Replication
of Mammalian
Parvoviruses

Replication of Mammalian Parvoviruses

EDITED BY

David C. Ward
Yale University

Peter Tattersall
Imperial Cancer Research Fund, Mill Hill Laboratories

COLD SPRING HARBOR LABORATORY / 1978

Replication of Mammalian Parvoviruses
© *1978 by Cold Spring Harbor Laboratory*
All rights reserved
International Standard Book Number 0-87969-120-4
Library of Congress Catalog Card Number 77-90839
Cover and book design by Emily Harste
Printed in United States of America

Cover: Particles of minute virus of mice, approximately 250 Å in diameter (electron micrograph courtesy of Peter Tattersall).

Contents

Transcription, Translation, and Virus Maturation

APPENDIX

Preface

Detailed studies of the structure and replication of bacteriophage genomes, such as those of øX174 and λ, have revealed many fundamental principles of prokaryotic biology. In a similar fashion, in-depth analyses of simple animal virus systems should help to unravel the molecular complexities of the eukaryotic cell. One such group of potential probes are the parvoviruses. This family of small, icosahedral viruses contains linear single-stranded DNA genomes of only 1.4–2.2×10^6 daltons and infects hosts as diverse as man and moth. This volume deals with the mammalian parvoviruses which comprise both nondefective viruses and the defective adeno-associated viruses.

The picture emerging of the parvovirus chromosome is that it encodes a very small number of genes, perhaps only the one for its coat protein, and, in addition, all the control signals required to redirect its host cell to the predominant, if not exclusive, function of producing progeny virus. The limited genetic capacity of these viruses is reflected in their extreme dependence upon host-cell or helper-virus functions. This makes them particularly suitable for the detailed investigation of eukaryotic DNA replication and of the control of gene expression during differentiation and neoplastic transformation.

Part I of this book provides an introduction to the mammalian parvoviruses and briefly reviews some interesting facets of their biology, structure, and mode of replication. Part II contains a comprehensive collection of previously unpublished papers most of which were presented during the Cold Spring Harbor meeting on Parvoviruses in May, 1977. Topics covered in these articles include the specificity of virus-cell interactions, the role of host-cell or helper-virus functions in the replication process, genome structure, mechanism of DNA replication, transcription, translation, properties of the structural polypeptides, and virion maturation.

We thank Dr. J. D. Watson for making the conference and publication facilities of the Cold Spring Harbor Laboratory available to us and for providing financial assistance for the meeting. We also gratefully acknowledge the funds provided by the International Cancer Research

Data Bank Programme of the National Cancer Institute, National Institutes of Health, under contract No. NOI-CO-65341 with the International Union Against Cancer. We express our sincere appreciation to those who reviewed research articles, to Paul Bethge for his help with editing, and to Nancy Batter and Brenda Marriott for secretarial assistance. Finally, we would like to acknowledge the stimulating contributions to the meeting of those participants whose work is not represented here because of our emphasis on the molecular biology of the mammalian viruses.

David Ward
Peter Tattersall

Replication of Mammalian Parvoviruses

PART I

REVIEW ARTICLES

The Parvoviruses— An Introduction

PETER TATTERSALL
DAVID WARD

Recent studies of the replication of the mammalian parvoviruses are the major subject of this book. The development of in vitro cell-culture systems for these viruses over the last few years has made possible considerable advances in the understanding of their multiplication at the molecular level and of some unique aspects of their interaction with their animal hosts in vivo.

The purpose of this introductory essay is to summarize briefly the major common characteristics of the parvoviruses. We hope also to provide a rationale for the growing conviction that the continued study of these particular viruses will yield important new insights into basic eukaryotic biology and the mechanisms of virus-induced disease. The article is not intended as an exhaustive treatise referring to all published data on parvoviruses, but rather as a preamble to the three detailed reviews which follow and which introduce the original research contributions.

For those who wish to acquaint themselves with aspects of parvovirus biology not covered in this book, a list of more comprehensive reviews has been included at the end of this article.

THE PARVOVIRUS GROUP

During the 1960s a considerable number of viruses were isolated which shared enough characteristics for the International Committee on the Nomenclature of Viruses (ICNV) to recognize them as a distinct group, giving them the family name "Parvoviridae." The numerous viruses now accepted as members of this family are listed in Table 1, along with their original sources and dates of isolation. Strictly, the name "parvovirus"

4

TABLE 1
The Parvovirus Group

	Acronym	Full name	Source and date of original isolation
NONDEFECTIVE SUBGROUP (GENUS *Parvovirus*)			
	KRV (or RV)	Kilham rat virus	rat-liver sarcomas and transplantable leukemia (1959)
	H-3	—	human HEP-3 transplantable tumor (1960)
	X-14	—	X-ray- and MC-treated rats (1963)
a	L-S	Lum-Schreiner virus	rat choroleukemic tumor (1963)
	HER	hemorrhagic encephalopathy virus of rats	CNS of cyclophosphamide-treated rats (1967)
	KIRK	—	human Detroit-6 tumor cells (1970)
a	H-1	—	human HEP-1 transplantable tumor (1960)
	HT	—	human placentas and embryos (1960)
b	FPV (or LV)	feline panleukopenia virus (or leopard virus)	leopard spleen (1965)
	MEV	mink enteritis virus	mink liver and spleen (1952)
a	PPV	porcine parvovirus	hog cholera virus stock (1966)
	KBSH	—	human KB tumor cells (1971)
	MVM	minute virus of mice	mouse adenovirus stock (1966)

BPV	bovine parvovirus[c]	calf feces (1961)
TVX	—	human permanent tumor cell lines (1971)
HB	—	human placentas and tumors (1964)
MVC	minute virus of canines	dog feces (1968)
LuIII	—	human Lu106 permanent cell line (1971)
RTV	—	rat AT permanent cell line (1967)
GHV	goose hepatitis virus	gosling liver and heart (1967)

DEFECTIVE SUBGROUP (GENUS *Adeno-associated virus*)

AAV1	adeno-associated virus type 1	simian virus 15 stock (rhesus monkey) (1965)
AAV2	adeno-associated virus type 2	human adenovirus 12 stock (H, M strains) (1966)
AAV3	adeno-associated virus type 3	human adenovirus 7 stock (H, K and T strains) (1966)
AAV4	adeno-associated virus type 4	simian virus 15 stock (African green monkey) (1967)
AAV-X7	bovine adeno-associated virus	bovine adenovirus 1 stock (1970)
AAAV	avian adeno-associated virus	quail bronchitis virus stock (1967)
CAAV	canine adeno-associated virus	canine hepatitis virus stock (1969)

ARTHROPOD SUBGROUP (GENUS *Densovirus*)

DNV1	densonucleosis virus 1	*Galleria mellonella* larvae (1964)
DNV2	densonucleosis virus 2	*Junonia coenia* larvae (1972)

[a] Brackets include antigenically cross-reacting viruses.
[b] These cross-react antigenically and are probably the same virus.
[c] Originally known as Haden—hemadsorbing enteric virus of calves.

describes only one of the three genera in this family proposed by the ICNV, but the name is commonly used to describe the entire family, and it is in this sense that the term will be used in this book. Of the three genera, the *Parvovirus* and *Adeno-associated-virus* (AAV) subgroups cover the viruses of vertebrates, whereas the genus *Densovirus* (densonucleosis virus, DNV) comprises arthropod agents. As all of the research papers in this book are concerned specifically with the mammalian parvoviruses, little further reference will be made to the insect subgroup.

The vertebrate parvoviruses are divided into two genera or subgroups on the basis of their requirement for helper viruses. Viruses of the largest subgroup, the genus *Parvovirus*, are usually referred to (perhaps colloquially but certainly more precisely) as the nondefective or autonomous parvoviruses. These are capable of productive replication without the aid of a helper virus in the vast majority of host cells studied to date. On the other hand, all members of the *Adeno-associated-virus* subgroup are defective and are (as their name implies) entirely dependent upon adenovirus coinfection for their own replication in all systems examined so far.

The viruses are strikingly similar in physicochemical properties, growth characteristics in vitro, and, in general, pathogenic potential. They are among the smallest known DNA viruses, being spherical, nonenveloped particles 20–25 nm in diameter. In the electron microscope these particles appear to possess an icosahedral symmetry and comprise 32 morphological capsomers. The infectious particle contains the genome, which is a predominantly single-stranded, linear DNA molecule with a molecular weight of 1.35–1.70×10^6. This DNA accounts for about 25% of the total virion mass; the balance is protein, with no detectable RNA, carbohydrate, or lipid. The viruses are extremely stable with regard to heat, freezing and thawing, lipid solvents, desiccation, detergents, and even low concentrations of chaotropic agents.

The vertebrate viruses show hemagglutinating activity, with the exception of AAV serotypes 1, 2, and 3. They are mostly antigenically distinct, as measured by hemagglutinin inhibition and complement fixation, apart from the cross-reacting groups indicated in Table 1. Within these cross-reacting groups individual viruses can be distinguished by the species of red blood cell they will agglutinate, by host range, and in some cases by pathogenic potential.

Most viral antigen is extracted from infected cells as assembled particles, which can be separated into four main classes by physical techniques. Particles which band at 1.30–1.32 g/cm³ in cesium chloride equilibrium gradients and have sedimentation constants of 60–70S appear to be empty capsids devoid of DNA. Infectivity is predominantly associated with particles which band at 1.38–1.43 g/cm³ in cesium chloride and sediment at approximately 110S. Another class of particles, with lower specific infectivity, is found at 1.46–1.48 g/cm³ in cesium chloride. These particles appear to be immature forms of the infectious virion and have the same sedimentation characteristics as the mature virus. A somewhat heterogeneous class of particles banding at 1.32–1.38 g/cm^3 in cesium chloride and sedimenting at 70–110S is also observed in some parvovirus preparations and appears to comprise capsids containing DNA molecules of less than genome length. More detailed consideration of the structure of the parvovirus genome and virion are given in the subsequent review articles.

In spite of their common physicochemical characteristics, the nondefective and defective subgroups do exhibit important differences in addition to helper-virus requirement. Whereas members of the autonomous subgroup package one unique strand of DNA (probably the minus strand with respect to transcription), populations of AAV and DNV particles contain approximately equal numbers of plus and minus strands, packaged in separate virions. As will be seen in later articles, this difference between the subgroups has significantly affected the approaches that investigators have used in studying the molecular biology of these viruses. For instance, the availability of both DNA strands as hybridization probes has made possible considerable progress in analyzing the transcription process of AAV. On the other hand, until recently it has proved easier to investigate the mechanism of DNA replication for the nondefective parvoviruses, where the analysis is not complicated by the concomitant replication of a helper-adenovirus genome.

In addition to these differences in encapsidation and helper requirement, another important difference exists between the parvovirus subgroups. The AAV do not appear to be associated per se with any particular disease in their hosts, although they may affect the course of disease induced by their companion adenoviruses. The nondefective parvoviruses, however, elicit a unique and fascinating spectrum of pathological conditions in their hosts, which will be discussed later.

PARVOVIRUS BIOLOGY—IN VITRO STUDIES

Cell-culture systems for the assay of parvovirus infectivity and for the study of single-cycle growth kinetics under controlled conditions have made possible detailed analysis of virus-host interactions at the cellular and molecular levels. A considerable body of information has been accumulated on various aspects of virus multiplication, such as DNA replication, transcription, and virion maturation, which will be discussed in detail in the subsequent review articles. A major part of the effort in the in vitro investigation of parvovirus biology has been concentrated on the examination of the dependence of these small viruses on host functions and, in the case of AAV, on the functions supplied by the helper virus.

Kinetic studies initially indicated that AAV is dependent upon a late function of adenovirus. In the absence of helper virus, adsorption and penetration occur, but no AAV-specific synthesis of DNA, RNA, or antigen can be detected, and no infectious virus is formed. Recent studies using temperature-sensitive adenovirus mutants as helpers have shown that, although the helper function appears to be late, it is not dependent upon adenovirus DNA synthesis for its expression. An indication that AAV might be defective in more than one function essential for replication came from the finding that members of the herpesvirus group could provide an incomplete helper function leading to the formation of AAV antigen but not infectious virus. Subsequently it was established that DNA synthesis, transcription, and antigen formation are normal in herpesvirus-coinfected cells, and that therefore the block appears to be in the packaging of progeny DNA into capsids. In cells coinfected with adenovirus and AAV, successful replication of one virus type appears to result in inhibition of the other. This interference may be the mechanism behind both the inhibition by AAV of the transformation of cells by adenovirus 12 in vitro and the inhibition of Ad12 oncogenicity in hamsters.

Several investigators have reported the stimulation of nondefective-parvovirus replication in vitro by coinfection with oncogenic DNA viruses such as Ad12 and polyoma virus. In the case where H-1 was found to be incapable of productive replication in a human cell culture, coinfection with Ad12 overcame the block. In parallel with the helper effect of adenovirus for AAV, H-1 also inhibited the growth of Ad12 in such coinfections. In addition, H-1 was subsequently shown to inhibit the oncogenicity of Ad12 in hamsters.

During the early in vitro studies of nondefective-parvovirus replication it was observed that these viruses grew most efficiently in rapidly dividing cell cultures. This correlated strongly with their predilection for mitotically active tissues in vivo (discussed below), which had been described by experimental pathologists some years earlier, and stimulated numerous studies, many employing synchronous cell cultures, which demonstrated that the nondefective parvoviruses require a cellular function expressed transiently during the late S phase of the cell cycle. The infection of mitotically quiescent cell cultures indicated that, unlike the oncogenic DNA viruses, the parvoviruses are unable to stimulate cells to enter S phase, and therefore noncycling cells cannot support lytic infection. This requirement, coupled with the limited genetic capacity of the viral genome, has led to the suggestion that the virus has to "wait" for a particular part of the cell's DNA-synthesizing apparatus to be activated as part of the normal traverse of the cell cycle. Only then can the virus usurp this cellular machinery for the replication of its own genome.

Thus, the parvovirus group exhibits a broad range of defectiveness: from the AAV (totally dependent upon adenovirus in all systems) to the autonomous subgroup (dependent upon adenovirus in some cells, but in others dependent upon host functions whose programming is entirely under the cell's control).

PARVOVIRUS BIOLOGY—IN VIVO STUDIES

In the ten years or so following the first parvovirus isolations, there were numerous studies of the induction of disease in experimental animals with these agents. The outstanding findings of these investigations were that the viruses cause fetal and neonatal disease and that, in general, the adult animal is resistant to the induction of overt disease. Figure 1 attempts to show diagrammatically the pathogenic spectrum of the "typical" nondefective parvovirus.

H-1 was discovered because of its ability to cause a characteristic "mongoloidlike" deformity when inoculated intracerebrally into newborn hamsters. Subsequently, KRV, H-3, HER, HT, X14, and HB were also shown to exhibit this syndrome and were grouped together with H-1 as the "hamster osteolytic viruses." Such infections, resulting in craniofacial lesions and microcephaly, are characterized by the proclivity of these agents towards dividing cells in developing dental and skeletal tissues. Unlike Down's syndrome in man, the condition is not associated with chromosomal abnormalities, nor is it hereditary, as these mongoloid animals can be raised to breeding age and produce nondeformed offspring.

Another striking result of intracerebral inoculation of neonates is the occurrence of cerebellar hypoplasia, often resulting in chronic ataxia. This was shown to be the result of selective viral attack on the rapidly proliferating cell population of the cerebellar granular cortex. The production

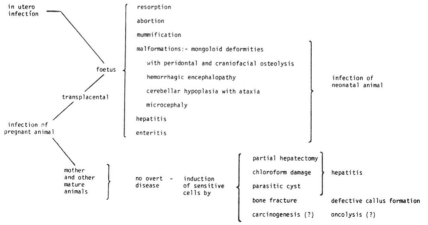

FIGURE 1
Pathogenic spectrum of "typical" nondefective parvovirus.

of this cerebellar lesion is not confined to the "hamster osteolytic" group but appears to be a general and quite specific feature of parvovirus pathology.

It was soon discovered that intrauterine infections of the developing embryo, resulting in these pathological conditions, can be established in many instances by parenteral injection of the mother. This shows that some of the viruses are capable of crossing both the placenta and the blood-brain barrier, depending on the host species. Infection during early gestation with large amounts of a placenta-crossing parvovirus often gives rise to a more generalized and devastating disease involving much of the mesodermal tissue of the embryo and resulting in resorption, abortion, or mummification of the fetus. Other diseases often associated with fetal and neonatal infection are hemorrhagic encephalopathy, neonatal hepatitis, and hemorrhagic enteritis. Again, the involvement of cell populations with high mitotic activity at the time of infection is a consistent feature of these diseases.

Overall, the studies made by the experimental pathologists led them to describe the parvoviruses as mitolytic agents, because they had a particular affinity (perhaps a requirement) for dividing cells. As we have seen, subsequent in vitro studies demonstrated that the viruses require the host cell to traverse the S phase in order to initiate lytic virus replication.

Interestingly, in most intrauterine infections the mother remains resistant to the induction of any overt disease. Even though she may harbor fatally diseased embryos, which presumably act as a continuous source of large amounts of infectious virus, usually her only response is to raise antibody to the agent used. This reinforces the general observation that most parvoviruses are unable to cause detectable illness in adult animals. As if to indicate that the reason for the resistance of mature animals is not trivial, a dramatic exception to this "rule" exists. Although feline panleukopenia virus (FPV) is the agent responsible for spontaneous ataxia in kittens, causing characteristic cerebellar lesions in infected neonates, it is also extremely pathogenic for adult cats. It has been shown to cause, in addition to infectious panleukopenia, feline enteritis, which is probably the most deadly disease of domestic cats.

Normally resistant adult animals, however, can be rendered susceptible to parvoviruses by inducing some tissue to undergo an abnormal proliferative response. Thus, hepatitis due to H-1 can be induced in adult rats by partial hepatectomy. The sites of viral attack are the regenerating margins of the hepatic lobes. Similar susceptibility can be brought about by inducing mitotic activity in the liver by carbon tetrachloride damage or by infection with the parasite Cysticercus fasciolaris. In addition, H-3 has been shown to infect healing osseous wounds, causing defective callus formation in normally resistant adult hamsters.

PARVOVIRUSES AND CANCER

The ability of parvoviruses to multiply in normally resistant tissues after the tissue has been induced to engage in "unscheduled" mitotic activity leads us to speculate about the relationship between parvoviruses and neoplastic disease. The circumstances of the first isolation of parvoviruses, mostly from tumors (Table 1), suggested that they might have some association with cancer in their hosts. Indeed, after the original isolations of KRV and H-1, control attempts to isolate these viruses from nonmalignant tissues by the same procedures were unsuccessful.

Although the study of animals surviving parvovirus infection at birth has indicated that such animals may carry the virus as a prolonged asymptomatic or latent infection, the fact emerges quite conclusively that these agents cannot be regarded as oncogenic viruses. How then can one explain the correlation between the isolation and distribution of these viruses and the existence of tumors in their hosts? The simplest explanation is that such tumors, because of their relatively high mitotic activity, provide a suitable environment for the proliferation of endogenous parvoviruses. This interpretation is supported by the extremely high incidence of contamination of transplantable tumors with parvoviruses—for instance, in one study minute virus of mice (MVM) was demonstrated in 40% of transplantable murine tumors examined.

In animals where a persistent, though latent, parvovirus infection has been established, carcinogenesis leading to cell proliferation might be expected to activate replication of the virus. This activation might lead in some cases to the arrest of tumor development by direct cell lysis, or to the inhibition of some function essential for cell division, such as DNA replication. The term "viral surveillance" has been suggested for such an equilibrium between virus replication and oncogenic transformation. In one long-term study, involving large numbers of hamsters, those which had survived H-1 infection at birth without deformity had a significantly lower (by a factor of 5) spontaneous tumor rate than the uninfected control group, and the rate for the mongoloidlike survivors was lower still (by a factor of 20–25). Is it possible that such a "viral surveillance" mechanism was operating in this situation?

Although FPV shows the same cell-cycle requirement for replication in vitro as other nondefective parvoviruses, as mentioned previously it has the ability to cause extreme, often fatal, disease in adult animals. This indicates that the inability of the other viruses to cause overt disease in adults is not simply a lack of dividing cells as targets, but that further constraints are placed on virus replication in the mature animal. It also suggests that these constraints are not merely the onset of immune competence or activation of the interferon response, but may be some subtle change in susceptibility at the cellular level related to the developmental stage of the animal. Recent in vitro experiments have shown that the ability to support lytic parvovirus replication is indeed controlled by a differentiation-dependent event which is not related to the ability of the cells to proliferate. The onset of the permissive state appears to be the result of a stable change in cellular gene expression which is inherited within the clone of differentiated cells. It is possible that, in some cases, this block to replication might also be lifted as a consequence of carcinogenesis as well as differentiation. This would render a tumor cell and its progeny susceptible to parvovirus infection, whereas its normal neighbors would be resistant regardless of their mitotic activity. This could explain the association of parvoviruses with tumors in their hosts and their ability, in some instances, to suppress oncogenesis.

In-depth analysis of virus and host components required for lytic parvovirus infection will, hopefully, further our understanding of differentiation and carcinogenesis, and may even enable us to control certain neoplastic diseases.

GENERAL BIBLIOGRAPHY

Berns, K. I. 1974. Molecular biology of the adeno-associated viruses. *Curr. Top. Microbiol. Immunol.* **65**:1.

Kilham, L. and G. Margolis. 1975. Problems of human concern arising from animal models of intrauterine and neonatal infections due to viruses: A review. I. Introduction and virologic studies. *Prog. Med. Virol.* **20**:113.

Kurstak, E. 1972. Small DNA densonucleosis virus. *Adv. Virus Res.* **17**:207.

Margolis, G. and L. Kilham. 1975. Problems of human concern arising from animal models of intrauterine and neonatal infections due to viruses: A review. II. Pathologic studies. *Prog. Med. Virol.* **20**:144.

Rose, J. A. 1974. Parvovirus reproduction. In *Comprehensive virology* (ed. H. Fraenkel-Conrat and R. R. Wagner), vol. 3, p. 1. Plenum, New York.

Siegl, G. 1976. The parvoviruses. *Virol. Monogr.* **15**:1.

Toolan, H. W. 1972. The parvoviruses. *Prog. Exp. Tumor Res.* **16**:410.

Parvovirus DNA Structure and Replication

KENNETH I. BERNS
WILLIAM W. HAUSWIRTH

Department of Immunology and Medical Microbiology
College of Medicine
University of Florida
Gainesville, Florida 32610

Parvovirus genomes are linear single-stranded DNA molecules of molecular weight $1.2–2.2 \times 10^6$ and are among the smallest viral genomes known. Indeed, it has been speculated that the DNA codes for but a single structural gene. Because parvoviruses replicate in the nuclei of infected cells, the DNA is an attractive model system for studying in detail the molecular mechanisms of DNA replication and transcription in eukaryotic cells. The system is made more attractive by the linearity of the DNA, a feature which may make relatively easy the localization, and thus the characterization, of regions involved in regulation—especially regulation of transcription.

The family Parvoviridae contains three genera: *Parvovirus*, the *Adeno-associated-virus* group, and *Densovirus*. However, since this book deals exclusively with mammalian parvoviruses, members of the last (arthropod) genus will not be discussed further. Because there are significant differences among the physical properties of the DNAs of viruses from the first two genera, as well as among their biological properties, studies of the fine structure and the replication of these viral genomes will be considered separately below. It should be noted, however, that, with reference to DNA replication, all of the mammalian parvoviruses have a number of significant features in common. Our overall aim is to review parvovirus DNA structure and replication, with an emphasis on relating recent findings to the general problem of linear DNA replication.

As the mature viral DNA is the final product of DNA replication, any replication model derived from studies of replicative intermediates isolated from infected cells must be able to account for those particular nucleotide-sequence arrangements (e.g., terminal repetitions or palindromes) found in virion DNA. Conversely, these same arrangements of nucleotide sequences in the mature viral DNA may offer significant clues as to the mode of viral DNA replication.

DNA STRUCTURE

Adeno-Associated Virus

AAV is an exception among viruses with single-stranded genomes in that plus and minus strands of the DNA are packaged in separate virions. From an experimental point of view this is an enormous aid in studying the fine structure of a single-stranded genome, because several of the various tests of structure utilize enzymes which require double-stranded DNA as a substrate. Such duplex DNA is easily obtained by annealing the plus and minus strands purified from virions. On the other hand, the plus and minus strands can readily be separated by various techniques so that properties of the individual single strands may also be studied.

Initially the AAV genome was considered to be double-stranded DNA because the plus and minus strands had annealed during the extraction procedure then used (Rose et al. 1966). After the initial proposal by Crawford et al. (1969) that the virion could not accommodate the initially reported duplex genome (m.w. 3.6×10^6), Rose et al. (1969) were able to demonstrate in a density-shift experiment that the complementary single strands of duplex AAV DNA originated from different virions (i.e., 50% of the duplex DNA isolated from a mixture of virions containing either unsubstituted or bromodeoxyuridine-substituted DNA had a hybrid density in CsCl gradients). Mayor et al. (1969) and Berns and Rose (1970) isolated single-stranded DNA from virions under conditions which do not denature double-stranded AAV DNA. Finally, Berns and Adler (1972) physically separated the virions containing minus strands from those with plus strands.

Double-stranded AAV DNA initially was characterized as a linear duplex molecule with a molecular weight of about 3×10^6 by sedimentation in neutral sucrose gradients and electron microscopy. Density in CsCl, melting temperature (T_m), and chemical analysis of base composition were all in accord with a completely duplex molecule with a GC content of 53% (Parks et al. 1967; Rose et al. 1968). However, the profile of the annealed DNA in neutral sucrose gradients always had a leading shoulder, which indicated the presence of other than unit-length linear double-stranded molecules. Indeed, several duplex molecular species could be detected by gel (agarose or acrylamide) electrophoresis (Gerry et al. 1973). These additional species were identified by electron microscopy as linear double-stranded molecules of monomeric and dimeric length and circular double-stranded monomeric molecules (Gerry et al. 1973; Koczot et al. 1973). Circles and dimers could be converted to linear monomers by heating at temperatures below the T_m of the DNA; thus, it was concluded that circles and dimers were joined by short, cohesive, single-stranded regions at the ends of linear duplex molecules. These cohesive regions were estimated to represent less than 5% of the genome. However, about half of all duplex AAV DNA molecules lacked such cohesive regions at the termini (i.e., would not circularize or form oligomers under annealing conditions). At that time it was suggested that the DNA contained two nucleotide-sequence permutations with starting points close together on the genome. Annealing of complementary strands with the same starting point would yield duplex termini, whereas complementary strands with different starting points would yield double-stranded DNA with cohesive termini.

As stated above, 50% of duplex linear monomers did not have cohesive termini; thus, it was concluded that these molecules had normal duplex structure at the ends. Limited digestion of these duplex molecules with exonuclease III (exonucleolytic digestion from the 3' termini of double-stranded DNA) created cohesive, single-stranded, 5' termini. After this treatment the DNA would circularize when exposed to annealing conditions. Thus, it was concluded that AAV DNA has a natural terminal-nucleotide-sequence repetition (12 - - - 12, where 1 and 2 represent nucleotide sequences). From the extent of digestion required for circle formation to occur, the natural terminal repetition was estimated to represent about 1% of the DNA.

The complementary strands of AAV DNA may be separated in two ways. The thymidine (T) content of the two strands is unequal [minus strand 27% T, plus strand 21% T (Rose and Koczot 1971)], so substitution of T by BUdR allows separation in CsCl gradients (Berns and Rose 1970). More recently, the strands have been separated directly on agarose-acrylamide composite gels after denaturation (Hauswirth and Berns 1977). Koczot et al. (1973) found that the separated single strands of AAV DNA form single-stranded circles whose ends are held together by base pairing of short regions, thus establishing that AAV DNA also has an inverted terminal-nucleotide-sequence repetition (12 - - - 2'1', where 2' and 1' represent sequences complementary to 2 and 1). Using the electron microscope, Berns and Kelly (1974) were able to visualize the base-paired terminal regions as projections from the single-stranded circles formed after self-annealing of linear minus strands. From measurements of the projections, the inverted terminal nucleotide sequence was estimated to represent 1–2.5% of the AAV genome.

A model was proposed to explain the existence of both a natural and an inverted terminal repetition in AAV DNA. Gerry et al. (1973) suggested that the terminal-nucleotide-sequence repetition was a palindrome of the type 122'1'---122'1'. This type of terminal sequence has properties of both a natural and an inverted terminal repetition. Evidence to support this model has been reported recently. Fife et al. (1977) digested AAV DNA, labeled internally with [^{3}H]thymidine and at the 5' termini with ^{32}P, with endoR·HindIII, which makes a single cut at 0.39 units. The two fragments were separated, heat-denatured, quick-chilled, and digested with a single-strand-specific nuclease. The internal ^{3}H-label was solubilized much more rapidly than the 5' terminal ^{32}P. These are the results that would be expected if the terminal repetition were a palindrome. After denaturation and quick chilling, the self-complementary regions of the terminal repetition could "snap back" to form a base-paired hairpin structure resistant to the single-strand-specific nuclease.

As isolated from virions, all AAV single-polynucleotide chains are linear (Gerry et al. 1973); however, it is difficult to prove conclusively that the DNA is linear in the virion. Data which suggest that this is the case include the following: (1) DNA released by treatment of the virions with proteolytic enzymes and ionic detergents or by treatment with alkali is linear. (2) The nucleotide sequence is not detectably permuted (within 5%). (3) All 5' and 3' nucleotide sequences are identical for at least 40 nucleotides. Fife et al. (1977) have reported that both 5' termini have the same nucleotide sequence except for the terminal one or two nucleotides (35% TTGGCCA, 50% TGGCCA, 15% GGCCA). Berns et al. (this volume) have extended the unique sequence to 40 nucleotides using the method of Maxam and Gilbert (1977). The reason for the one- or two-nucleotide heterogeneity at the termini is unknown. The heterogeneity may be caused by an exonuclease, but why the activity would be so limited is unclear. In any event, if an initially circular DNA within the virion is cleaved by an endonuclease to create the purified linear DNA, the enzyme must be site-specific.

The portion of the DNA that is transcribed into stable mRNA has also been determined (Carter et al. 1976; Carter, this volume). Only the minus strand is transcribed into a stable messenger, which is coded for by the region 0.18–0.90 on the genome. This was determined by annealing messenger to separated fragments of AAV DNA produced by digestion of the DNA with bacterial restriction endonucleases. These data suggest that the terminal regions of the DNA (0–0.18 and 0.90–1.0) control initiation, termination, and perhaps processing of the AAV transcript; thus, these regions are being studied intensively. Similarly, Hauswirth and Berns (1977) have demonstrated that the origin and termination of DNA replication occur within the terminal repetitions (discussed more fully below).

Recently, longer AAV transcripts have been found in the Hirt pellet (high-salt precipitate of infected KB-cell lysates) and have been assumed to be associated with chromatin (Jay et al., this volume). These transcripts map from approximately 0.05 to 0.95, suggesting that transcription may begin and terminate near the internal limits of the terminal repetitions.

There are four AAV serotypes from primates, three isolated from humans and one from monkeys (Hoggan et al. 1966). The extent of nucleotide-sequence homology among these four serotypes has been estimated to be 33–50% (Rose et al. 1968; Koczot 1975). Examination in the electron microscope of hybrid duplex molecules formed by annealing of plus and minus strands from different AAV serotypes has revealed that the homology exists primarily at the terminal regions of the molecule (Koczot 1975). However, the extent of homology observed extends beyond the purported regulatory regions into the region transcribed into stable mRNA.

In order to characterize the fine structure of the AAV genome in more detail, the DNA has been mapped extensively using a variety of restriction enzymes. Enzymes for which cleavage sites have been mapped include EcoRI, SalI, PstI, BamHI, HindII, HindIII, HpaII, HaeII, HaiIII, AluI, and HhaI (Berns et al. 1975; Carter and Khoury 1975; Carter et al. 1975; Denhardt et al. 1976; de la Maza and Carter 1976; Fife 1977; Spear 1977) (Fig. 1). HpaI has no cleavage sites. The mapping procedure has been straightforward for internal fragments, which supports the absence of significant nucleotide-sequence permutation. However, structural heterogeneity has been observed for terminal fragments. Those enzymes that make a cut outside the terminal repetition produce two terminal fragments from each end of the molecule which can be resolved on acrylamide gels. Enzymes that cleave within the terminal repetition produce multiple (>2) terminal fragments resolvable on acrylamide gels.

Details of a model to account for the terminal heterogeneity observed in duplex AAV DNA are presented by Berns et al. (this volume). Briefly, the terminal repetition of AAV2 DNA has been determined to be 140 nucleotides long, and a tentative sequence has been determined for all but 6 of these nucleotides. Nucleotides 1–41 are complementary to nucleotides 120–80, which accounts for the palindromic nature of the terminal repetition. Nucleotides 42–79 are not palindromic, although there is an internal palindrome of 16 nucleotides within this sequence. Since analysis of the terminal repetition after a partial digest with HpaII revealed twice the number of HpaII cleavage sites required to account for the fragments known to come from this region, we have suggested that the first 120 nucleotides of AAV2 DNA can exist either in the orientation 1–120 or in the reverse orientation, 120–1 (i.e., a "flip-flop" model for the first 120 nucleotides of the terminal repetition). A possible origin for the flip-flop is discussed later in the section on replication models.

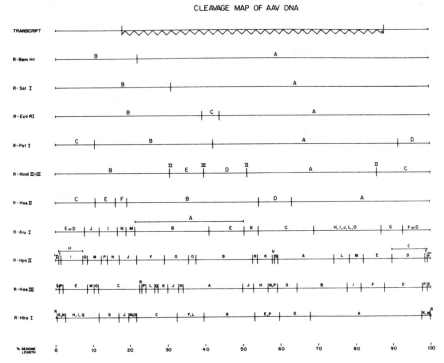

FIGURE 1
Restriction-endonuclease cleavage maps for AAV2 DNA.

Autonomous Parvoviruses

The autonomous parvoviruses differ significantly from AAV in that they do not require helper virus and only the minus strand (i.e., that transcribed into stable mRNA) is encapsidated into virions. [About 1 in 400 virions may contain the plus strand (Rhode 1977b).] Thus, purified autonomous parvovirus DNA does not self-anneal. Detailed studies of the structure of the DNA have been hampered by difficulty in purifying large amounts of virus, but recently this has been overcome, allowing rapid progress in structural studies. Additional technical advances furthering such studies have included isolation of significant amounts of RF DNA and synthesis of the complementary plus strand in vitro, allowing restriction-enzyme analysis (Bourguignon et al. 1976; Rhode 1977a,b).

DNA purified from several autonomous parvoviruses was shown to be linear, predominantly single-stranded, molecules by a variety of physicochemical techniques, including melting-profile and sedimentation-property studies and electron microscopy, as well as by susceptibility to single-strand-specific DNases (Crawford et al. 1969; McGeoch et al. 1970; Salzman et al. 1971; Bourguignon et al. 1976).

Another major difference between AAV DNA and that of the autonomous parvoviruses is that the autonomous-parvovirus DNA does not contain an inverted terminal-nucleotide-sequence repetition. Bourguignon et al. (1976) were able to demonstrate in parallel experiments that conditions which lead to the formation of hydrogen-bonded single-stranded circles of AAV DNA do not cause the circularization of MVM DNA.

In spite of the apparent lack of an inverted terminal repetition, the DNA does contain sequences at both termini with the properties of palindromes (Bourguignon et al. 1976; Rhode 1977; Salzman 1977; Lavelle and Mitra, this volume). Bourguignon et al. (1976) digested MVM DNA labeled at the 5' termini with ^{32}P and internally with [^{3}H]thymidine with S1 single-strand-specific nuclease and found that the ^{32}P was resistant to solubilization under conditions which solubilized the internal ^{3}H, much the same as has subsequently been demonstrated for AAV DNA. Initial attempts to show the same properties at the 3' terminus were unsuccessful. However, the 3' terminus was able to function as a primer for DNA synthesis, showing indirectly that it could fold back and base-pair with an internal sequence. More recently it has been possible to demonstrate directly, using a different preparation of the S1 nuclease, that 3' terminal ^{32}P is resistant to cleavage by a single-strand-specific DNase (Chow and Ward, this volume). As determined by gel mobilities, the 5' terminal palindrome was estimated to be ~130 nucleotides long and the 3' palindrome to be ~110 nucleotides long.

Several other approaches have been used to demonstrate that the terminal nucleotide sequences of parvovirus DNA can fold back to self-base-pair. Kilham rat virus (KRV) DNA labeled at the 5' termini with ^{32}P eluted from BND-cellulose as single-stranded DNA, yet, after S1-nuclease digestion, the ^{32}P was associated with acid-precipitable material which eluted as double-stranded DNA (Salzman 1977). In addition, about 5–6% of uniformly ^{32}P-labeled KRV DNA was resistant to digestion by S1 nuclease, and this material also eluted as double-stranded DNA from BND-cellulose. Finally, a double-stranded form of KRV DNA can be isolated from infected cells. Limited treatment of uniformly ^{32}P-labeled duplex KRV DNA with exonuclease III reduced by half the fraction of DNA resistant to S1 digestion after denaturation. Thus, Salzman (1977) concluded that both the 5' and 3' termini of KRV DNA can fold back to self-anneal. From the fraction of DNA resistant to S1 nuclease and from its sedimentation rate in neutral sucrose gradients, she estimated the palindromic region at each end of the molecule to be 90–150 nucleotides long.

In parallel experiments, Lavelle and Mitra (this volume) demonstrated that single-stranded DNA eluted from hydroxyapatite at a position consistent with the existence of double-stranded regions in the DNA. Single-stranded KRV DNA is resistant to exonuclease I (single-strand-specific 3' exonuclease) digestion, which demonstrates directly the duplex nature of the 3' terminus. Pretreatment of the DNA with exonuclease III (double-strand-specific 3' exonuclease) renders the DNA susceptible to exonuclease-I digestion. Two single-strand-specific endonuclease-resistant fragments of 135 and 110 nucleotides were present on gels. Pretreatment of the DNA with exonuclease III prior to single-strand-specific endonuclease digestion caused the disappearance of the 135-nucleotide fragment; hence, it was concluded that this represented the 3' terminal palindrome. Although the estimated sizes of these endonuclease-resistant fragments are similar to those reported by Chow and Ward (this volume), their assignments to specific termini differ. Lavelle and Mitra (this volume) have used the resistance of KRV DNA to exonuclease I to conclude that the DNA has a duplex structure at the 3' terminus, and yet the susceptibility of parvovirus DNA to exonuclease I was initially cited as evidence (as discussed above) that the DNA was linear. Although the reason for this discrepancy is not clear, it suggests that the enzyme used in previous studies contained a contaminating nuclease active on duplex DNA.

Restriction-enzyme analyses of double-stranded forms of H-1 DNA (Rhode 1977b) and MVM DNA (Ward and Dadachanji, this volume) provide further evidence that both the 3' and 5' termini of these genomes have palindromic sequences. Interestingly, restriction-endonuclease digests yielded pairs of terminal fragments from both ends of the DNA similar to those observed for duplex AAV DNA. The analysis is not yet as detailed as that of AAV DNA, but the intriguing possibility exists that the same type of nucleotide-sequence heterogeneity described above for the termini of AAV DNA may also exist for autonomous-parvovirus DNAs. This suggestion is strengthened by the variety of terminal structures observed when duplex H-1 DNA isolated from infected cells is denatured and reannealed in the presence of virion DNA labeled at the 5' termini (Rhode 1977b).

DNA REPLICATION

Adeno-Associated Virus

Because AAV produces and packages both minus and plus strands, DNA replication must generate approximately equal numbers of complementary strands, most likely in a single-strand form for direct encapsidation to avoid annealing in vivo. Purified duplex AAV DNA is infectious in the presence of an adenovirus helper (Hoggan et al. 1968; Boucher et al. 1971). Although virion proteins do not appear to be necessary for initiation of DNA replication, their role in chain elongation and packaging remains open, particularly in view of the possible packaging and replication functions of virion proteins for the autonomous parvoviruses (Rhode 1976; Tattersall and Ward 1976; Richards et al. 1977; Tattersall et al. 1977).

The infectivity of either plus or minus single strands alone is unclear at present. Separation of plus- and minus-strand virions by BUdR substitution and density centrifugation afforded insufficient virion purity for unequivocal strand-infectivity experiments. The recent separation of unsubstituted AAV DNA strands by gel electrophoresis (Hauswirth and Berns 1977) should lead to low levels of cross-contamination and a direct assessment of strand infectivity.

The precise nature of AAV defectiveness remains unresolved. Coinfection with either adenovirus or herpes simplex virus results in AAV DNA replication, but only after the onset of helper-virus replication [reviewed by Rose (1974) and Berns (1974)]. However, helper-virus DNA replication is not necessary because DNA-negative temperature-sensitive (ts) adenovirus mutants will still help AAV (Ito and Suzuki 1970; Ishibashi and Ito 1971; Mayor and Ratner 1973; Handa et al. 1976; Straus et al. 1976a). Although AAV DNA replication is undoubtedly dependent on host and/or helper functions, no specific role of host or helper-virus gene products in AAV DNA replication is known at present.

Use of ts adenovirus helpers defective in their own DNA synthesis has allowed analysis of AAV replicative intermediates in the absence of contaminating Ad DNA. After a short radiolabel pulse of KB cells coinfected with Ad5 (ts125) at a nonpermissive temperature, Straus et al. (1976b) isolated covalently linked plus and minus AAV DNA strands which can "snap back" to self-anneal via a hairpin structure and which can be chased to unit-length single strands. Concatemers up to four times unit length were also observed, although single-strand dimers of one plus strand and one minus strand were the predominant species. Handa et al. (1976) have also observed single- and double-stranded replicative intermediates larger than unit length after coinfection with AAV1 and Ad31 (tsA13) at the nonpermissive temperature of 40°C. Such DNA structures do not appear to be artifacts of the temperature-sensitive nature of the helper viruses, because coinfection with a wild-type Ad2 helper also produces concatemeric AAV DNA molecules which chase to unit length (Hauswirth and Berns, this volume).

Analysis of labeled duplex AAV DNA monomers completed during a short [^{3}H]thymidine pulse using restriction endonuclease HpaII shows a bimodal pattern of label incorporation with maximum incorporation in a fragment 24 base pairs long which maps 42 or 56 base pairs inside both molecular ends (Hauswirth and Berns, this volume). This is within the region of terminal-nucleotide-sequence repetition and is consistent with DNA replication terminating in sequences 42–56 base pairs inside either end.

This pattern is very similar to data for Ad2 DNA replication obtained by analysis of pulse-label distribution or by molecular hybridization studies (Winnacker 1974; Schilling et al. 1975; Horwitz 1974, 1976; Tolun and

Pettersson 1975; Lavelle et al. 1975; Bourgaux et al. 1976; Flint et al. 1977; Weingartner et al. 1977). In contrast with AAV DNA, however, the most heavily pulse-labeled regions in Ad2 DNA are contained in the terminal restriction fragments. This difference between Ad2 and AAV may be artifactual, since the terminal restriction fragments used for Ad2 DNA analysis are so large (800–1500 base pairs) that a small region of unlabeled terminal nucleotides (42–56 base pairs), as was observed for AAV DNA, cannot be detected.

Straightforward interpretation of the AAV data locates the termination of AAV DNA replication at least 50–60 nucleotides from the end of the molecule. There are several problems inherent in this direct interpretation. Given the palindromic nature of AAV DNA termini, the most obvious way to reproduce the pulse-label pattern is to terminate DNA replication at a parental 5' hairpin and transfer this unlabeled terminus to the newly formed progeny strand (Hauswirth and Berns 1977). However, the best estimate as to the extent of the palindromic terminal sequence is 120 base pairs (see above), and were the entire parental palindromic sequence transferred, maximum label would be subterminal by this 120 base pairs, not 42–56 as we observed. Three explanations seem worthy of consideration: (1) Perhaps maximum pulse-label density actually is 120 base pairs subterminal to either end but was missed by analysis using *Hpa*II restriction fragments. The distance to the next *Hpa*II site inward from the 24-base-pair fragment is either 230 (fragment I) or 380 (fragment D) base pairs. Hence, pulse-label incorporation around base pair 120 would be in *Hpa*II fragments I and D, and termination at base pair 120 may have been missed because only the average label density was determined for each restriction fragment. (2) During the pulse-label period newly incorporated bases may remain associated with the replicative form. If the AAV has hairpin termini, which seems likely during at least part of its replicative cycle, restriction-enzyme analysis may miss terminal fragments because of aberrant mobilities on polyacrylamide gels. Hence, peak pulse-label in subterminal fragments may be artifactual because of underrepresentation of normal terminal fragments in replicative forms. (3) Terminal genome segments may be underrepresented in replicating molecules if replication slowed considerably near the genome termini, because, for instance, a 5' hairpin needed to be displaced to complete replication or a virion packaging event required a replication delay at this point. These three possibilities are being investigated.

Similar restriction analysis of pulse-labeled AAV DNA for each complementary single strand reveals a unimodal gradation of pulse-label on both strands, each highest near the 3' end and lowest near the 5' end (Hauswirth and Berns 1977 and this volume). Although label in fragments from the terminal 2% could not be determined, the highest activity in each strand is close to the 3' end, with a downward gradient toward each 5' end. Hence, summation of plus- and minus-strand label patterns reproduces the observed bimodal duplex DNA label gradient as far as the data allow. The unimodal label patterns are consistent with replication originating within or near this same genome region and proceeding unidirectionally in the 5'→3' direction from both parental 3' ends.

In addition, products of apparently abortive replication are composed of only those sequences near either terminus of full-length AAV DNA (Hauswirth and Berns, this volume; de la Maza and Carter, this volume). If these molecules are initiated normally, restriction fragments in the highest molarity serve to locate the origin of replication. Such analyses suggest that AAV DNA replication initiates in the same genome regions where it terminates, near the ends of the terminal repetitions.

Autonomous Parvoviruses

Optimal yields of autonomous parvoviruses are obtained in cultures of actively dividing cells (Johnson 1967; Mayr et al. 1968; Tennant et al. 1969; Parker et al. 1970) during or immediately after cellular DNA synthesis, in late S or G_2 phase (Hampton 1970; Tennant and Hand 1970; Tattersall 1972; Siegl and Gautschi 1973; Rhode 1973). Thus, a potential viral DNA replication dependence on host-cell factors made during S phase was suggested. This is supported indirectly by a concomitant rise in the cellular DNA polymerase α activity and bovine parvovirus (BPV) infectivity in synchronous cell cultures (Bates et al. 1977 and this volume). The fact that Bates and coworkers could find no virion-associated DNA polymerase activity in highly purified infectious BPV, H-1, or LuIII virions is consistent with their suggestion that a host DNA polymerase is involved in viral replication. It is not known whether or not modification of host polymerase activity is required for viral DNA replication.

The fact that only the minus strand of autonomous parvoviruses is encapsidated raises an immediate problem for DNA replication. It must be either highly asymmetric with regard to strand synthesis, perhaps analogous to single-stranded DNA phages, or highly wasteful, if both plus and minus strands are made in equal amounts as in the case of AAV. In either case it seems imperative that the replicating pool contain at least some plus-strand-template molecules. Recently a number of workers have characterized the structural forms present in such replicating pools of several autonomous parvoviruses. Tattersall et al. (1973) observed that, on the basis of retention on hydroxyapatite, a portion of in-vivo-labeled MVM DNA was in a duplex monomer form. In addition, about 35% of this duplex MVM DNA underwent spontaneous, unimolecular renaturation after heat denaturation, which suggested that a significant proportion of the duplex MVM existed as covalently linked viral (v) and complementary (c) strands. Rhode (1974a,b) also isolated a double-stranded replicative form (RF) of the parvovirus H-1. On the basis of the self-annealing (snap-back) characteristics of the left end (viral strand, 3' end) contained in an EcoRI 0/0.22 fragment, 10% of unit-length duplex molecules are covalently linked v–c-strand hairpins (Rhode 1977a). A small proportion of RF molecules sediment as dimer-length duplex molecules, some of which are also composed of covalently linked v and c strands. Further restriction-enzyme analysis of covalently linked v and c strands suggests that most, if not all, linkage between strands occurs at the 3' ends of the v strands. This conclusion was strengthened by the observation that all v-c linkage at the viral 5' end could be melted out at appropriate formaldehyde concentration (Singer and Rhode 1977a). Similar, but less well characterized, unit-length RFs have been reported for KRV (Salzman and White 1973; Gunther and May 1976; Lavelle and Li 1977) as well as for KRV oligomeric forms (Gunther and May 1976). The similarity of the structures of autonomous and defective parvoviruses is striking and suggests the likelihood of parallel replication processes, as discussed more completely below.

Replicative intermediate (RI) forms can be operationally defined as the immediate precursors of progeny viral strands. For the autonomous parvoviruses, RIs may be distinguished from replicative forms (RFs) by the fact that they preferentially incorporate labeled nucleotides and usually contain a monomer duplex structure with additional variable lengths of single- or double-stranded viral genomes, giving them a Y-shaped appearance in the electron microscope. Such RIs have been detected for MVM (Tattersall et al. 1973), KRV (Lavelle and Li 1977; Li et al., this volume), LuIII (Siegl and Gautschi 1973 and this volume), and H-1 (Singer and Rhode 1977a,b). In each case except for H-1, evidence has been obtained for some degree of single-strandedness in the RI, which has been interpreted as representing partial displacement of a progeny strand from an RF molecule. For H-1, however, only Y-shaped molecules with all arms giving the appearance in the electron microscope of double-stranded DNA were reported. Perhaps the use of a ts H-1 mutant defective in progeny-strand synthesis but not in RF or RI synthesis at the restrictive temperature (Rhode 1976) has led to complementary-strand synthesis on the normally rapidly encapsidated single-stranded viral progeny strand.

Singer and Rhode (1977a,b) have extended their electron-microscope analysis of RI molecules from H-1 in order to locate the origin of DNA replication. Using partial-denaturation patterns, Y-shaped molecules could be oriented. The replication origin was thus located by the shortest Y-shaped molecule to within 300 base pairs of the viral 5' end. This conclusion is supported by pulse-labeling experiments in which the viral 5' end of complete RF molecules contained the least label (Rhode 1977a). One replication origin at the viral 5' end for H-1 suggests unidirectional replication of the v strand. If DNA synthesis is primed by a 3' hairpin on the c strand, this process may be analogous to the mechanism by which AAV produces both plus and minus strands from origins at both molecular ends primed by 3' hairpins as discussed below.

REPLICATION MODELS

The biological differences between autonomous and defective parvoviruses do not apear to extend to DNA structure and its general mode of replication. All mature virion DNA is single-stranded. In all parvovirus DNAs investigated thus far, both termini of the viral strands are able to form duplex hairpin structures. The fact that AAV terminal sequences are repetitious, whereas those in autonomous-parvovirus DNA are not, may well relate directly to the respective forms of infectious DNA. Assuming that the 3' terminal hairpin structure of viral strands is utilized in vivo as primer for initiating RF formation and progeny-DNA synthesis, AAV might be expected to have identical termini so that both strands can be synthesized equally. Conversely, autonomous parvoviruses would require nonidentical termini so that a distinction could be made between initiation of v-strand and c-strand syntheses. In addition, a specific terminal sequence or structure may be required for strand recognition during encapsidation, in view of the recent suggestion that viral strand packaging may drive the displacement of progeny strands from in vivo RFs (Tattersall and Ward 1976; Richards et al. 1977; Tattersall et al. 1977).

The structure of replicative forms is also basically identical for these two classes of parvovirus. All RFs are duplex molecules of monomer length made up of both unit-length strands and, to a lesser degree, hairpin single-strand dimers comprising one minus and one plus (viral and complementary) strand linked covalently. Larger than unit-length duplex RF forms have also been noted for both parvovirus classes. Such RF molecules have been shown to chase into progeny viral strands. For autonomous parvoviruses, RIs containing a partially displaced progeny strand on an otherwise duplex RF have been isolated.

The origin and termination of DNA replication have been located near the molecule ends. For AAV DNA, both ends apparently serve to initiate and terminate progeny DNA synthesis; for the parvovirus H-1, initiation is only at the 5' end of the viral strand. In addition, AAV DNA replication has been shown to be unidirectional, 5'→3', on both strands. These observations are all consistent with the ability of parvovirus DNA strands to form 3' hairpins and thus serve as primer and template for cellular DNA polymerases.

However, the most compelling evidence for self-primed DNA synthesis at a 3' hairpin comes from the terminal-sequence heterogeneity deduced for AAV DNA. The first 120 nucleotides of both ends of AAV

DNA exist with equal probability in one of two inverted orientations, either 1 to 120 or 120 to 1. Terminal hairpin structures created during the replicative process could directly account for this unusual terminal-sequence arrangement.

In spite of the fact that recently proposed parvovirus-replication models have not had the benefit of much of the above experimental data, they remain largely consistent with these new findings. From their identification of concatemeric plus and minus AAV DNA strands as intermediates in DNA replication and their finding that unit-length duplex DNA is an immediate precursor to progeny strands, Straus et al. (1976b) proposed a replication model patterned after that of Cavalier-Smith (1974). It utilized a 3'-OH hairpin to prime DNA synthesis and a site-specific nicking to create a duplex structure with a 5' overhang such that the 3'-OH end could be filled in, thus avoiding loss of sequences due to the hairpinning process. On the basis of replicating structures (Tatersall et al. 1973) and in vitro self-priming experiments (Bourguignon et al. 1976) observed for MVM, Tattersall and Ward (1976) have proposed a very similar replication model. In the case of nondefective parvoviruses, however, only the minus DNA strand is encapsidated; thus, the internal portions of the plus strand are unused. The observation that the origin of progeny-DNA strand synthesis for H-1 virus is at the 5' end of the viral strand (adjacent to the 3' end of the complementary strand in a hairpinned RF molecule) suggests that only the viral strand is synthesized from autonomous-parvovirus RFs (Rhode 1977b).

Utilizing the basic features of these models, we have attempted to incorporate all recent experimental data into a parvovirus replication model (Fig. 2). From a 3' hairpin primer a duplex replicative form is produced (Fig. 2A,B). To conserve terminal sequences, this duplex must be nicked at a specific site and opened, and the gap must be filled so that the original parental strand is extended to full length (Fig. 2C,D). With this duplex RF, the same processes of 3' hairpin-primer formation and displacement synthesis create either progeny plus and minus single strands (Fig. 2E,E'), in the case of AAV, or progeny viral strands only (Fig. 2E), in the case of the autonomous parvoviruses. The mechanism of this latter strand-specific process is conjectural, but it may involve specific polymerase priming of the unique complementary-strand 3' hairpin and/or specific viral-strand 5' sequence recognition by the encapsidation apparatus. Inversion of terminal nucleotide sequences (the flip-flop model) occurs as a natural consequence of site-specific nicking and gap-filling (e.g., compare the 5' terminal-sequence arrangement of the displaced strand in step E' with the original strand and, similarly, the 3' terminal sequence in the displaced strand from step E). This may occur for both defective and autonomous parvoviruses. Oligomeric RFs are produced by extension of the original hairpin RF (Fig. 2G,H). Identical

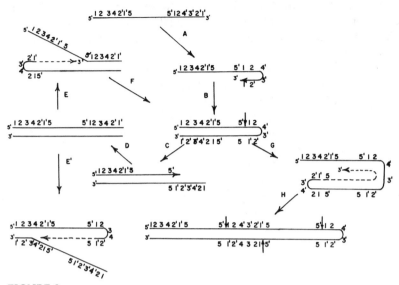

FIGURE 2
Proposed replication schemes for parvovirus DNA. Numerals 1, 2, 3, 4, and 5 represent nucleotide sequences within the terminal regions. Primed numerals represent the corresponding complementary sequence. Depicted are the terminal-nucleotide-sequence repetitions deduced for AAV DNA. See text for additional details.

site-specific nicking creates duplex intermediates whose 3' ends can be extended to full length so that no loss of sequence occurs, but again terminal-sequence inversions will occur.

This model is intended to be consistent with all presently available data on parvovirus-DNA structure and replication using a minimal number of processes. As such it should form a framework for further experimentation.

SUMMARY

Studies of the structure of parvovirus DNAs have now reached the point where truly detailed analysis of the rather small regions of the genome implicated in the initiation and termination of both DNA replication and transcription is possible. Clearly, work got off to an earlier start with AAV, mostly because it is easy to produce and because both strands are present in virions, but the regions of interest have also been identified in autonomous-parvovirus DNA and are being directly sequenced for both AAV and the autonomous parvoviruses. Already, the presence of palindromic nucleotide sequences at or near the origin of DNA replication has suggested an alternative to an initiating RNA primer (Tattersall and Ward

1976; Straus et al. 1976b). Not only should the sequences of these potential regulatory regions be of interest, especially in comparison with the sequences of similar regions from the DNAs of other viruses and cells, but the regions themselves should prove to be good probes for potential regulatory proteins, polymerase binding sites, etc., from infected cells.

As indicated, studies of parvovirus-DNA replication have been directly linked to structural studies. In the case of autonomous parvoviruses, the packaging of strands of only one polarity allowed somewhat easier detection of duplex replicative forms in vivo. The significant point, however, is that all parvoviruses share similar DNA structure, replicative forms, and locations for initiating and terminating DNA synthesis. This has led to consideration of a common replication mechanism, the details of which may well have application to the problem of linear-DNA replication in general.

REFERENCES

Bates, R. C., C. P. Kuchenbuch, J. T. Patton, and E. R. Stout. 1977. DNA polymerase activity associated with purified bovine parvovirus. *Am. Soc. Microbiol. Abstr.* S11.

Berns, K. I. 1974. Molecular biology of the adeno-associated viruses. *Curr. Top. Microbiol. Immunol.* **65**:1.

Berns, K. I. and S. Adler. 1972. Separation of two types of adeno-associated virus particles containing complementary polynucleotide chains. *J. Virol.* **9**:394.

Berns, K. I. and T. J. Kelly, Jr. 1974. Visualization of the inverted terminal repetition in adeno-associated virus DNA. *J. Mol. Biol.* **82**:267.

Berns, K. I. and J. A. Rose. 1970. Evidence for a single-stranded adenovirus-associated virus genome: Isolation and separation of complementary single strands. *J. Virol.* **5**:693.

Berns, K. I., J. Kort, K. H. Fife, W. E. Grogan, and I. Spear. 1975. Study of the fine structure of adeno-associated virus DNA with bacterial endonucleases. *J. Virol.* **16**:712.

Boucher, D. W., J. Melnick, and H. Mayor, 1971. Non-encapsidated infectious DNA of adeno-satellite virus in cells coinfected with herpes virus. *Science* **173**:1243.

Bourgaux, P., L. Delbecchi, and D. Bourgaux-Ramoisy. 1976. Initiation of adenovirus type 2 DNA replication. *Virology* **72**:89.

Bourguignon, G. J., P. J. Tattersall, and D. C. Ward. 1976. DNA of minute virus of mice: Self priming, nonpermuted, single-stranded genome with a 5' terminal hairpin duplex. *J. Virol.* **20**:290.

Carter, B. J. and G. Khoury. 1975. Specific cleavage of adenovirus-associated virus DNA by restriction endonuclease R. *Eco*R1—Characterization of cleavage products. *Virology* **63**:523.

Carter, B. J., G. Khoury, and D. T. Denhardt. 1975. Physical map and strand polarity of specific fragments of adenovirus-associated virus DNA produced by endonuclease R.*Eco*R1. *J. Virol.* **16**:559.

Carter, B. J., K. H. Fife, L. M. de la Maza, and K. I. Berns. 1976. Genome localization of adeno-associated virus RNA. *J. Virol.* **19**:1044.

Cavalier-Smith, T. 1974. Palindromic base sequences and replication of eukaryotic chromosome ends. *Nature* **250**:467.

Crawford, L. V., E. A. C. Follet, M. G. Burdon, and D. J. McGeoch. 1969. The DNA of a minute virus of mice. *J. Gen. Virol.* **4**:37.

de la Maza, L. M. and B. J. Carter. 1976. Cleavage of adeno-associated virus DNA with SalI, PstI and HaeII restriction endonucleases. *Nucleic Acids Res.* **3**:2605.

Denhardt, D. T., G. Eisenberg, K. Bartok, and B. J. Carter. 1976. Multiple structures of adeno-associated virus DNA: Analysis of terminally-labelled molecules with endonuclease R.HaeIII. *J. Virol.* **18**:672.

Fife, K. H. 1977. "Structure and nucleotide sequence studies of adeno-associated virus DNA." Ph.D. thesis, Johns Hopkins University, Baltimore, Maryland.

Fife, K. H., K. I. Berns, and K. Murray. 1977. Structure and nucleotide sequence of the terminal regions of adeno-associated virus DNA. *Virology* **78**:475.

Flint, S. J., S. Bergot, and P. Sharp. 1977. Characterization of single-stranded viral DNA sequences present during replication of adenovirus types 2 and 5. *Cell* **10**:559.

Gerry, H. W., T. J. Kelly, Jr., and K. I. Berns. 1973. Arrangement of nucleotide sequences in adeno-associated virus DNA. *J. Mol. Biol.* **79**:207.

Gunther, M. and P. May. 1976. Isolation and structural characterization of monomeric and dimeric forms of replicative intermediates of Kilham rat virus DNA. *J. Virol.* **20**:86.

Hampton, E. G. 1970. H-1 virus growth in synchronized rat embryo cells. *J. Microbiol.* **16**:266.

Handa, H., H. Shimojo, and K. Yamaguchi. 1976. Multiplication of adeno-associated virus type 1 in cells coinfected with a temperature-sensitive mutant of human adenovirus type 31. *Virology* **74**:1.

Hauswirth, W. W. and K. I. Berns. 1977. Origin and termination of adeno-associated virus DNA replication. *Virology* **78**:488.

Hoggan, M. D., N. R. Blacklow, and W. P. Rowe. 1966. Studies of small DNA viruses found in various adenovirus preparations: Physical, biological and immunological characteristics. *Proc. Natl. Acad. Sci.* **55**:1467.

Hoggan, M. D., A. Shatkin, N. Blacklow, F. Koczot, and J. Rose. 1968. Helper-dependent infectious deoxyribonucleic acid from adenovirus-associated virus. *J. Virol.* **2**:850.

Horwitz, M. S. 1974. Location of the origin of DNA replication in adenovirus type 2. *J. Virol.* **13**:1046.

———. 1976. Bidirectional replication of adenovirus type 2 DNA. *J. Virol.* **18**:307.

Ishibashi, M. and M. Ito. 1971. The potentiation of type 1 adeno-associated virus by temperature-sensitive conditional lethal mutants of CELO virus at the restrictive temperature. *Virology* **45**:317.

Ito, M. and E. Suzuki, 1970. Adeno-associated satellite virus growth supported by a temperature-sensitive mutant of human adenovirus. *J. Gen. Virol.* **9**:243.

Johnson, R. H. 1967. Feline panleucopaenia virus. IV. Methods for obtaining reproducible in vitro results. *Res. Vet. Sci.* **8**:256.

Koczot, F. J. 1975. "Analysis of genetic relatedness among three adeno-associated virus serotypes by competition hybridization and electron heteroduplex mapping." Ph.D. thesis, University of Maryland.

Koczot, F. J., B. J. Carter, C. F. Garon, and J. A. Rose. 1973. Self-complementarity of terminal sequences within plus or minus strands of adenovirus-associated virus DNA. *Proc. Natl. Acad. Sci.* **70**:215.

Lavelle, G. C. and A. T. Li. 1977. Isolation of intracellular replication forms and progeny single strands of KRV DNA in sucrose gradients containing guanidine hydrochloride. *Virology* **76**:464.

Lavelle, G., C. Patch, G. Khoury, and J. Rose. 1975. Isolation and partial characterization of single-stranded adenoviral DNA produced during synthesis of adenovirus type 2 DNA. *J. Virol.* **16**:775.

Mayor, H. D and J. Ratner. 1973. Analysis of adeno-associated satellite virus DNA. *Biochim. Biophys. Acta* **299**:189.

Mayor, H. D., K. Torikai, J. Melnick, and M. Mandel. 1969. Plus and minus single-stranded DNA separately encapsidated in adeno-associated satellite virions. *Science* **166**:1280.

Mayr, A., P. A. Bachmann, G. Siegl, H. Mahnel, and B. E. Sheffy. 1968. Characterization of a small porcine DNA virus. *Arch. gesamte Virusforsch.* **25**:38.

Maxam, A. M. and W. Gilbert. 1977. A new method for sequencing DNA. *Proc. Natl. Acad. Sci.* **74**:560.

McGeoch, D. J., L. V. Crawford, and E. A. C. Follett. 1970. The DNAs of three parvoviruses. *J. Gen. Virol.* **6**:33.

Parker, J. C., M. J. Collins, S. S. Cross, Jr., and W. P. Rowe. 1970. Minute virus of mice. II. Prevalence, epidemiology and natural occurrence as a contaminant of transplanted tumors. *J. Natl. Cancer Inst.* **45**:305.

Parks, W. P., M. Green, M. Pina, and J. L. Melnick. 1967. Physiochemical characterization of adeno-associated satellite virus type 4 and its nucleic acid. *J. Virol.* **1**:980.

Rhode, S. L. III. 1973. Replication process of the parvovirus H-1. I. Kinetics in a parasynchronous cell system. *J. Virol.* **11**:856.

―――. 1974a. Replication process of the parvovirus H-1. II. Isolation and characterization of H-1 replicative form DNA. *J. Virol.* **13**:400.

―――. 1974b. Replication process of the parvovirus H-1. III. Factors affecting H-1 RF DNA synthesis. *J. Virol.* **14**:791.

―――. 1976. Replication process of the parvovirus H-1. V. Isolation and characterization of temperature-sensitive H-1 mutants defective in progeny DNA synthesis. *J. Virol.* **17**:659.

―――. 1977a. Replication process of the parvovirus H-1. VI. Characterization of a replication terminus of H-1 replicative-form DNA. *J. Virol.* **21**:694.

―――. 1977b. Replication process of the parvovirus H-1. IX. Physical mapping studies of the H-1 genome. *J. Virol.* **22**:446.

Richards, P., R. Linser, and R. W. Armentrout. 1977. Kinetics of assembly of a parvovirus, minute virus of mice, in synchronized rat brain cells. *J. Virol.* **22**:778.

Rose, J. A. 1974. Parvovirus reproduction. In *Comparative virology* (ed. H. Fraenkel-Conrat and R. Wagner), vol. 3, p. 1. Plenum, New York.

Rose, J. A. and F. Koczot. 1971. Adenovirus-associated virus multiplication. VI.

Base composition of the deoxyribonucleic acid strand species and strand-specific in vivo transcription. *J. Virol.* **8**:771.

Rose, J. A., M. D. Hoggan, and A. J. Shatkin. 1966. Nucleic acid from an adeno-associated virus: Chemical and physical studies. *Proc. Natl. Acad. Sci.* **56**:86.

Rose, J. A., K. I. Berns, M. D. Hoggan, and F. J. Koczot. 1969. Evidence for a single-stranded adenovirus-associated virus genome: Formation of a DNA density hybrid on release of viral DNA. *Proc. Natl. Acad. Sci.* **64**:863.

Rose, J. A., M. D. Hoggan, F. Koczot, and A. J. Shatkin. 1968. Genetic relatedness studies with the adenovirus-associated viruses. *J. Virol.* **2**:999.

Salzman, L. A. 1977. Evidence for terminal S_1-nuclease-resistant regions on single-stranded linear DNA. *Virology* **76**:454.

Salzman, L. A. and W. White. 1973. In vivo conversion of the single-stranded DNA of Kilham rat virus to a double-stranded form. *J. Virol.* **11**:299.

Salzman, L. A., W. L. White, and T. Kakefuda. 1971. Linear, single-stranded deoxyribonucleic acid isolated from Kilham rat virus. *J. Virol.* **7**:830.

Schilling, R., B. Weingartner, and E. Winnacker. 1975. Adenovirus type 2 DNA replication. II. Termini of DNA replication. *J. Virol.* **16**:767.

Siegl, G. and M. Gautschi. 1973. The multiplication of parvovirus LuIII in a synchronized culture system. I. Optimum conditions for virus replication. *Arch. gesamte Virusforsch.* **40**:105.

Singer, I. I. and S. L. Rhode III. 1977a. Replication process of the parvovirus H-1. VII. Electron microscopy of replicative-form DNA synthesis. *J. Virol.* **21**:724.

————. 1977b. Replication process of the parvovirus H-1. VIII. Partial denaturation mapping and localization of the replication origin of H-1 replicative-form DNA with electron microscopy. *J. Virol.* **21**:724.

Spear, I. 1977. "The use of bacterial restriction endonucleases in elucidating the fine structure of AAV DNA." Ph.D. thesis, Johns Hopkins University, Baltimore, Maryland.

Straus, S. E., H. Ginsberg, and J. Rose. 1967a. DNA-minus temperature-sensitive mutants of adenovirus type 5 help adenovirus-associated virus replication. *J. Virol.* **17**:140.

Straus, S. E., E. Sebring, and J. Rose. 1976b. Concatemers of alternating plus and minus strands are intermediates in adenovirus-associated virus DNA synthesis. *Proc. Natl. Acad. Sci.* **73**:742.

Tattersall, P. 1972. Replication of the parvovirus MVM. I. Dependence of virus multiplication and plaque formation on cell growth. *J. Virol.* **10**:586.

Tattersall, P. and D. C. Ward. 1976. The rolling hairpin: A model for the replication of parvovirus and linear chromosomal DNA. *Nature* **263**:106.

Tattersall, P., A. J. Shatkin, and D. C. Ward. 1977. Sequence homology between the structural polypeptides of minute virus of mice. *J. Mol. Biol.* **111**:375.

Tennant, R. W. and R. E. Hand, Jr. 1970. Requirement of cellular synthesis for Kilham rat virus replication. *Virology* **42**:1059.

Tennant, R. W., K. R. Layman, and R. E. Hand, Jr. 1969. Effect of cell physiological state on infection by rat virus. *J. Virol.* **4**:872.

Tolun, A. and U. Pettersson. 1975. Termination sites for adenovirus type 2 DNA replication. *J. Virol.* **16**:759.

Weingartner, B., E. Winnacker, A. Tolun, and U. Pettersson. 1977. Two complementary strand specific termination sites for adenovirus DNA replication. *Cell* **10**:259.

Winnacker, E. L. 1974. Origins and termini of adenovirus type 2 DNA replication. *Cold Spring Harbor Symp. Quant. Biol.* **39**:547.

Parvovirus Transcription

BARRIE J. CARTER

Laboratory of Experimental Pathology
National Institute for Arthritis, Metabolism, and Digestive Diseases
National Institutes of Health
Bethesda, Maryland 20014

Gene transcription in eukaryotic cells is extremely complex and is likely to be controlled by a variety of mechanisms. Analysis of the properties of bulk populations of cellular RNA species can be expected to reveal only the most general principles. The determination of the precise mechanisms involved in transcription of any particular gene requires the availability of specific molecular probes. The most convenient and readily obtainable probes for eukaryotic systems are viral genomes. Since the DNA genome of the parvoviruses is among the smallest known (4000–5000 nucleotide bases in length), these viruses offer attractive experimental systems for analysis of certain aspects of eukaryotic transcription.

At the present time, detailed studies of parvovirus transcription have been carried out only with the defective adeno-associated virus (AAV). This is primarily because both plus and minus strands of AAV DNA are readily available in purified virions and because, fortunately for the purposes of transcription mapping, the strands can be separated readily on a preparative (microgram) scale. In addition, it is relatively easy to obtain milligram quantities of purified AAV DNA. Furthermore, over 100 bacterial restriction-endonuclease cleavage sites have been accurately mapped on the AAV2 genome (Carter and Khoury 1975; Berns et al. 1975; Carter et al. 1975, 1976; Denhardt et al. 1976; de la Maza and Carter 1976, 1977; de la Maza et al., this volume). Until very recently, transcription studies of the nondefective parvoviruses have been hindered because only the viral DNA strand was available in virions and because it was difficult to obtain the complementary DNA strand either from RF DNA molecules synthesized in vivo or by in vitro synthesis using the viral strand and purified polymerases. This current paucity of information concerning the nondefective parvoviruses restricts the main focus of this review to the discussion of studies with AAV.

AAV RNA TRANSCRIPTION AND THE VIRUS GROWTH CYCLE

Synthesis of AAV RNA during the virus growth cycle has been studied in KB-cell suspension cultures which were infected with AAV and a helper adenovirus (Carter et al. 1973). Viral DNA or RNA was detected by pulse-labeling and hybridization to viral DNA immobilized on nitrocellulose filters. After simultaneous infection with AAV2 and adenovirus 2, both AAV and Ad DNA synthesis were detected at 6–7 hr, whereas AAV RNA synthesis was first detected at 9–10 hr after infection. This lag between the onset of AAV DNA and RNA synthesis was eliminated if the cells were infected with the helper adenovirus for 10 hr prior to infection with AAV. Furthermore, in the adenovirus preinfection, both AAV DNA and RNA synthesis were detected much earlier (within 3–4 hr of AAV infection) than in the simultaneous infection. Analogous results were obtained in other, more indirect experiments in which AAV-antigen production was assayed rather than transcription (Blacklow et al. 1967). Similar kinetics of AAV DNA and RNA synthesis were observed in simultaneous-infection experiments using a DNA-negative adenovirus mutant as the helper at the nonpermissive temperature (Straus et al. 1976a). The time of onset of AAV RNA synthesis was also confirmed using the more sensitive assay of hybridization in solution between unlabeled RNA from infected cells and trace amounts of highly radioactive AAV DNA (Carter 1976). The presence of FUdR for 3 hr prior to infection prevented synthesis of AAV RNA and adenovirus "late" RNA but not adenovirus "early" RNA. This observation, together with the fact that DNA-negative mutants of adenovirus (Ito and Suzuki 1970; Ishibashi and Ito 1971; Mayor and Ratner 1972; Handa et al. 1975; Straus et al. 1976a,b) or herpesviruses (Drake et al. 1974) will help AAV RNA synthesis, indicates that AAV RNA synthesis is directly dependent upon synthesis of AAV DNA. Thus, AAV has no "early" RNA species analogous to those of most viruses. Similar conclusions have also been reached by Handa et al. (1976) from experiments using cycloheximide to inhibit any potential early AAV protein synthesis. The absence of an early function, together with the apparent lack of dependence on a specific S-phase host-cell function, as in the case of the nondefective parvoviruses (Tennant and Hand 1970; Tattersall 1972; Siegl and Gautschi 1973; Rhode 1973), provides a teleological rationale for the defectiveness of AAV. Also, since the infecting genome is single-stranded and infection with AAV shows single-hit kinetics (Blacklow et al. 1967), transcription of an early RNA would have implied synthesis from a single-strand template.

The explanation of the lag in onset of AAV RNA synthesis remains a mystery, as does the nature of the helper function(s) provided by adenovirus or herpesvirus. Our previous interpretation that the lag reflected the action of at least two independent adenovirus functions acting at the levels of AAV DNA synthesis and RNA synthesis respectively (Carter et al. 1973) is difficult to reconcile with the evidence using adenovirus DNA-negative helper viruses. Thus, any helper functions expressed or induced by adenovirus (or herpesvirus) for AAV DNA or RNA synthesis would appear to be expressed early after adenovirus infection. It is possible that the lag may result from a delay in availability of transcribable templates for AAV RNA synthesis rather than from intervention of another adenovirus helper function. Such a delay might result merely from competition by DNA replication enzymes or sequestering of AAV DNA progeny molecules synthesized in the earliest rounds of replication for further rounds. In a preinfection with adenovirus, the initial rounds of AAV DNA synthesis might occur more rapidly, or the available DNA replication sites might be saturated more rapidly, perhaps because of the prior replication of the adenovirus DNA. Other possibilities might include functional modifications of RNA polymerase molecules actively engaged in RNA synthesis following infection (Weinmann et al. 1976). It is interesting that the one-step growth cycle of AAV appears to exhibit strong phasing reminiscent of that for bacteriophages such as λ (Carter and Smith 1970; Dove 1971).

PHYSICAL CHARACTERIZATION AND MAPPING OF AAV RNA

Most of the physical studies of AAV RNA synthesized in vivo have employed RNA isolated from KB-cell spinner cultures infected with AAV2 and adenovirus as the helper. The overall objective of these studies has thus far been to obtain a transcription map of AAV DNA in order to define potential regulatory elements, such as sites for transcription initiation and termination and for post-transcriptional modification, in addition to defining informational (codigenic) regions of the genome. This has required construction of physical maps of AAV DNA in order to orient the transcription map. These general objectives have been achieved, at least in part, and AAV DNA now represents a well-characterized and sophisticated probe to be used in studying the molecular details of eukaryotic transcription.

The Size of AAV RNA

The size of AAV RNA synthesized in KB cells infected with AAV2 and Ad2 has been analyzed in some detail (Carter 1974, 1976; Carter and Rose 1972, 1974). In these studies RNA was labeled with [³H]uridine for varying periods 15–21 hr after infection, when viral RNA synthesis is maximal. The size distribution of isolated RNA molecules was analyzed in nonaqueous, denaturing solvents, either by velocity sedimentation in sucrose gradients containing 99% dimethyl sulfoxide (DMSO) or by acrylamide gel electrophoresis in 98% formamide. The use of strongly denaturing solvents was important to avoid various artifacts resulting from aggregation or conformational effects (Carter 1974). The ^{3}H-labeled AAV-specific RNA in gradient fractions or gel slices was detected by hybridizing to viral DNA immobilized on nitrocellulose filters. These analyses detected two populations of AAV RNA: a single, discrete, 20S species and a very heterogeneous population ranging in size from 4S to 18S (Carter and Rose 1974). Both populations were observed in the nucleus and the cytoplasm. In the cytoplasm, the small heterogeneous RNA was confined entirely to the nonpolysomal fraction (i.e., monosomes and cytoplasmic supernatant sedimenting at 80S or less). The 20S RNA was the only polysomal species and thus appears to be the only functional AAV mRNA species.

Analysis of RNA labeled in pulse-chase experiments showed that the small, heterogeneous AAV RNA was metabolically less stable than the 20S RNA (Carter and Rose 1974). The proportion of this RNA present in the nucleus or cytoplasm after a 4-hr label was frequently very low, whereas it was readily observed in a 30-min label. The significance of this heterogeneous AAV RNA is difficult to assess, and it was suggested that it might arise either by premature termination of transcription or from transcription of incomplete AAV DNA molecules, which constitute at least 10% of the total AAV DNA synthesized (de la Maza et al., this volume). Recent experiments (Jay et al., this volume) indicate that the unstable AAV RNA may represent, at least in part, nascent precursor RNA sequences. An additional distinguishing property of the two AAV RNA populations was that nearly all the 20S RNA contained poly(A), whereas the heterogeneous AAV RNA contained no poly(A) as assessed by binding to oligo(dT)-cellulose (Carter 1976).

After a 4-hr label with [^{3}H]uridine, the 20S species constituted 80–90% of the total cytoplasmic ^{3}H-labeled AAV RNA, and nearly all of this was polyadenylated. The remaining 10–20% of the cytoplasmic AAV RNA was the non-poly(A)-containing heterogeneous RNA. However, not more than about 45–55% of the total cytoplasmic RNA sedimented in a sucrose gradient with the polysomes. Thus, a significant proportion of the poly(A)-containing 20S mRNA, perhaps as much as one-third, was not associated with polysomes. Similar results (Philipson et al. 1973) have been observed with adenovirus RNA, and it was suggested that this may represent mRNA in ribonucleoprotein particles in the process of being transported to the cytoplasm.

Orientation of the AAV-DNA Physical Map

To obtain an RNA transcription map for a defined DNA genome it was necessary first to separate the complementary DNA strands and then to orient the DNA genome with respect to the DNA-strand polarity. Preparative separation of the complementary strands of AAV DNA was achieved (Berns and Rose 1970) by substitution of AAV DNA thymidine residues with BUdR, which resulted in a sufficient density difference in a CsCl density gradient at neutral pH. The strand which banded at a greater density in CsCl was defined as the minus strand because it is the strand that is transcribed into mRNA. The lighter strand was defined as the plus strand. By taking advantage of this finding, together with the ability to obtain specific fragments of DNA with bacterial restriction endonucleases, the AAV DNA genome was oriented with respect to DNA-strand polarity (Carter et al. 1975).

AAV2 DNA substituted with BUdR was labeled at the 5' termini with ^{32}P and then cleaved with endonuclease R·EcoRI (Carter and Khoury 1975). The 5' terminal ^{32}P-label was distributed equally in the terminal fragments A and B, which were then individually denatured and banded to equilibrium in CsCl density gradients. This showed that fragment A contained the 5' terminus of the minus strand and fragment B contained the 5' terminus of the plus strand. Thus, when the AAV DNA duplex is drawn in the way depicted in Figure 1, we define fragment B as the left terminus and fragment A as the right terminus (Carter et al. 1975). Thus, the 3' and 5' termini of the minus strand are at the left and right ends, respectively, of the duplex. This map of the AAV genome provided the basis for orienting the AAV transcription map and for obtaining additional AAV restriction maps. In addition, it provides a basis for orienting maps of the AAV genome with those of the nondefective parvoviruses in order to emphasize possible overall similarities of genome organization (see below).

Genome Localization of Stable AAV RNA

The DNA strand specificity of AAV RNA synthesized in vivo has been determined using two different types of molecular hybridization procedures. In one procedure (Rose and Koczot 1971) ^{32}P-labeled AAV

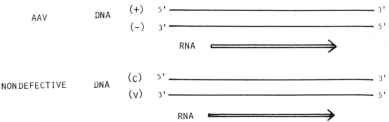

FIGURE 1
Orientation of transcription on parvovirus DNA. The polarities of AAV DNA plus (+) or minus (−) and nondefective parvovirus DNA viral (v) and complementary (c) strands are shown. The direction of transcription on minus- or v-strand DNA is from left to right, as indicated by the arrows.

RNA was first purified by filter hybridization to AAV DNA and then annealed in solution with an excess of separated plus or minus strands of [³H]BUdR AAV DNA. The resulting DNA-RNA hybrid was centrifuged to equilibrium in a CsCl density gradient. This experiment showed that the AAV RNA annealed only to the DNA minus strand. Also, the base composition of the purified AAV RNA was complementary to that of the AAV DNA minus strand. In addition, the lack of ribonuclease resistance in the self-annealed AAV RNA supported the absence of any transcript complementary to plus-strand DNA (Rose and Koczot 1971). This procedure relied on AAV RNA purified in the first step by hybridization selection and so may have failed to detect minor species. Also, it did not define how much of the DNA strand sequence was represented in the complementary RNA transcript.

An alternative procedure (Khoury and Martin 1972) which has now been used extensively to map AAV RNA (Carter et al. 1972, 1976; Jay et al., this volume) involves annealing in solution of trace amounts of high-specific-activity ³²P-labeled AAV DNA with an excessive or saturating amount of unlabeled AAV RNA. The amount of DNA contained in a hybrid molecule with RNA is then readily assessed by hydroxyapatite chromatography (Gelb et al. 1971) or digestion of nonhybridized DNA with the single-strand-specific nuclease S1 (Sutton 1971). This experimental approach has the following major advantages: (a) it is a rapid, one-step hybridization procedure, (b) it is significantly more sensitive than filter hybridization, (c) kinetic estimates of RNA concentration can be obtained, (d) the proportion of DNA that can be saturated by RNA can be easily and quantitatively determined, and (e) the viral RNA need not be purified from other cellular or viral RNAs present in the infected cells. One potential disadvantage (also a problem in the previous procedure) is that unstable RNA species with very short half-lives may not be detected.

Using this procedure, only 70–80% of the AAV DNA minus strand was saturated with AAV RNA contained in the total RNA extracted from infected cells (Carter et al. 1972). There was no significant reaction with the plus-strand DNA. Additional experiments of this type using AAV RNA fractionated from the nucleus, the cytoplasm, or the polysomes, or from poly(A)-containing or non-poly(A)-containing RNA obtained by oligo(dT)-cellulose chromatography, have failed to reveal any significant evidence for stable transcription from the DNA plus strand (Carter 1976; Carter et al. 1976; F. T. Jay, L. M. de la Maza, and B. J. Carter, manuscript in preparation).

It remains possible that plus-strand transcripts, if present at a very low concentration, might be sequestered by preferential hybridization with minus-strand transcripts and thus escape detection. To eliminate this possibility rigorously it will be necessary to perform experiments specifically designed to detect symmetric transcripts (Aloni 1972). It is clear, however, that the functional AAV mRNA is transcribed only from the minus strand, since symmetrical transcripts are not expected in polysomal mRNA. This, together with the known orientation of the AAV DNA map and the fact that RNA synthesis proceeds in the 5'→3' direction, defines the direction of transcription on the conventional AAV genome map as being from left to right (Carter et al. 1975). (See also Fig. 1 and Jay et al., this volume.)

The precise location of the stable AAV RNA on the minus strand was determined by hybridization of excess unlabeled AAV RNA with denatured ^{32}P-labeled AAV DNA restriction fragments under conditions where there was only minimal reannealing of the DNA strands (Carter et al. 1976). The data obtained with several sets of restriction fragments were internally consistent, and similar results were obtained when poly(A)-containing RNA was annealed with separated strands of the EcoRI fragments A and B. The stable AAV RNA annealed with a continuous region of the AAV DNA minus strand beginning at about 0.17–0.18 map units (i.e., 18%, or 700 nucleotides) from the left end of the molecule and ending at 0.88–0.9 units from the right end of the molecule. These determinations also showed that a total of about 70% of the minus strand was represented in the stable transcript. This is in accord with the finding that the stable 20S mRNA has a molecular weight of approximately 10^6, which is equivalent to 70% of the length of the DNA strand (Carter and Rose 1974). These observations and the fact that all regions of the AAV transcript were present at equal concentration are consistent with a single, contiguous AAV transcript. It should be noted that the procedures used in this work would not necessarily detect unstable RNA. At a minimum, the data provided an accurate map of the stable, and therefore presumably informational, AAV RNA.

Analysis of Unstable AAV RNA

The regions of the AAV DNA minus strand between 0–0.18 and 0.9–1.0 map units were not represented in stable mRNA and therefore contain potential transcriptional control elements. If there were no transcription at all in these regions, it would suggest that AAV RNA, in contrast with most other viral and eukaryotic mRNAs, was apparently not produced by post-transcriptional cleavage of a larger precursor. Alternatively, transcripts from these regions might be very rapidly degraded or perhaps preferentially lost during extraction because of their attachment to the DNA template. For instance, it is possible that the low-molecular-weight AAV RNA which was metabolically unstable (Carter and Rose 1974) might represent such sequences. Also, nuclear or total cytoplasmic RNA yielded a slightly higher saturation level of hybridization with the ^{32}P-labeled minus strand (76%) than did polysomal RNA (70%)

(Carter 1976). In the size analysis, a larger precursor to 20S RNA was not observed (Carter and Rose 1974), but a small amount of such a precursor would not be readily resolved in a DMSO-sucrose gradient and might be difficult to distinguish from residual aggregation in formamide gel electrophoresis (Carter 1974, 1976; Carter and Rose 1974). This problem has been reinvestigated using an RNA-extraction procedure different from the hot-phenol–detergent procedure used previously. In these recent experiments (Jay et al., this volume) RNA was fractionated into pellet RNA and supernatant RNA using the Hirt high-salt–SDS precipitation procedure (Jay et al., this volume). The RNA was then recovered by phenol extraction after extensive digestion with DNase and proteinase K.

When this RNA was analyzed in mapping experiments with ^{32}P-labeled AAV DNA, the supernatant RNA saturated approximately 70–76% of the minus strand and appeared to be equivalent to the stable RNA. The Hirt-pellet RNA, however, saturated approximately 90% of the minus strand. Neither fraction showed significant annealing to the plus strand. Thus, the chromatin-associated (or nascent) AAV RNA in the pellet contained sequences not present in the stable supernatant RNA. When restriction fragments of AAV DNA were used, the Hirt-supernatant RNA mapped between about 0.14 and 0.9 map units, as observed before for the stable (or 20S) AAV RNA, thus establishing the identity of these populations. The pellet RNA appears to contain all of this sequence plus contiguous regions at both the 5' end (between 0.06 and 0.14 map units) and the 3' end (between 0.90 and 0.96 map units). These experiments suggest, but do not prove, that 20S AAV mRNA may be derived by cleavage of both "leader" and "trailer" sequences from nascent transcripts.

The basis of the chromatin association of AAV RNA transcripts is not clear (Jay et al., this volume). The AAV DNA transcription template may be closely associated with the chromatin or perhaps even integrated. Alternatively, the AAV transcription template might be a higher-molecular-weight species of AAV DNA. At least half the AAV transcription activity can be solubilized as a 16–25S species (Jay et al., this volume); this suggests that this portion of the transcription template is not integrated. Furthermore, most of the AAV DNA replicating molecules, as well as the mature viral DNA, do not precipitate in the Hirt procedure, either with (Handa et al. 1976; Straus et al. 1976b) or without (Handa and Shimojo 1977) prior proteolytic digestion. This implies that the AAV transcription templates can be distinguished from the bulk of AAV replicating intermediates. Whether this is due to differences in physical structure or to cellular compartmentalization is not yet known. The possible close association of the most nascent AAV RNA transcripts with the cellular chromatin is of interest in view of recent evidence that both DNA replication and viral assembly of the nondefective parvoviruses LuIII and H-1 are also closely associated with cellular chromatin (Gautschi et al. 1976; Singer and Toolan 1975). Also, some replicating intermediates of LuIII DNA precipitate in the Hirt procedure (Siegl and Gautschi 1976). Thus, parvoviruses may prove to be useful probes for analysis of chromatin structure.

Post-Transcriptional Modification of AAV RNA

AAV RNA undergoes several types of post-transcriptional modifications. Polyadenylation of AAV RNA has been examined in detail (Carter 1976). Most of the 20S AAV RNA contained a polyadenylate tail at the 3′ end, and the lower-molecular-weight, unstable AAV RNA contained little, if any, poly(A). To estimate the size of the poly(A), AAV RNA labeled with [³H]adenosine was purified by hybridizing to AAV DNA and then digested with ribonuclease under conditions in which poly(A) was resistant to cleavage. The size of the poly(A) tract was deduced both from the proportion of [³H]adenosine-label which was resistant to ribonuclease digestion and by polyacrylamide gel electrophoresis of the limit digest. The two procedures gave similar results. Cytoplasmic RNA labeled for 4 hr contained a poly(A) sequence about 180–200 nucleotides long, which is about 6% of the total molecular mass. The poly(A) in nuclear AAV RNA was slightly larger than cytoplasmic poly(A) and also decreased in size with the length of the labeling period, as has also been reported for poly(A) in other mRNA molecules (Sheiness and Darnell 1973).

Cytoplasmic poly(A)-containing AAV RNA is also methylated both at the 5′ end and internally (B. Carter, A. Gershowitz, and B. Moss, unpublished experiments). Cytoplasmic AAV RNA with ³H-labeled methyl methionine and ¹⁴C-labeled uridine has been purified by poly(U) Sepharose chromatography and hybridization, then analyzed for its methylated nucleoside content. The predominant 5′ cap was a type-I structure and a lesser proportion were type-II structures. Since some type-II cap structures result from cytoplasmic methylation, the decreased amount of this structure in AAV RNA may merely reflect the length of the labeling period. The 20S AAV RNA also contains four to six residues per molecule of internally located N-6-methyl adenylic acid.

The biological functions of these post-transcriptional modifications are not clearly understood. Poly(A) is not required for translation (Greenberg 1975) but may be involved in determining mRNA stability, although this is controversial. The functions and positions of the internal methylations are not yet known. The function of the 5′ methyl cap is at present also controversial. For both vesicular stomatitis virus (VSV) and reovirus mRNAs, the methyl cap appears to be obligatory for ribosome binding and translation in a plant-cell system derived from wheat germ (Both et al.

1975) but is completely unnecessary for translation of VSV mRNA in the animal cell-free system derived from rabbit reticulocytes (Lodish and Rose 1977). Very recent experiments indicate that the blocking (but not the methylation) of the 5' end may promote mRNA stability by preventing 5'-exonucleolytic degradation (Furiuchi et al. 1977).

The presence of "chromatin-associated" AAV RNA sequences in the Hirt pellet in addition to those found in the stable mRNA suggests that the AAV mRNA may be derived by post-transcriptional cleavage. This cleavage may involve both the 5' and 3' regions of the putative precursor. However, we do not yet have any definitive evidence that the additional AAV sequences are contained in transcripts that are precursors of polysomal mRNA. The necessity for caution in interpretation of this information is related to a major problem in eukaryotic transcription: the significance of hnRNA. Although it is generally assumed that eukaryotic mRNA arises by cleavage and modification of higher-molecular-weight hnRNA, this faith is supported by remarkably little rigorous evidence (Davidson and Britten 1973). Analysis of the problem is hampered because of the large number of eukaryotic mRNA species, because of the difficulty of performing adequate pulse-chase studies with isotopic labels, and because at most only a very small proportion of the hnRNA is transported from the nucleus. Even for the more limited cases of viral mRNAs there is little rigorous evidence about the relevance of hnRNA species as precursors, although recent studies with adenovirus provide stronger evidence (Flint 1977).

Location of Potential Control Regions in AAV DNA

On the basis of the transcription data presently available, it appears that the informational portion of the AAV gene is located between 0.14–0.18 and 0.88–0.9 map units. Also, since additional sequences to the left and right of this region are transcribed, these two sites may contain important recognition sequences for post-transcriptional cleavage systems.

Since there is only one final mRNA end product, it seems reasonable to suppose that there may be only one promoter for RNA transcription. The location of this is not immediately defined, except that it must be at or to the left of (i.e., upstream of) about 0.06 map units. This places the promoter potentially very close to the origin or termination site of DNA replication, since the promoter cannot be more than 200 nucleotides from the 5' terminus, and the replication origin at the left end is about 40–50 nucleotides from the 5' terminus (see Hauswirth and Berns, this volume). Unambiguous definition of the in vivo AAV transcription promoter requires identification of the 5' triphosphate terminus of the nascent RNA strand, and this is now being attempted. The termination site for AAV RNA transcription is also difficult to define solely on the basis of hybridization experiments because of the instability of RNA from this region. In contrast with the promoter site, there is no simple, unambiguous marker for precise definition of the in vivo transcription-termination site. It is probably located at or to the right of (i.e., downstream of) about 0.95 map units. This is potentially within 150–200 nucleotides of the right origin or termination site of DNA replication (Hauswirth and Berns, this volume).

One interesting feature is that even the most nascent AAV RNA transcripts detected thus far do not appear to extend into the inverted repetitious sequences in AAV DNA, which comprise approximately the terminal 2% (i.e., about 80–100 nucleotides) (Koczot et al. 1973; Berns and Kelly 1974; de la Maza and Carter 1977; Berns et al., this volume). This raises some interesting possibilities for coordination of the control of the viral growth cycle (Jay et al., this volume).

Effect of Helper Virus or Host Cell on AAV RNA Transcription

The analysis of AAV2 transcription has been performed mostly in the permissive system using human cells as the host and human adenovirus as the helper. In several other systems using other host cells or helper viruses, AAV multiplication is abortive, but none of these systems characterized thus far appear to have any defect specifically at the level of AAV RNA transcription. For instance, herpesviruses are partial helpers of AAV to the extent that AAV capsid antigen is made (Atchison 1970; Blacklow et al. 1970), and in some cases empty particles are formed, but no infectious virions are produced. In this system, AAV DNA is synthesized and is claimed to be infectious (Boucher et al. 1971).

The AAV2-specific RNA molecules synthesized in KB cells using either herpes simplex type I or adenovirus type 2 as a helper were identical as compared by size in sedimentation experiments or by nucleotide sequence in RNA/DNA hybridization experiments (Carter and Rose 1972; Carter et al. 1972). The defect in herpesvirus helper cells appears to be at the level of DNA encapsidation.

A number of studies have indicated that temperature-sensitive, DNA-negative mutants of human adenoviruses and herpes simplex virus will efficiently help AAV at the nonpermissive temperature. This implies that the helper function is produced early after infection and might therefore be present in adenovirus-transformed or herpesvirus-transformed cells. However, various adeno-transformed cell lines studied so far do not help AAV. AAV did not multiply in Ad12-transformed cells (Hoggan et al. 1966), in the Ad2-transformed rat-embryo cell line 8517 of Freeman et al. (1967), which contains most of the early regions of the genome, or in the Ad2-transformed rat-embryo cell line F18 of Gallimore et al. (1974), which contains only the left region of the adenovirus genome. In the 8517 and F19 cell lines, AAV DNA and

RNA synthesis were not detected even when the cells were superinfected with adenovirus (B. Carter, unpublished).

AAV2 will multiply in human-embryo kidney (HEK) cells transformed by either a fragment of Ad5 DNA (Graham et al. 1974) or SV40 [the NB cell line of Shein and Enders (1962)], but only if adenovirus is added as a helper (C. A. Laughlin and B. Carter, unpublished). Also, it has been reported recently that AAV protein (and therefore presumably AAV RNA) is synthesized in some herpesvirus-transformed cell lines in the absence of any additional helper virus (Blacklow 1975). However, it is not clear whether these herpesvirus-transformed cell lines contained free herpesvirus DNA.

In the primary cells from the African green monkey kidney (AGMK), growth of human adenoviruses is severely restricted, but it is enhanced by coinfection of the cells with the simian virus SV40. AAV will multiply in AGMK cells, but only if they are coinfected with both a human adenovirus and SV40 (Blacklow et al. 1967). In AGMK cells infected with AAV and human adenovirus type 7 or type 5, both AAV DNA and RNA are synthesized but there is no synthesis of AAV capsid protein. In the absence of SV40 as a helper there is no synthesis of AAV DNA or RNA. The AAV DNA replication in the presence of Ad2 alone appears to be normal, and the AAV RNA synthesized is the normal 20S species. Thus, the defect overcome by SV40 appears to be a post-transcriptional host-cell restriction (presumably at the level of translation) analogous to the restriction for adenovirus in these cells (Salzman and Khoury 1974). The nature of this host-cell restriction in not clear, and it is the subject of some controversy because, for adenovirus in AGMK cells, there is a deficiency of some particular late mRNA species (Klessig and Andersen 1975). The Vero cell line, which was derived from an African green monkey kidney, is permissive for human adenoviruses, and this system is also permissive for production of intact AAV virions without the necessity for SV40 as a second helper (B. Carter, unpublished). Primary AGMK cells are permissive for the simian adenoviruses SV15 and SA7, and in these cells both viruses help multiplication of the simian virus AAV4. However, in the same cells SV15 but not SA7 will help AAV serotypes 1, 2, and 3 (Boucher et al. 1969). The basis of these restrictions and whether they are host-cell-specified have not been investigated. However, it is interesting that AAV4 is of African green monkey origin, AAV1 is believed to be of rhesus monkey origin, and AAV2 and AAV3 are human viruses.

In Vitro Transcription of AAV

We have begun analysis of AAV RNA synthesis in vitro (see Jay et al., this volume), since this offers a potentially simple system, having a well-defined single mRNA end product, for studying eukaryotic transcription.

When purified nuclei from cells infected with AAV and Ad2 are lysed with Sarkosyl in the presence of 0.1 M salt according to the procedure of Gariglio and Mousset (1975), at least 50% of the AAV-specific transcription activity is released from infected cells as a soluble viral transcription complex (VTC) sedimenting in sucrose at 16–26S. This complex elongates but does not reinitiate predominantly AAV RNA transcripts. However, these complexes offer an approach to purifying AAV RNA and to determining the AAV transcription template as discussed above. Synthesis of AAV RNA by the VTC was inhibited with intermediate levels (1–2 μg/ml) of α-amanitin (Jay et al., this volume), which indicated that AAV RNA, like most other viral and cellular mRNAs, is probably synthesized by an RNA polymerase II (Roeder 1976). A similar conclusion was reached from analysis of α-amanitin inhibition of AAV RNA in a system employing isolated nuclei from infected cells (Bloom and Rose 1977).

Transcription of purified AAV2 DNA with RNA polymerase II isolated from uninfected or adenovirus-infected KB cells has been studied, and results were similar with the enzyme from either source (see Jay et al., this volume). These experiments suggested that the purified enzyme initiates in vitro transcription randomly on either strand. These findings imply that the strand-specific AAV transcription observed in vivo requires other factors in addition to RNA polymerase II. It is still possible, but much less likely, that strand-specific transcription in vivo results from post-transcriptional control by very rapid degradation of the plus-strand transcript. The prokaryotic E. coli RNA polymerase also transcribed purified AAV DNA symmetrically (Rose and Koczot 1971).

TRANSCRIPTION OF NONDEFECTIVE-PARVOVIRUS GENOMES

Studies on transcription of the nondefective parvoviruses have lagged far behind those on AAV, but on the basis of the sparse information available it is tempting to suggest that the overall transcription scheme of all the parvoviruses may be similar.

Salzman and Redler (1974) studied transcription of Kilham rat virus (KRV) in randomly growing rat nephroma cells using the viral (v) strand of KRV DNA as a hybridization probe. They detected KRV-specific RNA complementary to the viral DNA as early as 2 hr after infection. The KRV RNA labeled between 5 and 20 hr after infection was analyzed in DMSO-sucrose gradients and showed a major peak sedimenting at the rate of 18S rRNA. This is equivalent in size to a molecule of about 50% the length of the viral DNA. Since the proportion of this DNA strand represented in the viral RNA was not measured, it is not known how many RNA species the 18S RNA might represent. No information is available on whether any RNA is synthesized from the nonviral complementary (c) strand of DNA. There was some indication of KRV RNA transcripts as large as 26S

(Salzman and Redler 1974) analogous to those seen for AAV (Carter and Rose 1974; Carter 1974), but for similar reasons their significance could not be assessed. Preliminary analysis of RNA transcribed from MVM DNA reveals the presence in infected cells of polyadenylated RNA-containing sequences complementary to approximately 70% of the DNA v strand (D. C. Ward, personal communication).

The preliminary characterization of KRV and MVM stable RNA is reminiscent of that from AAV with respect to size and proportion of the viral strand sequence represented. However, it is not yet known whether any RNA strand is transcribed from the viral DNA c strand. The viral strands of KRV and MVM DNAs are analogous to the minus strand of AAV in that they are also rich in thymidine. These facts begin to point to a possible general scheme for parvovirus transcription. It is suggested, therefore, that for the purposes of ready comparison the transcription and genome maps of all these viruses be drawn according to an analogous scheme based on that for AAV, as shown in Figure 1. Thus, the AAV minus strand is drawn with its 3' end at the left and the nondefective-parvovirus v strand is drawn with its 3' end at the left. Transcription on either minus-strand or v-strand DNA will then proceed from left to right.

The temporal relationship between KRV RNA synthesis and DNA synthesis in rat nephroma cells is not clear. While KRV RNA synthesis was apparently detected as early as 2 hr after infection (Salzman and Redler 1974), progeny DNA synthesis, as measured by FUdR inhibition of infectious-virus production, did not appear to begin until 8 hr after infection (Salzman et al. 1972). On the other hand, it was reported (Salzman and White 1973) that the infecting KRV genome became double-stranded within 60 min after infection. Interpretation of these findings is difficult for several reasons. In all this work, the rat nephroma cells were not synchronized for growth, and it is well known that the nondefective parvoviruses are dependent upon a specific phase of the cell cycle. The experiments of Salzman et al. (1972) measured only the onset of packaging of progeny DNA strands into virions and not the onset of DNA synthesis. The double-stranded KRV DNA present within 60 min of infection might be due to the presence in the inoculum of virus particles containing the complementary DNA strand rather than the viral DNA strand (Bourguignon et al. 1976).

It is now a relatively straightforward task to obtain the complementary DNA strand of the nondefective parvoviruses by self-primed in vitro DNA synthesis on the viral strand. This now allows detection of RNA which may be transcribed from the c-strand DNA. It will also allow determination by molecular hybridization of the temporal relationships between DNA and RNA synthesis.

Parvovirus Transcription as Related to Translation and Genome Structure

Electrophoretic analysis of disrupted virions in several laboratories indicated that both AAV and the nondefective parvoviruses contain three polypeptides, A, B, and C, having molecular weights of about 90K, 70K, and 60K, respectively. For AAV, the A and B proteins each contained about 10% and C contained 80% of the total virion protein (Rose et al. 1971; Johnson et al. 1971). Similar observations were made for the nondefective parvoviruses, except that in some cases only two proteins were seen (i.e., A and B or C) or the relative amounts of B and C showed an inverse relationship in "heavy" and "light" virions (see review by Tattersall, this volume). These observations suggested a coding problem, since a parvovirus genome approximately 4500 nucleotides long could not specify three proteins in the 60–90K range unless translation occurred in more than one reading frame. [Recent studies on φX174 by Sanger et al. (1977) show that it is possible to specify two proteins from different reading frames on the same stretch of nucleotide sequence.] This problem was highlighted more when it was shown that AAV mRNA is a single 20S mRNA species which can specify only one protein of up to about 100K, and it was therefore argued that AAV must contain only one gene, presumably for its capsid protein (Carter and Rose 1974). This protein might appear as three polypeptide species as a result of different translation initiation or termination sites or because of post-translational cleavage or processing. Concomitantly, it was shown by peptide mapping that the three polypeptides of MVM are related and presumably derive from a single translation product.

The purified 20S mRNA of AAV has recently been translated in vitro by Buller and Rose (1977) (see also Buller and Rose, this volume). In a reticulocyte cell-free system, only the 90K protein is synthesized, whereas in a KB cell-free system all three proteins, 90K, 70K, and 60K, are synthesized. In vivo experiments using the amino acid analog L-canavanine, which abolishes the 60K protein, and in vitro experiments with formyl-[^{35}S]L-methionine to label the protein amino terminus indicate that the three polypeptides result from post-translational proteolytic cleavage rather than from multiple sites for initiation or termination of translation. ›

The present information clearly suggests that AAV contains only one gene: that for its coat protein. The less complete information available for the nondefective parvoviruses, such as MVM, H-1, LuIII, and KRV, argues for an analogous situation. The basis for the biological difference between the defective AAV and nondefective parvoviruses appears to lie at the level of DNA synthesis and is presumably related to the different types of terminal-sequence arrangement for the two virus groups. Thus, the nondefective parvovirus DNA can be replicated by the cellular machinery present in S phase, whereas AAV replicates only in cells somehow modified by a coinfecting helper adenovirus or herpesvirus.

SUMMARY

The basic elements of AAV transcription now appear to be fairly well defined. It seems clear that AAV synthesizes only one mRNA species and contains only one gene. The AAV mRNA appears to be synthesized in a fashion analogous to that of most eukaryotic mRNA species with respect to both post-transcriptional cleavage and subsequent modification. The general organization of the AAV genome in fact resembles that of eukaryotic genomes to the extent that it is a single unique gene (codigenic) sequence bounded by reiterated sequences which might play some role in transcriptional control in addition to their role in replication. Therefore, AAV now offers a useful experimental system for analyzing the molecular mechanisms of eukaryotic transcription and post-transcriptional processing in more detail. Initial objectives for future studies include a more precise determination of the AAV transcription promoter and terminator and definition and characterization of the transcription template.

In addition, AAV may be useful as a probe to study the association of transcriptional templates with the cellular chromatin. It may also be possible to establish a variety of in vitro systems to assay for protein or other factors, in addition to RNA polymerase, which probably are required for transcription initiation, elongation, or termination and for controlling strand specificity of transcription.

It is anticipated that, with the availability of both DNA strands of several nondefective parvoviruses and with work currently proceeding, similar information will soon be available for these viruses as well as for AAV.

ACKNOWLEDGMENTS

A large part of the work on AAV transcription which I have reviewed here is the result of many enjoyable collaborations with Luis de la Maza, Francis Jay, Cathy Laughlin, Bill Cook, Jim Rose, Bernie Moss, Dave Denhardt, Ken Berns, Ken Fife, and George Khoury. I thank Mark Buller, Steve Straus, Ken Berns, and Dave Ward for communicating unpublished results.

REFERENCES

Aloni, Y. 1972. Extensive symmetrical transcription of simian virus 40 DNA in virus yielding cells. *Proc. Natl. Acad. Sci.* **69**:2024.

Atchison, R. W. 1970. The role of herpesviruses in adeno-associated virus replication in vitro. *Virology* **42**:155.

Berns, K. I. and T. J. Kelly, Jr. 1974. Visualization of the inverted terminal repetition in adeno-associated virus DNA. *J. Mol. Biol.* **82**:267.

Berns, K. I. and J. A. Rose. 1970. Evidence for a single-stranded adenovirus-associated virus genome: Isolation and separation of complementary single strands. *J. Virol.* **5**:693.

Berns, K. I., J. Kort, K. H. Fife, W. Grogan, and I. Spear. 1975. Study of the fine structure of adeno-associated virus DNA with bacterial restriction endonucleases. *J. Virol.* **16**:712.

Blacklow, N. R. 1975. Potentiation of an adenovirus-associated virus by herpes simplex virus type 2 transformed cells. *J. Natl. Cancer Inst.* **54**:241.

Blacklow, N. R., M. D. Hoggan, and M. S. McClanahan. 1970. Adenovirus-associated viruses: Enhancement by human herpesviruses. *Proc. Soc. Exp. Biol. Med.* **134**:952.

Blacklow, N. R., M. D. Hoggan, and W. P. Rowe. 1967. Immunofluorescent studies of the potentiation of an adenovirus-associated virus by adenovirus 7. *J. Exp. Med.* **125**:755.

Bloom, M. E. and J. A. Rose. 1977. Transcription of AAV RNA in isolated KB cell nuclei. *Am. Soc. Microbiol. Abstr.*, p. 281

Both, G. W., Y. Furiuchi, S. Muthukrishnan, and A. J. Shatkin. 1975. Ribosome binding to reovirus mRNA in protein biosynthesis requires 5' terminal 7-methyl guanosine. *Cell* **6**:185.

Boucher, D. W., J. L. Melnick, and H. D. Mayor. 1971. Non-encapsidated infectious DNA of adeno-satellite virus in cells coinfected with herpesvirus. *Science* **173**:1243.

Boucher, D. W., W. P. Parks, and J. L. Melnick. 1969. Failure of a replicating adenovirus to enhance adeno-associated satellite virus replication. *Virology* **39**:932.

Bourguignon, G. J., P. J. Tattersall, and D. C. Ward. 1976. DNA of minute virus of mice: Self-priming, non-permuted, single-stranded genome with a 5'-terminal hairpin duplex. *J. Virol.* **20**:290.

Buller, R. M. L. and J. A. Rose. 1977. In vivo and in vitro translation of AAV messenger RNA. *Am. Soc. Microbiol. Abstr.*, p. 302

Carter, B. J. 1974. Analysis of parvovirus mRNA by sedimentation and electrophoresis in aqueous and nonaqueous solutions. *J. Virol.* **14**:834.

―――. 1976. Intracellular distribution and polyadenylate content of adeno-associated virus RNA sequences. *Virology* **73**:273.

Carter, B. J. and G. Khoury. 1975. Specific cleavage of adenovirus-associated virus DNA by restriction endonuclease R.*Eco*R1—Characterization of cleavage products. *Virology* **63**:523.

Carter, B. J. and J. A. Rose. 1972. Adenovirus-associated virus multiplication. Analysis of the in vivo transcription induced by complete or partial helper viruses. *J. Virol.* **10**:9.

―――. 1974. Transcription in vivo of a defective parvovirus: Sedimentation and electrophoretic analysis of RNA synthesized by adenovirus-associated virus and its helper adenovirus. *Virology* **61**:182.

Carter, B. J. and M. G. Smith. 1970. Intracellular pools of bacteriophage λ deoxyribonucleic acid. *J. Mol. Biol.* **50**:713.

Carter, B. J., G. Khoury, and J. A. Rose. 1972. Adenovirus-associated virus multiplication. Extent of transcription of the viral genome in vivo. *J. Virol.* **10**:1118.

Carter, B. J., L. M. de la Maza, and F. T. Jay. 1977. The structure of the adeno-associated virus genome. In *DNA insertion elements, plasmids, and episomes* (ed. A. I. Bukhari, J. A. Shapiro, and S. L. Adhya), p. 477. Cold Spring Harbor Laboratory, Cold Spring Harbor, New York.

Carter, B. J., G. Khoury, and D. T. Denhardt. 1975. Physical map and strand polarity of specific fragments of adenovirus-associated virus DNA produced by endonuclease R.*EcoR1. J. Virol.* **16**:559.

Carter, B. J., K. H. Fife, L. M. de la Maza, and K. L. Berns. 1976. Genome localization of adeno-associated virus RNA. *J. Virol.* **19**:1044.

Carter, B. J., F. J. Koczot, J. Garrison, J. A. Rose, and R. Dolin. 1973. Separate helper functions provided by adenovirus for adenovirus-associated virus multiplication. *Nat. New Biol.* **244**:71.

Davidson, E. H. and R. J. Britten. 1973. Organization, transcription and regulation in the animal genome. *Q. Rev. Biol.* **48**:565.

de la Maza, L. M. and B. J. Carter. 1976. Cleavage of adeno-associated virus DNA with *Sal*I, *Pst*I, and *Hae*II restriction endonucleases. *Nucleic Acids Res.* **3**:2605.

————. 1977. Adeno-associated virus DNA. Restriction endonuclease maps and arrangement of terminal sequences. *Virology* (in press).

Denhardt, D. T., S. Eisenberg, K. Bartok, and B. J. Carter. 1976. Multiple structures of adeno-associated virus DNA: Analysis of terminally labeled molecules with endonuclease R.*Hae*III. *J. Virol.* **18**:672.

Dove, W. F. 1971. Biological inferences. In *The bacteriophage lambda* (ed. A. D. Hershey), p. 297. Cold Spring Harbor Laboratory, Cold Spring Harbor, New York.

Drake, S., P. A. Schaffer, J. Esparza, and H. D. Mayor. 1974. Complementation of adeno-associated satellite viral antigens and infectious DNA by temperature-sensitive mutants of herpes simplex virus. *Virology* **60**:230.

Flint, J. 1977. The topography and transcription of the adenovirus genome. *Cell* **10**:153.

Freeman, A. E., P.H. Black, E. A. Vanderpool, P. H. Henry, J. B. Austin, and R. J. Huebner. 1967. Transformation of primary rat embryo cells by adenovirus type 2. *Proc. Natl. Acad. Sci.* **58**:1205.

Furiuchi, Y., A. Lafiandra, and A. J. Shatkin. 1977. 5'-Terminal structure and mRNA stability. *Nature* **266**:235.

Gallimore, P. H., P. A. Sharp, and J. Sambrook. 1974. Viral DNA in transformed cells. II. A study of the sequences of adenovirus DNA in nine lines of transformed rat cells using specific fragments of the viral genome. *J. Mol. Biol.* **89**:49.

Gariglio, P. and S. Moussett. 1975. Isolation and partial characterization of a nuclear RNA polymerase–SV40 DNA complex. *FEBS Lett.* **56**:149.

Gautschi, M., G. Siegl, and G. Kronauer. 1976. Multiplication of parvovirus LuIII in a synchronized culture system. IV. Association of viral structural polypeptides with the host cell chromatin. *J. Virol.* **20**:29.

Gelb, L. D., D. E. Kohne, and M. A. Martin. 1971. Quantitation of SV40 sequences in African green monkey, mouse and virus-transformed cell genomes. *J. Mol. Biol.* **57**:129.

Graham, F. L., A. J. van der Eb, and H. L. Heijneker. 1974. Size and location of the transforming region in human adenovirus type 5 DNA. *Nature* **251**:687.

Greenberg, J. 1975. Messenger RNA metabolism in animal cells. *J. Cell Biol.* **64**:269.

Handa, H. and H. Shimojo. 1977. Viral DNA synthesis in vitro with nuclei isolated from adenovirus-associated virus type-1 infected cells. *Virology* **77**:424.

Handa, H., H. Shimojo, and K. Yamaguchi. 1976. Multiplication of adeno-associated virus type 1 in cells coinfected with a temperature-sensitive mutant of human adenovirus type 31. *Virology* **74**:1.

Handa, H., K. Shiroki, and H. Shimojo. 1975. Complementation of adeno-associated virus growth with temperature-sensitive mutants of human adenovirus types 12 and 5. *J. Gen. Virol.* **29**:239.

Hoggan, M. D., N. R. Blacklow, and W. P. Rowe. 1966. Studies of small DNA viruses found in various adenovirus preparations: Physical, biological and immunological characteristics. *Proc. Natl. Acad. Sci.* **55**:1467.

Ishibashi, M. and M. Ito. 1971. The potentiation of type I adeno-associated virus by temperature-sensitive conditional-lethal mutants of CELO virus at the restrictive temperature. *Virology* **45**:317.

Ito, M. and E. Suzuki. 1970. Adeno-associated satellite virus growth supported by a temperature-sensitive mutant of human adenovirus. *J. Gen. Virol.* **9**:243.

Johnson, F. B., H. L. Ozer, and M. D. Hoggan. 1971. Structural proteins of adenovirus associated virus type 3. *J. Virol.* **8**:860.

Khoury, G. and M. A. Martin. 1972. A comparison of SV40 DNA transcription in vivo and in vitro. *Nat. New Biol.* **238**:4.

Klessig, D. F. and C. W. Anderson. 1975. Block to multiplication of adenovirus serotype 2 in monkey cells. *J. Virol.* **16**:1650.

Koczot, F. J., B. J. Carter, C. F. Garon, and J. A. Rose. 1973. Self-complementarity of terminal sequences within plus or minus strands of adenovirus-associated virus DNA. *Proc. Natl. Acad. Sci.* **70**:215.

Lodish, H. F. and J. K. Rose. 1977. Relative importance of 7-methylguanosine in ribosome binding and translation of vesicular stomatitis virus mRNA in wheat germ and reticulocyte cell-free systems. *J. Biol. Chem.* **252**:1181.

Mayor, H. D. and J. Ratner. 1972. Conditionally defective helper adenoviruses and satellite virus replication. *Nat. New Biol.* **239**:20.

Philipson, L., U. Lindberg, T. Persson, and B. Vennström. 1973. Transcription and processing of adenovirus RNA in productive infection. *Adv. Biosci.* **11**:167.

Rhode, S. L. III. 1973. Replication process of the parvovirus H-1. Kinetics in a parasynchronous cell system. *J. Virol.* **11**:586.

Roeder, R. G. 1976. Eukaryotic nuclear RNA polymerases. In *RNA polymerase* (ed. R. Losick and M. Chamberlin), p. 285. Cold Spring Harbor Laboratory, Cold Spring Harbor, New York.

Rose, J. A. and F. J. Koczot. 1971. Adenovirus-associated virus multiplication. Base composition of the deoxyribonucleic acid strand species and strand-specific in vivo transcription. *J. Virol.* **8**:771.

Rose, J. A., J. V. Maizel, Jr., J. K. Imman, and A. J. Shatkin. 1971. Structural proteins of adenovirus-associated viruses. *J. Virol.* **8**:766.

Salzman, L. A. and B. Redler. 1974. Synthesis of virus-specific RNA in cells infected with the parvovirus Kilham rat virus. *J. Virol.* **14**:434.

Salzman, L. A. and W. L. White. 1973. In vivo conversion of the single-stranded DNA of Kilham rat virus to a double-stranded form. *J. Virol.* **11**:299.

Salzman, L. A., W. L. White, and L. McKerlie. 1972. Growth characteristics of Kilham rat virus and its effect on cellular macromolecular synthesis. *J. Virol.* **10**:573.

Salzman, N. P. and G. Khoury. 1974. Reproduction of papovaviruses. In *Comprehensive virology* (ed. H. Fraenkel-Conrat and R. R. Wagner), vol. 3, p. 63. Plenum, New York.

Sanger, F., G. M. Air, B. G. Barrell, N. L. Brown, A. R. Coulson, J. C. Fiddes, C. A. Hutchison III, P. M. Slocombe, and M. Smith. 1977. Nucleotide sequence of bacteriophage φX174 DNA. *Nature* **265**:687.

Shein, H. M. and J. F. Enders. 1962. Multiplication and cytopathogenicity of simian vacuolating virus 40 in cultures of human tissues. *Proc. Soc. Exp. Biol.* **109**:495.

Sheiness, D. and J. E. Darnell. 1973. Polyadenylic acid segment in mRNA becomes shorter with age. *Nat. New Biol.* **241**:265.

Siegl, G. and M. Gautschi. 1973. The multiplication of parvovirus LuIII in a synchronized culture system. II. Biochemical characteristics of virus replication. *Arch. gesamte Virusforsch.* **40**:119.

———. 1976. Multiplication of parvovirus LuIII in a synchronized culture system. III. Replication of viral DNA. *J. Virol.* **17**:841.

Singer, I. I. and H. W. Toolan. 1975. Ultrastructural studies of H-1 parvovirus replication. I. Cytopathology produced in human NB epithelial cells and hamster embryo fibroblasts. *Virology* **65**:40.

Straus, S. E., H. S. Ginsberg, and J. A. Rose. 1976a. DNA-minus temperature-sensitive mutants of adenovirus type 5 help adenovirus-associated virus replication. *J. Virol.* **17**:140.

Straus, S. E., E. D. Sebring, and J. A. Rose. 1976b. Concatemers of alternating plus and minus strands are intermediates in adenovirus-associated virus DNA synthesis. *Proc. Natl. Acad. Sci.* **73**:742.

Sutton, W. D. 1971. A crude nuclease preparation suitable for use in DNA reassociation experiments. *Biochim. Biophys. Acta* **240**:522.

Tattersall, P. 1972. Replication of the parvovirus MVM. Dependence of virus multiplication and plaque formation on cell growth. *J. Virol.* **10**:586.

Tennant, R. W. and R. E. Hand, Jr. 1970. Requirement of cellular synthesis for Kilham rat virus replication. *Virology* **42**:1054.

Weinmann, R., J. A. Jaehning, H. J. Raskas, and R. G. Roeder. 1976. Viral RNA synthesis and levels of DNA-dependent RNA polymerases during replication of adenovirus 2. *J. Virol.* **17**:114.

Parvovirus Protein Structure and Virion Maturation

PETER TATTERSALL

Imperial Cancer Research Fund
Mill Hill Laboratories
London NW7 1AD, England

When parvovirus particles extracted from infected cells are centrifuged to equilibrium in cesium chloride, four main density classes are usually observed (Rose 1974). Particles banding at 1.30–1.32 g/cm³ appear in the electron microscope to be empty capsids. Labeling studies with radioactive thymidine have shown that these particles are essentially devoid of DNA. The particles banding at 1.35–1.37 g/cm³ are fairly heterogeneous in density distribution and appear to contain DNA molecules shorter than genome length (Torikai et al. 1970; Siegl 1972). The origin of this DNA (whether cellular or viral) and the mechanism of its formation are not clear for many parvoviruses, but recent studies suggest defective viral genomes generated by aberrant replication (de la Maza et al., this volume; Müller et al., this volume). The major peak of infectious virions occurs at 1.39–1.42 g/cm³; a minor peak occurs at 1.45–1.47 g/cm³.

POLYPEPTIDE COMPOSITION OF NONDEFECTIVE PARVOVIRUSES

The polypeptide compositions of some of these different virion forms have been examined for a number of parvovirus serotypes by polyacrylamide gel electrophoresis in sodium dodecyl sulfate (SDS). The results published to date are summarized in Table 1. At first sight the data look somewhat confusing. The observed polypeptides have been grouped into four size classes in an attempt to rationalize the differences among viruses and, indeed, among different reports concerning the same virus. The largest virion protein is a minor component with a molecular weight

TABLE 1
Parvovirus Structural Polypeptides

Virus	Particle type[a]	Particle density[b]	Structural polypeptides				Reference
			A	B	C	D	
KRV	LF	1.40–1.41	72[c] (13)[d]		62 (76)	55 (11)	1
H-1	F	nr	92 (15)		72 (75)	56 (10)	2
BPV	LF	1.40–1.42	86 (10)	77 (8)	67 (83)	—	3
	E	1.30–1.31	86 (10)	77 (8)	67 (83)	—	3
LuIII	LF	1.41	74 (16)		62 (85)	—	4
	I	1.35	74 (14)	[68 (4)][e]	61 (82)	—	4
	E	1.32	76 (9)	[69 (4)][e]	63 (87)	—	4
FPV	LF	1.41	73 (10)		60 (86)	40 (3–6)	5
H-1	F	nr	92 (nr)	72 (nr)	69 (nr)	56 (nr)	6
H-1	F	nr	92 (20)	72 (80)	—	—	7
MVM	HF	1.46	92 (12)	72 (86)	69 (2)	—	8
	LF	1.41	92 (12)	72 (27)	69 (61)	—	8
	E	1.32	92 (14)	72 (79)	69 (7)	—	8
MVM	LF-HF	1.42–1.47	83 (14)	64 (15–76)	61 (7–70)	50 (vtr)	9
	E	1.32	84 (17)	64 (83)	—	—	9
X14	LF	1.40–1.43	82 (14)		68 (86)	—	10
AAV1	LF	1.38–1.42	87 (8)	73 (5)	62 (86)	—	10

54

AAV2	LF	1.38–1.42	87 (8)	73 (5)	≤? (86)	—	11
AAV3	LF	1.38–1.42	87 (8)	73 (5)	62 (86)	—	11
AAV3H	LF	1.40–1.41	92 (11)	78 (11)	66 (79)	—	12
	HF	1.47	93 (7)	80 (7)	69 (85)	—	12
AAV3H	E	1.34	92 (10)	80 (10)	66 (80)	—	13
AAV1	LF	1.38–1.40	84 (10)	70 (12)	56 (78)	—	10
AAV4	LF	1.43	83 (7)	71 (10)	58 (83)	—	10

nr: Not reported.
vtr: Variable; trace amounts detected in some virus preparations (Tattersall et al. 1976).
[a] LF: "light" full; HF: "heavy" full; I: intermediate; E: empty; F: full particle of unspecified density.
[b] Density in CsCl (g/cm³).
[c] Molecular weight × 10⁻³.
[d] Figures in parentheses are weight percents of total virion protein.
[e] Probable host contaminant (Gautschi and Siegl 1973).

References

[1] Salzman and White 1970.
[2] Kongsvik and Toolan 1972.
[3] Johnson and Hoggan 1973.
[4] Gautschi and Siegl 1973.
[5] Johnson et al. 1974.
[6] Kongsvik et al. 1974b.
[7] Kongsvik 1974a.
[8] Clinton and Hayashi 1975.
[9] Tattersall et al. 1976.
[10] Salo and Mayor 1977.
[11] Rose et al. 1971.
[12] Johnson et al 1971.
[13] Johnson et al. 1975.

55

variously reported to be between 92,000 and 73,000. This polypeptide species, which has been demonstrated in all density classes of both nondefective-parvovirus and AAV virions, will be referred to here as the A protein. It is consistently a minor component, comprising between 7% and 20% of the total virion protein.

The major differences between reports on parvovirus proteins concern the two middle size classes. Early studies showing the middle polypeptide of the nondefective parvoviruses as a single polypeptide species (Salzman and White 1970; Kongsvik and Toolan 1972; Gautschi and Siegl 1973) are almost certainly incorrect. The reason for this is that the SDS-phosphate gel system of Shapiro et al. (1967) used by these investigators, combined with conventional gel-slicing techniques for the detection of radiolabeled protein, does not resolve the two components in this protein band, which differ in molecular weight by less than 5%. This doublet is often resolved in stained SDS-phosphate gels, but more convincing evidence is obtained with the SDS-Tris-glycine gel system described by Laemmli (1970) used as the slab-gel modification (Anderson et al. 1973; Studier 1973); the doublet is detected by staining or autoradiography. The doublet, containing what will be referred to here as polypeptides B and C, represents the majority of the virion protein mass. The doublet nature of this protein band has been reported for H-1 (Kongsvik et al. 1974b) and minute virus of mice (MVM) (Clinton and Hayashi 1975; Tattersall et al. 1976) and has recently also been demonstrated for Kilham rat virus (KRV), H-3, LuIII, and bovine parvovirus (BPV) (see Peterson et al., this volume; Bates et al., this volume). Which polypeptide is the major species in the doublet depends on the particle density class examined, as will be explained later.

An additional lower-molecular-weight protein, easily separated from the others, has been reported for a number of nondefective parvoviruses. This minor component has a molecular weight between 56,000 and 40,000 (Salzman and White 1970; Kongsvik and Toolan 1972; Johnson et al. 1974; Tattersall et al. 1976) and comprises less than 11% of the total protein. For convenience this will be referred to here as polypeptide D, although it was originally thought to be the third and smallest component of the virion and was denoted polypeptide C (Salzman and White 1970; Kongsvik and Toolan 1972). The origin of polypeptide D is not clear. Kongsvik et al. (1974a) have shown that the equivalent polypeptide in H-1, which they have called VP3, was not present in virus grown in synchronized NB cells, but was found in virus propagated in randomly growing hamster-embryo cells. Tattersall et al. (1976) have reported that polypeptide D appears consistently in MVM grown in the RL5E line of murine-sarcoma-virus-transformed rat cells, but not in virus grown in BHK21 hamster cells or in A-9 or EA mouse cells. The presence or absence of this protein did not affect the infectivity of the virus preparation. It is not clear whether this protein is a specific cellular contaminant picked up from some host cells but not from others, or a breakdown product due to the action on virion proteins of some protease particularly active in some host cells but not in others. Whatever its origin, it does not appear to be an essential component of the infectious virion.

Another protein species migrating between polypeptides A and B has been described for LuIII (Gautschi and Siegl 1973). This appears to be a cellular protein picked up during purification (Gautschi and Siegl 1973; Gautschi et al. 1976). Recent studies suggest that a host-cell polypeptide of similar molecular weight can be stably associated with a number of parvoviruses, notably BPV, depending upon the procedure used for virus purification (Bates et al., this volume). This may be the same as the minor B polypeptide of molecular weight 77,000 reported by Johnson and Hoggan (1973) for BPV.

Although Table 1 indicates considerable disparity in molecular weight and distribution between nondefective-parvovirus structural proteins, this is probably a result of the different analytical systems used by the various investigators. The major discrepancies appear to be due to inadequate molecular-weight calibration of the gel system used, usually because too few standard protein markers were used to cover the range in which the viral polypeptides fall. When different nondefective parvoviruses are purified in the same way and run in the same gel, they show polypeptide patterns that are very similar, but not necessarily identical (Bates et al., this volume; Peterson et al., this volume).

POLYPEPTIDE COMPOSITION OF ADENO-ASSOCIATED VIRUSES

The polypeptide patterns for all four AAV serotypes have been reported (Rose et al. 1971; Johnson et al. 1971; Salo and Mayor 1977). Three structural proteins have been identified, the largest of which is the A polypeptide described previously. The consistent difference between AAV and the nondefective parvoviruses is that component B is a minor species similar in weight percent to the A polypeptide, whereas polypeptide C (readily separable from B even in SDS-phosphate gels) is the major component, comprising 79–86% of the total virion mass. Unfortunately, no analysis of AAV polypeptides on high-resolution SDS-Tris-glycine gels has been reported, so it is not absolutely clear whether each of the protein components identified is indeed a homogeneous species. Only one report has been published comparing AAV polypeptides with one of the nondefective parvoviruses, in this case BPV (Johnson and Hoggan 1973). In that study the A polypeptide of AAV3 was shown to be larger than that of BPV, although the major component of each virus, i.e., polypeptide C, was almost coincident. This supports the general conclusion, drawn from the published molecular-weight data shown in Table 1, that the molecular-weight differences between the proteins of AAV are consistently greater than those observed for the nondefective parvoviruses.

The amino acid analysis of the total protein of AAV2 full particles has been reported (Rose et al. 1971), as have the analyses of H-1 full particles (Kongsvik and Toolan 1972) and MVM full and empty particles (Tattersall et al. 1976). The mole percent values for each of the amino acids are exceedingly similar for all four particle types, although the actual composition does not show any remarkable features. There is no evidence, for instance, for a preponderance of arginine, as was found for several adenoviruses (Polasa and Green 1967) and shown to be an important feature of adenovirus internal proteins (Prage and Pettersson 1971).

ENZYMATIC ACTIVITY

A DNA-polymerase activity associated with KRV purified from infected rat nephroma cells has been described by Salzman (1971). This polymerase had been purified from virions and has been characterized further. Salzman and McKerlie (1975) reported that the enzyme has a molecular weight of 75,000 and has no exact counterpart in the uninfected cell. Rhode (1973) attempted to demonstrate a similar activity in H-1 and KRV propagated in hamster cells but had no success. Recent studies suggest that the polymerase may be picked up from the host cell by the virion during purification, perhaps in a modified form (Bates et al., this volume).

GENETIC ORIGIN OF THE STRUCTURAL POLYPEPTIDES

Early in the investigation of parvovirus structural proteins Rose et al. (1971) pointed out an anomaly which came to be known as the "coding problem." The problem was that the sum of the molecular weights of the polypeptide species in the AAV virion exceeded (by some 30%) the theoretical coding capacity of the viral genome. Subsequently this has been shown to be the case for the great majority of parvoviruses, and several possible explanations have been advanced. The first is simply that molecular weights of the viral proteins as determined in SDS-polyacrylamide gels are overestimates of actual polypeptide-chain length, owing to some secondary modification such as glycosylation. Another possibility is that one or more of the polypeptides are coded for by the host cell, either as pre-existing proteins or as polypeptides induced during infection. However, if the molecular weights are correct and the proteins are all coded for by the viral genome, they may contain regions of overlapping amino acid sequence. Recent studies with bacteriophage φX174 suggest a further intriguing possibility: Two proteins might be derived from the same nucleotide sequence by translation in different reading frames (Barrell et al. 1976).

Some of these alternatives have been explored for a few parvoviruses. First, it was important to establish that the apparent molecular weights as measured in SDS gels are reasonable estimates of the true polypeptide molecular weights. Many glycoproteins, for instance, show aberrant SDS binding (Pitt-Rivers and Impiombato 1968), and consequently do not give correct molecular weights, in terms of polypeptide-chain length, in SDS-polyacrylamide gels (Schubert 1970; Bretscher 1971). The poly-peptides of MVM have been examined in a number of gel systems, including transverse-gradient gels (Tattersall et al. 1976). These studies failed to detect any abnormalities in migration as a function of gel concentration that might be indicative of a side chain on any of the viral polypeptides. In addition, no carbohydrate was detected in any MVM protein by direct periodic acid-Schiff staining or by the incorporation of radioactive glucosamine. Although recent results (Peterson et al., this volume) demonstrate that all MVM virion polypeptides are phos-phorylated, this modification does not appear to result in aberrant elec-trophoretic mobility. In the absence of any information to the contrary, therefore, we must assume that the molecular weights as measured in gels are true measures of chain length.

Second, comparisons of polypeptide patterns from infected and unin-fected cells have indicated that the viral polypeptides are synthesized after infection and are not present in the uninfected cell (Kongsvik et al. 1974a; Tattersall et al. 1976; Gautschi et al. 1976), with the exception of the polypeptide VP3 mentioned earlier, which migrates between the A and B components of LuIII (Gautschi et al. 1976). When MVM was grown in cells prelabeled with radioactive amino acids, label was incorporated into the viral polypeptides with 2–3% of the efficiency of label supplied during infection (Tattersall et al. 1976). This incorporation was assumed to be due to turnover of the prelabel, because no viral polypeptide was pref-erentially labeled, as in the case of polyoma virus, which incorporates pre-existing host histones (Frearson and Crawford 1972). In addition, the molecular weights and distributions of MVM polypeptides A, B, and C were shown to be independent of the species of host cell in which the virus was propagated (Tattersall et al. 1976). This indicates that, were the virus to incorporate a host protein induced during infection, this host protein would have to be evolutionarily conserved among several rodent species. However, as mentioned previously, the appearance of poly-peptide D seems to depend upon the host cell in which the virus is grown (Kongsvik et al. 1974b; Tattersall et al. 1976). In the absence of coupled in vitro transcription and translation or a suitable chain-termination mutant system to test the hypothesis directly, the evidence so far suggests strongly that the three polypeptides A, B, and C found in infectious parvovirus particles are coded for by the viral genome. How then are these three polypeptides synthesized, when one considers the "coding problem" mentioned earlier?

VIRAL PROTEIN SYNTHESIS

The kinetics and location of viral protein synthesis have been followed by fluorescent-antibody staining and by hemagglutinin production. Both of these techniques essentially measure the accumulation of assembled particles. This occurs in the nucleus starting approximately 10–12 hr after infection and is complete by 18 hr in synchronized cells infected at the beginning of the S phase (Siegl and Gautschi 1973a; Rhode 1973; Parris and Bates 1976). In randomly dividing cell populations the kinetics of hemagglutinin synthesis and the appearance of antigen-positive nuclei are somewhat less synchronous (Tennant et al. 1969; Fields and Nicholson 1972; Tattersall et al. 1973).

The initiation of synthesis of viral protein is blocked by inhibitors of DNA synthesis (Parker et al. 1970; Siegl and Gautschi 1973b; Rhode 1973). Studies with synchronized cells have identified a DNA-synthesis event occurring late in S phase which is essential for the initiation of H-1 hemagglutinin synthesis (Rhode 1973). Using antibody prepared against separated SDS-denatured AAV proteins, Johnson et al. (1972) showed that the major component, polypeptide C, could be detected in the cytoplasm about 14 hr after infection and was present in both cytoplasm and nucleus by 18 hr. The two minor proteins were detected at about the same time in the nucleus, but each had a distinct and different staining distribution. Interestingly, these antisera did not cross-react with anti-serum against intact virions. The cytoplasmic fluorescence observed with anti-C serum was postulated to be staining of nascent, unfolded C-polypeptide chains. In support of this, the antipolypeptide sera were found to detect viral antigen in infected cells some 2 hr before intact virus could be detected by its homologous antibody. Kongsvik et al. (1974a) labeled H-1 growing in synchronized NB cells with radioactive amino acids for 8-hr periods until 24 hr after infection and then used a chase of unlabeled amino acids until the virus was harvested and purified 30 hr after infection. This experiment showed that the maximum rate of synthesis of virion polypeptides occurred 8–16 hr after infection, and that the A and B polypeptides were both synthesized during the same period. Gautschi et al. (1976) have examined the synthesis of the A and B polypeptides of LuIII in synchronized HeLa cells. The viral proteins were detected in the nucleus at 11–12 hr after infection, after the cell had traversed the S phase. Using chromatin fractionation techniques coupled with pulse-chase labeling they showed that most of the intranuclear viral polypeptides became associated with the chromatin acidic proteins within 20–30 min of being synthesized. The A and B polypeptides were present in the same ratio as in purified virions, and over 90% of the label had the buoyant density and sedimentation characteristics of empty particles. In addition, Gautschi et al. (1976) found that 90–95% of the replicative-form viral DNA was also chromatin-associated at this time after infection, and they postulated that the chromatin was the site of synthesis of progeny DNA and the maturation of complete full particles.

In the majority of nondefective-parvovirus systems examined to date, all of the hemagglutinin can be extracted from infected cells as assembled capsids, predominantly as empty particles (Payne et al. 1964; Robinson and Hetrick 1969; Salzman and Jori 1970; Tattersall 1972; Clinton and Hayashi 1975; Tattersall et al. 1976). The particles of MVM banding in the empty-capsid region (1.30–1.32 g/cm^3) in CsCl were shown to have very low contents of polypeptide C (Clinton and Hayashi 1975). Studies of several highly purified empty-particle preparations led Tattersall et al. (1976) to conclude that the empty virions contain only polypeptides A and B and no polypeptide C. Since the majority of viral protein in these MVM-infected cells was in particles of this type, it is not surprising that polypeptide C was not detected in whole-cell lysates, although it was present in full particles purified from the same batch of infected cells (Tattersall et al. 1976). The ratio of full particles to empty particles varies among preparations and virus-host systems, although the empty particles are usually present in 2- to 50-fold excess. The considerable stability of the full particles suggests that empty particles are unlikely to arise from the degradation of complete virions. In addition, thin sections of cells infected with H-1 show large numbers of empty particles, notably associated with the nucleolus (Al-Lami et al. 1969; Singer and Toolan 1975; Singer and Rhode, this volume). It is not clear at present whether empty capsids are themselves precursors for full particles, whether they are derived during isolation from unstable intermediates in full-virus assembly, or whether they are merely the product of a dead-end pathway of assembly due to overproduction of viral protein.

Clinton and Hayashi (1975) separated the full particles of MVM isolated at 24 hr after infection into two density classes in CsCl, one banding at 1.42 g/cm^3 and the other at 1.47 g/cm^3. These particles both sediment at 110S in sucrose velocity gradients, have the same protein-to-DNA ratio, and contain a genome-length DNA molecule. As shown in Table 1, they are markedly different in polypeptide composition. The "heavy" full particle (1.47 g/cm^3) contains polypeptide B as its major component, whereas the "light" full particle (1.42 g/cm^3) contains predominantly polypeptide C. Furthermore, when infected cells were labeled with radioactive thymidine for 24 hr and the progeny virus was purified, the label was predominantly in "heavy" full particles. However, if at the end of the labeling period the label was removed, the infected cells incubated under chase conditions, and virus purified at various times afterwards, the label chased from "heavy" fulls into "light" fulls almost quantitatively over a 48-hr period. Tattersall et al. (1976) found essentially the same phenomenon when they purified virus from the nuclei of MVM-infected cells at various times after radioactive methionine had been incorporated. When total nuclear full virions (1.41–1.47 g/cm^3) were examined in SDS-polyacrylamide gels, a change was seen in the major capsid protein of the particle, again from predominantly B to predominantly C with increasing time. During the same interval the composition of the empty particles, containing only polypeptides A and B, remained constant.

These results suggested a precursor-product relationship between the B and C polypeptides of the full virion (Clinton and Hayashi 1975; Tattersall et al. 1976). Subsequently, Clinton and Hayashi (1976) demonstrated an activity in infected-cell-conditioned medium which would convert "heavy" fulls to "light" fulls. They postulated that this activity was a protease which specifically cleaved about 30 amino acids from the B protein to create polypeptide C. Such a relationship between polypeptides B and C would suggest that the viruses may only code for A and B. Although this was within the coding capacity of the viral genome (in most cases only marginally), it did appear that a much higher proportion of the viral genetic information of nondefective parvoviruses was expended on coat protein than was the case for most helper-independent viruses.

SEQUENCE HOMOLOGY AMONG THE CAPSID PROTEINS

If polypeptide C were a cleavage product of polypeptide B, then the entire sequence of C would be contained within B. The first indication that sequences might be shared between parvovirus coat proteins was that antiserum raised against AAV3 component C cross-reacted slightly with polypeptide-A and -B antigens in immunodiffusion assays (Johnson et al. 1972). Subsequently, Tattersall et al. (1977) compared all of the polypeptides of MVM empty and full particles by tryptic and chymotryptic fingerprinting, after radioiodination of their tyrosyl residues in vitro. This analysis showed that the equivalent proteins in full and empty particles were indistinguishable, and also that the precursor-product relationship between polypeptides A and C in full virions was very likely correct since their peptide maps were almost identical.

An unexpected sequence relationship also emerged from these studies. The tryptic and chymotryptic fingerprints showed quite clearly that the entire sequence of polypeptide B, and therefore also that of C, was contained within polypeptide A. Of the 52 chymotryptic and 45 tryptic peptides identified for polypeptide A, 44 and 36, respectively, were identical to those present in fingerprints of polypeptide B. All of the peptides in chymotryptic or tryptic maps of polypeptide B were present in the equivalent maps of A. The difference in peptide number between the two polypeptides was not exactly in agreement with their molecular-weight difference as measured in SDS-polyacrylamide gels, which suggested that perhaps the frequency of tyrosine residues in the A-specific portion of the A polypeptide was relatively low (Tattersall et al. 1977).

The presence of the B sequence in polypeptide A raises the possibility that the A polypeptides found in virions are merely uncleaved precursor molecules for polypeptide B, and then for C. A number of observations suggest that this is not the case, and that A plays a unique, but so far unknown, functional role in the assembly or structure of the virion. First, approximately the same number of molecules of A is found in all density classes of virions. This number is independent of the time of infection at which the virions are purified (Tattersall et al. 1976) and is constant under pulse-chase conditions (Kongsvik et al. 1974a; Clinton and Hayashi 1975). Second, the amount of polypeptide A in the virion does not change during cleavage in vitro by trypsin or chymotrypsin, proteases which do affect the relative proportions of polypeptides B and C (Clinton and Hayashi 1976; Tattersall et al. 1977). Finally, as shown in Table 1, a polypeptide species of similar molecular weight and proportion of total virion protein has been demonstrated for all of the parvoviruses examined to date, prepared from many different cell types by a number of different purification techniques and in several laboratories independently. These observations indicate that molecules of A incorporated into virions are not subsequently processed to either polypeptide B or C. Polypeptide A may, however, be a precursor for B prior to the assembly of virions. If such a precursor-product relationship exists for polypeptides A and B, the processing of A to B must be coupled tightly to the synthesis of polypeptide A, as both proteins are present in the infected cell in approximately the same proportions as in purified virions (Kongsvik et al. 1974a; Gautschi et al. 1976; Tattersall et al. 1976).

Fingerprints of the iodotyrosyl chymotryptic peptides derived from the A polypeptide of MVM showed three A-specific peptides which were very basic (Tattersall et al. 1977). This was thought to be due to a preponderance of arginyl and lysyl residues in the A-specific region, since the A-specific tryptic iodopeptides, which would presumably only contain one C-terminal arginine or lysine each, were no more positively charged than the tryptic peptides common to all three proteins. The exceedingly basic nature of the A-specific region is also suggested by the finding that although the sequences of polypeptides A and B are over 75% the same, the isoelectric point of A is more than one pH unit higher than that of B or C (Peterson et al., this volume). Tattersall et al. (1977) postulated that the basic A-specific region might be located internally and interact with the DNA in full particles, much like what has been suggested for the host histone molecules incorporated into polyoma (Frearson and Crawford 1972) and simian virus 40 virions (Lake et al. 1973). Charge-neutralizing and packaging functions in the parvovirus virion (if any are required) must be fulfilled by the virus-coded capsid proteins, because, as we have seen, no histones are incorporated into the completed full particle. If polypeptides A and B (or C) have separate functions, this would be another striking example of the economy of viral genomes, where two distinct molecules are coded in the same reading frame from a common sequence. Such an arrangement has been described for the two A-gene products of bacteriophage φX174 (Linney et al. 1972) and the two subunits of avian oncornavirus reverse transcriptase (Gibson and Verma 1974). The two minor structural proteins VP2 and VP3 of polyoma virus also have overlapping amino acid sequences (Gibson 1974; Hewick et al. 1975), but it is not clear whether these two polypeptides have different functions.

The total number of protein molecules in the parvovirus virion, variously reported to be 6–9 of A and 50–60 of B, is close to 60, the largest number of equivalent units which can be arranged on the surface of a sphere in such a way that each is situated identically (Caspar and Klug 1962). As the sequences of polypeptides A and B (or C) of MVM are mostly identical, Tattersall et al. (1977) suggested that these nominally different species could interact equivalently in the construction of such a spherical virion. However, as they pointed out, several reports suggest that the parvoviruses—at least those in the nondefective subgroup—have an icosahedral symmetry and comprise 32 morphological capsomers (Vasquez and Brailovsky 1965; Karasaki 1966; Mayor and Jordan 1966). This would suggest that the morphological subunits are dimers of the structural polypeptides. If these capsomers (and thus their constituent polypeptides) are arranged in an icosahedral fashion, it is not clear how 6–9 molecules of polypeptide A might be symmetrically disposed within such a capsid structure.

PROTEOLYTIC CLEAVAGE AND THE MATURATION OF THE VIRION

The similarity of the fingerprints of iodopeptides derived from polypeptides B and C reported by Tattersall et al. (1977) supports the precursor-product relationship between them suggested by previous kinetic studies (Clinton and Hayashi 1975; Tattersall et al. 1976). These fingerprints also demonstrate a reciprocal difference between the two polypeptides involving one pair of peptides. The demonstration of a component-C-specific peptide in both tryptic and chymotryptic maps suggested that the in vivo cleavage did not occur at either a normal tryptic or a normal chymotryptic site within the B polypeptide. Moreover, Tattersall et al. (1977) proposed that, were chymotrypsin able to cleave at the carboxyl side of iodotyrosine, the presence of such a C-specific peptide in chymotryptic maps would imply that the in vivo cleavage site must be at the N terminal of this peptide. Therefore, they tentatively placed the in vivo cleavage site about 30 amino acids in from the N terminal of polypeptide B.

The substantial difference in density between MVM full particles containing B as their major polypeptide and those containing predominantly the C protein suggests that the B-C cleavage may induce a considerable rearrangement of the surface structure and hydration of the particle. Treatment with trypsin in vitro appears to mimic the in vivo cleavage, since the majority of the B protein in full particles is converted to polypeptide which coelectrophoreses with C (Clinton and Hayashi 1976; Tattersall et al. 1977). Chymotrypsin also cleaves the B polypeptide in full particles, but in this case the product is intermediate in molecular weight between B and C. The B polypeptide in empty particles is completely resistant to cleavage by either enzyme (Tattersall et al. 1977), although peptide mapping indicates that it has a sequence identical to that of polypeptide B in full particles. However, Clinton and Hayashi (1976) have shown that trypsin-treated "heavy" full particles do not convert to the "light" full density in cesium chloride even though the B protein is converted to the C-like polypeptide species. In addition, they have shown that "heavy" full particles shift to the "light" full density after treatment with medium conditioned by infected cells, and that this conversion is not blocked by bovine pancreatic trypsin inhibitor. This indicates that the cleavage may have to be exact to result in the correct conformational change and supports the conclusion drawn from peptide fingerprinting that the protease responsible for the B-C cleavage in vivo does not have a trypsinlike specificity (Tattersall et al. 1977).

Interestingly, the A polypeptide of MVM does not appear to be susceptible to in vivo proteolysis (or indeed to trypsin or chymotrypsin cleavage in vitro) in any particle type (Tattersall et al. 1977), although the sequence homology between the polypeptides indicates that the B-C cleavage site is present in the sequence of the A protein. In this respect, the conformation of polypeptide A in all particle types resembles the conformation adopted by polypeptide B in the empty particle, and the two proteins may not interact equivalently in the "heavy" full particle.

An essential role for the postassembly cleavage of a structural protein has been demonstrated in Sendai-virus maturation (Homma 1971, 1972; Scheid and Choppin 1974). Structural polypeptide processing by proteolytic cleavage has also been shown to occur during the maturation of a number of animal-virus and bacteriophage virions (Casjens and King 1975; Hershko and Fry 1975). In an attempt to correlate what is now known about the structure and synthesis of the various forms of parvovirus particles, Tattersall et al. (1977) proposed a general scheme for the maturation of the nondefective-parvovirus virion, using MVM as an example, which is shown diagrammatically in Figure 1.

According to this scheme, polypeptides A and B are synthesized (presumably from a common sequence in the genome) and rapidly assembled into empty virions (step 1), the form in which almost all of the viral protein is found upon extraction at various times after infection (Tattersall et al. 1976). Virtually nothing is known about the mechanism of synthesis of polypeptides A and B, or of its control. Because no iodopeptides were detected in fingerprints of polypeptide B that did not also occur in fingerprints of polypeptide A, Tattersall et al. (1977) were unable to position the sequence of B within that of A, even tentatively. Analysis of parvovirus transcription (Carter, this volume) has identified a single mRNA species, large enough to code for the A polypeptide, copied from the genomes of AAV (Carter 1974) and KRV (Salzman and Redler 1974). This suggests that the production of polypeptides A and B is controlled at the translational or post-translational level, rather than by

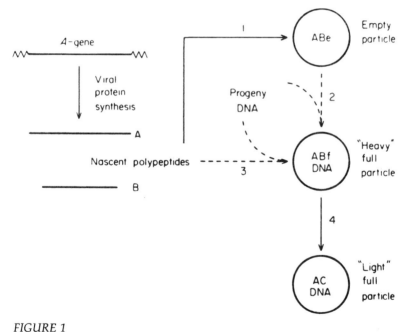

FIGURE 1
Scheme for the maturation of the nondefective parvovirus virion. A, B, and C denote the three viral polypeptides. Suffixes denote the B-C cleavage-resistant (e) or cleavage-sensitive (f) conformations of the particle. Empty, "heavy" full, and "light" full describe virions, which in the case of MVM band in cesium chloride at 1.32, 1.47, and 1.42 g/cm³, respectively. (Reproduced, with permission, from Tattersall et al. 1977).

the synthesis of two mRNA molecules with overlapping sequences. It remains for future research to determine whether these two molecules are differentially translated from separate initiation sites by premature termination of polypeptide A, or by post-translational cleavage of the A polypeptide or even a larger precursor molecule. It is possible that rapid incorporation of newly synthesized polypeptides into empty particles may play an important role in the coordinate synthesis of proteins A and B in the required ratio.

The empty particle, as extracted from the infected cell, is in the conformation resistant to B-C cleavage (the e state). It has a density of 1.32 g/cm^3 in cesium chloride. It binds to red blood cells and the host cell, and therefore hemagglutinates; however, because it lacks the viral genome, it is noninfectious. Two possible pathways for the incorporation of DNA into the virion have been proposed: (step 2) the genome might be packed directly into pre-existing empty particles, or (step 3) the viral DNA could condense directly with the nascent polypeptides. Whichever packaging process operates, the result is a full particle containing polypeptides A and B. This full particle is in a conformation (the f state) where the B polypeptide is susceptible to cleavage in vitro by trypsin and chymotrypsin (Clinton and Hayashi 1976; Tattersall et al. 1977) and by the proteolytic activity (whatever its specificity) which processes polypeptide B to C in vivo (Clinton and Hayashi 1976; Tattersall et al. 1976). In this conformation the MVM virion has a density of 1.47 g/cm^3 in cesium chloride, that is, the so-called "heavy" full particle. It has a lower specific hemagglutinating activity and a somewhat lower rate of absorption to host cells than the empty particle or the mature virion (Clinton and Hayashi 1976; P. Tattersall, P. J. Cawte, and D. C. Ward, unpublished results; Kongsvik et al., this volume), probably reflecting an altered affinity for both the red-blood-cell receptors and the host-cell receptors.

Tattersall et al. (1977) proposed that the conformational change in the particle responsible for the decrease in receptor-binding activity and for the exposure of the B-C cleavage site is a direct result of the packaging process. It had been suggested (Tattersall and Ward 1976) that the packaging of progeny single-stranded genomes into preformed capsids might drive the asymmetric-displacement synthesis of the viral strand from duplex replicative-intermediate forms of intracellular viral DNA (Tattersall et al. 1973; Rhode 1974; Siegl and Gautschi 1976; Gunther and May 1976) and that the energy for such a packaging reaction might be supplied by the conformational change in the particle. Recent studies have provided more direct evidence for involvement of packaging in progeny DNA synthesis (Richards et al.; Gautschi et al.; Rhode; Gunther and Revet; all this volume). The packaging process of bacteriophage T4 also involves proteolytic cleavage and changes in the conformation of the viral capsid. The major component of the T4 head is cleaved as a first step in maturation, resulting in a conformational change in the viral shell. This is followed by two further cleavage events which are intimately linked to the packaging of the viral DNA (Laemmli and Favre 1973; Laemmli et al. 1976).

As a final step in the scheme for parvovirus maturation, Tattersall et al. (1977) proposed that the conformational change in the particle induced by packaging is reversed by the in vivo cleavage of polypeptide B to C (step 4). The cleavage allows a rearrangement of the capsid structure to take place, which results in a shift of density in cesium chloride to that of the "light" full particle (in the case of MVM, 1.42 g/cm³). This structural rearrangement concomitantly restores maximum receptor binding potential. Thus, the particle assumes the full infectivity and hemagglutinating activity of the mature parvovirus virion.

SUMMARY

The nondefective parvovirus virion contains three species of structural polypeptide, A, B, and C, which have overlapping sequences. The A protein, which is the largest, is a minor species which appears to be metabolically stable during maturation. Polypeptide C, however, is derived by proteolytic cleavage of B molecules in the assembled, DNA-containing virion. The purpose of this maturation event appears to be correction of some distortion in the capsid caused by the packaging of the DNA. The adeno-associated-virus capsid also comprises three types of structural polypeptide, which again appear to contain homologous sequences. Evidence to date, however, suggests that both the A and B proteins are stable, minor species and that the bulk of the virion is made up of polypeptide-C molecules.

It is not now known how these sequence-related polypeptides are synthesized, or how they are arranged in the construction of the virion.

REFERENCES

Al-Lami, F., N. Ledinko, and H. Toolan. 1969. Electron microscope study of human NB and SMH cells infected with the parvovirus H-1: Involvement of the nucleolus. *J. Gen. Virol.* **5**:485.

Anderson, C. W., P. R. Baum, and R. F. Gesteland. 1973. Processing of adenovirus 2-induced proteins. *J. Virol.* **12**:241.

Barrell, B. G., G. M. Air, and C. A. Hutchison III. 1976. Overlapping genes in bacteriophage φX174. *Nature* **264**:34.

Bretscher, M. S. 1971. Major human erythrocyte glycoprotein spans the cell membrane. *Nat. New Biol.* **231**:229.

Carter, B. J. 1974. Analysis of parvovirus mRNA by sedimentation and electrophoresis in aqueous and nonaqueous solution. *J. Virol.* **14**:834.

Casjens, S. and J. King. 1975. Virus assembly. *Annu. Rev. Biochem.* **44**:555.

Caspar, D. L. D. and A. Klug. 1962. Physical principles in the construction of regular viruses. *Cold Spring Harbor Symp. Quant. Biol.* **27**:1.

Clinton, G. M. and M. Hayashi. 1975. The parvovirus MVM: Particles with altered structural proteins. *Virology* **66**:261.

————. 1976. The parvovirus MVM: A comparison of heavy and light particle infectivity and their density conversion in vitro. *Virology* **74**:57.

Fields, H. A. and B. L. Nicholson. 1972. The replication of the Kilham rat virus (RV) in various host systems: Immunofluorescent studies. *Can. J. Microbiol.* **18**:103.

Frearson, P. M. and L. V. Crawford. 1972. Polyoma virus basic proteins. *J. Gen. Virol.* **14**:141.

Gautschi, M. and G. Siegl. 1973. Structural proteins of parvovirus LuIII. Evidence for only two protein components within infectious virions. *Arch. gesamte Virusforsch.* **43**:326.

Gautschi, M., G. Siegl, and G. Kronauer. 1976. Multiplication of parvovirus LuIII in a synchronized culture system. IV. Association of viral structural polypeptides with the host cell chromatin. *J. Virol.* **20**:29.

Gibson, W. 1974. Polyoma virus proteins: A description of the structural proteins of the virion based on polyacrylamide gel electrophoresis and peptide analysis. *Virology* **62**:319.

Gibson, W. and I. M. Verma. 1974. Studies on the reverse transcriptase of RNA tumor viruses. Structural relatedness of two subunits of avian RNA tumor viruses. *Proc. Natl. Acad. Sci.* **71**:4991.

Gunther, M. and P. May. 1976. Isolation and structural characterization of monomeric and dimeric forms of replicative intermediates of Kilham rat virus DNA. *J. Virol.* **20**:86.

Hershko, A. and M. Fry. 1975. Post-translational cleavage of polypeptide chains: Role in assembly. *Annu. Rev. Biochem.* **44**:775.

Hewick, R. M., M. Fried, and M. D. Waterfield. 1975. Nonhistone virion proteins of polyoma: Characterisation of the particle proteins by tryptic peptide analysis by use of ion-exchange columns. *Virology* **66**:408.

Homma, M. 1971. Trypsin action on the growth of Sendai virus in tissue culture cells. I. Restoration of the infectivity for L-cells by direct action of trypsin on L-cell-borne Sendai virus. *J. Virol.* **8**:619.

————. 1972. Trypsin action on the growth of Sendai virus in tissue culture cells. II. Restoration of the hemolytic activity of L-cell-borne Sendai virus by trypsin. *J. Virol.* **9**:829.

Johnson, F. B. and M. D. Hoggan. 1973. Structural proteins of Haden virus. *Virology* **51**:129.

Johnson, F. B., N. R. Blacklow, and M. D. Hoggan. 1972. Immunological reactivity of antisera prepared against the sodium dodecyl sulfate-treated structural polypeptides of adenovirus-associated virus. *J. Virol.* **9**:1017.

Johnson, F. B., H. L. Ozer, and M. D. Hoggan. 1971. Structural proteins of adenovirus-associated virus type 3. *J. Virol.* **8**:860.

Johnson, R. H., G. Siegl, and M. Gautschi. 1974. Characteristics of feline panleucopaenia virus strains enabling definitive classification as parvoviruses. *Arch. gesamte Virusforsch.* **46**:315.

Karasaki, S. 1966. Size and ultrastructure of the H-viruses as determined with the use of specific antibodies. *J. Ultrastruct. Res.* **16**:109.

Kongsvik, J. R. and H. W. Toolan. 1972. Capsid components of the parvovirus H-1. *Proc. Soc. Exp. Biol. Med.* **139**:1202.

Kongsvik, J. R., J. F. Gierthy, and S. L. Rhode III. 1974a. Replication process of the parvovirus H-1. IV. H-1 specific proteins synthesized in synchronized human NB kidney cells. *J. Virol.* **14**:1600.

Kongsvik, J. R., I. I. Singer, and H. W. Toolan. 1974b. Studies on the red cell and antibody-reactive sites of the parvovirus H-1: Effect of fixatives. *Proc. Soc. Exp. Biol. Med.* **145**:763.

Laemmli, U. K. 1970. Cleavage of structural proteins during the assembly of the head of bacteriophage T4. *Nature* **227**:680.

Laemmli, U. K. and M. Favre. 1973. Maturation of the head of bacteriophage T4. I. DNA packaging events. *J. Mol. Biol.* **80**:575.

Laemmli, U. K., L. A. Amos, and A. Klug. 1976. Correlation between structural transformation and cleavage of the major head protein of T4 bacteriophage. *Cell* **7**:191.

Lake, R. S., S. Barban, and N. P. Salzman. 1973. Resolutions and identification of the core deoxynucleoproteins of the simian virus 40. *Biochem. Biophys. Res. Commun.* **54**:640.

Linney, E. A., M. N. Hayashi, and M. Hayashi. 1972. Gene A of φX174. I. Isolation and identification of its products. *Virology* **50**:381.

Mayor, H. D. and E. L. Jordan. 1966. Electron microscopic study of the rodent "picodnavirus" X14. *Exp. Mol. Pathol.* **5**:580.

Parker, J. C., S. S. Cross, M. J. Collins, Jr., and W. P. Rowe. 1970. Minute virus of mice. I. Procedures for quantitation and detection. *J. Natl. Cancer Inst.* **45**:297.

Parris, D. S. and R. C. Bates. 1976. Effect of bovine parvovirus replication on DNA, RNA and protein synthesis in S phase cells. *Virology* **73**:72.

Payne, F. E., T. F. Beals, and R. E. Preston. 1964. Morphology of a small DNA virus. *Virology* **23**:109.

Pitt-Rivers, R. and F. S. A. Impiombato. 1968. The binding of sodium dodecyl sulphate to various proteins. *Biochem. J.* **109**:825.

Polasa, H. and M. Green. 1967. Adenovirus proteins. I. Amino acid composition of oncogenic and nononcogenic human adenoviruses. *Virology* **31**:565.

Prage, L. and U. Pettersson. 1971. Structural proteins of adenoviruses. VII. Purification and properties of an arginine-rich core protein from adenovirus type 2 and type 3. *Virology* **45**:364.

Rhode, S. L. III. 1973. Replication process of the parvovirus H-1. I. Kinetics in a parasynchronous cell system. *J. Virol.* **11**:856.

————. 1974. Replication process of the parvovirus H-1. II. Isolation and characterization of H-1 replicative form DNA. *J. Virol.* **13**:400.

Robinson, D. M. and F. M. Hetrick. 1969. Single-stranded DNA from the Kilham rat virus. *J. Gen. Virol.* **4**:269.

Rose, J. A. 1974. Parvovirus reproduction. In *Comprehensive virology* (ed. H. Fraenkel-Conrat and R. R. Wagner), vol. 3, p. 1. Plenum, New York.

Rose, J. A., J. V. Maizel, Jr., J. K. Inman, and A. J. Shatkin. 1971. Structural proteins of adenovirus-associated viruses. *J. Virol.* **8**:766.

Salo, R. J. and H. D. Mayor. 1977. Structural polypeptides of parvoviruses. *Virology* **78**:340.

Salzman, L. A. 1971. DNA polymerase activity associated with purified Kilham rat virus. *Nat. New Biol.* **231**:174.

Salzman, L. A. and L. A. Jori. 1970. Characterization of the Kilham rat virus. *J. Virol.* **5**:114.

Salzman, L. A. and L. McKerlie. 1975. Characterization of the deoxyribonucleic acid polymerase associated with Kilham rat virus. *J. Biol. Chem.* **250**:5583.

Salzman, L. A. and B. Redler. 1974. Synthesis of viral-specific RNA in cells infected with the parvovirus Kilham rat virus. *J. Virol.* **14**:434.

Salzman, L. A. and W. L. White. 1970. Structural proteins of Kilham rat virus. *Biochem. Biophys. Res. Commun.* **41**:1551.

Scheid, A. and P. W. Choppin. 1974. Identification of biological activities of paramyxovirus glycoproteins. Activation of cell fusion, hemolysis and infectivity by proteolytic cleavage of an inactive precursor protein of Sendai virus. *Virology* **57**:475

Schubert, D. 1970. Immunoglobulin biosynthesis. IV. Carbohydrate attachment to immunoglobulin subunits. *J. Mol. Biol.* **51**:287.

Shapiro, A. L., E. Vinuela, and J. V. Maizel, Jr. 1967. Molecular weight estimation of polypeptide chains by electrophoresis in SDS polyacrylamide gels. *Biochem. Biophys. Res. Commun.* **28**:815.

Siegl, G. 1972. Parvoviruses as contaminants of human cell lines. V. The nucleic acid of KBSH virus. *Arch. gesamte Virusforsch.* **37**:267.

Siegl, G. and M. Gautschi. 1973a. The multiplication of parvovirus LuIII in a synchronized culture system. I. Optimum conditions for virus replication. *Arch. gesamte Virusforsch.* **40**:105.

———. 1973b. The multiplication of parvovirus LuIII in a synchronized cell system. II. Biochemical characteristics of virus replication. *Arch. gesamte Virusforsch.* **40**:119.

———. 1976. Multiplication of parvovirus LuIII in a synchronized culture system. III. Replication of viral DNA. *J. Virol.* **17**:841.

Singer, I. I. and H. W. Toolan. 1975. Ultrastructural studies of H-1 parvovirus replication. I. Cytopathology produced in human NB epithelial cells and hamster embryo fibroblasts. *Virology* **65**:40.

Studier, F. W. 1973. Analysis of bacteriophage T7 early RNAs and proteins on slab gels. *J. Mol. Biol.* **79**:237.

Tattersall, P. 1972. Replication of the parvovirus MVM. I. Dependence of virus multiplication and plaque formation on cell growth. *J. Virol.* **10**:586.

Tattersall, P. and D. C. Ward. 1976. Rolling hairpin model for replication of parvovirus and linear chromosomal DNA. *Nature* **263**:106.

Tattersall, P., L. V. Crawford, and A. J. Shatkin. 1973. Replication of the parvovirus MVM. II. Isolation and characterization of intermediates in the replication of the viral deoxyribonucleic acid. *J. Virol.* **12**:1446.

Tattersall, P., A. J. Shatkin, and D. C. Ward. 1977. Sequence homology between the structural polypeptides of minute virus of mice. *J. Mol. Biol.* **111**:375.

Tattersall, P., P. J. Cawte, A. J. Shatkin, and D. C. Ward. 1976. Three structural polypeptides coded for by minute virus of mice, a parvovirus. *J. Virol.* **20**:273.

Tennant, R. W., K. R. Layman, and R. E. Hand, Jr. 1969. Effect of cell physiological state on infection by rat virus. *J. Virol.* **4**:872.

Torikai, K., M. Ito, L. E. Jordan, and H. D. Mayor. 1970. Properties of light particles produced during growth of type-4 adeno-associated satellite virus. *J. Virol.* **6**:363.

Vasquez, C. and C. Brailovsky. 1965. Purification and fine structure of Kilham's rat virus. *Exp. Mol. Pathol.* **4**:130.

PART II

RESEARCH ARTICLES

BIOLOGY AND
CELL-VIRUS INTERACTION

Isolation and Characterization of KB Cell Lines Latently Infected with Adeno-Associated Virus Type 1

HIROSHI HANDA
KAZUKO SHIROKI
HIROTO SHIMOJO

Institute of Medical Science
University of Tokyo
Tokyo 108, Japan

Adeno-associated virus (AAV) is a defective parvovirus which is capable of multiplying only in cells coinfected with adenovirus (Atchison et al. 1965; Hoggan et al. 1966; Smith et al. 1966; Parks et al. 1967). Despite its requirement for adenovirus functions for the production of infectious particles, AAV can occasionally be isolated from human-embryo kidney (HEK) cells in culture in the absence of demonstrable adenovirus (Hoggan 1970). This phenomenon could be explained by the presence of an undetected helper virus or by the ability of AAV to establish a helper-independent carrier state in these cells. The latter possibility is suggested by more recent studies showing that AAV carrier lines of Detroit-6 cells (Hoggan et al. 1973) contain AAV-specific nucleotide sequences in their DNA (Berns et al. 1975). This helper-independent carrier state offers a system for studying the requirements for maintaining a defective genome in stable association with its host cell.

METHODS

DNA Extraction

Cellular DNA was extracted by a slightly modified version of the method described by Varmus et al. (1973a). Cells were washed with phosphate-buffered saline (PBS), scraped off, suspended in RSB [10 mm NaCl, 10 mm Tris-HCl (pH 7.4), 1.5 mm MgCl$_2$] at 10^7 cells/ml (Yamashita et al. 1975), and incubated in an ice bath for 15 min. Nuclei were collected by disruption of these swollen cells with a Dounce homogenizer and by low-speed centrifugation. Cellular DNA was extracted from the nuclei. Network DNA was prepared as described by Varmus et al. (1973b). The preparation of ^{32}P-labeled AAV1 DNA was performed as described by Rose et al. (1966).

Induction by BUdR or IUdR

Subconfluent cultures were treated with BUdR (5–20 μg/ml) or IUdR (5–10 μg/ml) in the dark for 8 hr. After the treatment, the medium was removed and the cultures were washed once with maintenance medium [Eagle's minimum essential medium (MEM) with 2% calf serum]. The cells were infected with wild-type human adenovirus type 31 (Ad31) at 30 plaque-forming units (PFU) per cell.

RESULTS

Isolation of KB-Cell Clones Latently Infected with AAV1

KB cells were infected at 500 fluorescent infectious units (FIU) per cell (Blacklow et al. 1967) with AAV1, which had been purified by banding in CsCl (Rose et al. 1966) and heated at 60°C for 10 min. These AAV1-infected KB cells were cloned by the limit dilution technique and recloned by picking up cells in a colony with the use of a stainless steel cylinder (Handa et al. 1977a). Clones of cells positive for AAV1 antigen after superinfection with Ad31 were selected. We have established two such clones out of 120 clones tested. These clones, designated KB13 and KB302, remain inducible for AAV1 antigen and infectious virus by Ad31 superinfection, even after 45 and 39 passages, respectively, in the absence of helper virus (see Table 1). No infectious virus or AAV1 antigen was detectable in those cells before helper-adenovirus infection. Similarly, no infectivity or antigen could be detected in control KB cells infected with Ad31 alone. These results suggest that the AAV1 genome is continuously maintained in KB13 and KB302 cells in a stable, but masked, state.

*Time Course for Induction of AAV1 Infectious Progeny Virus in
KB13 Cells After Infection with Ad31*

The infectivity of AAV1 induced in KB13 cells was titrated at intervals of 4 hr post-infection (p.i.) with Ad31 (Fig. 1). The titer of infectious AAV1 in KB13 cells began to increase after 16 hr p.i. with Ad31, whereas the titer in

TABLE 1
Induction of AAV1 Antigens and Infectious Progeny Virus in AAV1-Carrier Cells

Clone	No. of passages	Infection with Ad31	Ad31 capsid antigen[a]	No. of AAV1- antigen-positive cells per cover slip[b]	AAV1 titer[c] log (FIU/ 0.2 ml)
KB	—	+	+	0	< 1
	—	—	—	0	< 1
KB13	2	+	+	98	nd[d]
	13	+	+	196	nd
	31	+	+	101	nd
	45	+	+	146	3.7
	45	—	—	0	< 1
KB302	3	+	+	38	nd
	11	+	+	50	nd
	25	+	+	43	nd
	39	+	+	58	3.2
	39	—	—	0	< 1

[a] Cells on cover slips (about 2×10^5 cells) were infected with Ad31 at 30 PFU/cell, harvested at 24 hr p.i., fixed and stained with the anti-Ad31 conjugate. +: More than 50% of cells were antigen-positive. −: No antigen-positive cells were detected.
[b] Cells on cover slips were stained with the anti-AAV1 conjugate. The numbers of cells positive for AAV1 antigen per cover slip were counted. "0" indicates no AAV1-antigen-positive cells.
[c] AAV1 was titrated as described previously (Handa et al. 1976) and expressed as log (FIU/0.2 ml).
[d] nd: Not done.

KB cells coinfected with Ad31 (30 PFU/cell) and AAV1 (15 FIU/cell) began to increase after 12 hr p.i. Thus, infectious AAV1 progeny virus appeared 4 hr later in Ad31-infected KB13 cells than in KB cells coinfected with Ad31 and AAV1.

Detection of the AAV1 DNA Sequence in KB13 and KB302 Cell DNA by DNA Reassociation Kinetics

AAV1 DNA sequences were detected in KB13 cell DNA, and the number of AAV1 genomes per cell was estimated by DNA reassociation kinetics (Gelb et al. 1971). Acceleration of the reassociation kinetics of ^{32}P-labeled AAV1 DNA was demonstrated in the presence of unlabeled DNA from KB13 cells (Fig. 2a). When KB13 cell DNA was fractionated into network DNA, both the network DNA and the DNA remaining in the supernatant

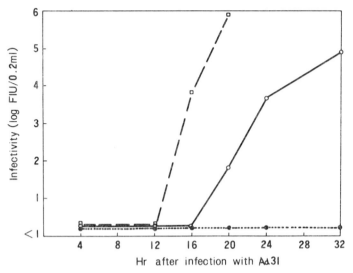

FIGURE 1

Induction of AAV1 in KB13 cells by infection with Ad31. KB or KB13 cells grown in small bottles (3 × 10⁵ cells per bottle) were infected with Ad31 at 30 PFU/cell and incubated at 37°C in maintenance medium after adsorption for 1 hr. At intervals of 4 hr p.i., the cultures were frozen and thawed three times, and FIU in the supernatants after low-speed centrifugation were determined in HEK cells. o——o: KB13 cells infected with Ad31; ●- - - - -●: KB cells infected with Ad31; □- - - -□: KB cells coinfected with Ad31 and AAV1.

showed the same capacity as total cellular DNA to accelerate the reassociation kinetics (Fig. 2a). When network DNA was prepared from [³H]thymidine-labeled KB or KB13 cells, 81% and 79% of the total acid-insoluble radioactivity were found in the network DNA fractions from KB and KB13 cells, respectively. When a mixture of ³H-labeled KB cell DNA and ³²P-labeled AAV1 DNA was fractionated similarly, most of the AAV1 DNA (92%) was recovered in the supernatant. These results suggest that most of the AAV1 DNA exists in a form integrated into cellular DNA. However, the existence of AAV1 DNA in a free (unintegrated) form cannot be excluded.

The hybridization data shown in Figure 2 indicate that the entire nucleotide sequence of the AAV1 genome is present in KB13 cells. It was calculated that KB13 cells contained 6.0 AAV1 genomes per cell. It was also calculated that KB302 cells contained 4.1 AAV1 genomes per cell (data not shown).

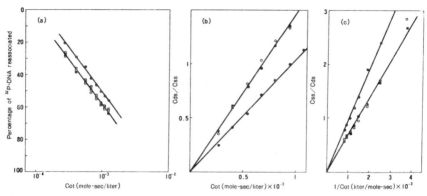

FIGURE 2

Reassociation of ^{32}P-labeled AAV1 DNA in the presence of cellular DNA. C_{ss} and C_{ds} represent the concentrations (moles per liter) of single- and double-stranded DNA, respectively. ^{32}P-labeled AAV1 DNA (13 ng/ml, 8.1×10^6 cpm/ml) was sonicated, denatured, and reassociated in the presence of KB cell DNA (●), unfractionated KB13 cell DNA (○), KB13 network DNA (■), and KB13 supernatant DNA (□) (cellular DNA sample adjusted to 1.1 mg/ml) at 68°C in 250 μl of 0.4 M phosphate buffer (pH 6.8). The fraction of reassociated ^{32}P-labeled AAV1 DNA was measured by hydroxyapatite chromatography as described previously (Shiroki et al. 1976). (a) C_0t vs. percentage of ^{32}P-labeled DNA reassociated, (b) C_0t vs. C_{ds}/C_{ss}, (c) $1/C_0t$ vs. C_{ss}/C_{ds}.

Effect of BUdR or IUdR on Induction of AAV1 Antigens in AAV1-Carrier Cells by Infection with Ad31

It has been reported that virus can be induced from cells transformed with simian virus 40 (Rothschild and Black 1970) and Epstein-Barr virus (Gerber 1972) by BUdR or IUdR treatment. KB13 or KB302 cells were exposed to BUdR or IUdR in order to examine the effect of these chemical agents on induction of AAV1. Treatment with BUdR (5–20 μg/ml) or IUdR (5–10 μg/ml) enhanced the induction of AAV1 in KB13 and KB302 cells by infection with Ad31 (Tables 2 and 3). No AAV1 antigen was developed in control KB cells treated in the same way. The treatment of KB13 or KB302 cells with BUdR or IUdR alone without Ad31 infection did not induce AAV1 antigens in AAV1-carrier cells. The effect of BUdR on the efficiency of adenovirus infection was examined by comparing the numbers of adenovirus-antigen-positive cells in Ad31-infected AAV1-carrier cells treated and untreated with BUdR. The results showed that nearly equal fractions of the cells were positive for adenovirus capsid antigens in the two cases (data not shown), indicating that the enhancing effect of BUdR on the induction of AAV1 antigens was not due to the increased sensitivity of drug-treated cells to adenovirus infection.

TABLE 2
Effect of BUdR on Induction of AAV1 Antigens in AAV1-Carrier Cells After Infection with Ad31

Cell culture	Infection with Ad31	Concentration of BUdR (µg/ml)	Average number of AAV1-antigen-positive cells per cover slip
KB	+	0	0[a]
	+	5	0
	+	10	0
	+	20	0
KB13	+	0	170
	+	5	7973
	+	10	6502
	+	20	1862
	−	5	0
KB302	+	0	36
	+	5	169
	+	10	117
	+	20	70
	−	5	0

[a] "0" indicates no AAV1-antigen-positive cells.

The effect of BUdR on the lag time for development of AAV1 antigens in KB13 cells by infection with Ad31 was also examined. Detection of AAV1 antigens began after 16 hr p.i. with Ad31 in KB13 cells either treated or untreated with BUdR (Fig. 3). Although treatment with BUdR did not shorten the lag time for appearance of AAV1 antigen in these cells, the number of cells positive for AAV1 antigen after Ad31 infection was always increased 10- to 20-fold by prior BUdR treatment.

DISCUSSION

The establishment of AAV1-carrier KB cell lines is reported herein. KB cells were infected with AAV1 alone, and AAV1-carrier-cell clones, which are capable of producing AAV1 antigens or infectious progeny virus by infection with adenovirus, were selected.

The AAV genome is maintained in these cells for more than 45 passages (with fivefold cell dilution at each passage) in a stable form which is activated only upon infection with helper adenovirus. The network DNA from AAV-carrier cells was as rich in viral sequences as the unfractionated DNA, and the supernatant DNA was not enriched further (Fig. 2). A reconstruction experiment indicated, however, that more than 90%

TABLE 3

Effect of IUdR on Induction of AAV1 Antigens in AAV1-Carrier Cells After Infection with Ad31

Cell culture	Infection with Ad31	Concentration of IUdR (μg/ml)	Average number of AAV1-antigen-positive cells per cover slip
KB	+	0	0[a]
	+	5	0
	+	10	0
KB13	+	0	105
	+	5	1802
	+	10	1501
	−	5	0
KB302	+	0	86
	+	5	265
	+	10	513

[a] "0" indicates no AAV1-antigen-positive cells.

FIGURE 3
Time course of development of AAV1 antigens in KB13 cells treated or untreated with BUdR after infection with Ad31. KB13 cells grown on cover slips (2 × 10⁵ cells) in small bottles were pretreated with BUdR (5 μg/ml) for 8 hr in the dark. The cells were infected with Ad31 at 30 PFU/cell. At intervals of 4 hr p.i., the cells were stained and the number of cells positive for AAV1 antigens was counted. o——o: KB13 cells untreated with BUdR; •----•: KB13 cells treated with BUdR.

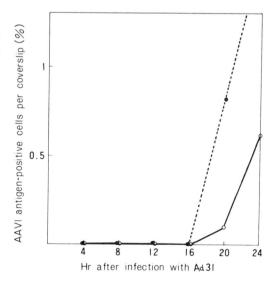

of free AAV1 DNA appears in the supernatant. These results suggest that most AAV1 DNA is covalently integrated into the cellular DNA of AAV1-carrier cells. However, the existence of a small number of AAV1 genomes present in the cell in a free form cannot be excluded.

Presumably the AAV1 genome must be in a free form for the production of progeny infectious virus in AAV1-carrier cells superinfected with adenovirus. Three possible mechanisms for the production of free AAV1 genomes are (1) the existence of a small number of free AAV1 genomes replicating autonomously in the cell as episomes, (2) excision of the integrated AAV1 genome induced by infection with adenovirus, and (3) spontaneous excision of the integrated AAV1 genome at low frequency.

The results presented here tend to support the last proposition. AAV1 antigen was detected only in a small number of adenovirus-infected cells (0.6% or less), even at a multiplicity of infection (30 PFU/cell) high enough to ensure the infection of almost all cells (Table 1, Fig. 3). The kinetics of the appearance of infectious progeny AAV1 (Fig. 1) and antigen-positive cells (Fig. 3) is slower and less synchronous than that found for cells coinfected with AAV1 and adenovirus. This suggests that new free AAV1 genomes are becoming available for replication randomly throughout the adenovirus infection cycle. In addition, pretreatment of carrier cells with halogenated pyrimidines, such as BUdR or IUdR, causes the enhancement of AAV1 induction in these carrier cells by adenovirus superinfection (Tables 2 and 3). We postulate that BUdR increases the frequency of excision of the AAV1 genome from cellular DNA in the absence of Ad31 infection.

BUdR-substituted DNA is more sensitive to breakage by visible light than unsubstituted DNA (Puck and Kao 1967; Ben-Hur and Elkind 1972), and a direct relationship between length of exposure to visible light and yield of SV40 induced from transformed cells has also been reported (Kaplan et al. 1975). AAV1-carrier cells treated with BUdR (5–20 μg/ml) were exposed to visible light as described by Fogel (1973) in order to determine whether such a correlation also exists for the induction of AAV1 antigens. However, the results showed that exposure of BUdR-treated AAV1-carrier cells to several doses of visible light failed to affect the induction of AAV1 antigen by adenovirus superinfection (Handa et al. 1977a).

The effect of DNA strand breakage by UV irradiation on Ad31 induction of AAV1 antigen was also examined. Subconfluent cultures of KB13 cells were infected with Ad31 after irradiation with various doses of UV light (50–300 ergs). The number of cells positive for AAV1 antigens was counted at 24 hr p.i. No enhancing effect on induction of AAV1 antigens was observed at any UV dose (data not shown). Halogenated pyrimidines have shown several biological effects, such as mutation (Freese 1959),

DNA strand breaking (Hsu and Somers 1961), and alteration of protein binding to DNA (Lin and Riggs 1972). The above results suggest that the enhancing effect of BUdR or IUdR on induction of AAV1 antigens is due to some mechanism other than DNA strand breaking, in agreement with previous results with other BUdR inductive systems (Rutter et al. 1973; Aaronson and Stephenson 1976). Further studies are also necessary to clarify the enhancing mechanism with halogenated pyrimidines on AAV1 induction.

REFERENCES

Aaronson, S. A. and J. R. Stephenson. 1976. Endogenous type-C RNA viruses of mammalian cells. *Biochim. Biophys. Acta* **458**:323.

Atchison, R. W., B. C. Casto, and W. M. Hammon. 1965. Adenovirus-associated defective virus particles. *Science* **149**:754.

Ben-Hur, E. and M. M. Elkind. 1972. Damage and repair of DNA in 5-bromodeoxyuridine-labeled Chinese hamster cells exposed to fluorescent light. *Biophys. J.* **12**:636.

Berns, K. I., T. C. Pinkerton, G. F. Thomas, and M. D. Hoggan. 1975. Detection of adeno-associated virus (AAV)-specific nucleotide sequences in DNA isolated from latently infected Detroit 6 cells. *Virology* **68**:556.

Blacklow, N. R., M. D. Hoggan, and W. P. Rowe. 1967. Immunofluorescent studies of the potentiation of an adenovirus-associated virus by adenovirus 7. *J. Exp. Med.* **125**:755.

Fogel, M. 1973. Induction of polyoma virus synthesis by fluorescent (visible) light in polyoma-transformed cells pretreated with 5-bromodeoxyuridine. *Nat. New Biol.* **241**:182.

Freese, E. J. 1959. The specific mutagenic effect of base analogues on phage T4. *J. Mol.Biol.* **1**:87.

Gelb, L. D., D. E. Kohne, and M. A. Martin. 1971. Quantitation of simian virus 40 sequences in African green monkey, mouse and virus-transformed cell genomes. *J. Mol. Biol.* **57**:129.

Gerber, P. 1972. Activation of Epstein-Barr virus by 5-bromodeoxyuridine in "virus-free" human cells. *Proc. Natl. Acad. Sci.* **69**:83.

Handa, H., H. Shimojo, and K. Yamaguchi. 1976. Multiplication of adeno-associated virus type 1 in cells coinfected with a temperature-sensitive mutant of human adenovirus type 31. *Virology* **74**:1.

Handa, H., K. Shiroki, and H. Shimojo. 1977a. Establishment and characterization of KB cell lines latently infected with adeno-asssociated virus type 1. *Virology* **82**:84.

———. 1977b. Helper factor(s) for growth of adeno-associated virus in cells transformed by adenovirus 12. *Proc. Natl. Acad. Sci.* **74**:4508.

Hoggan, M. D. 1970. Adenovirus-associated viruses. *Prog. Med. Virol.* **12**:211.

Hoggan, M. D., N. R. Blacklow, and W. P. Rowe. 1966. Studies of small DNA viruses found in various adenovirus preparations: Physical, biological, and immunological characteristics. *Proc. Natl. Acad. Sci.* **55**:1467.

Hoggan, M. D., G. F. Thomas, and F. B. Johnson. 1973. Continuous "carriage" of adenovirus-associated virus genome in cell culture in the absence of helper adenovirus. In *Possible episomes in eukaryotes* (ed. L. G. Silvestri), p. 243. North-Holland, Amsterdam.

Hsu, T. C. and C. E. Somers. 1961. Effect of 5-bromodeoxyuridine on mammalian chromosomes. *Proc. Natl. Acad. Sci.* **47**:396.

Kaplan, J. C., S. M. Wilbert, J. J. Collins, T. Rakusanova, G. B. Zamansky, and P. H. Black. 1975. Isolation of simian virus 40-transformed inbred hamster cell lines heterogeneous for virus induction by chemicals or radiation. *Virology* **68**:200.

Lin, S. and A. Riggs. 1972. *Lac* operator analogues: Bromodeoxyuridine substitution in the *lac* operator affects the rate of dissociation of the *lac* repressor. *Proc. Natl. Acad. Sci.* **69**:2574.

Parks, W. P., J. L. Melnick, R. Rongey, and H. D. Mayor. 1967. Physical assay and growth cycle studies of a defective adeno-satellite virus. *J. Virol.* **1**:171.

Puck, T. T. and F. Kao. 1967. Genetics of somatic mammalian cells. V. Treatment with 5-bromodeoxyuridine and visible light for isolation of nutritionally deficient mutants. *Proc. Natl. Acad. Sci.* **58**:1227.

Rose, J. A., M. D. Hoggan, and A. J. Shatkin. 1966. Nucleic acid from an adeno-associated virus: Chemical and physical studies. *Proc. Natl. Acad. Sci.* **64**:863.

Rothschild, H. and P. H. Black. 1970. Analysis of SV40 induced transformation of hamster kidney tissue in vitro. VII. Induction of SV 40 from transformed hamster cell clones by various agents. *Virology* **42**:251.

Rutter, W. J., R. J. Pictet, and P. W. Morris. 1973. Toward molecular mechanisms of developmental processes. *Annu. Rev. Biochem.* **42**:601.

Shiroki, K., H. Shimojo, K. Sekikawa, K. Fujinaga, J. Rabek, and A. J. Levine. 1976. Suppression of the temperature-sensitive character of adenovirus 12 early mutants in monkey cells transformed by an adenovirus 7–simian virus 40 hybrid. *Virology* **69**:431.

Smith, K. O., W. D. Gehle, and J. F. Thiel. 1966. Properties of a small virus associated with adenovirus type 4. *J. Immunol.* **97**:754.

Varmus, H. E., J. M. Bishop, and P. K. Vogt. 1973a. Appearance of virus-specific DNA in mammalian cells following transformation by Rous sarcoma virus. *J. Mol.Biol.* **74**:613.

Varmus, H. E., P. K. Vogt, and J. M. Bishop. 1973b. Integration of deoxyribonucleic acid specific for Rous sarcoma virus after infection of permissive and nonpermissive hosts. *Proc. Natl. Acad. Sci.* **70**:3067.

Yamashita, T., M. Q. Arens, and M. Green. 1975. Adenovirus deoxyribonucleic acid replication. II. Synthesis of viral deoxyribonucleic acid in vitro by a nuclear membrane fraction from infected KB cells. *J. Biol. Chem.* **250**:3273.

Interactions of Adeno-Associated Viruses with Cells Transformed by Herpes Simplex Virus

NEIL R. BLACKLOW
GEORGE CUKOR

Departments of Medicine and Microbiology
University of Massachusetts Medical School
Worcester, Massachusetts 01605

SIDNEY KIBRICK
GERALD QUINNAN

Boston University School of Medicine
Boston, Massachusetts 02118

Adeno-associated viruses (AAV) are unconditionally defective human viruses requiring a morphologically and immunologically unrelated helper adenovirus for production of infectious progeny (Atchison et al. 1965; Hoggan et al. 1966; Smith et al. 1966; Parks et al. 1967; Blacklow et al. 1967). Adenovirus-transformed cells that contain T antigen fail to supply necessary helper functions for detectable AAV replication (Hoggan et al. 1966). However, AAV inhibits adenovirus oncogenicity in hamsters (Kirschstein et al. 1968). It has also been found that herpes-group viruses can incompletely complement AAV replication. AAV immunofluorescent-stainable antigen and nonencapsidated AAV DNA are detectable when herpes simplex virus (HSV) and AAV are used to coinfect cells, but complete infectious AAV particles are not produced (Atchison 1970; Blacklow et al. 1970; Boucher et al. 1971; Dolin and Rabson 1973; Mayor and Young, this volume). Some temperature-sensitive mutants of HSV have also been shown to complement AAV in the production of viral antigens and DNA at the nonpermissive temperature (Drake et al. 1974).

The purpose of this report is to define interactions between AAV and cells transformed by HSV type 2. Ultraviolet-irradiated HSV2 has been shown by Duff and Rapp (1971a,b) to transform hamster-embryo fibroblasts, which then produce tumors accompanied by metastases in hamsters (Duff et al. 1973). These HSV2-transformed cells have been studied for their ability to support AAV immunofluorescent-stainable antigen formation and have been examined for the effect of AAV on their oncogenicity.

MATERIALS AND METHODS

Cells

Hamster-embryo fibroblasts transformed by HSV2 (line 333-8-9), obtained from Duff and Rapp (1971a,b), were shown to be free from infectious herpesviruses and adenoviruses (Blacklow 1975) and were used in our laboratory at passage levels of 17–35. Simian-virus-40-induced hamster tumor TT101 and baby hamster kidney 21 (BHK-21) cells were obtained from Flow Laboratories. A tumorigenic variant of BHK-21 cells transformed by dimethylnitrosamine (DMN) (line DMN4B) was supplied by di Mayorca et al. (see di Mayorca et al. 1973). All cells were grown at 35°C in an atmosphere of 5% CO_2 with Eagle's minimum essential medium (MEM) supplemented with 10% heat-inactivated fetal-calf serum (FCS).

Viruses

The AAV3(H) strain was grown in HEK cells with adenovirus type 7 (Ad7) as helper (Johnson et al. 1972). Stock pools of AAV1(H), supplied by Hoggan et al. (see Hoggan et al. 1966), were prepared in KB cell cultures with adenovirus type 2 (Ad2) as helper (Johnson et al. 1971, 1972). The virus stocks were heated at 56°C for 30 min immediately before use. This procedure destroys adenovirus infectivity with minimal inactivation of AAV and therefore makes it possible to use AAV free of infectious adenovirus (Hoggan et al. 1966). Gradient-purified AAV1(H), kindly provided by M. D. Hoggan, was prepared by three serial bandings in CsCl and was shown to be free of adenovirus by both complement-fixation and infectivity tests (Hoggan et al. 1966). Ad2, shown to be free of AAV (Johnson et al. 1972), and HSV1 were used as described previously (Blacklow 1975).

Immunofluorescence Techniques

Cells were grown on glass cover slips in plastic petri dishes to about 75% confluency. The medium was removed, and the washed cover slips were inoculated with virus at a multiplicity of 10 $TCID_{50}$/cell (Hoggan et al. 1966). Two hours later the cells were washed three times with serum-free MEM and then renewed with MEM supplemented with 10% FCS. Twenty-four hours after virus inoculation cover slips were removed, rinsed in isotonic saline (pH 7.4), and fixed in cold acetone. The indirect fluorescent-antibody (FA) staining method using hyperimmune guinea-pig sera was employed to detect AAV virion and polypeptide antigens. The hyperimmune guinea-pig sera have been described in detail by Johnson et al. (1972). In brief, they include antibodies (designated anti-AAV1 and anti-AAV3 virion sera) prepared against whole AAV1 and AAV3 and antibodies (designated anti-VP1, anti-VP2, and anti-VP3 sera) prepared against the three structural polypeptides of AAV3 which had been treated with sodium dodecyl sulfate (SDS).

Test for Effect of Virus on Oncogenicity

The HSV2-transformed hamster cell line 333-8-9 and control lines TT101 and DMN4B were grown to confluency in plastic culture dishes (Cukor et al. 1975). A culture containing 5×10^5 cells was infected at a multiplicity of 5 $TCID_{50}$/cell with a heated AAV1 preparation or 50 $TCID_{50}$/cell with gradient-purified AAV1. Control cultures were either inoculated at a multiplicity of 5 $TCID_{50}$/cell with heat-inactivated, AAV-free Ad2 or treated with MEM. After a 24-hr incubation with MEM containing 10% FCS, each flask was washed twice with Hanks's balanced salt solution and trypsinized. The cells were resuspended in MEM containing 1% Lakeview Syrian hamster serum to the required concentration of viable cells. Weanling (3–4-week-old) Lakeview Syrian hamsters were inoculated subcutaneously with 0.2 ml of the cell suspension. Ten to twelve hamsters were inoculated from each culture; they were checked daily for palpable tumors.

Detection of Lung Metastases

Hamsters were sacrificed 10 weeks after tumor inoculation, and in order to make pulmonary metastases visible, the lungs were insufflated with India ink according to the method of Wexler (1966). After the lungs had been washed in a bleach preservative, tumor implants became discretely visible as white spots on a black background. Histological examinations ·confirmed that areas of lung that failed to perfuse with India ink were tumor metastases and that fully perfused lungs did not contain tumors.

Studies on Clones of 333-8-9 Cells

In order to study the properties of subpopulations of the 333-8-9 line, a suspension of 10^3 cells was placed in the central chamber of a Costar

Cuprak dish (Cuprak 1975) containing multiple discrete microwells, and deionized water was added to the outer chamber for humidification. After a 4-hr incubation, the medium was removed by gentle aspiration, leaving a small bead of medium in each well. Wells containing single cells were identified under an inverted phase-contrast microscope. The plates were reincubated, and each day for the following week the cells in each selected well were counted and the media changed. In about half of the wells that initially had contained single cells, growth failed to occur. By 96 hr after plating, those wells in which exponential cell growth was taking place could be identified readily. At this point, cells were either infected with 10^3 $TCID_{50}$/cell of gradient-purified AAV1 or mock-infected with medium alone. Monitoring of growth rates in the preselected wells continued after virus inoculation.

RESULTS

Enhancement of AAV by 333-8-9 Cells

Table 1 summarizes results obtained when 333-8-9 cells were inoculated with AAV3 alone and in combination with Ad2 or HSV1 helpers. AAV3-inoculated cells produced FA-stainable antigens in the absence of helper virus that were detectable by the VP1, VP3, and AAV3 virion antisera, but no fluorescence was observed with the VP2 antiserum. Addition of either

TABLE 1
Synthesis of AAV3 FA-Stainable Antigens by 333-8-9 and BHK-21 Cells

Virus inoculated[a]			FA-stainable antigen detected in 333-8-9 cells with indicated antiserum[b]				FA-stainable antigen detected in BHK-21 cells with indicated antiserum			
AAV3	Ad2	HSV1	VP1	VP2	VP3	AAV3 virion	VP1	VP2	VP3	AAV3 virion
+	−	−	+	−	+	+	−	−	−	−
+	+	−	+	+	+	+	+	+	+	+
+	−	+	+	+	+	+	+	+	+	+
−	+	−	−	−	−	−	−	−	−	−
−	−	+	−	−	−	−	−	−	−	−
−	−	−	−	−	−	−	−	−	−	−

Data from Blacklow (1975).
[a] All cover slips were fixed for FA tests 24 hr after virus inoculation.
[b] All antisera were used at a concentration containing 8 FA-staining units.

Ad2 or HSV1 helper to AAV3-inoculated 333-8-9 cells permitted the development of FA-stainable antigens detected by all of the AAV antisera. Uninfected 333-8-9 cells and 333-8-9 cells infected solely with Ad2 or HSV1 helper failed to react with any of the AAV antisera.

Figure 1 shows the staining patterns obtained with the antisera to VP1, VP2, and VP3 of AAV3 in 333-8-9 cells inoculated with AAV alone. The cytoplasmic fluorescence observed previously with the anti-VP1 serum was produced, as was the primarily intranuclear staining observed previously with the anti-VP3 serum. Not shown is the typical nuclear granular fluorescence, accompanied by some cytoplasmic staining, seen with anti-AAV3 virion serum in all AAV3-infected 333-8-9 cells with or without added helper.

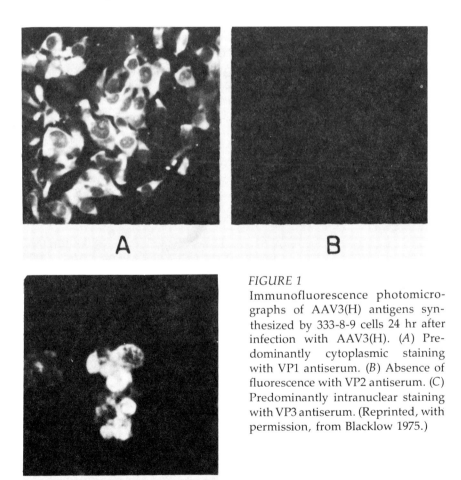

FIGURE 1
Immunofluorescence photomicrographs of AAV3(H) antigens synthesized by 333-8-9 cells 24 hr after infection with AAV3(H). (*A*) Predominantly cytoplasmic staining with VP1 antiserum. (*B*) Absence of fluorescence with VP2 antiserum. (*C*) Predominantly intranuclear staining with VP3 antiserum. (Reprinted, with permission, from Blacklow 1975.)

Infectious AAV3 could not be isolated from 333-8-9 cells that were initially inoculated with AAV3 alone and then carried through 14 serial culture passages (Blacklow 1975).

In contrast with 333-8-9 cells, BHK-21 cells (a virus-free hamster line) failed to enhance the synthesis of any AAV3 FA-stainable antigens in the absence of added Ad2 or HSV1 helper (see Table 1).

Line 333-8-9 cells inoculated with AAV1 alone produced both cytoplasmic and nuclear fluorescence when reacted with anti-AAV1 virion serum.

Inhibition of 333-8-9-Cell Oncogenicity in Hamsters by AAV1

As shown in Table 2, we examined the specific effect of prior AAV1 infection of 333-8-9 cells on their subsequent oncogenic potential after inoculation into hamsters. In all three experiments, a statistically significant inhibition of mean palpable tumor latency period was observed after hamsters had been inoculated with 333-8-9 cells initially infected with a crude preparation of AAV1 rather than with uninfected 333-8-9 cells. In the second experiment, mean survival time after inoculation of animals that received AAV1-infected cells was prolonged a statistically significant period over the survival time of animals that received uninfected 333-8-9 cells; furthermore, two animals in the AAV group underwent tumor regressions. In the third experiment, an additional control, 333-8-9 cells inoculated with AAV-free, heat-inactivated Ad2, failed to cause a significant change in tumor latency, in contrast with 333-8-9 cells inoculated with crude infectious AAV1 that had first undergone similar heat treatment.

Additional control experiments for the specificity of the AAV inhibitory effect on herpesvirus-transformed cells were performed (fourth and fifth experiments in Table 2). Inoculation of SV40-transformed TT101 cells or DMN-transformed DMN4B cells with AAV1 before injection into hamsters failed to alter significantly the mean tumor latency period of either cell line.

Effect of Purified AAV1 on Metastases

We have performed a preliminary study of the effect of AAV pretreatment of 333-8-9 cells on the incidence of lung metastases that are known to occur with 333-8-9-cell tumors (Duff et al. 1973). Each of ten weanling hamsters was inoculated with 10^3 333-8-9 cells; a similar group received the same number of cells pretreated with gradient-purified AAV1. After 10 weeks all animals were sacrificed and examined for the presence of lung metastases. Table 3 shows the effect of purified AAV1 on the development of primary and metastatic 333-8-9-cell tumors. Pre-

TABLE 2
Effect of AAV on Oncogenicity of Transformed Hamster Cells

Experiment[a]	Cell type	Transformed by	Number of cells given	Cells inoculated with	Mean tumor latency (days±s.e.)	Mean survival after inoculation (days±s.e.)
1	333-8-9	HSV2	5×10^2	media	26.7 ± 1	
				AAV	49.8 ± 4 $(P<0.001)$	82.5 ± 5.5
2	333-8-9	HSV2	1×10^3	media	20.1 ± 1	
				AAV	51.3 ± 6 $(P<0.001)$	$122.5^{b}\pm9.3$ $(P<0.01)$
3	333-8-9	HSV2	1×10^3	media	19.6 ± 2	
				AAV	31.2 ± 3 $(P<0.01)$	
				heated adenovirus	22.0 ± 2 (NS)[c]	
4	TT101	SV40	1×10^3	media	15.2 ± 0.1	
				AAV	17.2 ± 0.7 (NS)[c]	
5	DMN4B	DMN	1×10^3	media	24.8 ± 1	
				AAV	27.5 ± 2 (NS)[c]	

Data from Cukor et al. (1975).

[a] Ten to twelve animals/group.

[b] Two animals had tumor regressions in this group. One subsequently redeveloped a tumor and died; the other lived and was tumor-free.

[c] Difference not significant with respect to media control by the nonpaired t-test.

93

TABLE 3
Effect of Purified AAV1 on the Development of Primary and Metastatic Tumors

	Mean primary tumor latency (days±s.e.)	Incidence of lung metastases (%)
333-8-9 Cells alone[a]	29.9±4	50
AAV1-treated 333-8-9 cells[a]	54.0±6 ($P<0.01$)	0

[a] Ten hamsters received this inoculum.

treatment with AAV1 increased mean tumor latency significantly. Moreover, half of the animals receiving untreated tumor cells developed multiple lung metastases. This observation is similar to that of Duff et al. (1973), who found that 57% of untreated tumor-bearing animals had lung metastases. In contrast, none of the hamsters in the group inoculated with AAV-treated tumor cells showed pulmonary metastases.

It appears that the absence of lung metastases in the AAV-treated group did not simply reflect a delayed rate of development of the primary tumor. We have observed in over 50 hamsters inoculated with untreated 333-8-9 cells (N. R. Blacklow, G. Cukor, S. Kibrick, and G. Quinnan, unpublished observations) that those animals bearing primary tumors greater than 7 ml in volume at 10 weeks also showed lung metastases. However, of the hamsters receiving AAV-treated 333-8-9 cells in the experiment described in Table 3, some possessed primary tumors larger than 7 ml in volume which appeared just as early, yet they had no lung metastases.

Effect of AAV1 on Growth Characteristics of 333-8-9 Cells in Culture

The inhibitory effect of AAV on tumor development was shown not to be caused by direct toxicity of the virus preparation on 333-8-9 cells in culture. The heated AAV1 preparation, as well as the heat-inactivated Ad2 stock, was not directly toxic to 333-8-9, TT101, or DMN4B cells, as determined by assay of viable cells that excluded erythrocin B and by measurement of the rate of protein synthesis by cell cultures 24 hr after treatment (Cukor et al. 1975). However, these observations did not exclude the possibility that AAV treatment eliminated a small but critical subpopulation of 333-8-9 cells. Therefore, we studied the effect of AAV on the growth characteristics of individual clones of 333-8-9 cells. Cells were plated in Cuprak dishes, and after the 4 hr required for attachment, wells containing single cloned cells were selected for observation. Figure

FIGURE 2
Semilogarithmic plot of the growth curves of 333-8-9-cell clones observed in Cuprak dishes. (*A*) The mean±s.d. of 15 clones; 10^3 TCID$_{50}$/cell of gradient-purified AAV1 was added to the cultures 96 hr after plating. (*B*) The mean±s.d. of 28 untreated control clones.

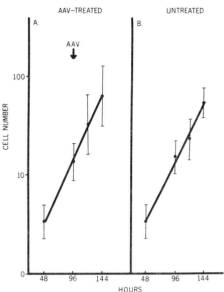

2 shows that the cells exhibited exponential growth in the interval between 48 and 144 hr after plating. No deviation from the logarithmic growth rate was noted in any of the clones after treatment with purified AAV1.

DISCUSSION

The oncogenic transformation of hamster-embryo fibroblasts by UV-irradiated HSV2 (Duff and Rapp 1971a,b) has provided a system for studying the effect of AAV on the oncogenicity of HSV2. Our data suggest that the 333-8-9 cell line, at the passage levels employed, contains enough HSV genetic material to enhance the formation of two of the three AAV structural polypeptides, whereas complete infectious herpes simplex virions potentiate the synthesis of all three AAV proteins (Johnson et al. 1972). In contrast with 333-8-9 cells, cells of the virus-free hamster line BHK-21 failed to enhance the synthesis of AAV FA-stainable antigens without added adenovirus or herpesvirus helper.

We have also demonstrated a striking inhibition by AAV of the oncogenicity of HSV2-transformed cells in syngenic hamsters. When 333-8-9 cells were treated with AAV prior to inoculation into animals, the appearance of palpable tumors was delayed by 50–150% as compared with the development of tumors in animals receiving untreated cells. On the other hand, AAV treatment of two non-herpes-transformed cell lines caused no significant delay in mean tumor latency. Further, it appears that the metastasizing of the primary tumor is specifically restricted in animals inoculated with AAV-treated 333-8-9 cells.

We did not observe direct toxicity of AAV to cultures of 333-8-9 cells or to clones derived from that cell line. The exponential growth rate continued unaltered after AAV treatment. One hypothesis to account for the inhibition of 333-8-9-cell oncogenicity could involve the AAV-specified antigens expressed on the surface of infected tumor cells. These foreign antigens could serve as a target for an increased immune response against tumor cells by the syngenic animal. The added inhibition of metastases also fits nicely with this hypothesis, since it is known that the occurrence of 333-8-9-cell metastases may be a sensitive indicator of the immunological status of the host (Duff et al. 1973). It has also been shown that metastases produced by HSV1-transformed cells are inhibited by the immune potentiator drug Levamisole, although no effect on primary tumors is seen (Sadowski and Rapp 1975). A second hypothesis to account for the inhibitory effects produced by AAV could involve the direct interaction of AAV with HSV2 genetic material in the tumor cell, which would render the 333-8-9 cell less oncogenic. Analysis of our 333-8-9-cell clones for their oncogenic potential as well as for their ability to complement AAV may help to elucidate the mechanism of tumor inhibition by AAV.

Recently, much effort has been directed at understanding the potential oncogenic effects of HSV2 in man. However, little attention has been given to the interaction of other human viruses with HSV2. Our data indicate that AAV, a ubiquitous defective virus of man (Blacklow et al. 1967, 1968; Parks et al. 1970), can influence the malignancy of HSV2-transformed cells.

ACKNOWLEDGMENTS

We are grateful to Dr. M. David Hoggan for kindly providing gradient-purified AAV1 and to Dr. Shirley Zajdel for histologic studies. We thank Frank Capozza, Roberta Ferriani, and Ingrid C. Swan for technical assistance.

This work was supported by grant IM21 from the American Cancer Society and by grants 1F22CA01761, BRSG RRO5712, and CA13397 from the National Institutes of Health.

REFERENCES

Atchison, R. W. 1970. The role of herpesviruses in adenovirus-associated virus replication in vitro. *Virology* **42**:155.

Atchison, R. W., B. C. Casto, and W. M. Hammon. 1965. Adenovirus-associated defective virus particles. *Science* **149**:754.

Blacklow, N. R. 1975. Potentiation of an adenovirus-associated virus by herpes simplex virus type 2 transformed cells. *J. Natl. Cancer Inst.* **54**:241.

Blacklow, N. R., M. D. Hoggan, and M. S. McClanahan. 1970. Adenovirus-associated viruses: Enhancement by human herpesviruses. *Proc. Soc. Exp. Biol. Med.* **134**:952.

Blacklow, N. R., M. D. Hoggan, and W. P. Rowe. 1967. Isolation of adenovirus-associated viruses from man. *Proc. Natl. Acad. Sci.* **58**:1410.

————. 1968. Serologic evidence for human infection with adenovirus-associated viruses. *J. Natl. Cancer Inst.* **40**:319.

Boucher, D. W., J. L. Melnick, and H. D. Mayor. 1971. Nonencapsidated infectious DNA of adeno-satellite virus in cells co-infected with herpesvirus. *Science* **173**:1243.

Cukor, G., N. R. Blacklow, S. Kibrick, and I. C. Swan. 1975. Effect of adeno-associated virus on cancer expression by herpesvirus-transformed cells. *J. Natl. Cancer Inst.* **55**:957.

Cuprak, L. J. 1975. An improved method for single cell cloning and observation. *TCA Manual* **1**:49.

di Mayorca, G., M. Greenblatt, T. Trauthen, A. Soller, and R. Giordano. 1973. Malignant transformation of BHK_{21} clone 13 cells in vitro by nitrosamines—A conditional state. *Proc. Natl. Acad. Sci.* **70**:46.

Dolin, R. and A. S. Rabson. 1973. *Herpesvirus saimiri*: Enhancement of adenovirus-associated virus. *J. Natl. Cancer Inst.* **50**:205.

Drake, S., P. A. Schaffer, J. Esparza, and H. D. Mayor. 1974. Complementation of adeno-associated satellite viral antigens and infectious DNA by temperature-sensitive mutants of herpes simplex virus. *Virology* **60**:230.

Duff, R. and F. Rapp. 1971a. Oncogenic transformation of hamster cells after exposure to herpes simplex virus type 2. *Nat. New Biol.* **233**:48.

————. 1971b. Properties of hamster embryo fibroblasts transformed in vitro after exposure to ultraviolet-irradiated herpes simplex virus type 2. *J. Virol.* **8**:469.

Duff, R., E. Doller, and F. Rapp. 1973. Immunologic manipulation of metastases due to herpesvirus transformed cells. *Science* **180**:79.

Hoggan, M. D., N. R. Blacklow, and W. P. Rowe. 1966. Studies of small DNA viruses found in various adenovirus preparations: Physical, biological and immunological characteristics. *Proc. Natl. Acad. Sci.* **55**:1467.

Johnson, F. B., N. R. Blacklow, and M. D. Hoggan. 1972. Immunological reactivity of antisera prepared against the sodium dodecyl sulfate-treated structural polypeptides of adenovirus-associated virus. *J. Virol.* **9**:1017.

Johnson, F. B., H. L. Ozer, and M. D. Hoggan. 1971. Structural proteins of adenovirus-associated virus type 3. *J. Virol.* **8**:860.

Kirschstein, R. L., K. O. Smith, and E. A. Peters. 1968. Inhibition of adenovirus 12 oncogenicity by adeno-associated virus. *Proc. Soc. Exp. Biol. Med.* **128**:670.

Parks, W. P., J. L. Melnick, R. Rongey, and H. D. Mayor. 1967. Physical assay and growth cycle studies of a defective adenosatellite virus. *J. Virol.* **1**:171.

Parks, W. P., D. W. Boucher, J. L. Melnick, L. Taber, and M. Yow. 1970. Seroepidemiological and ecological studies of the adenovirus-associated satellite viruses. *Infect. Immun.* **2**:716.

Sadowski, J. and F. Rapp. 1975. Inhibition by Levamisole of metastases by cells transformed by herpes simplex virus type 1. *Proc. Soc. Exp. Biol. Med.* **149**:219.

Smith, K. O., W. D. Gehle, and J. F. Thiel. 1966. Properties of a small virus associated with adenovirus type 4. *J. Immunol.* **97**:754.

Wexler, H. 1966. Accurate identification of experimental pulmonary metastases. *J. Natl. Cancer Inst.* **36**:641.

Expression of Helper Function for Adeno-Associated Virus in Adenovirus-Transformed Cells

HIROSHI HANDA
KAZUKO SHIROKI
HIROTO SHIMOJO

Institute of Medical Science
University of Tokyo
Tokyo 108, Japan

The helper function(s) for adeno-associated-virus (AAV) growth induced by Ad31 in human cells appears as early as 6 hr after infection (Handa et al. 1976), at the same time as T antigen. The presence of adenovirus T antigen alone is not sufficient for the helper effect, since AAV does not grow in T-antigen-positive Ad2-transformed rat-embryo cells (B. Carter, personal communication). The ability of AAV to grow in cells coinfected with DNA-negative mutants (Ito and Suzuki 1970; Ishibashi and Ito 1971; Mayor and Ratner 1972; Handa et al. 1975; Straus et al. 1976) suggests strongly that the helper effect is an early adenovirus function and therefore might be present in transformed cells. Transformed cells may lack helper function because of a defect in adenovirus early gene expression in these cells or simply because AAV requires cellular factors not present in these cells (that is, AAV has a host range similar to that of its helper adenovirus). To distinguish between these two possibilities we have examined the infection of heterokaryons between adenovirus-transformed cells and normally permissive cells.

MATERIALS AND METHODS

Virus Infection and Cell Fusion

KB cells and Ad12-transformed cells were cocultured in plastic petri dishes in low-calcium (1 mM) Eagle's minimum essential medium (MEM) supplemented with 10% fetal-calf serum. The cells were infected with AAV1 (Handa et al. 1975) at 10 fluorescent infectious units (FIU) per cell (Blacklow et al. 1967) after cocultivation at 37°C for 4 hr. After adsorption for 2 hr at 37°C, the cells were incubated in medium containing 100–200 neutralizing units of anti-AAV1 guinea-pig serum for 30 min. Cell fusion was performed by the method of Dr. Y. Okada (personal communication). The cells were washed with Ca-BSS (0.0114 M NaCl, 0.054 M KCl, 0.34 M Na_2HPO_4, 0.44 mM KH_2PO_4, 0.01 M Tris-HCl, 1 mM $CaCl_2$, pH 7.6), and UV-inactivated Sendai virus strain Z (Okada 1962) suspension (500–3000 hemagglutinating units per dish) was added to the cells. The dishes were kept on ice for 10 min and then washed with Ca-BSS. Low-calcium (1 mM) Eagle's MEM supplemented with 2% fetal-calf serum was added, and the cultures were then incubated at 37°C in a CO_2 incubator.

Cell Lines

The use of KB cells has been described before (Handa et al. 1975). A rat cell line, 3Y1, established from a Fischer rat embryo (Kimura et al. 1975), was kindly provided by Dr. G. Kimura. The following Ad12-transformed cell lines were used: WY3 cells [3Y1 cells transformed by the whole Ad12 DNA (Yano et al. 1977)], CY1 cells [3Y1 cells transformed by the *Eco*RI C fragment of Ad12 DNA (Yano et al. 1977)], GY1 cells [3Y1 cells transformed by the *Hin*dIII G fragment of Ad12 DNA (Shiroki et al. 1977)], and JtsA3 cells [strain JAR rat-embryo cells transformed by a temperature-sensitive mutant (tsA) of Ad12 (K. Shiroki, unpublished)]. HT4 cells were derived from an Ad12-induced tumor in a hamster (Nakajima et al. 1973). W-3Y-23 cells [3Y1 cells transformed by wild-type simian virus 40 (SV40) (Segawa et al. 1977)] were kindly provided by Dr. N. Yamaguchi. H5 cells [monkey cells transformed by an Ad7-SV40 hybrid virus (Shiroki and Shimojo 1971)] were used. H-NRK cells [nonproducer normal-rat-kidney (NRK) cells infected and transformed by Harvey murine sarcoma virus (Levy 1977)] were kindly provided by Dr. Y. Yuasa.

Virus Titration

The infected cells and medium were frozen and thawed three times. After centrifugation, 0.2 ml of the supernatant was titrated by production of AAV1-antigen-positive cells on cover-slip cultures as described previously (Handa et al. 1976). Virus titer is expressed as FIU per 0.2 ml.

Immunofluorescent Staining and Autoradiography

KB cells labeled with [³H]thymidine (0.5 μCi/ml, 22 Ci/mmole) for 48–72 hr before cocultivation were used to produce heterokaryons. At 20 hr after fusion, the heterokaryons on cover slips were fixed with acetone. Immunofluorescent staining was done by a direct method similar to that described by Ishibashi (1970). After examination by fluorescence microscopy, the cover slips were washed with phosphate-buffered saline, treated with 2% perchloric acid, coated with Sakura NR-M2 emulsion, and kept in the dark for 10 days. The samples were developed with Konidol X, stained with Giemsa, and examined.

Labeling, Extraction, and Analysis of Viral DNA

At 11 hr after cell fusion, the culture was labeled with [^{3}H]thymidine (10 μCi/ml) for 1 hr. Viral DNA was selectively extracted by the Hirt procedure (Hirt 1967). The extracted DNA in Hirt supernatants was analyzed in 5–20% neutral sucrose gradients containing 0.01 M Tris-HCl (pH 7.2), 0.1 M NaCl, 0.01 M EDTA, and 0.1% Sarkosyl by centrifugation in an SW41 rotor at 4°C for 20 hr at 25,000 rpm. Fractions were assayed for radioactivity as described previously (Handa et al. 1976).

RESULTS

Microscopic Examination of AAV1 Growth

KB and JtsA3 cells were used to produce heterokaryons. After infection of the cocultured cells with AAV1 and subsequent cell fusion, the formation of AAV1 antigens was detected by immunofluorescence and the differentiation between KB- and JtsA3-cell nuclei was made by autoradiography. The nucleus of the KB cell in the heterokaryon was packed with grains (Fig. 1A). The results of autoradiography demonstrated that the larger KB-cell nuclei could be differentiated from JtsA3-cell nuclei on the basis of size. The formation of heterokaryons which contained 2–20 nuclei was confirmed by autoradiography. AAV1 virion antigens were detected only in heterokaryons, and never in homokaryons or in mononucleate cells. However, the proportion of AAV1-antigen-positive heterokaryons was small (approximately 1% of all the heterokaryons). AAV1 virion antigens were detected in nuclei derived from one or the other parental cell or in nuclei of both types in the same heterokaryon, as shown in Figure 1B.

Growth of AAV1 in Heterokaryons of KB Cells and Ad12-Transformed Cells

KB and JtsA3 cells were cocultured in different proportions, infected with AAV1, and fused with UV-inactivated Sendai virus. Cultures were harvested at 20 hr after fusion, and the virus yield in each culture was titrated (Table 1). The maximum virus yield resulted from the fusion of KB and

FIGURE 1
Microscopic examination of AAV1 growth in heterokaryons. (A) An autoradiogram of a heterokaryon. The KB-cell nucleus is shown by an arrow. (B) Immunofluorescence of AAV1 virion antigens in a heterokaryon.

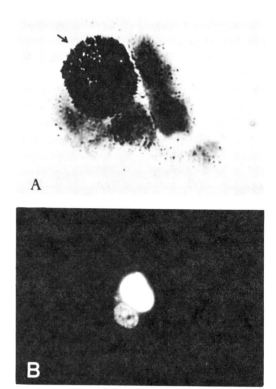

A

B

JtsA3 cells in the proportion 3:1. This ratio of parental cell types was used in the following experiments.

Time Course of AAV1 Growth After Cell Fusion

Virus yields in heterokaryons of KB and JtsA3 cells were titrated at intervals of 4 hr after cell fusion (Fig. 2). An increase in virus titer was detected at 4 hr and reached its maximum at 12 hr after cell fusion. No virus growth was observed in the absence of UV-inactivated Sendai virus.

Analysis of DNA Synthesized in Heterokaryons

The AAV1-infected heterokaryons of KB and JtsA3 cells were pulse-labeled for 1 hr with [³H]thymidine (10 µCi/ml) at 11 hr after cell fusion. Hirt supernatants prepared from the labeled cells were analyzed in neutral sucrose gradients. As a control, a similar experiment was performed without cell fusion (Fig. 3). A peak of [³H]thymidine-label cosedi-

TABLE 1
Growth of AAV1 in Heterokaryons Formed by Fusion of KB Cells with JtsA3
Cells in Different Proportions

	KB cells: JtsA3 cells[a]				
	10:1	3:1	1:1	1:3	1:10
AAV1 titer, log (FIU/0.2 ml)	3.6	3.9	3.8	3.3	3.0

[a] Proportion of KB cells and JtsA3 cells used for fusion.

menting with ^{32}P-labeled AAV1 DNA was detected in the coculture
treated with UV-inactivated Sendai virus, but not in the control coculture. This DNA profile is characteristic of Hirt supernatants prepared
from infected cells, as shown previously (Handa et al. 1976). The results
indicate that AAV1 DNA is synthesized in heterokaryons without coinfection of cells with adenovirus. Similar results were obtained when GY1
cells were used instead of JtsA3 cells.

*Growth of AAV1 in Heterokaryons Formed with Various Lines of Transformed
Cells and in Homokaryons*

The growth of AAV1 was examined in heterokaryons of KB cells and
different Ad12-transformed cells (Table 2). The results showed that all the
heterokaryons of KB cells with WY3, CY1, GY1, or JtsA3 cells helped

FIGURE 2
Growth of AAV1 after cell
fusion. Cocultures of KB and
JtsA3 cells with (o——o) or
without (●---●) treatment
with UV-inactivated Sendai
virus.

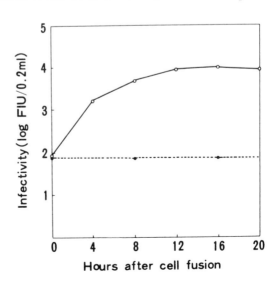

FIGURE 3

Analysis of DNA synthesized in heterokaryons. Cocultures of KB and JtsA3 cells with (o——o) or without (●---●) treatment with UV-inactivated Sendai virus were infected with AAV1 and labeled with [³H]thymidine at 11 hr p.i. The Hirt supernatants extracted from the labeled cells were analyzed in sucrose gradients as described in the text. The position of ³²P-labeled AAV1 DNA (a marker) is shown by an arrow.

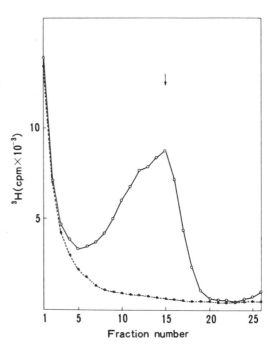

TABLE 2

Growth of AAV1 in Different Types of Heterokaryons and Homokaryons

Combination of cells		AAV1 virion antigens[a]	AAV1 titer log (FIU/0.2 ml)
KB	WY3	+	3.9
KB	CY1	+	nd[b]
KB	GY1	+	3.8
KB	JtsA3	+	3.9
KB	HT4	+	nd
KB	3Y1	−	nd
KB	KB	−	1.9
JtsA3	JtsA3	−	1.8
KB	W–3Y–23	−	1.8
KB	H–NRK	−	nd
KB	H5	−	nd

[a]+: AAV1-antigen-positive cells were detected. −: No AAV1-antigen-positive cells were detected.
[b]nd: Not done.

AAV growth to the same extent. Similar results were obtained with HT4 cells, cells derived from a tumor induced in a hamster with Ad12. However, no AAV1 growth was found in heterokaryons of KB cells with 3Y1, W-3Y-23, or H-NRK cells. No AAV1 growth was detected in homokaryons of either KB or JtsA3 cells fused with UV-inactivated Sendai virus. No AAV1 virion antigen was detected in heterokaryons of KB and H5 cells, which were SV40 T-antigen-positive and Ad12 T-antigen-negative (Shiroki and Shimojo 1971).

DISCUSSION

The adeno-associated viruses are defective parvoviruses which can replicate only in cells coinfected with a helper adenovirus. We have shown previously (Handa et al. 1976) that adenovirus induces a helper factor(s) for AAV growth early after infection of human cells with adenovirus. This observation led us to examine whether or not the helper factor(s) is present in Ad12-transformed cells. The growth of AAV1 in heterokaryons of KB cells and Ad12-transformed rodent cells was shown by formation of the AAV1 antigen, by increase in AAV1 titers, and by synthesis of AAV1 DNA. These results indicate that a helper factor(s) is present in Ad12-transformed cells and that AAV1 can replicate in heterokaryons without coinfection with adenovirus.

AAV1, the virus used in this experiment, is originally a human virus; it replicates in human cells coinfected with human adenovirus. AAV1 does not replicate in either rat cells coinfected with human adenovirus or rodent cells transformed by adenovirus (H. Handa, unpublished data). This restriction of AAV1 growth in rodent cells may be due to the lack of a host-range factor; however, this matter has not yet been clarified. The restriction was overcome by fusion of Ad12-transformed rodent cells with human KB cells, which supply a host-range factor. It would appear, therefore, that the growth of AAV1 is dependent upon two factors, a host-range factor controlled by the cell and an adenovirus-induced helper factor.

The nature of the adenovirus-induced helper factor is not known. However, it was shown previously that the time of the appearance of the helper factor was quite similar to that of the Ad12-specific T antigen in Ad31- or Ad12-infected cells (Handa et al. 1976; Shimojo et al. 1967). This observation suggests that the helper factor is closely related to the Ad12-specific T antigen. This suggestion is supported by the present observations that heterokaryons of KB cells and either CY1 or GY1 cells supported the growth of AAV1. CY1 and GY1 cells are rat cells transformed by transfection with the EcoRI C fragment (16% of the left end) (Yano et al. 1977) and with the HindIII G fragment (7.2% of the left end) (Shiroki et al. 1977) of Ad12 DNA, respectively. It is probable that CY1 or GY1 cells contain no Ad12-specific early proteins other than T antigen, the putative product of the Ad12 transforming gene. The helper factor for the AAV1 growth therefore may be the Ad12-specific T antigen, which may influence AAV1 growth either directly or indirectly. Although T antigen is believed to play an important role in transformation of cells by adenovirus, its biological function is still unknown. The helper function for AAV1 growth therefore provides a useful tool for analysis of the biological function(s) of adenovirus T antigen.

ACKNOWLEDGMENTS

We are grateful to Dr. Yoshio Okada for kind instruction in cell fusion and to Dr. Barrie J. Carter for critical review of the manuscript. This work was supported by grants from the Ministry of Education, Science, and Culture, Japan, and from the Princess Takamatsu Fund for Cancer Research.

REFERENCES

Blacklow, N. R., M. D. Hoggan, and W. P. Rowe. 1967. Immunofluorescent studies of the potentiation of an adenovirus-associated virus by adenovirus 7. *J. Exp. Med.* **125**:755.

Handa, H., H. Shimojo, and K. Yamaguchi. 1976. Multiplication of adeno-associated virus type 1 in cells coinfected with a temperature-sensitive mutant of human adenovirus type 31. *Virology* **73**:1.

Handa, H., K. Shiroki, and H. Shimojo. 1975. Complementation of adeno-associated virus growth with temperature-sensitive mutants of human adenovirus types 12 and 5. *J. Gen. Virol.* **29**:239.

Hirt, B. 1967. Selective extraction of polyoma DNA from infected mouse cell cultures. *J. Mol. Biol.* **26**:365.

Ishibashi, M. 1970. Retention of viral antigen in the cytoplasm of cells infected with temperature-sensitive mutants of an avian adenovirus. *Proc. Natl. Acad. Sci.* **65**:304.

Ishibashi, M. and M. Ito. 1971. The potentiation of type 1 adeno-associated virus by temperature-sensitive conditional-lethal mutants of CELO virus at the restrictive temperature. *Virology* **45**:317.

Ito, M. and E. Suzuki. 1970. Adeno-associated satellite virus growth supported by a temperature-sensitive mutant of human adenovirus. *J. Gen. Virol.* **9**:243.

Kimura, G., A. Itagaki, and J. Summers. 1975. Rat cell line 3Y1 and its virogenic polyoma- and SV40-transformed derivatives. *Int. J. Cancer* **15**:694.

Levy, J. A. 1971. Demonstration of differences in murine sarcoma virus foci formed in mouse and rat cells under a soft agar overlay. *J. Natl. Cancer Inst.* **46**:1001.

Mayor, H. D. and J. Ratner. 1972. Conditionally defective helper adenoviruses and satellite virus replication. *Nat. New Biol.* **239**:20.

Nakajima, S., C. Hamada, and H. Uetake. 1973. Alternate changes of surface antigen(s) in adenovirus type 12-transformed and tumor cells. *Japan. J. Microbiol.* **17**:303.

Okada, Y. 1962. Analysis of giant polynuclear cell formation caused by HVJ virus from Ehrlich's ascites tumor cells. *Exp. Cell Res.* **26**:98.

Segawa, K., N. Yamaguchi, and K. Oda. 1977. Simian virus 40 gene A regulates the association between a highly phosphorylated protein and chromatin and ribosomes in simian virus 40-transformed cells. *J. Virol.* **22**:679.

Shimojo, H., H. Yamamoto, and C. Abe. 1967. Differentiation of adenovirus 12 antigens in cultured cells with immunofluorescent analysis. *Virology* **31**:748.

Shiroki, K. and H. Shimojo, 1971. Transformation of green monkey kidney cells by SV40 genome: The establishment of transformed cell lines and the replication of human adenoviruses and SV40 in transformed cells. *Virology* **45**:163.

Shiroki, K., H. Handa, H. Shimojo, S. Yano, S Ojima, and K. Fujinaga. 1977. Establishment and characterization of rat cell lines transformed by restriction endonuclease fragments of adenovirus 12 DNA. *Virology* (in press).

Straus, S. E., H. S. Ginsberg, and J. A. Rose. 1976. DNA-minus temperature-sensitive mutants of adenovirus type 5 help adenovirus-associated virus replication. *J. Virol.* **17**:140.

Yano, S., S. Ojima, K. Fujinaga, K. Shiroki, and H. Shimojo. 1977. Transformation of a rat cell line by the adenovirus type 12 DNA fragment. *Virology* **82**: 207.

Complementation of Adeno-Associated Virus by Temperature-Sensitive Mutants of Human Adenovirus and Herpesvirus

HEATHER D. MAYOR
JAMES F. YOUNG

Department of Microbiology and Immunology
Baylor College of Medicine
Houston, Texas 77030

The adeno-associated satellite viruses (AAV) are unconditionally defective members of the parvovirus group. Their defectiveness necessitates the presence of replicating adenovirus in the same host cell for mature infectious progeny virus to be produced (Parks et al. 1967; Ito et al. 1967; Blacklow et al. 1967; Torikai and Mayor 1969). Herpesviruses exhibit an incomplete helper function for AAV, potentiating viral antigens, detectable by immunofluorescence (Atchison 1970; Blacklow et al. 1970), and infectious viral DNA (Boucher et al. 1971), but no infectious virions or complement-fixing antigens. AAV therefore appears to be defective in both its replication and maturation processes in the absence of a helper virus.

Numerous temperature-sensitive mutants of adenoviruses and herpesviruses known to be defective in viral DNA synthesis and/or other virus-specific functions have been isolated and characterized. These mutants are proving to be useful tools in dissecting the basis and strategy of AAV defectiveness.

MATERIALS AND METHODS

Cells and Viruses

All stocks of herpes simplex virus type 1 (HSV1) wild type (wt) and its temperature-sensitive mutants were grown in a continuous line of African green monkey kidney cells (Vero). Stocks were also assayed for infectivity and monitored for leakiness by plaque titration in Vero cells at 34°C (permissive temperature) and 39°C (nonpermissive temperature). For all mutants, the virus yield was consistently smaller at the non-permissive temperature; the factors were $\leq 10^{-5}$.

Stocks of human adenovirus-31 (Ad31) wild type and its temperature-sensitive (ts) mutants were grown and titered in cultures of human-embryo kidney (HEK) cells at 34°C (permissive temperature) and 40°C (nonpermissive temperature). Titers at the permissive temperature ranged from $10^{7.5}$ $TCID_{50}$/ml for wild-type virus and ts mutants 7, 9, and 13 to $10^{6.5}$ $TCID_{50}$/ml for ts mutants 38 and 94. Mutant ts94 had a titer of $10^{2.5}$ $TCID_{50}$/ml at the nonpermissive temperature (40°C), which indicated a small degree of leakiness. The other ts mutants had titers less than 10^2 $TCID_{50}$/ml at 40°C. Wild-type virus had a titer of $10^{6.9}$ $TCID_{50}$/ml at 40°C, which indicated some temperature sensitivity.

Adeno-associated virus type 1 (AAV1), titer $10^{7.4}$ complement-fixing units per ml, was grown in CV1 cells with simian adenovirus SV15 as helper. Complementation tests and immunofluorescence were per-formed as described previously (Drake et al. 1974; Mayor et al. 1977).

Infectivity Titration of AAV and AAV DNA

AAV1 infectivity was titrated by complement-fixation production using AAV-free adenovirus 2 (Ad2) as helper virus in HEK cells (Mayor et al. 1977). DNA infectivity was assayed using DEAE-dextran and fluorescent-antibody techniques as described by Boucher et al. (1971).

Extraction and Analysis of DNA

Low-molecular-weight DNA was selectively extracted from control and infected cells using the method of Hirt (1976). DNA was further analyzed by CsCl and sucrose gradients (neutral and alkaline) as described pre-viously (Mayor et al. 1974, 1977).

Vero cells coinfected with either AAV1 and SV15 or AAV1 and HSV were labeled with 5 μCi/ml [^{3}H]thymidine and extracted with 0.25% Triton X-100 (pH 7.9) 24 hr after infection. Cell debris was removed, and supernatants were adjusted to 0.04 M and applied to hydroxyapatite columns. Eluates were collected with 0.1 M phosphate buffer (single-stranded-DNA region) and with 0.5 M buffer (double-stranded-DNA region) and further categorized by velocity sedimentation in neutral sucrose gradients and by enzyme digestion.

RESULTS AND DISCUSSION

Complementation of AAV by HSV mutants

Members of seven complementation groups of HSV1 strain KOS ts mutants as categorized by Schaffer (1975) were used in the present study (Table 1). The partial helper function provided by HSV to AAV production of AAV structural antigens and infectious AAV DNA was studied in mixed infections of AAV1 and HSV mutants in Vero cells at the nonpermissive temperature (39°C). In agreement with the results obtained with wild-type HSV1 and those obtained in a previous study using AAV4 (Drake et al. 1974), mutants in complementation groups A, D, and G, which either were DNA-negative or showed minimal viral DNA production, and groups E and O, which were DNA-positive, complemented AAV-antigen synthesis at the nonpermissive temperature. In addition, infectious AAV DNA, migrating as a 16S species in neutral sucrose gradients, was isolated from cells coinfected with AAV1 and these mutants. However, members of complementation groups B and C, which were DNA-negative, failed to complement AAV-antigen synthesis, and it was also shown that ts847, a member of group B, failed to stimulate production of AAV DNA. Thus, the ability of DNA-negative ts mutants to complement AAV structural-antigen synthesis and DNA production established that these functions are not dependent on HSV DNA synthesis. The fact that two distinct DNA-negative complementation groups, B and C, were defective in potentiating AAV antigens and (at least in the case of group B) defective in stimulating synthesis of AAV DNA indicates that at least two HSV cistrons may be involved in potentiating AAV-antigen production. However, the defects in the B and C mutants may indirectly prevent the expression of the HSV helper function. It is of interest that members of these two groups of mutants failed to synthesize wild-type amounts of their major capsid protein VP154 (Table 1; also see Schaffer 1975 for review). In fact, mutants of groups B and C are among the most defective and exhibit the greatest alterations in polypeptide synthesis of any of the HSV mutants (Courtney et al. 1976). However, the failure to complement AAV appears to be independent of the mutant's inability to replicate its own genome, or to express the virally coded DNA polymerase (Aron et al. 1975).

Complementation of AAV by Adenovirus Mutants

Human adenovirus type 31 is highly oncogenic in hamsters and induces a T antigen identical to that of adenovirus type 12 (Ad12). We have studied

TABLE 1
Characteristics of ts Mutants of HSV1 (Strain KOS) at Nonpermissive Temperature

Complementation group*	AAV-antigen synthesis[a]	AAV-DNA production[b]	Herpes DNA[c]*	Herpes DNA polymerase*	Herpes thymidine kinase*	Herpes viral antigens (IF)*	Selected HSV polypeptides (VP)[d]*
wt	100	100	100	+	+	+	
A	100 ± 5	100 ± 10	<1	+	–	+	↓ VP154 ↓ VP175
B	<1	<10	<1	±	–	+	↓ VP154 →
C	<1	ND	<1	–	+	+	↑ VP134 ↓ VP154
D	100 ± 5	100 ± 10	<1	–	ND	+	↓ VP154
E	100 ± 5	100 ± 10	70	+	+	+	= wt
G	100 ± 5	100 ± 10	<10	+	–	+	= wt
O	100 ± 5	100 ± 10	<10	+	+	+	↓ VP154

* These portions of table adapted from Schaffer 1975.
wt: Wild type.
VP: Values given as m.w. × 10³.
+: Present; –: absent.
ND: No data yet available.
↓: Depressed; ↓↓: markedly depressed.
[a] Results expressed as percentage of fluorescing cells in wt HSV-AAV coinfections.
[b] Results expressed as percentage of infectivity titer by IF DNA assay under standard conditions with wt HSV and AAV.
[c] [³H]thymidine was incorporated into DNA in mutant- and wt-infected cells. Infected-cell lysates were banded in neutral CsCl gradients. Results are expressed as percentage of viral DNA in wt-virus-infected cultures (Aron et al. 1975).
[d] Data abstracted from Courtney et al. 1976.

the ability of temperature-sensitive mutants of Ad31 (categorized by Suzuki et al. 1972) and Ad12 (kindly supplied by Dr. Nada Ledinko and categorized by her; see Ledinko 1974) to support AAV replication. Wild types of both adenoviruses, Ad31 and Ad12, and all the ts mutants tested (type 31: ts7, 9, 13, 38 and 94; type 12: ts401, 406, and 409) were able to complement AAV-antigen production as detected by immunofluorescence in both HEK and KB cells at both permissive and nonpermissive temperatures (Table 2). However, the results of AAV infectivity titrations from cultures coinfected with AAV and the various mutants revealed that Ad31 mutant ts94 was consistently defective in its ability to potentiate the production of complete AAV virions at the nonpermissive temperature. Mutant ts94 is not a DNA-negative mutant; in fact, it appears to be competent in all adenovirus functions except maturation (Table 2).

AAV DNA was synthesized in ts94 coinfections. The majority of this DNA was apparently not encapsidated, as no virus particles were seen in electron microscopy (data not shown). DNA present in the Hirt supernatant fraction (Hirt 1967) was subjected to velocity sedimentation in neutral sucrose gradients to separate the Ad31 DNA from the AAV DNA. Electron-microscope examination of DNA with a sedimentation constant of less than 25S revealed that this DNA, although heterogeneous in size, contained no molecules of greater than unit AAV genome length (~1.5 μm). Ad12 mutant ts409 also appeared to be impaired in its ability to potentiate AAV infectivity and DNA at the nonpermissive temperature. AAV antigen was produced, as measured by immunofluorescence (IF), but the pattern was markedly cytoplasmic, possibly indicating a block in the transport of AAV polypeptides to the nucleus. In contrast with Ad31 ts94 and Ad12 ts409 mutants, Ad31 mutant ts38 appeared to have an enhanced ability to potentiate AAV replication at the nonpermissive temperature.

Isolation of AAV Nucleoprotein Complexes

It might be productive to follow the potential for encapsidating AAV DNA through the examination of intermediates released from infected cells. Preliminary work to characterize AAV maturation through biochemical examination of the in vitro products synthesized during coinfections of AAV with either adenovirus or herpesvirus are being undertaken. We have previously used the method of Hirt (1967) in preparing viral DNA from infected cells (Mayor et al. 1974, 1977). This procedure utilizes sodium dodecyl sulfate, pronase, and 1 M NaCl followed by phenol extraction, probably resulting in the removal of any lipid or protein that may have been bound to the DNA. This method, although excellent for analysis of naked nucleic acid, gives no indication of the

TABLE 2
Characteristics of ts Mutants of Ad31 and Ad12 at Nonpermissive and Permissive Temperatures

Adenovirus type or mutant	AAV-antigen synthesis[a]	AAV-DNA production[b]	AAV infectivity[c] 40°C/34°C $\log_{10}$ ID$_{50}$	Adenovirus DNA[d]	Adenovirus structural proteins[e]			Adenovirus antigens (IF)
					hexon	penton	fiber	
Type 31 wt	100	100	7.4/7.4	100	+	+	+	+
ts13	100 ± 5	100 ± 10	7.0/7.0	0	−	−	−	−
ts7	50 ± 3	ND	6.5/6.5	70	−	+	−	< wt
ts9	50 ± 3	ND	6.5/6.5	100	−	→	−	< wt
ts38	125 ± 10	125 ± 15	7.0/5.5	90	+	+	→	< wt
ts94	100 ± 5	100 ± 10	4.4/7.4	100	+	+	+	+
Type 12 wt	100	100	7.0/7.0	100	ND	ND	ND	+
ts401	100 ± 5	100 ± 10	6.0/6.0	40	ND	ND	ND	−
ts406	100 ± 5	100 ± 10	5.5/5.5	30	ND	ND	ND	−
ts409	100 ± 5	ND	5.5/7.0	100	ND	ND	ND	< wt

wt: Wild type.
+: Present; −: absent.
↓: Depressed.
ND: No data available.
[a] Results expressed as percentage of fluorescing cells in wt Ad-AAV coinfections.
[b] [³H]thymidine incorporated into DNA in cells coinfected with mutants and AAV. Hirt-supernatant 16S neutral sucrose fractions. Results expressed as percentage of infectivity titer by IF DNA assay under standard conditions with wt Ad and AAV1.
[c] Infectivities are shown as $\log_{10}$ CFU (complement-fixing units)/ml at the nonpermissive and permissive temperatures in HEK cells. Harvests were heated at 60°C for 15 min to inactivate any adenovirus activity. Surviving AAV, if present, was titrated at 37°C in 35-mm petri-dish cultures of HEK cells coinfected with 5 PFU (plaque-forming units)/cell AAV-free Ad2 helper.
[d] [³H]thymidine incorporated into DNA in mutant- and wt-infected cells. Hirt supernatants were banded in neutral sucrose. Results expressed as percentage of viral (32S) DNA in wt-virus-infected cultures.
[e] Data abstracted from Suzuki et al. 1972.
(Table modified from Mayor et al. 1977.)

actual state of the DNA as it exists in virus-infected cells. We are currently using a different approach (Green et al. 1971) known to maintain the integrity of protein-DNA complexes. This method has been used to extract complexes of polyoma (Goldstein et al. 1973) and simian virus 40 (Meinke et al. 1975). The technique involves the use of a lysing solution containing 0.01 M ethylenediaminetetraacetate, 0.01 M Tris-HCl (pH 8.9) containing 0.25% Triton X-100. This treatment releases the nucleoprotein complexes gently from infected cells, apparently by imposing a leakiness on the nuclear membrane. This leakiness must not be very drastic, as neither adenovirus DNA nor herpesvirus DNA has yet been found to be released in our experiments.

Initial experiments using this method indicate that a protein-DNA complex does exist in Vero cells coinfected with AAV and simian adenovirus SV15. This complex elutes from hydroxyapatite (Bernardi 1969) with 0.1 M phosphate buffer, a salt concentration similar to that required to elute free single-stranded DNA (Fig. 1). Of the material in the 0.1 M

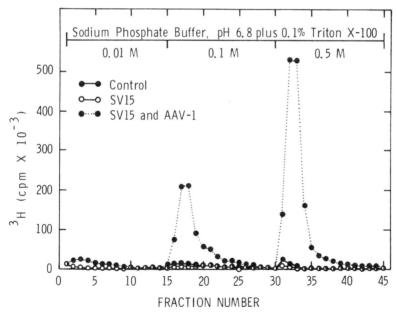

FIGURE 1
Hydroxyapatite chromatography of AAV nucleoprotein complexes. Complexes were eluted under pressure with 15 1-ml portions of 0.1 M sodium phosphate (pH 6.8) containing 0.1% Triton X-100. Double-stranded AAV DNA was eluted with 15 1-ml portions of 0.5 M sodium phosphate (pH 6.8) containing 0.1% Triton X-100.

eluant, 33% was found to have a sedimentation coefficient of around 65S (Fig. 2) in sucrose (pH 7.9); the remainder of the radioactivity was at the top of the gradient, possibly complexed to lipid. After extraction with phenol, 80% of the 65S complex then exhibited a sedimentation constant of 16S; 20% of the radiolabel sedimented at 27S (Fig. 3). These values are consistent with double- and single-stranded AAV DNA, respectively. The high level of double-stranded DNA observed probably reflects the reassociation of separated AAV strands after deproteinization, because prior to phenol extraction the DNA is sensitive to single-strand-specific nucleases (data not shown).

The material from the 0.5 M eluate was found to have a sedimentation profile consistent with naked double-stranded AAV DNA (16S) (Fig. 2). By comparing the relative amounts of material in the 0.1 M (65S) and 0.5 M (16S) eluates, it was found that after 12 hr of labeling with [³H]thymidine, beginning at 12 hr post-infection, 35% of the total AAV DNA extracted by this procedure was found in the form of 65S complexes. Although this represents a relatively high percentage of the AAV DNA, such a complex has not as yet been found in cells coinfected with herpes simplex virus and AAV. We plan to examine these complexes by immune electron microscopy to determine if AAV capsid material is associated with the DNA. This would be of great interest in light of the observation that a 65S complex is not found in HSV-AAV systems in which complete AAV maturation does not take place.

FIGURE 2
Velocity sedimentation in 5–20% sucrose gradients (pH 7.9) of [³H]thymidine-labeled AAV DNA, isolated by the method of Green et al. (1971), after chromatography on hydroxyapatite columns. The samples analyzed were fraction 17 (●) and fraction 32 (○) from the hydroxyapatite column (Fig. 1). Gradients were run at 36,000 rpm for 3 hr at 4°C in an SW 50.1 rotor. Fractions were collected by bottom puncture and assayed for radioactivity.

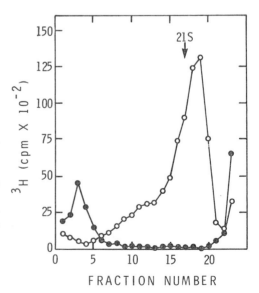

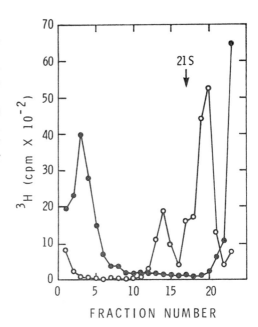

FIGURE 3

Velocity sedimentation in 5–20% sucrose gradients (pH 7.9) of 0.1 M eluate material from hydroxyapatite columns before (●) and after (○) phenol extraction. Gradients were run at 36,000 rpm for 3 hr at 4°C in an SW 50.1 rotor. Fractions were collected by bottom puncture and assayed for radioactivity.

ACKNOWLEDGMENTS

This research was supported by grant Q-398 from the Robert A. Welch Foundation, Houston, Texas, and grant CA 14618 from the National Cancer Institute, U.S. Public Health Service.

REFERENCES

Aron, G. M., D. J. M. Purifoy, and P. A. Schaffer. 1975. DNA synthesis and DNA polymerase activity of herpes simplex virus type 1 temperature-sensitive mutants. *J. Virol.* **16**:498.

Atchison, R. W. 1970. The role of herpes virus in adeno-associated virus replication in vitro. *Virology* **42**:155.

Bernardi, G. 1969. Chromatography of nucleic acids on hydroxyapatite. I. Chromatography of native DNA. *Biochim. Biophys. Acta* **174**:423.

Blacklow, N. R., M. D. Hoggan, and M. S. McClanahan. 1970. Adenovirus-associated viruses: Enhancement by human herpes virus. *Proc. Soc. Exp. Biol. Med.* **134**:952.

Blacklow, N. R., M. D. Hoggan, and W. P. Rowe. 1967. Immunofluorescent studies of the potentiation of an adenovirus-associated virus by adenovirus 7. *J. Exp. Med.* **125**:755.

Boucher, D. W., J. L. Melnick, and H. D. Mayor. 1971. Nonencapsidated infectious DNA of adeno-satellite virus in cells coinfected with herpesvirus. *Science* **173**:1243.

Courtney, R. J., P. A. Schaffer, and K. L. Powell. 1976. Synthesis of virus-specific polypeptides by temperature-sensitive mutants of herpes simplex. *Virology* **75**:306.

Drake, S., P. A. Schaffer, J. Esparza, and H. D. Mayor. 1974. Complementation of adeno-associated satellite virus antigens and infectious DNA by temperature sensitive mutants of herpes simplex virus. *Virology* **60**:230.

Goldstein, D. A., M. R. Hall, and W. Meinke. 1973. Properties of nucleoprotein complexes containing replicating polyoma DNA. *J. Virol.* **12**:887.

Green, M. H., H. I. Miller, and S. Handler. 1971. Isolation of a polyoma-nucleoprotein complex from infected mouse-cell cultures. *Proc. Natl. Acad. Sci.* **68**:1032.

Hirt, B. 1967. Selective extraction of polyoma DNA from infected mouse cell cultures. *J. Mol. Biol.* **26**:365.

Ito, M., J. L. Melnick, and H. D. Mayor. 1967. An immunofluorescence assay for studying replication of adeno-satellite virus. *J. Gen. Virol.* **1**:199.

Ledinko, N. 1974. Temperature-sensitive mutants of adenovirus type 12 defective in viral DNA synthesis. *J. Virol.* **14**:457.

Mayor, H. D., S. Carrier, and L. Jordan. 1977. Complementation of adeno-associated satellite virus AAV by temperature-sensitive mutants of adenovirus type 31. *J. Gen. Virol.* (in press).

Mayor, H. D., S. Drake, and L. E. Jordan. 1974. Chemical and physical properties of adeno-associated satellite DNA produced during coinfection with herpes simplex virus. *Nucleic Acids Res.* **1**:1279.

Meinke, W., M. R. Hall, and D. A. Goldstein. 1975. Protein in intracellular simian virus 40 nucleoprotein complexes: Comparison with simian virus 40 core protein. *J. Virol.* **15**:439.

Parks, W. P., J. L. Melnick, J. Rongey, and H. D. Mayor. 1967. Physical assay and growth cycle studies of a defective adeno-satellite virus. *J. Virol.* **1**:171.

Schaffer, P. A. 1975. Temperature-sensitive mutants of herpesviruses. *Curr. Top. Microbiol. Immunol.* **70**:51.

Suzuki, E., H. Shimojo, and Y. Moritsugu. 1972. Isolation and a preliminary characterization of temperature-sensitive mutants of adenovirus 31. *Virology* **49**:426.

Torikai, K. and H. D. Mayor. 1969. Interference between two adeno-associated satellite viruses: A three component system. *J. Virol.* **3**:484.

Biological Properties of Rat-Virus Variants

M. DAVID HOGGAN
JOHNNA F. SEARS
GUNTER F. THOMAS
AUROBINDO ROY

Laboratory of Viral Diseases
National Institute of Allergy and Infectious Diseases
National Institutes of Health
Bethesda, Maryland 20014

Two distinct parvoviruslike agents have recently been isolated from two transplantable rodent tumors which were being studied because they exhibited strong immunosuppressive activity. These studies indicated that the primary inhibitory activity in each of the two tumors is due to these contaminating parvoviruses. One of the agents, which came from the chemically induced, thymus-derived murine tumor EL-4 (Gorer 1950), was shown to be a strain of minute virus of mice (MVM) (Bonnard et al. 1976). The second agent, isolated from the 13762 rat adeno-carcinoma, has been identified as Kilham rat virus. (Campbell et al. 1977). Subsequent studies in our laboratory showed that this agent reacted strongly when tested against our hyperimmune anti-rat-virus sera: >5120 in complement-fixation (CF) assays and >10,240 in hemagglutinin-inhibition (HAI) assays (see Table 1). Unexpectedly, the tumor rat-virus isolate (TRV) also reacted in HAI (326–640) with our hyperimmune anti-MVM sera, which had been shown to be highly specific for MVM when tested against numerous parvovirus strains (Hoggan 1971). In addition to TRV, the rat-virus (RV) strain 308, an original isolate of Kilham's, also reacted with our hyperimmune anti-MVM sera, although other RV strains did not. These include a strain designated MBA, which was obtained from Microbiological Associates, and an RV reference strain obtained through the Parvovirus Working

119

Team of the Program on Comparative Virology sponsored jointly by the World Health Organization and the Food and Agriculture Organization of the United Nations (WHO/FAO).

Furthermore, we found upon plaquing the 308 strain in rat nephroma cells (Babcock and Southam 1967; provided by L. A. Salzman) that two distinct plaque-size populations were produced, and that the larger plaques (usually measuring >5 mm) expressed the MVM antigen, whereas smaller plaques did not. Our studies designed to characterize and compare the biological and immunological properties of these RV308 variant populations, designated large-plaque rat virus (LPRV) and small-plaque rat virus (SPRV), are the subject of this presentation.

MATERIALS AND METHODS

Cells and Virus

Monolayer cultures of primary rat-embryo fibroblasts (REF) from parvovirus-antibody-free rats, rat nephroma cells (RN) (Babcock and Southam 1967), and rat brain-tumor cells (RT-7) (Richards et al. 1977; provided by R. W. Armentrout) were grown in 150-cm^2 plastic tissue-culture flasks in a 50-50 mixture of basal medium-Eagle's (BME) and 199 medium supplemented with 10% fetal-calf serum (FCS). Cultures were infected with the various viruses when 50% confluent and were harvested 5–6 days later by rapid freezing and thawing three times prior to storage in 2–3-ml aliquots at −70°C. Early passages of RV strain MBA and an MVM strain designated SC (obtained from J. C. Parker) were grown in REF. However, later pools of RV were grown in RN cells, whereas pools of MVM were grown in RT-7 cells. Large-plaque pool D19 was produced from the RV308 strain after the third serial plaque passage in RN cells. D20 is a small-plaque pool also produced from the RV308 strain after three serial plaque passages.

Virus Purification

For the production of hyperimmune guinea-pig antiparvovirus sera, the infected cells from 20 150-cm^2 flasks were scraped into the supernatant fluid after the cells showed a cytopathogenic effect of >3 (usually after 4–6 days). The cell debris containing over 99% of the virus was separated from the fluid by low-speed centrifugation (2000 rpm for 10 min). The cell pellet was then rapidly frozen and thawed three times and extracted overnight with 30 ml glycine buffer (pH 9). After clarification by a second low-speed centrifugation, 10 ml of cell extract was layered over 6 ml of 1.5 g/cm^3 CsCl in 0.01 м Tris buffer (pH 8) in a Beckman 27.1 rotor and spun for 5 hr at 25,000 rpm. The opalescent band just below the interphase was removed and then banded three times in isopycnic CsCl gradients (average density 1.4 g/cm^3).

RV Plaque Assay

RN suspension (5 ml), at a concentration of 5×10^5 cells/ml, was plated onto 60-mm plastic petri dishes in the BME–199 medium containing 10% FCS. They were incubated at 37°C until they reached 50% confluency (usually overnight). The growth medium was then aspirated and the plates were infected with 1.5 ml of virus diluted in serum-free medium. Virus was allowed to adsorb for 60 min at room temperature, with gentle shaking every 15 min, and the plates were then overlaid with 5 ml of medium comprised of modified Eagle's medium with 5% FCS and 0.9% Difco Noble agar. Plates were refed on day 4 with 2.5 ml of additional overlay and stained on day 6 by one of two methods. In the first method, used for picking plaques, the plates were stained by overlaying with 2.5 ml of overlay medium containing 2% Neutral Red. The second procedure consisted of fixing the plates with 10% formalin for 2–3 hr, removing the agar with a strong water jet, and staining for 1 hr with hematoxylin. The rinsed and air-dried plates provided an easily countable permanent record with excellent photographic qualities.

Antiserum Production

Reference antisera, provided by the Parvovirus Working Team of the WHO/FAO Program on Comparative Virology, were prepared in specific-pathogen-free Wistar rats. Each animal was given three intraperitoneal injections 14 days apart and was terminally exsanguinated 18 days later. The immunogen consisted of 1.5 ml of chloroform-treated, clarified, serum-free culture fluid from virus-infected rat-embryo cell cultures.

All antisera against parvoviruses were prepared in pretested guinea pigs using CsCl-gradient-purified virus which was injected with complete Freund's adjuvant into the rear foot pads of each animal as described previously (Hoggan et al. 1966). The hyperimmune anti-RV serum used for HAI, CF, immune electron microscopy (IEM), and serum neutralization was prepared against the MBA strain of RV and contained no demonstrable MVM antibodies in these tests. The hyperimmune anti-MVM serum was prepared using the SC strain of MVM.

Serological Techniques

All CF and HAI titrations were carried out using microtiter plates as described by Sever (1962). In order to remove nonspecific inhibitors, all rat and guinea-pig sera were treated with 200 units of receptor-destroying enzyme per ml, and hamster sera were treated by adsorbing with 12.5% kaolin. All sera were inactivated by heating at 56°C for 30 min, and those used in HAI assays were further treated by adsorption with 0.025 ml of packed guinea-pig red blood cells per ml of serum.

Serum Neutralization

Virus was diluted to contain $1–5 \times 10^6$ plaque-forming units (PFU) per ml. Large aggregates were minimized by low-speed centrifugation and filtering through a 450-μm Millipore filter. An equal volume of diluted virus was mixed with an equal volume of diluted antibody containing 64–128 HAI units and incubated at room temperature for 1 hr before being inoculated into 50%-confluent RN tube cultures. The cultures were observed daily for cytopathogenic effect and assayed by HAI and plaque induction. The IEM procedure of Feinstone et al. (1973) was modified as described previously (Thomas and Hoggan 1974).

Hamster Virulence Testing

Pregnant hamsters were obtained from NIH stock when near term and were monitored closely until they gave birth. Within 4 hr after birth the new litters were injected intraperitoneally with 0.1 ml of diluted virus or normal cell-culture fluid. They were then returned to their mothers and not disturbed for 5 days. The count at day 5 was used as a baseline. They were then periodically counted, weighed, examined for malformation, and bled for antibody determination.

RESULTS

Correlation of Plaque Size with the Presence of MVM Antigen

We found that when large RV308 plaques (>5 mm in diameter) were picked and grown up into new pools, these pools always contained the MVM antigen, whereas pools made from small plaques did not. It is important to point out that even though average plaque sizes may vary from experiment to experiment because of differences in rate of cell growth, temperature of incubation, and initial degree of cell confluency, the relative size difference of large and small plaques is so great that they can be distinguished easily in the same dish.

Of great interest to us was the observation that when replaquing LPRV we always found both large and small plaques, and again the large plaques always contained the MVM cross-reacting antigen but the small plaques did not. On the other hand, SPRV always bred true, and the small plaques produced never contained the cross-reacting antigen. Figure 1 demonstrates both the large plaques (LP) and the small plaques (SP) we saw when replaquing our large-plaque pool D19.

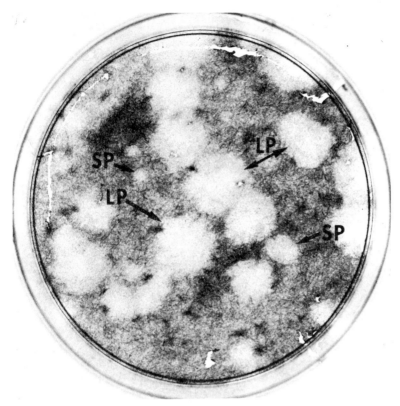

FIGURE 1
Large plaques (LP) and small plaques (SP) from LPRV pool.

In plaque titers of LPRV pools the number of small plaques near the terminal dilution always surpassed the number of large plaques, even though large plaques appeared to dominate at higher plaque concentrations. These observations led us to believe that the large-plaque particles are defective, requiring coinfection with small-plaque particles for productive infection.

Attempts to Verify Large-Plaque Defectiveness

If our defective-virus model is true, we should be able to detect such defectiveness by counting the number of large and small plaques at closely spaced dilutions near the terminal dilution. The results of one such attempt are shown in Figure 2, in which the number of plaques is plotted against the relative concentration. The open triangles represent the number of small plaques counted from LPRV pool D19 at different

FIGURE 2
Effect of concentration on the
production of small and large
plaques. △: small plaques from
LPRV; ○: large plaques from
LPRV; ▲: small plaques from
SPRV.

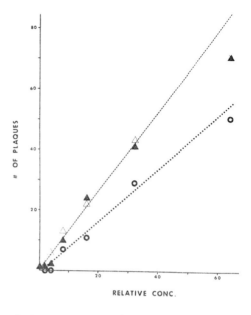

concentrations, and the open circles represent the number of large
plaques counted in the same plates. For comparison, the relative numbers
of small plaques from SPRV pool D20 at different concentrations are also
given (solid triangles).

Although this and similar experiments were somewhat disappointing
in that the differences were not as great as we had hoped, they do suggest
that only one SPRV particle is required for small-plaque production,
whereas more than one LPRV particle is required for large-plaque production. The small differences found may reflect our inability to prevent
aggregation completely.

More direct evidence for defectiveness of the large-plaque particles
came from our serum-neutralization studies. When either SPRV or LPRV
pools were treated with hyperimmune anti-RV serum, all cytopathogenic
effect was eliminated and no virus could be recovered. On the other
hand, when SPRV or LPRV was treated with cross-reacting hyperimmune anti-MVM serum, maximum cytopathogenic effect occurred at
approximately the same time as in the normal-serum controls. SPRV was
recovered from both preparations, but no LPRV could be recovered even
after two blind passages in RN cells.

*Demonstration of MVM Cross-Reacting Antigen Using Immune Electron
Microscopy*

An additional technique we have used to exploit further the MVM cross-
reacting antigen is immune electron microscopy. This technique has

allowed us to visualize the antigen-antibody reaction at the surface of virus particles.

Figure 3*A* demonstrates a typical field of an IEM preparation in which LPRV was reacted with an optimal dilution of cross-reacting MVM antiserum and examined in the electron microscope. As can be seen, both full and empty virus particles are connected with antibody, forming large

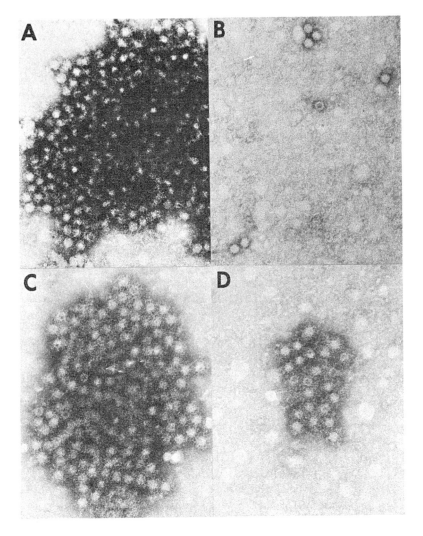

FIGURE 3
Demonstration of cross-reacting MVM antigen by immune electron microscopy. (*A*) LPRV vs. anti-MVM serum, (*B*) SPRV vs. anti-MVM serum, (*C*) LPRV vs. anti-RV serum, (*D*) SPRV vs. anti-RV serum.

aggregates. In contrast, Figure 3B is a typical field from an IEM preparation in which SPRV was reacted with the same concentration of cross-reacting MVM antiserum. It shows only widely spaced single virus particles or very small antibody-free groups of virus particles.

When hyperimmune anti-RV serum was used, both LPRV (Fig. 3C) and SPRV (Fig. 3D) formed massive (4+) antibody-containing virus aggregates. It is also important to point out that when RV antibody was used it reacted with virtually every particle, regardless of whether LPRV or SPRV was used.

Attempts at Physical Separation of LP and SP Particles by Isopycnic Centrifugation in CsCl

When LPRV or SPRV pools were centrifuged to equilibrium in isopycnic CsCl gradients, each usually formed four distinct bands. In either case, minor dense bands of DNA-containing particles were found in the density range 1.46–1.51 g/cm^3 and major infectious bands appeared in the range 1.40–1.42 g/cm^3. In addition, both preparations usually produced minor bands at 1.35–1.38 g/cm^3, comprised of noninfectious particles containing DNA molecules of less then genome length, and large empty-particle bands at 1.32–1.33 g/cm^3. The latter bands accounted for most of the total antigenic mass. All fractions from LPRV gradients which contained RV antigen also contained the MVM antigen, whereas no fractions from SPRV gradients contained the cross-reacting antigen.

From one such series of experiments the major infectious bands and the empty bands were rebanded twice in CsCl and further purified by running in 15–30% sucrose gradients. The peak hemagglutinin fractions were then used to hyperimmunize pretested guinea pigs. Table 1 gives the CF and HAI titers obtained by testing these hyperimmune guinea-pig sera against 8 antigen units and 8 hemagglutinin units, respectively, of SPRV, LPRV, and MVM.

As expected, both full and empty SPRV and LPRV produced good antibody titers against both rat-virus populations, but MVM cross-reacting antibodies were produced only in animals injected with large-plaque virus. Although no cross-reacting CF antibodies were found, this may only reflect the relative sensitivity of the CF test versus the HAI test. After all, the cross-reacting HAI titers found are quite low compared to homologous HAI titers [320 for LPRV antibody (Ab) vs. MVM compared to 20,480 for LPRV Ab vs. LPRV]. The reciprocal cross-reaction titers (MVM Ab vs. LPRV) are in the same range. Although the cross-reactivity is not high, the facts that it is reproducible and is found as a two-way cross suggest that it is real and due to an antigen present in both MVM and LPRV populations but not in SPRV.

TABLE 1
LPRV and MVM Cross-Reactivity

Virus	CF titer[a] vs. 8 antigen units			HAI titer[b] vs. 8 hemagglutinin units		
	SPRV	LPRV	MVM	SPRV	LPRV	MVM
SPRV (1.41 g/cm³)	320	160	< 10	20,480	5120	< 20
SPRV (1.32 g/cm³)	1280	320	< 10	40,960	5120	< 20
LPRV (1.41 g/cm³)	640	1280	< 10	5120	20,480	320
LPRV (1.32 g/cm³)	1280	640	< 10	10,240	40,960	80
MVM (1.42 g/cm³)	< 10	< 10	1280	< 20	320	5120

[a] Reciprocal of serum dilution that gives at least 3+ fixation of 2 full units of complement in the presence of 8 antigen units.
[b] Reciprocal of serum dilution that completely inhibits 8 hemagglutinin units of virus (Sever 1962).

Comparative Virulence of LPRV and SPRV

After noting a significant difference in the rate of spread of LPRV and SPRV as reflected by plaque size and considering the apparent defectiveness of LPRV, we wondered if any gross differences in their virulence could be detected in an animal model. We therefore carried out experiments designed to measure the relative virulence of these pools in newborn hamsters. Thus far, two experiments designed to answer this question have been completed. Table 2 summarizes the results from one such experiment. The SPRV pool is more virulent for newborn hamsters than the LPRV pool which contains both SPRV and LPRV. Although the differences are small, they are quite consistent whether measured by death, by runting, or by antibody production. Whether this reduction in virulence is a reflection of the defectiveness of the LPRV particles and/or whether they interfere with the virulence of the SPRV particles in the LPRV population is not yet known. In fact, it cannot be stated whether or not LPRV particles have any pathogenicity in and of themselves. Further studies are in progress to answer that question.

DISCUSSION

We have observed that two strains of RV (a field strain isolated from a rat tumor and strain 308) express an antigen common to MVM. This antigen can be detected with hyperimmune guinea-pig serum using HAI, virus neutralization, and IEM. When the 308 strain of RV is plaqued, two sizes of plaques are produced. The larger plaques (>5 mm) express the MVM

TABLE 2
Comparative Virulence of SPRV and LPRV for Newborn Hamsters

Virus dose (PFU/animal)	Day after injection	Deaths (%)		Runted (%)		Antibody-positive (%)		Mean HAI titer	
		SPRV	LPRV	SPRV	LPRV	SPRV	LPRV	SPRV	LPRV
1	10	29	33	0	0	—	—	—	—
1	30	29	33	0	0	40	0	3074	0
1	45	29	33	0	0	40	0	640	0
100	10	55	22	40	0	—	—	—	—
100	30	55	22	80	0	100	100	10,240	7862
100	45	55	22	80	0	100	100	9216	7131
1000	10	56	50	50	25	—	—	—	—
1000	30	56	50	75	25	100	100	10,240	1920
1000	45	56	50	75	0	100	100	20,480	7360
100,000	10	100	25	—	100	—	—	—	—
100,000	30	100	62	—	100	—	100	—	> 40,960
100,000	45	100	62	—	100	—	100	—	> 40,960

SPRV was compared with LPRV as measured by (1) percent deaths, defined as number of animals dead or missing on the day indicated divided by number of animals counted at day 5; (2) percent runted, or number weighing less than 75% of the controls, divided by number surviving on day indicated; (3) percent producing antibodies, or number with HAI ≥ 20, divided by number of surviving animals on day indicated; and (4) average HAI titer of each group on day indicated.

antigen; the smaller plaques do not. The large-plaque virus seems to be defective, as replaquing always results in the production of plaques of both sizes, whereas small plaques always breed true and are free of the antigen. Large-plaque populations are less virulent for newborn hamsters than small-plaque populations.

One of the first theories we considered to explain the reactivity between the T and 308 strains of rat virus and our hyperimmune anti-MVM sera was the possibility that these two strains were contaminated with small amounts of MVM which replicated poorly, if at all, in the cell-culture systems usually used for these viruses. Such a host-restricted MVM has been isolated by Bonnard et al. (1976). A virus of this type might replicate in rat nephroma cells only when coinfected with a helper such as RV. Attempts to isolate MVM from our LPRV pools using a wide variety of mouse cells, including mouse leukocytes, failed. We also ruled out the possibility of low levels of MVM contamination in LPRV and SPRV pools by subjecting them to the very sensitive antibody protection test, using mice, rats, guinea pigs, and hamsters. No MVM antibody was detected in any of the test animals.

One model we have entertained in attempting to explain our data postulates that LPRV pools are made up of a mixture of two virus subpopulations which are morphologically indistinguishable but differ in the following respects: The first is made up of particles which have at least two distinct antigenic sites on their surfaces, one which reacts with RV antibody and one which reacts with MVM antibody. These particular particles may contain some genetic information from MVM as well as RV genetic material, and appear to be incapable of replication without normal non-MVM-containing RV. The second subpopulation is made up of pure wild-type RV particles which only express normal RV antigens and are capable of replication without helper. The plaque data tend to support this hypothesis, but it is difficult to explain why the cross-reacting MVM antibody appears to attach to all particles in an LPRV pool when tested by IEM.

We should also point out that only hyperimmune guinea-pig serum has been used successfully to detect cross-reacting antigen. Neither the WHO/FAO reference sera nor infection sera produced in a variety of animals (including extremely high titer sera from runted hamsters infected with LPRV) can be used to detect the cross-reacting antigen.

An alternative hypothesis we have considered is that this cross-reacting antigen represents a common rodent parvovirus antigen which is normally not expressed at the surface of the parvovirions. Such an antigen may be identical to or related to the cross-reacting antigen common to RV, MVM, and H-1. This antigen cannot be detected by HAI or CF but can be clearly demonstrated in infected cells by immunofluorescence (Hoggan 1971; Cross and Parker 1972). Perhaps a variant RV has been selected which has an altered surface allowing for the expression of this "internal antigen."

ACKNOWLEDGMENTS

We thank Dr. Michael Collins of Microbiological Associates, Walkersville, Maryland, for carrying out the antibody protection tests. We also thank Mr. Lee Cline for carrying out the CF and HAI analyses.

REFERENCES

Babcock, V. I. and C. M. Southam. 1967. Stable cell line of rat nephroma in tissue culture. *Proc. Soc. Exp. Biol. Med.* **124**:217.

Bonnard, G. D., E. K. Manders, D. A. Campbell, Jr., R. B. Herberman, and M. J. Collins, Jr. 1976. Immunosuppressive activity of a subline of the mouse EL-4 lymphoma, evidence for minute virus of mice causing inhibition. *J. Exp. Med.* **143**:187.

Campbell, D. A., Jr., S. P. Staal, E. K. Manders, G. D. Bonnard, R. K. Oldham, L. A. Salzman, and R. B. Heberman. 1977. Inhibition of in vitro lymphoproliferative responses by in vivo passaged rat 13762 mammary adenocarcinoma cells. II. Evidence that Kilham rat virus is responsible for the inhibitory effect. *Cell. Immunol.* (in press).

Cross, S. S. and J. C. Parker. 1972. Some antigenic relationships of the murine parvoviruses: Minute virus of mice, rat virus, and H-1 virus. *Proc. Soc. Exp. Biol. Med.* **139**:105.

Feinstone, S., A. Kapikian, and R. Purcell. 1973. Hepatitis A: Detection by immune electron microscopy of a viruslike antigen associated with acute illness. *Science* **182**:1026.

Gorer, P. A. 1950. Studies in antibody response of mice to tumor inoculation. *Br. J. Cancer.* **4**:372.

Hoggan, M. D. 1971. Small DNA viruses. In *Comparative virology* (ed. K. Maramorosch and E. Kurstak), p. 43. Academic, New York.

Hoggan, M. D., N. R. Blacklow, and W. P. Rowe. 1966. Studies of small DNA viruses found in various adenovirus preparations: Physical, biological, and immunological characteristics. *Proc. Natl. Acad. Sci.* **55**: 1467.

Richards, R., P. Linser, and R. W. Armentrout. 1977. Kinetics of assembly of a parvovirus, minute virus of mice, in synchronized rat brain cells. *J. Virol.* **22**:778.

Sever, J. L. 1962. Application of a microtechnique to viral serological investigations. *J. Immunol.* **88**:320.

Thomas, G. F. and M. D. Hoggan. 1974. Immune electron microscopy (IEM) of canine adeno-associated virus. In *Proceedings of the 32nd Annual Meeting of the Electron Microscopy Society of America* (ed. C. J. Arcineaux), p. 256. Claitors, Baton Rouge, Louisiana.

Toolan, H. W. 1964. Studies on H-viruses. *Proc. Am. Assoc. Cancer Res.* **5**:64.

Susceptibility to Minute Virus of Mice as a Function of Host-Cell Differentiation

PETER TATTERSALL

Imperial Cancer Research Fund
Mill Hill Laboratories
London NW7 1AD, England

The destruction of rapidly proliferating tissues of fetal and neonatal animals is the hallmark of parvovirus pathology (Margolis and Kilham 1965; Toolan 1968). This effect has been ascribed to the dependence of the nondefective parvoviruses upon host function(s) expressed transiently during the S phase of the cell cycle (Tennant et al. 1969; Hampton 1970; Tattersall 1972; Rhode 1973; Siegl and Gautschi 1973).

The resistance of actively dividing cells of the early embryo to minute virus of mice (MVM) (Mohanty and Bachmann 1974) indicates that further constraints upon virus multiplication may operate as a function of development. In general, adult animals are also resistant to parvovirus infection (Kilham and Margolis 1975; Margolis and Kilham 1975). Although adult tissues are mitotically quiescent compared to the developing fetus and neonate, many contain dividing cells essential to the animal's survival, which one might expect to be targets for parvovirus attack. Most strains of parvovirus, however, do not cause any overt disease in the adult animal, unless some tissue has been induced to undergo an "unscheduled" proliferative response, for instance, liver regeneration following partial hepatectomy (Ruffolo et al. 1966; Kilham and Margolis 1975; Margolis and Kilham 1975). This again suggests that cell division is a necessary, but not sufficient, condition for susceptibility to parvovirus infection.

131

Recent advances in the culture of teratoma cells have provided a system for analysis of differentiation-related events in vitro. Teratomas are tumors arising in the testis or ovary which comprise a mixture of many differentiated tissue types. Malignant teratocarcinomas also contain pockets of undifferentiated, pluripotent stem cells. The tumors arise spontaneously in strain 129/Sv mice (Stevens 1958) and can be induced by implanting early embryos into extrauterine sites (Stevens 1968). Such implantation studies have suggested that the stem cells of the tumor are the primordial germ cells of the early embryo (Stevens 1967). Stem cells can be grown in culture and induced to differentiate in vitro (Martin 1975; Nicolas et al. 1975; Martin and Evans 1975a). This cell system has been used here to investigate a differentiation-dependent block in the replication of MVM.

MATERIALS AND METHODS

Virus and Cells

The isolation of the MVM(T) strain of minute virus of mice and its plaque assay in A-9 cells have been described previously (Tattersall 1972; Tattersall et al. 1976). Low-multiplicity-derived virus was grown in Ehrlich ascites cells as described before (Tattersall et al. 1976).

STO cells are from a 6-thioguanine- and ouabain-resistant cell line isolated by Dr. A. Bernstein of the Ontario Cancer Institute from fibroblasts of a Swiss inbred mouse. Clonal pluripotent stem-cell lines SCC-S2 and SCC-PSA4 (Martin and Evans 1975c) derived from the OTT-5568 transplantable 129/Sv teratocarcinoma (Stevens 1970) were cultured on feeder layers of mitomycin-C-blocked STO cells as described by Martin and Evans (1975a). The clonal pluripotent stem cell A-6 (Hogan 1976) derived from the OTT-6050 tumor (Stevens 1970) was also cultured on STO feeders as above. The clonal nullipotent stem-cell line nulli-SCC-S2 (Martin and Evans 1975c) derived from the LS-402C-1684 tumor (Stevens 1958) was grown on 1%-gelatin-treated culture dishes. WME 129 is an uncloned, established cell line derived from a culture of 16-day 129-strain whole mouse embryos, and was used in vitro at passage numbers between 22 and 30. The A-6 cells were kindly supplied by Dr. B. Hogan, and the other teratoma lines, the WME 129 line and the STO cells, were the generous gift of Dr. N. Teich. Monolayer cultures were grown in Dulbecco's modified Eagle's medium supplemented with 10% calf serum. For virus-growth experiments, medium was supplemented with 10% fetal-calf serum (FCS). Cells were detached by washing the monolayer twice with trypsin-EDTA [0.25% trypsin in phosphate-buffered saline (PBS) containing 0.5 mM EDTA], incubating at 37°C, then suspending in medium. Feeder cells and stem cells were separated by allowing this suspension to settle at 37°C for 30 min, duing which time the feeders rapidly and preferentially reattached to the dish. The medium containing the stem cells, with >90% of the feeders removed, was then gently transferred to another dish.

Differentiation In Vitro

In general, the techniques described by Martin and Evans (1975a) were used. These are summarized in Figure 1. Stem cells were induced to make three-dimensional colonies by passage on untreated tissue-culture dishes in the absence of feeders. Such colonies were then detached by gentle pipetting and grown in bacteriological dishes (to which they do not

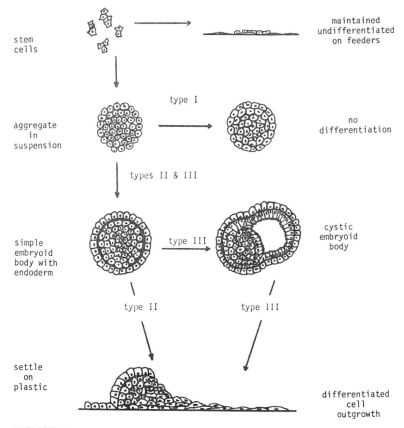

FIGURE 1
Diagrammatic representation of in vitro differentiation scheme. Types I–III refer to the three modes of differentiation described in "Results."

adhere) for at least 5 days, with one or two medium changes. Alternatively, single-cell suspenstions (~5 × 10⁵/ml) of stem cells from which the feeders had been removed were mixed in a sterile polystyrene container by rapid rotation at 37°C. After 18 hr, aggregates had formed; these were washed several times in fresh medium by allowing them to settle and were then carefully transferred to bacteriological dishes. Two to three days later a layer of endoderm was observed covering these aggregates, which are termed embryoid bodies (EB) (Pierce and Dixon 1959; Martin and Evans 1975b). After 5 days in suspension the EBs were transferred to untreated plastic tissue-culture dishes and allowed to attach for 3–4 days. The unattached EBs were then removed and the settled embryoid bodies (SEB) were fed with fresh medium every 3–4 days. Primary subcultures of differentiated cell types (SEB-der 1°) were produced from these differentiating cell masses by brief washing with trypsin-EDTA as described above. After ~30 min of incubation at 37°C, detached cells were gently pipetted off the dish into fresh medium, aggregates were allowed to settle out, and the cell suspension was seeded into 1%-gelatin-treated dishes. After 2 hr the medium was changed to a mixture of fresh medium and SEB-conditioned medium (1:1), and the unattached cells, including most of the stem cells released from the SEB culture, were discarded.

Infectious-Center Assay

Cells were seeded as semiconfluent monolayers on 1%-gelatin-treated dishes and incubated overnight in medium containing 10% FCS. The cells in the monolayer were then infected at the indicated multiplicity by the addition of a low-multiplicity-derived pool of MVM(T) diluted in medium containing 5% FCS. After 4 hr was allowed for adsorption and penetration of virus, the inoculum was removed and all extracellular virus was inactivated by incubation in medium containing 5% receptor-destroying enzyme, 5% FCS, and rabbit anti-MVM serum (hemagglutination inhibition = 1:800 against 4 hemagglutinating units per ml of MVM(T) antigen) added to 1%. After a further 4-hr incubation the medium was removed and the cells were detached by trypsin-EDTA wash and resuspended in medium containing 1% anti-MVM serum and 10% FCS. The cells were then centrifuged, resuspended in medium containing only 10% FCS, counted, and diluted in the same medium. Cell suspensions at various dilutions were mixed with an equal volume of the same medium containing 0.9% agar, and 5-ml aliquots were gently poured onto 50-mm dishes containing sparse A-9 monolayers prepared for the plaque assay as described elsewhere (Tattersall et al. 1976). In some assays a 1-mm-thick layer of medium containing 0.9% agar was poured on top of the A-9 monolayer before the addition of the infected cells. Assays were incubated 5 days and stained as described before (Tattersall 1972).

Heterokaryon Formation

The polyethylene glycol (PEG) method of cell fusion was used (Pontecorvo 1975; Hales 1977). Briefly, infected cells were prepared as for the infectious-center assay (described above) to the dilution stage. A mixture of 7.5–15 × 10^5 cells with an equal number of cells of the other partner of the fusion was prepared, centrifuged, and resuspended in 0.5 ml of medium containing 42.5% (w/v) PEG 1000 (Koch-Light) and 9% (w/v) dimethyl sulfoxide. After 1 min at 37°C, 0.5 ml of medium containing 15.3% (w/v) PEG 1000 was added and gentle mixing was continued for 2–3 min. Then 4 ml of medium alone was added and mixed in gently, and 1-ml aliquots were carefully pipetted into each well of a four-place Lab-Tek tissue-culture chamber slide. After 2 hr the medium was replaced with medium containing 10% FCS, and 22–24 hr later the fused cultures were processed for immunofluorescent analysis.

Immunofluorescent Staining

Cultures were grown in 1%-gelatin-treated Lab-Tek culture slides. Semiconfluent monolayers were infected at 10 plaque-forming units (PFU) per cell by the addition of 0.5 ml per well of virus suspension diluted in medium containing 5% FCS. After 1 hr the medium was changed. At 24 hr post-infection (p.i.) the cells were washed in PBS and fixed in methanol:acetone (3:1) for 10 min at room temperature, then air-dried. Fixed cells were incubated with rabbit anti-MVM serum (1:5 dilution in PBS of the antiserum described above, adsorbed overnight with methanol:acetone-fixed A-9 cells) for 30 min at 37°C, thoroughly washed in PBS, and incubated a further 30 min in fluorescein-conjugated goat anti-rabbit globulin (1:20 in PBS, Nordic). After further exhaustive washing, the samples were mounted in PBS-glycerol-polyvinyl alcohol (Rodriguez and Deinhardt 1960) and viewed in a Zeiss Photomikroscop III with epifluorescent optics.

RESULTS

Effect of MVM on the In Vitro Differentiation Process

Cultured teratoma stem cells can be divided into three types on the basis of their differentiation behavior in vitro (Martin and Evans 1975c). Type I, represented in this study by nulli-SCC-S2, are unable to differentiate in vitro, although they share biochemical and immunological characteristics with pluripotent stem cells. These cells do not differentiate in vivo either, but are transplantable as embryonal carcinomas. They will, however, aggregate in suspension and form "pre-embryoid bodies." When similar three-dimensional aggregates are formed with cells of types II and III, the outer cells respond by differentiating terminally to an endodermal cell with the biochemical characteristics of parietal yolk-sac endoderm. Cells of this type cover the surface of the aggregate to produce the simple embryoid body structure as shown diagramatically in Figure 1. Types II and III differ in their requirements for further differentiation. Type-II (e.g., SCC-S2) EBs must be allowed to attach to a surface before further differentiation will occur, whereas type-III (e.g., SCC-PSA4 and A-6) EBs will continue to differentiate in suspension, becoming complex structures containing many different cell types and developing a fluid-filled cyst often lined with cells resembling those of the visceral yolk sac of the midgestation embryo.

Stem cells were infected at a multiplicity of 10 PFU/cell with MVM(T) and examined for their growth and differentiation potential compared to uninfected cells. No difference was detected in survival, in ability to form simple EBs, or even (in the case of infected type-III cells) in the ability to form complex, cystic EBs. Furthermore, growth and differentiation in the continuous presence of MVM(T) (at an approximate multiplicity of 10 PFU/cell—renewed with each medium change) was apparently unaffected up to the stage of EB attachment and outgrowth (data not shown).

When simple or cystic EBs are allowed to attach to a surface, extensive outgrowth of differentiated cells occurs. Figure 2a shows a low-power view of a number of such SEB "colonies." Many differentiated cell types can be distinguished by standard histological procedures (several occur simultaneously within the same colony); these have been described extensively by Martin and Evans (1975c) and Sherman (1975). The cell types observed most frequently are those of smooth muscle, skeletal muscle (often observed as fused myotube), cardiac muscle (occasionally observed as rhythmically beating sheets of cells), neuronal networks, endoderm cells, fat-globule-containing cells, pigmented epithelium, keratinizing epithelium, and fibroblasts—a chaotic patchwork of apparently randomly differentiated cells derived from all three germ layers. Most noticeable, even under low power (Fig. 2a), are sheets of parallel spindle-shaped cells which under higher power resemble colonies of a contact-inhibited fibroblast such as the BHK-21 cell (Macpherson and Stoker 1962). These cells appear to migrate out from the SEB and form whorled patterns at the periphery (Fig. 2b). Where two SEBs are close to one another, the space between them is often filled with such cells, which are probably derived from both SEBs and which form a highly oriented monolayer, often with most cells lying at right angles to the line joining the SEB centers (Fig. 2c).

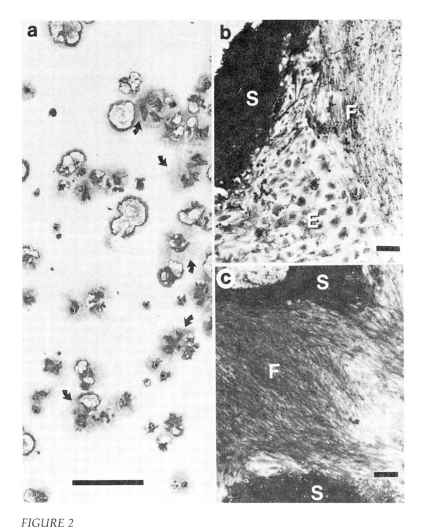

FIGURE 2
Settled embryoid bodies (SEB) with differentiated outgrowths, cultured in absence of MVM. Giemsa stain. (a) Low power. Arrows indicate typical areas of fibroblastlike outgrowth from SEBs. Bar represents 10 mm. (b) Higher-power view of typical outgrowth of endoderm (E) and fibroblastlike cells (F) at edge of SEB (S). Bar represents 100 μm. (c) As in (b), showing fibroblastlike cells (F) occupying space between two SEBs (S). Bar represents 100 μm.

In cultures of SEBs continuously exposed to MVM(T) it is immediately obvious, even at low power, that these "spindle" cells are missing (Fig. 3a). Examination of many such cultures has indicated the consistent

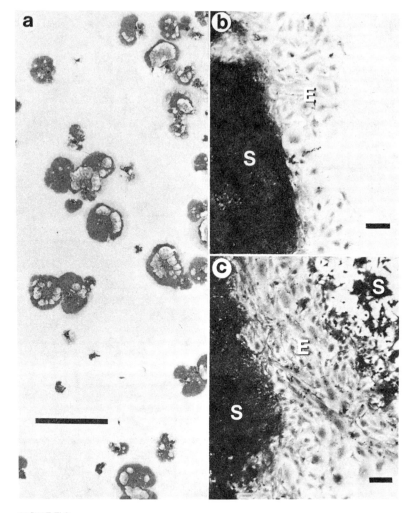

FIGURE 3
Settled embryoid bodies (SEB) cultured in continual presence of MVM at
~10 PFU/cell. Giemsa stain. (a) Low power. Bar represents 10 mm. (b)
Higher-power view showing endoderm (E) at edge of SEB (S). Bar represents 100 μm. (c) As in (b), showing space between two SEBs (S) occupied by
endodermal cells (E). Bar represents 100 μm.

absence of these cells, but no other major type of differentiated cell, as a
result of growth and differentiation in the presence of MVM(T). At low
power the margins of individual SEBs are sharply defined in the infected
culture (Fig. 3a), and the cells which form monolayers at the periphery of

these "colonies" are epithelioid (Fig. 3*b*,*c*). These cells are probably derived from the original endoderm layer of the EB which migrates off the SEB shortly after attachment. The higher-power micrographs show no parallel-oriented fibroblasts either at the edges of isolated SEBs (Fig. 3*b*) or filling the space between adjacent SEBs (Fig. 3*c*).

The appearance of these fibroblasts in uninfected SEBs was unaffected by the inclusion of HAT (hypoxanthine, aminopterin, and thymidine) in the medium (Szybalski et al. 1962). This ruled out the possibility that these cells were STO feeders, carried over in EB formation, which had "escaped" from the mitomycin-C block.

Culture of the Fibroblastlike Derivative

Surprisingly, infection of SEB cultures grown for several weeks in the absence of MVM did not lead to the rapid disappearance of this fibroblast cell type. Several different explanations for this were considered. The effect of virus infection might be to "shut down" the differentiation pathway to the fibroblastlike cell, or only a precursor to this cell type, rather than the fibroblast itself, may be sensitive to virus infection. A further possibility is simply that the majority of these cells are mitotically quiescent because of contact inhibition, and thus incapable of supporting lytic infection by MVM (Tattersall 1972). To decide which possibility was the most likely, attempts were made to culture these fibroblast cells from uninfected SEBs in the absence of stem cells and resistant differentiated cell types. All of the differentiated cell types have proved difficult to subculture, either because on their own they are incapable of producing all the nutrients or factors required for their growth or because they are quickly swamped by contaminating stem cells which use them as feeder layers. However, limited passage of differentiated cells released by gentle trypsin treatment was found to be possible in the presence of conditioned medium. Figure 4 compares the morphology of stem cells with that of a typical primary culture from extensively differentiated SEBs. Whereas the stem cell is small, compact, and stellate, with a low cytoplasm:nucleus ratio, and forms compact colonies (Fig. 4*a*), the SEB-derived culture contains cells with distinct bipolar, fibroblast-type morphology and extensive cytoplasm, which grow in oriented sheets (Fig. 4*b*).

When semiconfluent monolayers of these SEB-derived cells are infected and examined by immunofluorescence for viral capsid antigen 24 hr after infection, approximately 8% of the cells are found to have positive-staining nuclei (Fig. 4*e*,*f*), compared with 43% positive nuclei found in parallel-infected cultures of WME 129 cells (Fig. 4*c*,*d*). These results contrast with the extremely low (<0.1%) frequency of positive nuclei observed when stem cells are infected at similar multiplicities

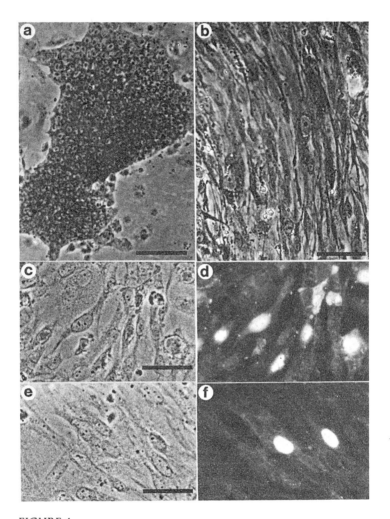

FIGURE 4

(a) Colony of SCC-PSA4 stem cells, stained with Giemsa. Bar represents 100 μm. (b) Primary culture derived from SCC-PSA4 settled embryoid bodies (SEB), stained with Giemsa. Bar represents 100 μm. (c) Culture of WME 129 cells 24 hr after infection with MVM(T) (10 PFU/cell). Phase contrast. Bar represents 50 μm. (d) Same field as in (c), showing anti-MVM capsid immunofluorescent staining. (e) SCC-PSA4 SEB-derived primary culture 24 hr after infection with MVM(T) (10 PFU/cell). Phase contrast. Bar represents 50 μm. (f) Same field as in (e), showing anti-MVM capsid immunofluorescent staining.

TABLE 1
Infection of Stem Cells and Differentiated Cells with MVM

Cell type	Multiplicity of infection	Infectious centers per cell	Virus yield per cell at 50 hr p.i. (PFU)	Ratio of PFU (50 hr:4 hr)
Nulli-SCC-S2	29	0.0074	7.4	13
SCC-S2	36	0.0013	0.47	0.11
SCC-PSA4	7	0.0007	0.97	0.56
PSA4 SEB-der 1°	17	0.11	25	19
WME 129	18	0.39	68	37

(Miller et al, 1977; P. Tattersall, unpublished results). Extensive cytopathogenic effect was observed by 48–72 hr p.i. in the SEB-derived cultures, indicating that a considerable fraction of these cells do support lytic infection by MVM. Since a number of mouse cell lines have been found to be killed by MVM without releasing infectious particles (P. Tattersall and P. J. Cawte, unpublished results; D. C. Ward, personal communication), it was of interest to determine the percentage of cells which formed infectious centers. Table 1 shows that the fraction of cells giving rise to infectious centers is equivalent to the fraction of capsid-antigen-positive nuclei for both the WME-129-derived and the SEB-derived cultures (Fig. 4). These infected cells also give rise to equivalent amounts of infectious progeny virus.

In addition, a small but significant fraction of each stem-cell line tested gave rise to infectious centers, and also, in the case of the nullipotent cell line, to a measurable rise in virus titers between 4 and 50 hr p.i. The low level of susceptibility of stem cells to infection was further shown to be multiplicity-dependent. As shown in Table 2, the fraction of infectious

TABLE 2
Effect of Multiplicity of Infection of SCC-S2 with MVM

Experiment	Multiplicity of infection	Infectious centers per cell
a	1.1	0.00004
b	12	0.00012
c	190	0.0038

centers for SCC-S2, the only line tested in this way, increased approximately in proportion to the multiplicity of infection.

Interaction of MVM with Stem Cells

It has so far proved impossible to establish stable virus-sensitive and virus-resistant clonal cell lines derived from teratoma stem cells in vitro with which to examine the nature of the block(s) in MVM replication. Instead, the interaction of virus with the stem cell itself was examined. An obvious first candidate for stem-cell resistance was a block in adsorption due to the lack of cell-surface receptors. In view of the low-multiplicity-dependent level of susceptibility, this explanation was considered unlikely, and indeed an immunofluorescent sandwich technique demonstrated that stem cells do have surface binding sites for MVM (data not shown). A less direct but more informative approach to this problem was to make heterokaryons between infected stem cells (from which all extracellular virus had been removed or inactivated) and a cell permissive for MVM replication, such as WME 129. When examined 24 hr after fusion, a considerable proportion of fused cells which could be identified as heterokaryons contained nuclei positive for MVM antigen. Typical positive heterokaryons are shown in Figure 5. In every case where each nucleus in the heterokaryon could be unambiguously assigned to one parent or the other, the only antigen-positive nuclei were those of the WME 129 permissive parent. A number of further fusion experiments were performed by making heterokaryons between SCC-S2 stem cells and STO cells. STO cells are permissive for the production of MVM antigen but produce little, if any, infectious progeny virus (data not shown). The advantage of this system is that DNA synthesis can be inhibited in the STO parent, without affecting the stem cell or any heterokaryon, by incubation in HAT-containing medium. This has the effect of suppressing viral antigen synthesis in an infected STO parent cell and thus enabling a small number of positive heterokaryons to be detected with relative ease. Figure 6 shows typical heterokaryons from crosses between infected stem cells and uninfected STO (Fig. 6a,b) and between infected STO and uninfected stem cells (Fig. 6c,d). In both types of cross the only positive nuclei identified in heterokaryons were those of the STO antigen-permissive parent.

In crosses where infected stem cells were fused with uninfected permissive or semi-permissive cells, no positive homokaryons or positive unfused parental cells of either type were detected. This indicates that activation of viral antigen synthesis was not due to the fusion process alone, or to the release into the medium of infectious virus which was subsequently able to infect the other parental cell of the cross.

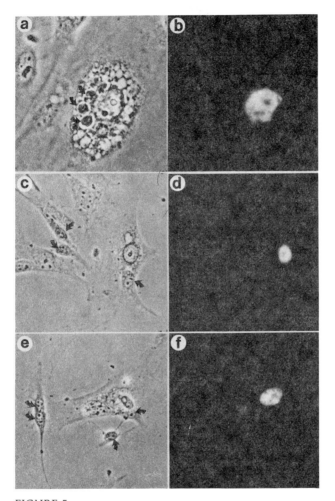

FIGURE 5

(*a*) Heterokaryon between MVM-infected (10 PFU/cell) SCC-S2 stem cells and uninfected WME 129 cells 24 hr after fusion. Phase contrast. (*b*) Same field as in (*a*), showing anti-MVM capsid immunofluorescent staining. (*c*) Heterokaryon from same experiment as described in (*a*). Phase contrast. (*d*) Same field as in (*c*), showing anti-MVM capsid immunofluorescent staining. (*e*) Heterokaryon from same experiment as described in (*a*). Phase contrast. (*f*) Same field as in (*e*), showing anti-MVM capsid immunofluorescent staining. Arrows indicate stem-cell nuclei.

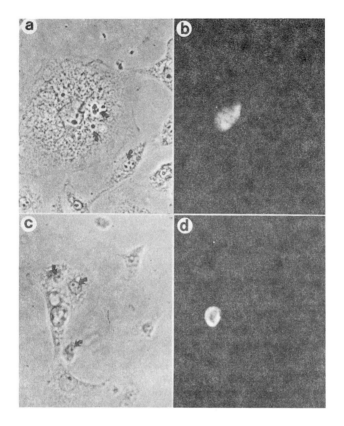

FIGURE 6

(a) Heterokaryon between MVM-infected (10 PFU/cell) SCC-S2 stem cells and uninfected STO cells, cultured in medium containing HAT, 24 hr after fusion. Phase contrast. (b) Same field as in (a), showing anti-MVM capsid immunofluorescent staining. (c) Heterokaryon between MVM-infected (10 PFU/cell) STO cells and uninfected SCC-S2 stem cells, cultured in medium containing HAT, 24 hr after fusion. Phase contrast. (d) Same field as in (c), showing anti-MVM capsid immunofluorescent staining. Arrows indicate stem-cell nuclei.

DISCUSSION

The infection of teratoma stem cells with MVM at relatively high multiplicity does not affect their ability to grow and differentiate in vitro. It is clear that most of the pathways of differentiation open to this stem cell proceed to the recognizable differentiated end cell without going through a virus-susceptible intermediate. The exception to this is a fibroblastlike cell which resembles morphologically the type of cell which grows out during continued passage of whole-embryo cultures or neonatal tissue cultures. Thus, the virus shows an extremely limited host range with respect to differentiated cell type, at least in this in vitro system. This contrasts with the pantropic effects observed in studies of neonatal infection reported by others (Kilham and Margolis 1970, 1975). In these studies, the original isolate MVM(CR) was used (Crawford 1966) and the virus was propagated in primary whole-mouse-embryo cultures or by passage in neonatal animals. The virus used in the present study was the clonal isolate MVM(T), which had been plaque-purified and propagated in A-9 cells (Tattersall 1972). MVM(T) therefore has been selected for its ability to infect lytically a laboratory strain of mouse cells and may not possess the entire host range of field strains of the virus. In support of this explanation is the isolation of a variant of MVM, believed to be the immunosuppressive agent carried in an EL-4 murine lymphoma cell line Bonnard et al. 1976). This variant grows in leukocyte cultures, but barely, if at all, in L cells (M. D. Hoggan, personal communication). This is in direct contrast with MVM(T), which grows in L cells but not at all in murine leukemic lymphocytes (D. C. Ward, P. J. Cawte, and P. Tattersall, unpublished results) or in concanavalin-A-stimulated splenic lymphoid cells (Miller et al. 1977). Thus it appears possible to select specific cytotropic variants of the pantropic field strain. Whether MVM in nature is a mixture of specific cytotropes or whether these arise spontaneously at low level and grow out preferentially under inadvertant selection conditions during in vitro propagation is an important question requiring further investigation.

The emerging picture of the parvovirus chromosome is that it contains little else than the information required for the control of expression and replication of a single gene. This makes the viruses especially attractive tools with which to probe the control processes of the cell during growth and differentiation. The present study shows that undifferentiated stem cells in the teratoma system, as well as the majority of their differentiated derivatives, express a block (or a series of blocks) to the successful infection and lytic replication of MVM. The block is lifted during differentiation down at least one pathway. The fact that virus replication (as measured by capsid-antigen synthesis) can be "rescued" in stem cells by fusion to uninfected permissive cells speaks against the possibility that this block is mediated through a diffusible inhibitor, or simply by the lack of cell-surface receptors for the virus.

A similar resistance to infection of stem cells, but not their differentiated derivatives, has been shown for polyoma and SV40 (Lehman et al. 1975; Swartzendruber and Lehman 1975) and Moloney murine leukemia virus (Teich et al. 1977). However, stem cells are permissive for other viruses, such as mengovirus (Lehman et al. 1975), adenovirus type 2 (Kelly and Boccara 1976), and vaccinia, encephalomyocarditis, sindbis, and vesicular stomatitis viruses (Teich et al. 1977).

Interestingly, the blocking of MVM replication in stem cells is not total, as a very small fraction of infected cells produce capsid antigen (Miller et al. 1977) and, as shown here, become infectious centers. The proportion of these apparently productive cells can be increased by increasing the multiplicity of infection. Further studies are required to show that these infectious centers are the result of cells producing a burst of normal infectious progeny. Apart from the possibility of spontaneous lysis releasing parental virus, there is the possibility that infectious centers are the result of a stem cell being infected with a virus variant with a different host range from the majority of MVM(T). This is unlikely, as the culture of stem cells in the continuous presence of MVM(T) does not appear to select for a variant capable of growing lytically in the culture. The production of infectious centers by spontaneous fusion of infected stem cells to permissive A-9 cells in the plaque-assay monolayer, or any other mechanism involving cell-cell contact, is excluded because infectious centers form with greater than 50% of control efficiency when a 1-mm-thick stiff agar layer is placed between the two cell types.

When infected stem cells are fused with uninfected permissive cells, the nucleus of the latter and the common cytoplasm become permissive for viral antigen synthesis. However, capsid components are apparently transported only into the permissive-cell nucleus. The recessive nature of stem-cell resistance as demonstrated in heterokaryons appears to be in disagreement with the results of Miller et al. (1977), who examined the infection of stable fusion hybrids between resistant teratoma stem cells and permissive Friend erythroleukemia cells. These studies showed that the hybrids, which resembled the teratoma parent biochemically and in their differentiation capacity, were also resistant to MVM infection, as measured by capsid-antigen synthesis.

These two results can be reconciled by considering that the stem-cell nucleus is resistant to MVM, and that this trait is dominant in the formation of the common hybrid nucleus. A possible mechanism for this type of resistance involves the transport of viral antigen or capsid to the nucleus. During lytic infection of a permissive cell, viral capsid antigen accumulates in the nucleus. This process probably involves active transport, as it appears to generate a considerable "chemical gradient" of antigen across the nuclear membrane. Appreciable amounts of viral antigen only appear in the cytoplasm when the nucleus disintegrates late in infection. Since it is likely that parvoviruses are too limited in their genetic capacity to code for their own transport system, their capsid proteins might mimic a cellular component which is normally transported to the nucleus. Such a cellular system may be switched off and on, depending upon the differentiated state of the cell. If parvoviruses depended upon such a system to reach the nucleus in order to initiate infection, cells which have this system switched off—say, teratoma stem cells and their stem-cell-like hybrids—would be resistant to infection. In this situation, however, heterokaryons with sensitive cells would be permissive, but would not transport viral antigen into the nucleus of the stem-cell parent.

ACKNOWLEDGMENTS

I wish to thank Deb Rowe for her assistance with cell culture, and Drs. B. Hogan and N. Teich for many useful discussions.

REFERENCES

Bonnard, G. D., E. K. Manders, D. A. Campbell, Jr., R. B. Herberman, and M. J. Collins, Jr. 1976. Immunosuppressive activity of a subline of the mouse EL-4 lymphoma. Evidence for minute virus of mice causing the inhibition. *J. Exp. Med.* **143**:187.

Crawford, L. V. 1966. A minute virus of mice. *Virology* **29**:605.

Hales, A. 1977. A procedure for the fusion of cells in suspension by means of polyethylene glycol. *Som. Cell Genet.* **3**:227.

Hampton, E. G. 1970. H-1 virus growth in synchronized rat embryo cells. *Can. J. Microbiol.* **16**:266.

Hogan, B. L. M. 1976. Changes in the behavior of teratocarcinoma cells cultivated in vitro. *Nature* **263**:136.

Kelly, F. and M. Boccara. 1976. Susceptibility of teratocarcinoma cells to adenovirus type 2. *Nature* **262**:409.

Kilham, L. and G. Margolis. 1970. Pathogenicity of minute virus of mice (MVM) for rats, mice and hamsters. *Proc. Soc. Exp. Biol. Med.* **133**:1447.

———. 1975. Problems of human concern arising from animal models of intrauterine and neonatal infections due to viruses: A review. I. Introduction and virologic studies. *Prog. Med. Virol.* **20**:113.

Lehman, J. M., I. B. Klein, and R. M. Hackenberg. 1975. The response of murine teratocarcinoma cells to infections with DNA and RNA viruses. In *Teratomas and differentiation* (ed. M. I. Sherman and D. Solter), p. 289. Academic, New York.

Macpherson, I. A. and M. G. P. Stoker. 1962. Polyoma transformation of hamster cell clones. An investigation of genetic factors affecting cell competence. *Virology* **16**:147.

Margolis, G. and L. Kilham. 1965. Rat virus, an agent with an affinity for the dividing cell. *N.I.N.D.B. Monogr.* **2**:361.

———. 1975. Problems of human concern arising from animal models of intrauterine and neonatal infections due to viruses: A review. II. Pathologic studies. *Prog. Med. Virol.* **20**:144.

Martin, G. R. 1975. Teratocarcinomas as a model system for the study of embryogenesis and neoplasia. *Cell* **5**:229.

Martin, G. R. and M. J. Evans. 1975a. Differentiation of clonal lines of teratocarcinoma cells: Formation of embryoid bodies in vitro. *Proc. Natl. Acad. Sci.* **72**:1441.

————. 1975b. The formation of embryoid bodies in vitro by homogeneous embryonal carcinoma cell cultures derived from isolated single cells. In *Teratomas and differentiation* (ed. M. I. Sherman and D. Solter), p. 169. Academic, New York.

————. 1975c. Multiple differentiation of clonal teratocarcinoma stem cells following embryoid body formation in vitro. *Cell* **6**:467.

Miller, R. A., D. C. Ward, and F. H. Ruddle. 1977. Embryonal carcinoma cells (and their somatic cell hybrids) are resistant to infection by the murine parvovirus MVM, which does infect other teratocarcinoma-derived cell lines. *J. Cell. Physiol.* **91**:393.

Mohanty, S. B. and P. A. Bachmann. 1974. Susceptibility of fertilized mouse eggs to minute virus of mice. *Infect. Immun.* **9**:762.

Nicolas, J. F., P. Dubois, H. Jakob, J. Gaillard, and F. Jacob. 1975. Tératocarcinome de la souris: Différenciation en culture d'une lignée de cellules primitives à potentialités multiples. *Ann. Microbiol. (Paris)* **126A**:3.

Pierce, G. B. and F. J. Dixon. 1959. Testicular teratomas. II. Teratocarcinoma as an ascitic tumor. *Cancer* **12**:584.

Pontecorvo, G. 1975. Production of mammalian somatic cell hybrids by means of polyethylene glycol. *Som. Cell Genet.* **1**:397.

Rhode, S. L. III. 1973. Replication process of the parvovirus H-1. I. Kinetics in a parasynchronous cell system. *J. Virol.* **11**:856.

Rodriguez, J. and F. Deinhardt. 1960. Preparation of a semipermanent mounting medium for fluorescent antibody studies. *Virology* **12**:316.

Ruffolo, P. R., G. Margolis, and L. Kilham. 1966. The induction of hepatitis by prior partial hepatectomy in resistant adult rats infected with H-1 virus. *Am. J. Pathol.* **49**:795.

Sherman, M. I. 1975. Differentiation of teratoma cell line PCC4:aza1 in vitro. In *Teratomas and differentiation* (ed. M. I. Sherman and D. Solter), p. 189. Academic, New York.

Siegl, G. and M. Gautschi. 1973. The multiplication of parvovirus LuIII in a synchronized culture system. I. Optimum conditions for virus replication. *Arch. gesamte Virusforsch.* **40**:105.

Stevens, L. C. 1958. Studies on transplantable testicular teratomas of strain 129 mice. *J. Natl. Cancer Inst.* **20**:1257.

————. 1967. Origin of testicular teratomas from primordial germ cells in mice. *J. Natl. Cancer Inst.* **38**:549.

————. 1968. The development of teratomas from intratesticular grafts of tubal mouse eggs. *J. Embryol. Exp. Morphol.* **20**:329.

————. 1970. The development of transplantable teratocarcinomas from intratesticular grafts of pre- and post-implantation mouse embryos. *Dev. Biol.* **21**:364.

Swartzendruber, D. E. and J. M. Lehman. 1975. Neoplastic differentiation: Interaction of simian virus 40 and polyoma virus with murine teratocarcinoma cells in vitro. *J. Cell. Physiol.* **85**:179.

Szybalski, W., E. H. Szybalski, and G. Ragni. 1962. Genetic studies with human cell lines. *Natl. Cancer Inst. Monogr.* **7**:75.

Tattersall, P. 1972. Replication of the parvovirus MVM. I. Dependence of virus multiplication and plaque formation on cell growth. *J. Virol.* **10**:586.

Tattersall, P., P. J. Cawte, A. J. Shatkin, and D. C. Ward. 1976. Three structural polypeptides coded for by minute virus of mice, a parvovirus. *J. Virol.* **20**:273.

Teich, N. M., R. A. Weiss, G. R. Martin, and D. R. Lowy. 1977. Virus infection of murine teratocarcinoma stem cell lines. *Cell* (in press).

Tennant, R. W., K. R. Layman, and R. E. Hand, Jr. 1969. Effect of cell physiological state on infection by rat virus. *J. Virol.* **4**:872.

Toolan, H. W. 1968. The picodna viruses: H, RV and AAV. *Int. Rev. Exp. Pathol.* **6**:135.

Binding of Minute Virus of Mice to Cells in Culture

PAUL LINSER
RICHARD W. ARMENTROUT

Department of Biological Chemistry
College of Medicine
University of Cincinnati
Cincinnati, Ohio 45267

The replication of the nondefective parvoviruses requires dividing cells (Rose 1974). It appears that some event occurring in late S phase of the host cell cycle is necessary for viral DNA synthesis (Rhode 1973). However, there are probably additional restraints on parvovirus replication in vivo which are poorly understood and which restrict lytic infection to certain specific classes of growing cells (Lipton and Johnson 1972). One mechanism for limiting virus replication to specific cell types is the requirement for a specific virus receptor on the cell surface to initiate infection. Cytotropism based on a specific virus receptor has been described for a number of different types of viruses (Dales 1973).

We present evidence here that the parvovirus MVM (minute virus of mice) initiates infection by binding to a specific receptor on the cell surface. Such a receptor may be a component in the cytotropism observed for parvovirus infection.

MATERIALS AND METHODS

Virus

Our initial plaque-purified MVM was the generous gift of Dr. Peter Tattersall. Infectious stocks and [³H]thymidine-labeled 110S virus were grown in monolayer culture and purified by sucrose velocity gradient centrifugation as described previously (Richards et al. 1977). Empty-capsid material (70S) was also isolated in sucrose gradients. The con-

151

centration and specific activity of radiolabeled virus were calculated from the optical density at 280 nm as described by Tattersall et al. (1976).

Cells and Culture

All cell lines used in this study are adapted for growth in suspension and are of murine origin. The lines used were the A-9 derivative of mouse L cells (Tattersall 1972) and an infection-resistant clone of A-9 cells designated 8-E (Linser et al. 1977). Culture conditions were as described previously (Linser et al. 1977).

Binding Assays

Binding assays were all performed on cells in suspension at 4°C. Briefly, the cells were suspended in phosphate-buffered saline (PBS), pH 7.2, and reacted with fixed quantities of virus for an appropriate length of time. After incubation, the cell suspension was filtered through 25-mm Nucleopore filters with 5-μm pores and washed with ice-cold buffer. After drying, the filter was solubilized with Soluene 100 (Packard) and the cell-associated radioactivity counted in a liquid scintillation counter.

CsCl Analysis of Bound Virus

To analyze the density classes of labeled 110S virus bound to cells as a function of time, binding assays were run for varying periods of time, the suspension was filtered as described above, and the filter was then treated with $Ca^{++}Mg^{++}$-free PBS containing 1 mM EDTA, which elutes MVM bound to cells. The resultant suspension was centrifuged to remove debris, and the supernatant was analyzed by CsCl equilibrium density gradient centrifugation as described previously (Richards et al. 1977).

Selecting Resistant Cells

A-9 cells resistant to MVM infection were cloned from cells surviving long-term infection in monolayer culture as described previously (Linser et al. 1977). The particular clonal derivative reported here was designated 8-E for reference purposes.

Electron Microscopy

Standard techniques for electron microscopy were used as described previously (Richards et al. 1977).

RESULTS

Virus Binding Assay

To determine whether MVM adsorbs to a specific surface receptor we have examined the kinetics of binding of virus to cells at 4°C. In Figure 1, it can be seen that adsorption of virus to cells follows a distinctly biphasic pattern. The initial binding component of this curve is linear and saturates at about 5–7×10^5 virus particles bound per cell. The second region of the binding curve occurs at high virus levels, is not saturable, and probably represents some form of low-affinity, nonspecific binding. Several technical points concerning this assay are worth noting: The virus used in the binding studies are DNA-containing particles internally labeled with [^{3}H]thymidine. Their interaction with the cells should closely reflect the initial stages of the infectious process. To ensure that the virus particles were not aggregated, only 110S full-virus particles freshly isolated from a sucrose gradient were used. The concentration of 110S virus which could be recovered from a sucrose gradient was limited, and this placed restraints on some of the competition experiments described below.

Experiments (not presented here) have shown that the binding of virus has an optimal pH of 7.2, is 75% complete within 15 min, and is not markedly temperature-dependent over the range 4°–37°C. The virus par-

FIGURE 1
Saturation curve for binding of MVM to A-9 cells in suspension. A-9 cells (2×10^5) suspended in 1 ml PBS at 4°C were reacted with the indicated multiplicity of [^{3}H]-thymidine-labeled MVM particles per cell (I.M.: input multiplicity) for 2 hr. Samples were then filtered and the cell-associated radioactivity was converted to B.M. (bound multiplicity) (see "Methods"). At all concentrations of input virus, filtration of cell-free control samples resulted in retention of less than 0.1% of the input radioactivity, whereas samples with cells retained up to 75%. Each point is the mean of three determinations. The vertical bars indicate the range of the data.

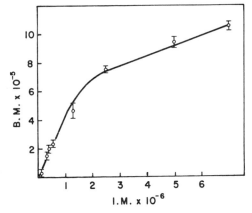

ticles adsorb readily to both glass and plastic but not to Nucleopore filters (Linser et al. 1977).

Although the virus-binding experiments were conducted at 4°C to minimize uptake into the cell, two pieces of evidence indicate that most of the measured binding was to the cell surface. First, 78% of the virus bound during a 2-hr incubation could be eluted from the cells by a brief exposure (5 min) to EDTA (0.001 M) (these conditions do not affect the viability of the cells). Second, it is possible to compete from the cell surface virus previously bound at 4°C by incubating the cell-virus complex with an excess of unlabeled virus in buffer (Fig. 2, left panel, curve B). These results indicate that the bulk of the virus is reversibly bound to the cell surface.

In order to test the specificity of the binding reaction, a fixed amount of labeled virus was mixed with increasing amounts of unlabeled virus prior to adsorption to cells. As shown in Figure 2 (left panel, curve A), the unlabeled virus competes effectively for the limited binding sites on the

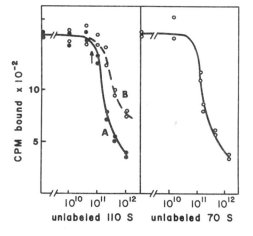

FIGURE 2

Competition analysis of the binding of 110S MVM and 70S empty-capsid material to A-9 cells in suspension. *Left panel:* Curve of competition between labeled and unlabeled 110S MVM. A-9 cells suspended in PBS at 4°C were reacted with a fixed small quantity of [³H]-thymidine-labeled MVM. In curve A (●——●) the labeled virus was mixed with increasing amounts of unlabeled virus prior to adsorption to cells. In curve B (o----o) the labeled virus was allowed to adsorb to the cells for 2 hr, and increasing amounts of unlabeled virus were subsequently added and incubated with cells for an additional hour prior to filtration. Arrow indicates concentration at which 5 × 10⁵ virus particles are bound per cell. *Right panel:* Same as curve A in left panel, but generated using 70S empty-capsid material as the unlabeled species in the reaction.

cell surface. As expected, there is no competition until the cellular binding sites are saturated. However, once the total bound virus exceeds the available binding sites, competition by unlabeled virus can be observerd. These results, along with experiments described below using virus-resistant cell strains, indicate that virus is binding to a limited number of specific surface sites on the cells.

The binding of virus to cells is complicated by the fact that the 110S material contains at least two distinct viral species: a dense particle (1.46 g/cm^3 in CsCl) and the major full-virus species (1.42 g/cm^3 in CsCl), to which the dense particle is a precursor. Clinton and Hayashi (1976) have shown that after a brief binding period the precursor particle is readily dissociated from the cells at 37°C by simple rinsing with buffer. This observation and the biphasic nature of our binding curves led us to determine that the two forms of virus were binding to cells under our conditions. As shown in Figure 3, both the 1.46- and the 1.42-g/cm^3 particles are found on the cell surface after adsorption at 4°C. Both particles appear to bind to the cells, and neither is preferentially released by washing at 4°C.

It should be noted that under these experimental conditions the cellular binding sites are in approximately two- to threefold excess over the input virus. The binding appears to be a reversible reaction, and only a portion of the input virus has bound in this case even though the cellular sites are in excess. With a higher ratio of cell receptors to virus (10–15-fold excess in a 1-ml volume) virtually all of the input virus (95%) becomes cell-associated within the 2-hr incubation period.

In addition to the two species of DNA-containing particles, empty capsids are produced in great excess during MVM infection. The empty capsid competes with labeled full virus for cellular binding sites with the same efficiency as unlabeled full virus (Fig. 2, right panel). Thus, the empty capsid appears to bind to the same cell receptor sites as the full virus.

To determine the relationship between virus binding and the infectious process we have isolated a number of clones of cells resistant to MVM infection. Several of these clones were found to bind a significantly lower number of virus particles. For example, clone 8-E derived from A-9 cells is extremely resistant to infection (Table 1) and binds virus at low levels with monophasic nonsaturable kinetics (Fig. 4). From this result and similar data from other resistant clones, it would appear that the higher-affinity saturable binding of virus leads to infection of the cell, whereas the nonsaturable form of virus adsorption is rarely productive. This correlation of infectivity with the high-affinity binding holds with other cell types as well. Murine-lymphocyte-derived L-1210 cells are resistant to infection and bind virus weakly in the nonsaturable mode. Friend erythroleukemia cells (strain 745) have 1–2 × 10^5 binding sites per cell (Linser et al. 1977) and can be productively infected (Miller et al. 1977).

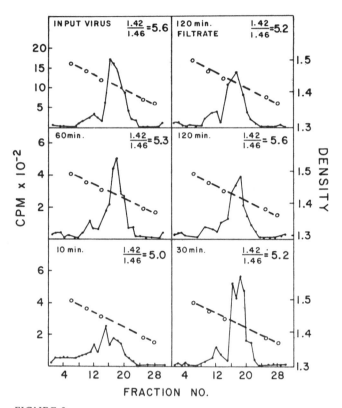

FIGURE 3

CsCl density gradient analysis of 110S MVM before and after binding to A-9 cells in suspension: 2×10^5 cells suspended in PBS were reacted with subsaturating quantities of [^{3}H]thymidine-labeled 110S MVM at 4°C for 10, 30, 60, or 120 min. Cells were filtered and washed as described in "Methods," and the cell-associated virus was eluted by treatment of the filter with Ca^{++}Mg^{++}-free PBS containing 1 mM EDTA. After low-speed centrifugation to remove debris, the eluant was adjusted to an average density of 1.46 g/cm^2 with CsCl and was centrifuged in a fixed-angle 40 rotor for 48 hr at 35,000 rpm and 4°C. Fractions were collected from the bottoms of the tubes, and acid-insoluble radioactivity was determined after solubilization of precipitates. The labeled virus used for binding (input virus) was analyzed, as was the filtrate (virus that did not become cell-associated). The CsCl gradient of only one of the filtrate samples (120 min) is presented. The other samples gave similar results. The ration of the 1.42-g/cm^3 virus to the 1.46-g/cm^3 particles was very similar (approximately 5:1) in all cases. ●——●: Radioactivity (cpm); o----o: density.

156

TABLE 1
Tissue-Culture Infectivity, A-9 vs. 8-E

Dilution of infectious stock	Hemagglutinin titers			
	A-9	mean	8-E	mean
10^{-2}	64; 64; 128	85	2; 4; 4	3.3
10^{-3}	128;128;256	170	0; 0; 0	0
10^{-4}	128;512;512	384	0; 0; 0	0
10^{-5}	64; 64; 64	64	0; 0; 0	0
10^{-6}	32; 8; 8	16	0; 0; 0	0
10^{-7}	2; 2; 2	2	0; 0; 0	0

Hemagglutination activity produced in a tissue-culture infectivity assay run on A-9 and 8-E cells in parallel. Cells (1×10^5) were seeded into 35-mm culture dishes. Dilutions of stock MVM were absorbed in 0.5 ml PBS at 37°C for 2 hr. The cultures were subsequently fed with growth media and refed daily. Four days following infection, mock-infected control cultures of both cell types had reached confluency. The lowest three dilution samples exhibited gross cytopathology in the A-9 cultures, whereas no growth inhibition or cytopathogenic effect was evident in any of the 8-E cultures. At this time cells were harvested into 0.01 M Tris, 0.005 M EDTA (pH 9.0) buffer and lysed by sonic treatment. Viral protein was assayed by standard hemagglutination assay.

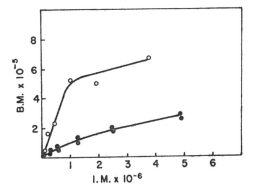

FIGURE 4
Comparison of MVM binding to virus-sensitive A-9 cells and to virus-resistant 8-E cells: 2×10^5 A-9 or 8-E cells suspended in 1 ml PBS at 4°C were reacted with the indicated multiplicity of [^{3}H]-thymidine-labeled MVM particles per cell (I.M.: input multiplicity) for 2 hr. Samples were then filtered, and the cell-associated radioactivity was measured and converted to B.M. (bound multiplicity) as described in "Methods." ○——○: A-9; ●——●: 8-E.

As a first step toward determining the sequence of virus uptake into cells we have examined by electron microscopy the location of virus binding to the cell surface. In general, virus appears on three regions of the cell surface when adsorbed at 4°C at the level of 2–3 × 10⁵ particles per cell. The particles are seen singly or in patches on the filopodia, on clefts with prominent internal and surface glycocalyx, and scattered about undefined regions of the cell surface (Fig. 5). The location of the virus receptor on the surface of the cell does not appear to be restricted. Under similar binding conditions, few particles were detected on the surfaces of approximately 30 8-E cells examined at random.

DISCUSSION

From these experiments we have obtained evidence that MVM initiates infection by binding to a specific receptor present on the cell surface. This surface component of virus-susceptible cells does not appear to be necessary for any obvious cellular function. Virus-resistant cell lines with greatly reduced levels of this receptor grow as rapidly and as densely in culture as the virus-susceptible cells from which they were derived.

The virus receptor is present in varying amounts on several different cultured cell lines. In addition, it is possible that the virus receptors are involved in the cytotropism of parvovirus infections observed in vivo. For example, when growing primary cultures of embryonic mouse brain cells are infected with MVM, we find that most cells eventually lyse, leaving a population of infection-resistant survivors. These surviving cells are of varied morphology and grow even in the presence of the mitolytic virus. While these observations are consistent with the loss (or masking) of a cell-surface receptor, they do not exclude the possibility that virus resistance is mediated by intracellular factors.

Of course, in addition to the cellular receptor, the virus particle plays an important role in the adsorption to the cell surface. It is known that one of the final steps in the formation of MVM particles is the proteolytic cleavage of the major capsid protein of the 1.46-g/cm³ particle (Clinton and Hayashi 1976; Tattersall et al. 1976; Richards et al. 1977). The precursor particle hemagglutinates red blood cells poorly compared to the product virus. Also, as Clinton and Hayashi have shown, the precursor particles adhere less tenaciously to cultured cells than the product particles and, as a result, are less infectious when cultures are rinsed after adsorption at 37°C (Clinton and Hayashi 1976). On the basis of these results, one might conclude that the processing of the virus particle by a proteolytic cleavage of the major capsid protein generates virus affinity for the cell surface. However, our results indicate that precursor and product virus particles bind equally well to cellular receptors at 4°C. It would seem likely that some temperature-dependent step following binding but preceding

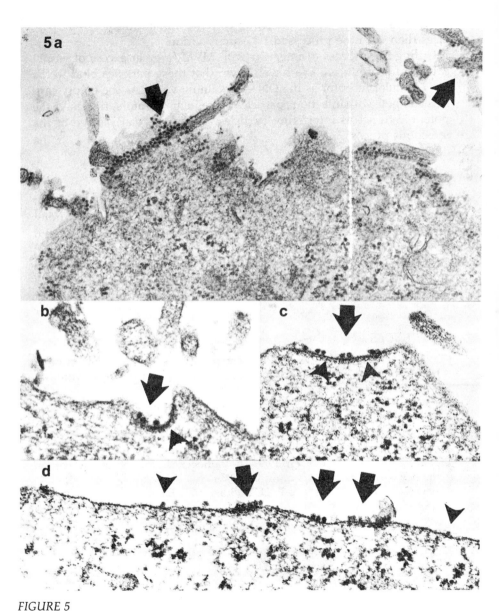

FIGURE 5

Electron micrographs of MVM bound to A-9 cells at the level of 2–3×10^5 per cell. (a) Portion of an A-9 cell with adherent virus in several locations, particularly on the filopodia (arrows), $52,000\times$. (b,c) Examples of virus (arrows) bound to specialized endocytotic clefts in the cell membrane which are characterized by an electron-dense inner-surface glycocalyx (arrowheads); (b) $65,000\times$ and (c) $60,000\times$. (d) MVM particles adsorbed to unspecialized regions of cell surface in patches (arrows) and as single particles (arrowheads), $60,000\times$.

internalization occurs less efficiently with the immature 1.46-g/cm³ particles than with the processed 1.42-g/cm³ virion.

During the process of infection with MVM, a great excess of empty capsids are produced. We have shown that these particles bind to the same cellular receptor as the DNA-containing virus. As the empty capsids compete with infectious particles for cellular receptors, they have the potential to act as interfering particles at this very early stage of the infectious process.

ACKNOWLEDGMENTS

We gratefully acknowledge helpful discussions with Dr. Peter Tattersall and Dr. David Ward during the course of this work. Ms. Helen Bruning provided expert technical assistance. This investigation was supported by grants 1 KO4 CA 00134 and 5 RO1 CA 16517 awarded by the National Cancer Institute, by grant 1-396 from the National Science Foundation and the March of Dimes, and by a grant from the United Fund Health Foundation of Canton, Ohio.

REFERENCES

Clinton, G. M. and M. Hayashi. 1976. The parvovirus MVM: A comparison of heavy and light particle infectivity and their density conversion in vitro. *Virology* **74**:57.

Dales, S. 1973. Early events in cell-animal virus interactions. *Bacteriol. Rev.* **37**: 103.

Linser, P., H. Bruning, and R. W. Armentrout. 1977. Specific binding sites for a parvovirus, minute virus of mice, on cultured mouse cells. *J. Virol.* (in press).

Lipton, H. L. and R. T. Johnson. 1972. The pathogenesis of rat virus infections in the newborn hamster. *Lab. Invest.* **27**: 508.

Miller, R. A., D. C. Ward, and F. H. Ruddle. 1977. Embryonal carcinoma cells (and their somatic hybrids) are resistant to infection by murine parvovirus MVM, which does infect other teratocarcinoma-derived cell lines. *J. Cell. Physiol.* **91**:393.

Rhode, S. L. 1973. Replication process of the parvovirus H-1. I. Kinetics in a parasynchronous cell system. *J. Virol.* **11**: 856.

Richards, R., P. Linser, and R. W. Armentrout. 1977. Kinetics of assembly of a parvovirus, minute virus of mice, in synchronized rat brain cells. *J. Virol.* **22**: 778.

Rose, J. A. 1974. Parvovirus reproduction. In *Comprehensive virology* (ed. H. Fraenkel-Conrat and R. R. Wagner), vol. 3, p. 1. Plenum, New York.

Tattersall, P. 1972. Replication of the parvovirus MVM. I. Dependence of virus multiplication and plaque formation on cell growth. *J. Virol.* **10**: 586.

Tattersall, P., P. J. Cawte, A. J. Shatkin, and D. C. Ward. 1976. Three structural polypeptides coded for by minute virus of mice, a parvovirus. *J. Virol.* **20**: 273.

Maternal Role in Susceptibility of Embryonic and Newborn Hamsters to H-1 Parvovirus

HELENE WALLACE TOOLAN

Institute for Medical Research
Putnam Memorial Hospital
Bennington, Vermont 05201

Since it was first learned that human maternal infection with rubella virus during the initial third of pregnancy could cause malformation of the fetus, an animal model has been sought for studies of maternally derived in utero infections resulting in death or deformity of the embryo. The parvovirus H-1 was reported to cause a "mongoloidlike" deformity, or death, if injected into newborn hamsters (Toolan 1960), and to traverse the placenta readily and produce the same deformity and/or death in embryos of subcutaneously injected pregnant hamsters. Subsequently a number of studies were performed to determine if infection with H-1 parvovirus during any particular period of the hamster's short 16-day gestation period was especially harmful to the embryo. On the basis of the work reported to date, it has been accepted that the hamster embryo is most susceptible to infection with H-1 on day 8 of gestation. This finding has been attributed to the particular developmental stage of the embryo at that time (Ferm and Kilham 1964, 1965) or to a hypothetical special permeability of the placenta at midgestation (Soike et al. 1976).

We have examined the immunological response of both mothers and offspring to parvovirus infection as a function of time during gestation in order to explore further the basis of the changes in sensitivity to virus during pregnancy.

MATERIALS AND METHODS

Virus Inoculation

H-1 and H-3 parvovirus preparations were distilled-water filtrates prepared from tissue pools of 5-day-old hamsters infected with H-1 or H-3 (Toolan 1968). Virus infectivity was measured by plaque assay on NB cells as described by Ledinko (1967).

Hamsters were obtained from Lakeview Hamster Colony (Charles River Labs); some were shipped as time-bred animals and some were obtained at the Lakeview Colony within a few hours of coitus. Pregnant animals were inoculated subcutaneously with 2×10^4 plaque-forming units (PFU) of H-1 (in one experiment the dose was 10^5 PFU). Three separate experiments gave uniform results, which have been combined in this article.

Mortality was expressed as the percent of offspring surviving at day 16 compared with the number present the day after delivery. Large litters of healthy animals were reduced to seven at day 16, after which time most normal individuals survive. When deformed babies were present, the entire litter was kept.

Serology

When a litter was 21 days old the mother was bled from the heart. Each baby was also bled from the heart, and sera from any one litter were pooled. All sera were kaolinized and tested for antibodies against 4–8 units of both H-1 and H-3 antigen by hemagglutination inhibition (HAI) as described previously (Toolan 1968). Serum titers are expressed as the reciprocal of the highest dilution causing inhibition in this assay. Spot tests of various samples showed that virus-infectivity-neutralizing and HAI titers were approximately the same.

Tissue-Culture Isolation of Virus

Embryos 10 and 13 days old were tested for the presence of virus by removing them aseptically from the uterus after the mother was first bled and killed. Groups of four to five whole embryos were finely chopped in a small volume of modified Eagle's medium containing 15% fetal-calf serum, diluted to approximately 20 ml with the same medium, and seeded in tissue-culture flasks. After 4 days at 37°C the medium and floating debris were removed and centrifuged, and the supernatant was tested for viral hemagglutinin (HA). The cells remaining were fed with fresh medium, and 1 ml was removed every second day thereafter for HA estimation. Surviving cultures were refed every 7–10 days.

Immunofluorescence Studies

Tissues from 10- or 13-day embryos were also examined by fluorescent-antibody staining as described by Tanagaki et al. (1967). The indirect staining procedure was used, employing anti-H-1 serum prepared in guinea pigs followed by fluorescein-conjugated IgG-fraction goat anti-guinea-pig serum (Cappel).

RESULTS

Infection of Pregnant Hamsters During the Preimplantation Period

Mortality and morbidity due to injection of H-1 from coitus up to implantation (on day 6) was studied. When five to seven pregnant hamsters were injected each day from 24 hr after coitus (day 1) through the subsequent 5 days (days 2–6) with 2×10^4 PFU of H-1, no significant mortality due to virus occurred among the resulting litters by the time of weaning, with the exception of a 20% mortality without any deformities in the offspring of females injected on day 3 of gestation. In litters from mothers injected on day 6, no mortality occurred, but five deformed hamsters (four out of four in one litter) were observed.

Embryonic Infection as Determined by Tissue-Culture Techniques

During the 1–6-day postcoitus period at least two additional mothers were injected on each day; one was bled and killed on day 10 of gestation and the other on day 13, and their embryos were removed, minced, and tested in tissue culture for the presence of virus. H-1 could be recovered from cultures of *all* embryos from mothers infected on day 2 or later. Only day 1 showed two negative results out of six litters minced and plated.

Fluorescent-Antibody Studies on Embryos from Mothers Injected on Days 1–6

Fluorescent-antibody studies on the 10- and 13-day embryos derived from mothers injected on days 1–6 of gestation are still in an early stage and have been limited to samples taken from various levels; serial sections have not yet been made. However, our results corroborate the presence of virus infection in all material examined, as noted previously by tissue-culture tests. The considerable amount of fluorescence we have observed in studies to date has all been found in nuclei of cells of mesenchymal origin, such as the perichondrial cells surrounding developing cartilage and the adjacent, still mitosing, young cartilage cells. Nuclei of the loose-connective-tissue cells of the dermis and cells outlining the hair follicles, though not those of the hair follicles themselves, have been shown to be positive. Epidermis and parenchymal cells of the liver failed to stain, but nuclei in a few kidney tubule cells (of mesodermal origin) were positive, as were many interstitial connective-tissue-cell nuclei of liver, lung, blood vessels, and muscle. Surprisingly, we have not seen any staining in embryonic brain or nerve tissues. This work is still preliminary; detailed study will be a future effort.

HAI Antibody Titers to H-1 of Mothers and Litters

There was no significant difference in maternal antibody titer to H-1 between mothers who had been infected during the implantation period (bled at weaning time, 21 days after giving birth) and normal adult hamsters injected with the same amount of virus—the average titer was approximately 1:640 to 1:1280 when tested with guinea-pig cells. Weanlings of this group had low HAI titers, with the majority of pooled litter sera (27 out of 36) showing titers of 1:40 or less. There were three particular exceptions. One of the day-3-injection litters, in which considerable mortality had occurred, had five animals remaining, with a pooled HAI titer of 1:5120, and the two litters from the day-6 group in which deformities were observed produced titers of 1:5120 (in the litter where one of seven weanlings was deformed) and 1:40,960 (in the litter where all of four were deformed). This last high HAI titer is typical of the mongoloidlike or "funny face" (FF) hamster (Toolan 1968, 1972).

Infection of Pregnant Hamsters on Days 7–15 of Gestation

In litters from mothers injected with H-1 after day 6 of pregnancy, considerable mortality and morbidity occurred (Fig. 1). A sharp rise in deaths was seen in litters from the day-7-injected mothers, and deaths continued to escalate with injection on days 9 and 10. Both mortality and deformities dropped in litters from mothers injected on days 11 and 12, and subsequently a peak of mortality was observed for day 13, with a second and precipitous fall to zero mortality for day 15. These results were the same in three separate experiments representing litters from 58 mothers injected. Figure 1 is a summary of the three.

HAI titers of mothers injected on days 7–15 of gestation were again within normal limits. Although two mothers that had raised litters containing all FF babies had titers of 1:10,240, the pooled titers of their litters were in each case 1:81,620. Previous work showed that mothers of FF litters (even those in which the babies were injected at birth and the mothers had never been inoculated) can develop very high HAI titers, though such titers are always 3–4 logs less than those of their offspring (Toolan 1968). On the other hand, mothers of 22 litters in this group where all babies died or were deformed had an average titer of 1:640. The reason for this difference is unknown. It may represent the amount of

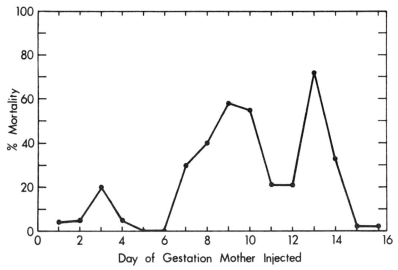

FIGURE 1

Mortality of babies from hamsters injected subcutaneously with 20,000 PFU of H-1 parvovirus on days 1–15 of gestation (parturition on day 16), expressed as percent of individuals surviving at 16 days after birth related to total number in litter the day after delivery.

cannibalism of dead and infected babies. The HAI antibody titers of the litters in the 7–15-day group varied from zero to 1:81,820, depending mainly upon the number of deformed animals in the litter. As noted previously, such FF's always have extremely high titers and maintain them throughout life (Toolan 1968, 1972).

Induction of Specific Tolerance in Hamsters Injected on Day 15 of Gestation

In a number of our experiments we found that pregnant hamsters injected with a parvovirus (either H-1 or H-3) on day 15 of gestation often showed no HAI antibody titer when bled 3 weeks after delivery, even though mothers and litters injected on other gestation days of the same experiment had good HAI responses; litters of day-15-injected mothers were also negative (Table 1). In general, such a lack of titer in mothers injected on day 15 occurred when the quantity of virus given was low (2×10^4 PFU or less); approximately 75% of the day-15 mothers injected with such low doses were negative. Yet occasionally, as in the experiments of Table 1, negative day-15 mothers were seen when the dose was fairly high (in this case 10^5 PFU). As noted previously, lack of response on the part of the day-15-injected mother, without apparent infection of her

TABLE 1
Data on Pregnant Hamsters Injected with H-1 During Last Three Days of Gestation

| | | Litter | | HAI titers[b] | | | |
| | | | | mother | | litter | |
Mother	Day injected[a]	no. of survivors	no. deformed	H-1	H-3	H-1	H-3
1	3	2	2	320	< 10	2560	< 10
2	3	6	6	160	< 10	2560	< 10
3	2	dead[c]		nd[d]	nd	—	—
4	2	dead[c]		nd	nd	—	—
5	2	5	5	1280	< 10	2560	< 10
6	1	8	0	< 10	< 10	< 10	< 10
7	1	9	0	< 10	< 10	< 10	< 10

[a] Days before delivery—mothers injected subcutaneously with 10^5 PFU.
[b] Determined at 21 days after delivery as described in "Methods."
[c] Babies all dead by 6 days after birth.
[d] nd: Not determined.

litter, is probably not due to insufficient time for placental passage of virus, since an injection of 10^6 PFU can produce death and/or deformity in a litter even if the mother is injected only 12 hr before delivery.

When these negative mothers and their litters were reinjected shortly after their first bleeding with the same parvovirus originally given to the mother, the babies readily developed antibodies but the mothers often failed to do so (Table 2). Even a third injection with a combination of H-1 and H-3 viruses can fail to produce an HAI response in the mother to the virus originally received, although she will then produce antibodies to the previously uninjected parvovirus. As shown in Table 2, the tolerance seen in the mothers was quite specific, since it could not be abrogated even with a related virus. An extreme example of such specificity maintained through continued reinjection and bleeding is seen in Table 3, where a mother first injected on gestation day 15 with H-3 parvovirus survived five virus reinjections, seven bleedings, and another pregnancy and yet remained tolerant to H-3, the virus she originally received; she was, however, readily susceptible to H-1.

High Level of Antibody in Deformed (FF) Hamsters

Hamsters deformed by infection with H-1 or H-3 parvovirus either in utero or at birth have extremely high levels of antibody to the virus injected (Toolan 1968, 1972). It is not unusual for them to have a titer of

TABLE 2
Induction of Tolerance in Hamster Mothers After Injection Just Prior to Delivery

Mother	Day[a] injected	Virus	Litter		HAI titer[b]				HAI titer[c]				HAI titer[d]			
			no. of survivors	no. deformed	mother		litter		mother		litter		mother		litter	
					H-1	H-3	H-1	H-3	H-1	H-3	H-1	H-3	H-1	H-3	H-1	H-3
1	2	H-1	3	2	320	< 10	2560	< 10	nd[e]	nd	nd	nd	nd	nd	nd	nd
2	2	H-1	4	3	160	< 10	2560	< 10	nd	nd	nd	nd	nd	nd	nd	nd
3	2	H-1	1	1	160	< 10	2560	< 10	nd	nd	nd	nd	nd	nd	nd	nd
4	2	H-1	9	9	80	< 10	2560	< 10	nd	nd	nd	nd	nd	nd	nd	nd
5	1	H-1	7	0	< 10	< 10	< 10	< 10	< 10	< 10	640	< 10	< 10	160	1280	320
6	1	H-1	8	0	< 10	< 10	< 10	< 10	< 10	< 10	1280	< 10	< 10	320	1280	640
7	1	H-3	6	0	< 10	< 10	< 10	< 10	< 10	< 10	< 10	640	640	< 10	640	640
8	1	H-3	7	0	< 10	< 10	< 10	< 10	< 10	< 10	< 10	320	320	< 10	640	320

All virus inoculations were 2 × 10⁴ PFU, injected subcutaneously.

[a] Days before delivery.
[b] Titers at 21 days after delivery.
[c] Reinjected with same virus on 23rd day after delivery and titers determined on 37th day after delivery.
[d] Injected with H-1 and H-3 on 39th day after delivery and titers determined on 53rd day after delivery.
[e] nd: Not determined.

TABLE 3

Permanent and Specific Tolerance Induced in a Hamster Mother Following H-3 Injection One Day Before Delivery

Day after delivery of first litter	Virus inoculated	mother		first litter		second litter	
		H-1	H-3	H-1	H-3	H-1	H-3
−1	H-3	—	—	—	—		
21	—	< 10	< 10	< 10	< 10		
24	H-3[a]	—	—	—	—		
40	—	< 10	< 10	< 10	2560		
42	H-1[b]	—	—				
56	—	640	< 10				
61	H-3[c]	—	—				
76	—	640	< 10				
84	—	640	< 10			< 10	< 10
92	H-3	—	—			—	—
104	—	2560	< 10			< 10	640
113	H-1 + H-3	—	—			—	—
133	—	2560	< 10			320	2560

All virus inoculations were 2×10^4 PFU, given subcutaneously.
[a] Mother and first litter reinjected.
[b] Mother only injected.
[c] Mother reinjected while pregnant, 2 days before delivery.

1×10^5 or more, which is maintained from the time of the single natal injection until their deaths. Such animals, which are often quite small, are fertile and when crossbred with one another produce offspring of normal size and appearance. However, for the first 5–6 weeks of life they do have an antibody titer to the particular parvovirus with which the mother was injected. Since this antibody is always at a much lower level than that of the mother and disappears shortly after weaning, it is presumed to be passive in nature.

Injection of Babies of FF Parents with H-1

In order to determine if the high levels of antibody in FF mothers might have an effect on the susceptibility of their young to parvovirus infection, we mated five pairs of deformed hamsters that had received H-1 virus at birth; all their litters were born within the same week. Five litters of normal hamsters were delivered during the same period. Three of the normal litters and three litters of the FF mothers were injected shortly after birth with 1–3 PFU of H-1 per animal, an amount of virus known to produce the usual deformity in approximately 50% of the litter (H. Toolan, unpublished). The other two litters in each group were sham-injected. When the babies were 3 weeks old, the mothers and babies were bled as usual and the anti-H-1 titers of their sera were determined. Results of these tests are given in Table 4. It is apparent that babies of the FF parents had the same antibody titer at 3 weeks whether or not they had been injected natally with the parvovirus—i.e., they apparently had passive antibodies only. This was borne out by the fact that babies from litters 4 and 5 lost their titers by 6 weeks of age. Surviving babies of litter 3 retained a low antibody level (the only instance we know of in our experience with over 5000 FF's where the antibody titer was below 1:5000). On the other hand, babies of litters 6, 7, and 8 developed the high antibody titers invariably seen in hamster babies deformed by a neonatal injection of H-1.

DISCUSSION

Effect of Day of Gestation on Course of H-1 Infection in Pregnant Hamsters

Our results indicate that the embryonic hamster is sensitive to in utero H-1 infection *throughout* the gestation period, even if the mother is infected in the 5 days prior to implantation, as shown by our tissue-culture tests and FA studies. However, with the possible suggestion of some mortality occurring in litters from mothers injected on day 3, no visible effects of the maternal infection were seen until deformities were observed in litters of mothers injected on day 6. The mortality in litters of mothers injected from days 7 to 15 produced a definite bimodal curve (Fig. 1). The width and maximum height of the first peak have varied slightly from experiment to experiment, but after day 10 we have always seen a considerable drop at day 11, which may extend to day 12 as shown. The most striking aspect of the whole curve is the sharp rise from a low at days 11 and 12 to a peak at day 13 or occasionally day 14, usually higher than any seen previously. There is then a considerable drop at or just after day 14 and finally a fall to zero level at day 15. It might be argued that the virus does not have time to traverse the placenta if injected on gestation day 15. However, if a large quantity of virus is given on this day, as high as 100% FF's can be obtained (H. Toolan, unpublished results).

One of the first questions that might arise would be why we have observed a bimodal curve when other investigators have described a single curve with a peak at day 8 and relatively little mortality before day 6 or after day 10 and have thereby deduced that the developmental stage of

TABLE 4
Effect of High Maternal Antibody on Response of Progeny to Neonatal H-1 Infection

Litter	Parents[a]	No. born	Treatment after birth[b]	Survivors at 21 days		HAI titer at 21 days[c]	
				total no.	no. deformed	mother	litter
1	FF × FF	4	H-1	1	1	163,840	640
2	FF × FF	5	H-1	0	0	327,680	—
3	FF × FF	4	H-1	3	3	40,960	320
4	FF × FF	6	sham	6	0	81,920	320
5	FF × FF	5	sham	5	0	40,960	640
6	N × N	8	H-1	6	5	< 10	40,960
7	N × N	8	H-1	6	2	< 10	20,480
8	N × N	6	H-1	6	3	40	81,920
9	N × N	8	sham	8	0	< 10	< 10
10	N × N	7	sham	7	0	< 10	< 10

[a] FF: "Funny face" mongoloid-type deformed hamster; N: normal hamster.
[b] Litters receiving H-1 were injected with 1–3 PFU/individual at birth. Sham injections were carried out as controls.
[c] All sera were tested for antibodies to H-3 in addition to being proved uniformly negative to this agent.

the embryo on day 8 causes it to be especially susceptible to virus infection at this time. We feel that there are two reasons for this: the amount of virus given and the frequency of testing. We have learned that the amount of virus given to the pregnant hamster is of the utmost importance. If it is too great, the embryos will die in utero or at birth and all details will be obscured by the overwhelming mortality. If the dose is too light, there may be no visible results. It is also necessary to test the pregnant hamster at least once and preferably the morning and evening of each day of pregnancy; otherwise, important peaks of morbidity and mortality may be missed since the hamster has so short a gestation period (16 days). The sharp peak at day 13 is an excellent example.

Possible Hormonal Effects

If the hamster is susceptible to H-1 infection throughout gestation, why should there be a distinct bimodal mortality pattern depending on the day of maternal infection? Differences in maternal antibody levels do not appear to be the answer to this question, although such antibodies may have been produced quickly enough (Toolan 1965) to offer some protection to the early embryos which we have learned contain H-1 virus. We have therefore looked for an explanation of this phenomenon not entirely concerned with susceptible developmental stages of the embryo, however important this factor might be. In examining the physiological changes of pregnancy as they occur in the hamster, we were impressed by the alterations in progesterone levels of the gravid hamster from coitus to parturition. Figure 2 shows hamster blood progesterone concentrations throughout gestation and early lactation according to Leavitt and Blaha (1970). As is evident, this is also essentially a bimodal curve, with the suggestion of a third peak shortly after coitus (day 2). The progesterone reaches its first major peak between days 4 and 9, falls at day 10, rises to a second higher and sharper peak on day 14, and drops precipitously thereafter to the day of delivery. The similarity to our mortality curve is obvious. A figure from another study of progesterone levels in hamster plasma during pregnancy as prepared by Lukaszewska and Greenwald (1970) is even more striking in that it shows the drop from day 9 to day 10, a flat low period from days 10 to 11, and again a sharp rise to day 14 and a drastic fall on day 15. It may be pertinent that the period around day 11, when our mortality data and the progesterone level are both relatively low, coincides with a sharp increase in the peripheral plasma level of estrone in the pregnant hamster (Baranczuk and Greenwald 1974). This is noteworthy, since Nicol et al. (1964) have called estrogens "the natural stimulant of body defense," as they strongly promote the phagocytic activity of the reticuloendothelial system (RES) of an infected animal; this is in contrast with progesterone, which ranks close to prednisone and

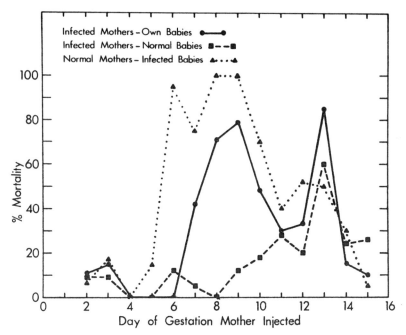

FIGURE 2

Blood progesterone concentrations (in μg/100 ml) during hamster pregnancy and early lactation. (Reprinted, with permission, from Leavitt and
Blaha 1970.)

cortisone as either having no effect on the RES or acting as a depressant.
In figures shown by these authors, the phagocytic activity of the RES
in the rat during pregnancy is also high during the implantation period
(days 1–6); peaks of RES activity were related by Nicol et al. to increments
in the urinary estrogen levels. Abel et al. (1975) have reported that
progesterone depresses macrophage activity both in vivo and in vitro. It
would thus appear that progesterone and estrogens are antagonistic in
their effect on the RES, and that the high levels of progesterone seen at
days 7–10 and at about day 13 of pregnancy in the hamster could hamper
seriously any maternal RES response to an infection.

Possible T-Cell Involvement

Other reports have noted an immunosuppressive effect of progesterone
as related to foreign skin or cell transplants (Andreson and Monroe 1962;
Turcotte et al. 1968; Watnick and Russo 1968; Moriyama and Sugawa
1972). Thymic cells (T lymphocytes) would therefore be involved (Nossal
and Ada 1971). Studies by several authors (Purtilo et al. 1972; Leiken 1972)
have shown that during pregnancy there is a depression of thymus-

dependent lymphocytic proliferation as measured after stimulation with T-cell mitogens; this effect was seen only if pregnancy serum was present, and was therefore attributed to a factor in such serum. In a recent paper, Siitari et al. (1977) propose that progesterone is this factor and that it "may be essential for maintenance of mammalian pregnancy by virtue of its ability to inhibit T-lymphocyte cell mediated responses involved in tissue rejection." An interesting note on the relationship of progesterone and the thymus has been reported by Weinstein et al. (1977), who have observed an enzyme to be present in the thymus that can metabolize progesterone; the authors have speculated that such an enzyme could be important as part of the defense mechanism mediated by the thymus if the progesterone level of the host is too high.

One must conclude, on the basis of the experiments cited, that progesterone can affect the immune response by immunosuppression of T lymphocytes and possibly the RES as well; in the latter case, the effect may be direct or it may be an indirect result of antagonism to the RES stimulant, estrogen.

It is even possible that parvoviruses per se may affect the T-cell response. A paper very pertinent to our own interests is a report by Schultz et al. (1976) noting that adult cats injected with the parvovirus of feline panleukopenia (FPLV) show a decreased lymphocyte response to the T-cell mitogens phytohemagglutinin and conconavalin A but no significant decrease in response to pokeweed mitogen, which stimulates both T and B cells. Neither primary nor secondary serum antibody responses were affected by the FPLV infection. They found, however, that all embryos infected at 35 days from one cat (gestation is approximately 63 days) and one embryo infected at 45 days from another animal showed a significant delay in rejection of skin allografts after birth which was not seen in control newborn kittens. Schultz et al. concluded that "FPLV has its primary immunosuppressive effect on T-cell activity."

How then do we view the role of a mother infected with H-1 or H-3 parvovirus in relation to infection of her embryos? It has long seemed logical to us that the exquisite sensitivity of the newborn hamster in vivo and of all embryonic and newborn hamster cells in vitro would argue against concluding that the developmental stage of the embryo is alone responsible for the fetal mortality curve. Our finding that cells of embryos of mothers injected in the period prior to implantation contain considerable virus shows that even the earliest embryos are susceptible, whether they become infected immediately or through the viremia of the mother after implantation occurs. As we have noted, such early preimplantation stages may be affected by a maternal parvovirus antibody response, which we have found is already present in a measurable quantity 4 days after injection (Toolan 1965). However, the presence of antibody does not seem to be a primary factor in postimplantation embryos.

The remarkable coincidence of maternal progesterone levels and the fetal mortality curves suggests a very real role of this hormone in fetal mortality due to parvovirus infection, especially since it is becoming more and more evident how effective progesterone is as a T-cell suppressor.

Induction of Maternal Tolerance

The type of tolerance seen in the day-15-injected mothers is unique. It is produced in the intact adult by a particulate rather than a soluble antigen, and has not to date been abrogated by injection of a related virus agent. At this time we can only speculate on the nature of the mechanism of this tolerance. For instance, it is possible that T cells capable of responding to virus in the pregnant animal were permanently modified in some manner during pregnancy by a hormonal event occurring just prior to delivery.

Effect of High Maternal Antibody on Resistance of Offspring to Parvovirus Infection

It seems apparent that the passively derived antibodies of litters 1–3 kept the babies from forming their own antibodies, with deleterious results, since only 4 of 13 survived while 18 of 22 progeny from normal mothers lived. (The χ^2 value for survival, corrected for continuity, was 7.065 in this experiment and 5.047 in an earlier and similar test, so the null hypothesis is very unlikely, as $P \ll 0.01$.) We can only conclude that in these experiments the high levels of maternal antibodies were generally disastrous rather than protective for nursing babies infected at birth with the agent to which the mother had shown such great reaction. The implications for human infections acquired during vaginal passage or at birth in babies born to mothers with high antibody titers to, e.g., herpesvirus or cytomegalovirus should possibly be considered.

In conclusion, we suggest the following on the basis of our experiments using the parvoviruses H-1 and H-3 to explore the maternal-fetal-newborn immunological relationship subsequent to maternal infection: (1) The status of maternal hormones (probably the progesterone-estrogen levels in particular) at time of infection is at least as important as the stage of fetal development. (2) The mother herself may be affected by infection at certain stages of pregnancy and develop a permanent tolerance to the agent injected. (3) High levels of FF maternal antibody to a parvovirus passively transferred to her offspring can negate subsequent immunological response of progeny to the same agent.

ACKNOWLEDGMENTS

We thank Bradley Davis for excellent technical assistance, Dr. K. A. O. Ellem for statistical analysis of Table 4, M. Sue Hopkins for the PFU determinations, R. Costantino for making the charts and photography, and V. Haas and J. Pratt for typing the manuscript.

This work was supported by U.S. Public Health Service Grants CA07826-12 from the National Cancer Institute and DE04512-01 from the National Institute of Dental Research.

REFERENCES

Abel, J. H., H. Komnick, H. Hahn, B. Wellek, W. Stockem, and C. Torbit. 1975. Progesterone antagonism of red cell engulfment by peritoneal exudate macrophages in vitro. *J. Cell Biol.* **67**:1a.

Andreson, R. H. and C. W. Monroe. 1962. Experimental study of the behavior of adult human skin homografts during pregnancy. *Am. J. Obstet. Gynecol.* **84**:1096.

Baranczuk, R. and G. S. Greenwald. 1974. Plasma levels of oestrogen and progesterone in pregnant and lactating hamsters. *J. Endocrinol.* **63**:125.

Ferm, V. H. and L. Kilham. 1964. Congenital anomalies induced in hamster embryos with H-1 virus. *Science* **145**:510.

————. 1965. Histopathologic basis of the teratogenic effects of H-1 virus on hamster embryos. *J. Embryol. Exp. Morphol.* **13**:151.

Leavitt, W. W. and J. C. Blaha. 1970. Circulating progesterone levels in the golden hamster during the estrus cycle, pregnancy and lactation. *Biol. Reprod.* **3**:353.

Ledinko, N. 1967. Plaque assay of the effects of cytosine arabinoside and 5-iodo-2'-deoxyuridine on the synthesis of H-1 virus particles. *Nature* **214**:1346.

Leiken, S. 1972. The immunosuppressive effects of maternal plasma. In *Proceedings of the Sixth Leukocyte-Culture Conference* (ed. M. R. Schwartz), p. 725. Academic, New York.

Lukaszewska, J. H. and G. S. Greenwald. 1970. Progesterone levels in the cyclic and pregnant hamster. *Endocrinology* **86**:1.

Moriyama, I. and T. Sugawa. 1972. Progesterone facilitates implantation of xenogenic cultured cells in hamster uterus. *Nat. New Biol.* **236**:150.

Nicol, T., D. Bilbey, L. Charles, J. Cordingley, and B. Vernon-Roberts. 1964. Oestrogen: The natural stimulant of body defense. *J. Endocrinol.* **30**:277.

Nossal, G. J. and G. L. Ada. 1971. *Antigens, lymphoid cells and the immune response.* Academic, New York.

Purtilo, D. T., H. M. Hallgren, and E. J. Yunis. 1972. Depressed maternal lymphocyte response to phytohaemagglutinin in human pregnancy. *Lancet* **i**:769.

Schultz, R. D., H. Mendel, and F. W. Scott. 1976. Effect of feline panleukopenia virus infection on development of humoral and cellular immunity. *Cornell Vet.* **66**:324.

Siitari, P. K., F. Febres, L. E. Clemens, R. J. Chang, B. Gondos, and D. Stites. 1977. Progesterone and maintenance of pregnancy: Is progesterone nature's immunosuppressant? *Ann. N. Y. Acad. Sci.* (in press).

Soike, K. F., M. Iatropoulis, and G. Siegl. 1976. Infection of newborn and fetal hamsters induced by inoculation of LuIII parvovirus. *Arch. Virol.* **51**:235.

Tanigaki, N., Y. Yagi, and D. Pressman. 1967. Application of the paired label radioantibody technique to tissue sections and cell smears. *J. Immunol.* **98**:274.

Toolan, H. W. 1960. Experimental production of mongoloid hamsters. *Science* **131**:1446.

―――. 1965. H-1 virus viremia in the adult hamster. *Proc. Soc. Exp. Biol. Med.* **119**:715.

―――. 1968. The picodna viruses; H, RV and AAV. *Int. Rev. Exp. Pathol.* **6**:135.

―――. 1972. The parvoviruses. *Prog. Exp. Tumor Res.* **16**:410.

Turcotte, J. G., R. F. Haines, G. Bordy, T. Meyer, and S. Schwartz. 1968. Immunosuppression with medroprogesterone acetate. *Transplantation* **6**:248.

Watnick, A. S. and R. A. Russo. 1968. Survival of skin homografts in uteri of pregnant and progesterone-estrogen treated rats. *Proc. Soc. Exp. Biol. Med.* **128**:1.

Weinstein, Y., H. R. Lindner, and B. Eckstein. 1977. Thymus metabolizes progesterone—Possible enzymatic marker for T lymphocytes. *Nature* **266**:632.

PARVOVIRUS DNA STRUCTURE

Terminal Structure of Adeno-Associated-Virus DNA

KENNETH I. BERNS
WILLIAM W. HAUSWIRTH

Department of Immunology and Medical Microbiology
College of Medicine
University of Florida
Gainesville, Florida 32610

KENNETH H. FIFE
ILENE S. SPEAR

Department of Microbiology
School of Medicine
Johns Hopkins University
Baltimore, Maryland 21205

The adeno-associated-virus (AAV) genome is a linear single-stranded DNA molecule with molecular weight 1.4×10^6 (Gerry et al. 1973). The existence of equal numbers of plus and minus strands encapsidated in separate particles (Mayor et al. 1969; Rose et al. 1969; Berns and Rose 1970; Berns and Adler 1972) has facilitated studies of the fine structure of the DNA. Such studies have revealed several interesting properties associated with the termini of the DNA. The DNA has properties consistent with both a natural terminal-nucleotide-sequence repetition (Gerry et al. 1973) and an inverted terminal-nucleotide-sequence repetition (Koczot et al. 1973; Berns and Kelly 1974). We have suggested that a terminal nucleotide sequence which is a palindrome (e.g., 122′1′-----122′1′, where 2′ and 1′ are sequences complementary to 2 and 1, respectively) would have both of these properties (Gerry et al. 1973; Berns and Kelly 1974). From these studies, the length of the terminal repetition was estimated to represent 1–2.5% of the genome. When the plus and minus strands of AAV DNA anneal to form a double helix, two distinct types of terminal structures are formed. In one case the termini appear to have normal duplex structure; molecules of this type remain as linear duplex monomers (m.w. 2.8×10^6). However, a second type of molecule may be formed with terminal regions in a single-stranded state available for further base pairing. These duplex molecules may then form circular duplex monomers or linear duplex dimers (m.w. 5.6×10^6) (Gerry et al. 1973; Koczot et al. 1973).

In this paper we present positive results of experiments testing the existence of a terminal nucleotide sequence with the properties of a palindrome and a tentative nucleotide sequence for large portions of the terminal repetition. In addition, we present a model to explain the several types of secondary structure observed at the termini of double-stranded molecules.

MATERIALS AND METHODS

CELLS AND VIRUSES. AAV2(H) (Hoggan et al. 1966) was grown on KB cells in suspension culture with adenovirus-2 helper as described previously (Berns et al. 1975).

VIRUS AND DNA PURIFICATION. Virus was purified by banding in CsCl after lysis of infected cells with trypsin and deoxycholate as described previously (Berns and Rose 1970). DNA labeled with ^{3}H or ^{32}P was purified by sedimentation through alkaline sucrose, and the resulting single strands were annealed as described elsewhere (Berns and Rose 1970; Berns et al. 1975).

ENZYMES. HpaII was purchased from Bethesda Research Labs. Some HaeIII was a gift of D. Brown and some was purchased from Bethesda Research Labs. BamHI was a gift of D. Shortle. Bacterial alkaline phosphatase was purchased from Worthington Biochemicals. T4 polynucleotide kinase was purified according to the method of Richardson (1965) or purchased from P-L Biochemicals. T4 DNA polymerase was provided by P. Englund. S1 nuclease was a gift from D. Ward.

GEL ELECTROPHORESIS. Electrophoresis on 1.4% agarose-6% or 8% polyacrylamide gels was as described previously (Berns et al. 1975; Spear 1977). High-resolution gels were composed of 20% acrylamide, 0.67% bisacrylamide, 7 M urea and were run in a buffer of 50 mM Tris-borate (pH 8.3), 1 mM EDTA at 15 V/cm (Gilbert et al. 1976; Maxam and Gilbert 1977).

TERMINAL LABELING. The 5' ends of the DNA molecules were labeled using polynucleotide kinase in the presence of $[\gamma\text{-}^{32}\text{P}]$ATP [synthesized according to the method of Glynn and Chappell (1964) as modified by Maxam and Gilbert (1977)] after removal of the terminal phosphate with bacterial alkaline phosphatase (Glynn and Chappell 1964).

For 3'-terminal labeling, 2–5 μg of linear duplex AAV DNA was incubated in 0.05 ml of reaction mixture containing 67 mM Tris-HCl (pH 8.1), 6 mM MCl$_2$, 6 mM 2-mercaptoethanol, 0.1 mM a-^{32}P-labeled deoxynucleoside triphosphate [some synthesized according to the method of Symons (1974) and some purchased from New England Nuclear], and 25 units of T4 DNA polymerase at 11°C for 15–30 min (Englund 1972). The reaction was terminated by the addition of EDTA to 0.05 M, and the reaction mixture was dialyzed extensively to remove free triphosphates.

S1-NUCLEASE DIGESTIONS. Terminal-labeled *Hin*dIII fragments were denatured by boiling for 10 min and then quick chilling. They were then digested in a reaction mixture (0.05 ml) containing 0.3 M NaCl, 0.002 M ZnCl$_2$, 0.03 sodium acetate (pH 4.5), 2 μg denatured salmon-sperm DNA, 2 μg sonicated calf-thymus DNA, and 10 units of S1 at 37°C (Ando 1966). Digestion was quantitated by taking samples, precipitating with cold 12% perchloric acid, and counting the acid-soluble and acid-precipitable material in a Beckman LS-230 liquid scintillation counter with Formula 950A (New England Nuclear) scintillation fluid.

RESTRICTION-ENZYME DIGESTIONS. Digestions with *Hpa*II and *Bam*HI were as described previously (Carter et al. 1976; Spear 1977).

CHEMICAL DEGRADATION OF TERMINAL FRAGMENTS. Terminal fragments containing ^{32}P-label at only one 5' terminus were subjected to partial base-specific cleavage according to the method of Maxam and Gilbert (1977).

SUCROSE GRADIENTS. Linear neutral sucrose gradients (10–30%) containing 0.05 M EDTA and 0.5 M Tris-HCl (pH 7.9) in 1 M NaCl were spun in the Beckman L2-65B ultracentrifuge at 36,000 rpm at 20°C for 18 hr. Fractions were collected through a hole punctured in the bottom of the tube.

RESULTS

Formation of Terminal Hairpin Structure

We have hypothesized that the terminal nucleotide sequence of AAV DNA is a palindrome. Under appropriate conditions this type of sequence ought to be able to fold back and self-anneal, creating a terminal hairpin structure resistant to a single-strand-specific DNase. To test this possibility, duplex AAV DNA labeled internally with [^{3}H]thymidine and at the 3' termini with ^{32}P was digested with *Bam*HI[1] (one cut at 0.22 on the genome map; see Carter et al. 1976). The two *Bam* fragments, repre-

---

[1] See Appendix to this volume for standardized nomenclature of restriction-enzyme fragments.

senting the right and left ends of the duplex molecule, were separated on a sucrose gradient. Each fragment was then heat-denatured, quick-chilled, and digested with S1 single-strand-specific nuclease. The internal ^{3}H-label became acid-soluble much more rapidly than the 3'-terminal [32]P (Fig. 1). We conclude from this experiment that the 3' termini of both plus and minus strands can loop back to base-pair stably with an internal sequence, as would be expected for a terminal palindrome.

Similar experiments for the 5' termini gave comparable results. The maximum size of each terminal hairpin structure was estimated to represent 4% of the genome (Fife et al. 1977).

Structural Heterogeneity at the Termini of Duplex AAV-DNA Molecules

Analyses of the fragments produced by digestion of AAV DNA with bacterial restriction endonucleases have revealed multiple forms of terminal fragments from either end of the molecule. The various forms are readily separable by polyacrylamide gel electrophoresis. For those restriction endonucleases which make a terminal cut outside the terminal repetition but within 25% of the end of the molecule, two classes of

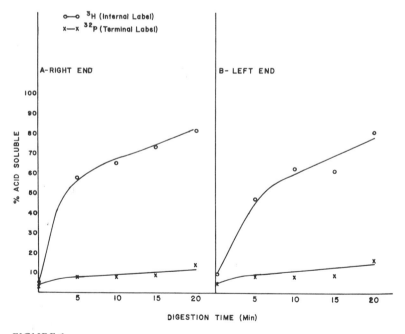

FIGURE 1
S1-nuclease digestion of denatured, 3'-terminal-labeled *Hin*dIII fragments. (See text for details.)

terminal fragments are resolved on acrylamide gels. Terminal fragments of this type (two species) include *Bam* B_1 and B_2 and *Hae*II C_1 and C_2 from the left end and *Hin*dII + III C_1 and C_2 from the right end. An example of the separation of *Bam* B_1 and B_2 is shown in Figure 2.

The resolution on gels of these terminal fragments into two classes could reflect differences in conformation, molecular weight, or both. To test this, *Bam* B_1 and B_2 were separated and then denatured and re-annealed. From this treatment of either B_1 or B_2 both forms were recovered, i.e., B_1 yielded both B_1 and B_2 and vice versa. We believe these data to be most consistent with the interpretation that *Bam* B_1 and B_2 represent differences in conformation. If B_1 and B_2 represented fragments of different molecular weight, more than two species should have resulted from denaturation and reannealing.

Digestion of AAV DNA labeled at the 5' termini with ^{32}P using a restriction endonuclease which cuts within the terminal repetition (thus

FIGURE 2
Terminal-labeled AAV DNA was digested with *Bam*HI and the products were fractionated on a 6% polyacrylamide gel. The two species of B fragments were recovered and were then denatured and reannealed and again run on a 6% polyacrylamide gel. Lane 1: B_1 denatured and reannealed. Lane 2: untreated B_1. Lane 3: mixture of B_1 and B_2. Lane 4: B_2 denatured and reannealed. Lane 5: B_2 untreated.

producing similar terminal fragments from both ends of the molecule) yields more than two classes of terminal fragments. For instance, *Hpa*II yields at least six types of terminal fragments (ters); some are heterogeneous and some are in higher yield, but all are reproducible (I. S. Spear, K. H. Fife, W. W. Hauswirth, and K. I. Berns, unpublished). We have determined from which of the two classes of terminal fragments produced by *Bam*HI, *Hae*II, or *Hin*dII + III (terminal cut outside the terminal repetition) each of the *Hpa*II ters originates. This was done by isolating *Hin*dII + III C_1 and C_2 (right end of the molecule), labeling them at the 5' termini with ^{32}P, and digesting with *Hpa*II (Fig. 3). The inboard terminus of both *Hin*dII + III C_1 and C_2 is represented by L'. Ters 1, 3, 4, and 5 come from C_1. Only ter 6 is produced from C_2, but ter 6 clearly has two components, which differ in length by 4 nucleotides (data not shown). Ters 1 and 2 extend beyond the terminal repetition, so that ter 2 is produced only from the left end of the molecule (I. S. Spear, K. H. Fife, W. W. Hauswirth, and K. I. Berns, unpublished). Some ter 6 is observed with *Hin*dII + III C_1, but it is considered to represent contamination of C_1 by C_2 in the original preparative gel (i.e., the trailing edge of C_2 overlaps C_1). Analysis of single strands produced by denaturation showed that ters 6a and 6b (46 and 42 nucleotides long, respectively) comprise a single species. However, multiple sizes of single strands (46, 55, 66–70, and longer) are produced by denaturation of ters 1–5, compatible with single-strand nicking of *Hin*dII + III C_1 by *Hpa*II (I. S. Spear, K. H. Fife, W. W. Hauswirth, and K. I. Berns, unpublished). These and other data (not shown) concerning dependence of mobility on gel concentration led us to conclude that half the termini of duplex AAV DNA have a normal double-helical structure, but that the other 50% have an aberrant secondary structure with possible hairpin loops and non-base-paired regions.

Products of Partial HpaII Digestion of Bam B_2

Smith and Birnstiel (1976) have described an alternative method for mapping restriction-enzyme cleavage sites. The labeled products of a partial digestion of a large DNA fragment labeled at one 5' terminus form a series of bands on a gel differing in apparent molecular weight by the distance between potential cleavage sites. From a partial digestion of *Bam* B_2 (normal secondary structure) with *Hpa*II (Fig. 4), *Hpa*II cleavage sites at 42, 46, 56, 66, 76, and 80 nucleotides from the original terminus of the molecules were determined. These data were paradoxical in that complete digestion reveals a fragment 24 nucleotides long (*Hpa*II T) contiguous with ter 6. The origin of *Hpa*II T and the resolution of the apparent paradox between the results of partial and complete *Hpa*II digestion are discussed below.

FIGURE 3

AAV DNA was digested with *Hin*dII + III and then labeled at the 5′ ends with ^{32}P using polynucleotide kinase. Fragments were separated on 6% polyacrylamide gels and recovered. The two *Hin*dII + III C fragments were each recovered and redigested with *Hpa*II. ^{32}P-labeled AAV DNA digested with *Hpa*II was used as marker.

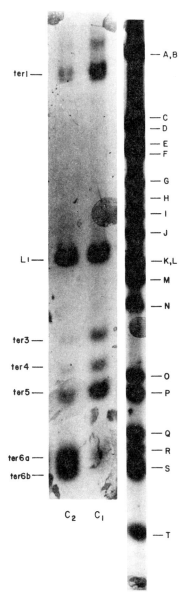

Hpa II
digest of
AAV

FIGURE 4

*Hpa*II partial-digestion products of the termini of AAV DNA. AAV2 DNA was labeled at the 5′ termini with ^{32}P using polynucleotide kinase. The labeled DNA was digested with *Bam*HI, and fragments were recovered from a preparative acrylamide gel and subjected to partial digestion with *Hpa*II. The products were then separated electrophoretically and identified by autoradiography. Only those partial-digestion products containing the original left terminus of the intact DNA contained label and were thus observed. Lane 1: *Hpa*II digestion of intact, uniformly ^{32}P-labeled AAV DNA. Lane 2: *Hpa*II partial digestion of *Bam* B₂. In lane 2 the numbers refer to the differences in base pairs between adjacent partial-digestion products.

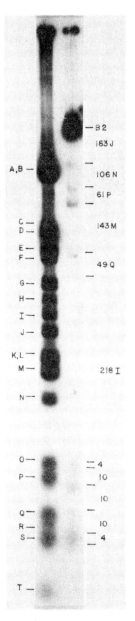

5′ Terminal Nucleotide Sequences of AAV DNA

Because of the evident complexity of the termini of AAV DNA and the evidence that the termini are the sites of the origin and termination of DNA replication (Hauswirth and Berns, this volume), we decided to determine the nucleotide sequence of this region of the DNA. The initial sequencing was done using the method described by Morrison and Murray (1974). These studies revealed that all 5′ sequences of AAV DNA (both plus and minus strands) are identical except for the terminal nucleotide (35% 5′ TTGGCCA, 50% 5′ TGGCCA, 15% 5′ GGCCA) (Fife et al. 1977). More extensive 5′ sequencing has been done using the method of Maxam and Gilbert (1977). These results are shown in Figure 5. Several conclusions can be drawn from this work: (1) the first 42 nucleotides are the same for all 5′ termini of AAV DNA. (2) The inverted terminal repetition is 140–141 nucleotides long—it ends 23 nucleotides in from the outboard ends of HaeIII E and D, and the outboard end of HaeIII E is 120–124 nucleotides from the end of the DNA as determined by the method of Smith and Birnstiel (1976) (K. H. Fife, I. S. Spear, M. C. Jones, W. W. Hauswirth, and K. I. Berns, manuscript in preparation). (3) The last 61 nucleotides of the terminal repetition are the same for all molecules. (4) There is extensive self-complementarity within the terminal repetition. A nucleotide sequence complementary to the first 42 nucleotides exists just outboard from the last 20 nucleotides of the terminal repetition.

Nucleotide-Sequence Homology between AAV DNA and Adenovirus-2 DNA

Previous hybridization assays have failed to reveal any nucleotide-sequence homology between the DNAs of AAV and helper adenovirus

```
                  10              20              30            ↓
5'(T) T G G C C A C T C C C T C T C T G C G C G C T C G C T C G C T C A C T G A G G C C
3'(A) A C C G G T G A G G G A G A G A C G C G C G A G C G A G C G A G T G A C T C C G G

                  50              60        ↓                ↓
5'   G G G C G A C C A A G G T C G C C C G A C G C C C G G - - - - - - C C G G C
3'   C C C G C T G G T T C C A G C G G G C T G C G G G C C - - - - - - G G C C G

      80           90           100           110           120
5'   G G C C T C A G T G A G C G A G C G A G C G C G C A G A G A G G G A G T G G C C A
3'   C C G G A G T C A C T C G C T C G C T C G C G C G T C T C T C C C T C A C C G G T

            130           140
5'   C T C C A T C A C T A G G G G T T C C T
3'   G A G G T A G T G A T C C C C A A G G A
```

FIGURE 5
Nucleotide sequence of the terminal repetition of AAV DNA using the orientation in Fig. 7b. The sequence has the following features: (1) The terminal nucleotide in parentheses (T) represents the 35% of all molecules which have an extra T at the 5′ terminus. (2) Nucleotides 1–41 can fold over to base-pair with nucleotides 80–120. (3) Nucleotides 42–49 are complementary to 52–59. The arrows represent HpaII cleavage sites within this orientation of the terminal repetition.

(Rose et al. 1968; Rose and Koczot 1972). Using the blotting technique of Southern (1975) we have been able to find evidence of such homology between AAV2 and Ad2 DNAs. AAV2 DNA digested with *Hae*II was blotted from an agarose gel onto a nitrocellulose filter. AAV2 DNA and adenovirus DNA were digested using *Hpa*II + *Hae*III or *Hpa*II, respectively. The 5' ends of the fragments were labeled with ^{32}P and then used as a probe to hybridize with filters containing the AAV2 *Hae*II bands. The entire *Hae*II pattern was observed using ^{32}P-labeled AAV2 fragments as a probe (control), but the ^{32}P-labeled Ad2 probe hybridized only to the *Hae*II AAV A and C bands (Fig. 6). We conclude that there is a limited amount of nucleotide-sequence homology in these regions of AAV2 DNA to an as yet unmapped region of adenovirus-2 DNA. It is of interest that *Hae*II A and C are the right and left terminal fragments of AAV2 DNA, respectively.

DISCUSSION

Analysis of the terminal structure of AAV DNA has proved to be complex. Earlier studies had described properties consistent with both a natural and an inverted terminal-nucleotide-sequence repetition. It was suggested that these were the result of a terminal-nucleotide-sequence palindrome. The physical studies and the nucleotide-sequence data reported in this paper support this hypothesis.

More perplexing has been the evidence for heterogeneity in the terminal secondary structure of duplex AAV DNA formed by annealing the complementary plus and minus strands. We have interpreted the data presented herein as follows. (1) Two classes of terminal structure are

FIGURE 6
Hybridization of ^{32}P probe DNA to a nitrocellulose blot of a *Hae*II digest of AAV DNA. (See text for details.) Lane 1: AAV DNA probe. Lanes 2 and 3: Adenovirus-2 DNA probe (each probe was made from a separate plaque-purified stock of adenovirus).

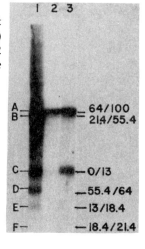

formed, (a) a normal double helix or (b) an aberrant secondary structure. In the case of the aberrant secondary structure, some *Hpa*II sites cannot be cleaved. (2) The two possible secondary structures reflect heterogeneity in the nucleotide sequence within the terminal repetition. Further, the heterogeneity in nucleotide sequence need reflect as few as two possible sequences, because 50% of all annealing events result in a terminus with a normal duplex structure (from the *Bam* B_1 and B_2 interconvertibility data). Two sets of data are consistent with the existence of only two different arrangements of nucleotide sequences in the terminal repetition. The partial-digestion data demonstrate that the greatest distance between adjacent *Hpa*II sites within the terminal repetition is 10 nucleotides. Yet complete *Hpa*II digestion of the terminal repetition produces *Hpa*II T (24 nucleotides), which is contained wholly within the terminal repetition. It could be argued that *Hpa*II cannot produce a fragment smaller than 24 nucleotides, but smaller fragments are apparent on analysis of *Hpa*II digests. A simple way to reconcile the partial-digestion data and the existence of *Hpa*II T is to have two sets of nucleotide sequences, each with its own *Hpa*II cleavage site. Thus, *Hpa*II ter 6a would represent the terminal fragment of one sequence and ter 6b would represent the terminal fragment of the other sequence.

The extent of the region within the terminal repetition in which nucleotide-sequence heterogeneity could exist is limited. From the nucleotide-sequence data we know that the sequences of the outboard 40 nucleotides and the inboard 23 nucleotides of the terminal repetition are unique. In addition, there may be as many as 40 more nucleotides near the inboard end of the terminal repetition which are complementary to the outboard 40 nucleotides and must therefore have their own unique sequence. Thus, only a stretch of about 40 nucleotides within the terminal repetition (total length: 141 nucleotides) may be involved in the proposed sequence heterogeneity. It is interesting that all the *Hpa*II sites within the terminal repetition occur within 38 nucleotides.

We have developed a model which we believe is consistent with all of the above data. We propose that the nucleotide-sequence heterogeneity observed within the terminal repetition results from the equal probability of nucleotides 1→120 occurring in that orientation or in the reverse orientation 120→1, i.e., the terminal 120 nucleotides may be "flip-flopped." This model serves to reconcile the apparently paradoxical results of partial and complete *Hpa*II digestion within the terminal repetition (Fig. 7).

A possible origin of such a "flip-flop" is illustrated in Figure 8. A terminal hairpin structure linked covalently to both a plus and a minus strand may occur within a replicative intermediate at either the origin or the termination of DNA replication. Such structures have been observed in AAV DNA replication in vivo (Straus et al. 1976) and in

FIGURE 7
Interpretation of *Hpa*II
partial-digestion patterns
within the terminal repetition
(Fig. 4) utilizing the "flip-
flop" model.

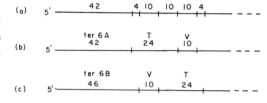

minute virus of mice DNA synthesis in vitro (Tattersall and Ward 1976).
The hairpin structure would be cleaved by a single-strand nick, which
could occur with equal probability between 1′ and X′ or between 1 and X.
A nick between 1 and X would result in the 5′ terminal structure
123452′1′X′, whereas a nick between 1′ and X′ would yield a 3′ terminal
structure 1′2′54321X, which would have a complementary 5′ terminal
structure 125′4′3′2′1′. Thus, effectively, only the region 345 would appear
to have been flipped.

We are currently engaged in further examination of the products of
partial *Hpa*II digestion to test the predictions of the "flip-flop" model.

The finding of a limited amount of nucleotide-sequence homology
between the terminal *Hae*II fragments of AAV DNA and adenovirus DNA
suggests the possibility that these regions are involved in the correction of
AAV defectiveness by adenovirus helper.

FIGURE 8
"Flip-flop" model to explain
nucleotide-sequence heterogeneity within the terminal
repetition of AAV DNA.
A hairpin structure (*a*) produced during DNA repli-
cation is nicked between 1 and
X or 1′ and X′ by a site-specific
endonuclease. A nick be-
tween 1 and X produces a 5′
terminal repetition on the
minus strand 123452′1′ (*b*),
whereas a nick between 1′ and
X′ produces a 3′ terminal
repetition on the plus strand
1′2′54321 (*c*). The com-
plementary sequence to (*c*)
would then be 125′4′3′2′1′- - -.

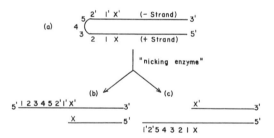

ACKNOWLEDGMENTS

The work reported here was supported by grant 1 P01 CA 16519–02 from the National Cancer Institute, U.S. Public Health Service.

REFERENCES

Ando, T. 1966. A nuclease specific for heat-denatured DNA isolated from a product of *Aspergillus oryzae*. *Biochim. Biophys. Acta* **114**:158.

Berns, K. I. and S. Adler. 1972. Separation of two types of adeno-associated virus particles containing complementary polynucleotide chains. *J. Virol.* **9**:394.

Berns, K. I. and T. J. Kelly, Jr. 1974. Visualization of the inverted terminal repetition in adeno-associated virus DNA. *J. Mol. Biol.* **82**:267.

Berns, K. I. and J. A. Rose. 1970. Evidence for a single-stranded adenovirus-associated virus genome: Isolation and separation of complementary single strands. *J. Virol.* **5**:693.

Berns, K. I., J. Kort, K. H. Fife, E. W. Groggan, and I. Spear. 1975. Study of the fine structure of adeno-associated virus DNA with bacterial restriction endonucleases. *J. Virol.* **16**:712.

Carter, B. J., K. H. Fife, L. M. de la Maza, and K. I. Berns. 1976. Genome localization of adeno-associated virus RNA. *J. Virol.* **19**:1044.

Englund, P. T. 1972. The 3′ terminal sequences of T7 DNA. *J. Mol. Biol.* **66**:209.

Fife, K., K. Murray, and K. Berns. 1977. Structure and nucleotide sequence of the terminal regions of adeno-associated virus DNA. *Virology* **78**:475.

Gerry, H. W., T. J. Kelly, Jr., and K. I. Berns. 1973. Arrangement of nucleotide sequences in adeno-associated virus DNA. *J. Mol. Biol.* **79**:207.

Gilbert, W., A. Maxam, and A. Mirzabekov. 1976. Contacts between *lac* repressor and DNA revealed by methylation. In *Control of ribosome synthesis* (Alfred Benzon Symposium IX) (ed. N. O. Kjelgaard and O. Maaløe), p. 139. Munksgaard, Copenhagen.

Glynn, I. M. and J. B. Chappell. 1964. A simple method for the preparation of ^{32}P-labeled adenosine triphosphate of high specific activity. *Biochem. J.* **90**:147.

Hoggan, M. D., N. R. Blacklow, and W. P. Rowe. 1966. Studies of small DNA virus found in various adenovirus preparations: Physical, biological and immunological characteristics. *Proc. Natl. Acad. Sci.* **55**:1467.

Koczot, F. J., B. J. Carter, C. F. Garon, and J. A. Rose. 1973. Self complementarity of terminal sequences within plus or minus strands of adenovirus-associated virus DNA. *Proc. Natl. Acad. Sci.* **55**:1467.

Maxam, A. M. and W. Gilbert. 1977. A new method for sequencing DNA. *Proc. Natl. Acad. Sci.* **74**:560.

Mayor, H. D., K. Torikai, J. L. Melnick, and M. Mandel. 1969. Plus and minus single-stranded DNA separately encapsidated in adeno-associated satellite virions. *Science* **166**:1280.

Morrison, A. and K. Murray. 1974. The behaviour of oligodeoxy-nucleotides on thin-layer chromatography on polyethyleneimine-cellulose and ion-exchange paper electrophoresis. *Biochem. J.* **141**:321.

Richardson, C. C. 1965. Phosphorylation of nucleic acid by an enzyme from T4 bacteriophage-infected *E. coli*. *Proc. Natl. Acad. Sci.* **54**:158.

Rose, J. A., K. I. Berns, M. D. Hoggan, and F. J. Koczot. 1969. Evidence for a single-stranded adenovirus-associated virus genome: Formation of a DNA density hybrid on release of viral DNA. *Proc. Natl. Acad. Sci.* **64**:863.

Smith, H. O. and M. L. Birnstiel. 1976. A simple method for DNA restriction site mapping. *Nucleic Acids Res.* **3**:2387.

Southern, E. M. 1975. Detection of specific sequences among DNA fragments separated by gel electrophoresis. *J. Mol. Biol.* **98**:503.

Spear, I. S. 1977. "The use of restriction endonucleases in elucidating the fine structure of AAV DNA." Ph. D. thesis, Johns Hopkins University, Baltimore, Maryland.

Straus, S. E., E. D. Sebring, and J. A. Rose. 1976. Concatemers of alternating plus and minus strands are intermediates in adenovirus-associated virus DNA synthesis. *Proc. Natl. Acad. Sci.* **73**:742.

Symons, R. H. 1974. Synthesis of [a-^{32}P]ribo- and deoxyribonucleotide 5' triphosphates. In *Methods in enzymology* (ed. L. Grossman and K. Moldave), vol. XXIX, p. 102. Academic, New York.

Tattersall, P. and D. Ward. 1976. The rolling hairpin: A model for the replication of parvovirus and linear chromosomal DNA. *Nature* **263**:106.

DNA Structure of Incomplete Adeno-Associated-Virus Particles

LUIS M. DE LA MAZA
BARRIE J. CARTER

Laboratory of Experimental Pathology
National Institute of Arthritis, Metabolism, and Digestive Diseases
National Institutes of Health
Bethesda, Maryland 20014

When lysates of cells infected with adeno-associated virus (AAV) and a helper adenovirus are banded to equilibrium in CsCl buoyant density gradients, a number of virus bands can be detected visually (Hoggan 1971). The major and minor infectious AAV components band at densities of 1.41 and 1.45 g/cm³, respectively. In the electron microscope, these virions appear to be "full" particles. Infected-cell lysates also contain a heterogeneous population of "light," noninfectious AAV particles which band at densities of less than 1.41 g/cm³. These light particles appear "empty" in the electron microscope and have been variously reported to contain either no DNA (Johnson et al. 1975) or DNA of less than genome length (Torikai et al. 1970).

We have examined the AAV DNA present in both "full" and "empty" particles. These studies revealed that all density classes of AAV virions appear to contain varying proportions of incomplete genomes. It is not clear whether there are any truly empty virions. Analysis of the physical and biological properties of these incomplete DNA molecules is a useful approach for studying the multiplication mechanism of AAV and the nature of its interaction with its helper virus or host cell.

We summarize here the physical properties of the incomplete AAV genomes. A more detailed description of their physical and biological properties will be reported elsewhere (L. M. de la Maza, F. T. Jay, and B. J. Carter, manuscript in preparation).

MATERIALS AND METHODS

VIRAL DNA. AAV2 was grown in KB3 spinner cells with Ad2 as the helper (Carter et al. 1973), and DNA was prepared as described previously (de la Maza and Carter 1977).

CLEAVAGE WITH RESTRICTION ENZYMES AND ELECTROPHORESIS. Restriction-endonuclease-digestion reactions and electrophoresis of the digested DNA in polyacrylamide gels were carried out as described previously (de la Maza and Carter 1977).

SUCROSE GRADIENTS. Neutral and alkaline sucrose sedimentation have been described previously (Carter and Khoury 1975).

RESULTS

Fractionation of Virus Particles in CsCl Gradients

Virus particles present in lysates of KB cells infected with AAV2 and Ad2 were fractionated by equilibrium density centrifugation in CsCl. The regions containing visible bands at densities of 1.45, 1.41, 1.35, and 1.32 g/cm^3 were individually pooled and centrifuged to equilibrium in CsCl two or three more times. Electron micrographs of negatively stained preparations showed that 95–98% of the AAV particles in the two heaviest bands appeared "full," whereas a large proportion of those present in the 1.35- and 1.32-g/cm^3 bands were "empty." Full adenovirus was found in the 1.35 band, and amorphous aggregates probably corresponding to disrupted adenovirus particles were observed in the 1.32-g/cm^3 band. When the areas between the visible bands were pooled and concentrated from large-volume preparations and then rebanded in CsCl, visible virus was observed, which indicated the presence of "incomplete" particles of different densities spread throughout the gradient.

Fractionation of Incomplete DNA from Purified Virions

DNA was released from virions by sedimentation in alkaline sucrose gradients, then reannealed in formamide and fractionated further in neutral sucrose gradients. When DNA was prepared from AAV virions of different densities, several classes of incomplete molecules were obtained (as summarized in Table 1). Specific classes of incomplete AAV DNA molecules were isolated in the following way: When virions from the

TABLE 1
Sedimentation Properties of Incomplete AAV-DNA Molecules

DNA	Density of virion[a] (g/cm³)	Sedimentation coefficient and molecular weight of isolated DNA[b]			
		alkaline sucrose		neutral sucrose	
		S	m.w. × 10⁻⁶	S	m.w. × 10⁻⁶
Intact	1.41	15.5	1.4	14.5	2.8
Type I	1.41	15.5	1.4	11.5	1.4
Type II	1.41	12.0	0.8	10.0	0.8
Type III (oligomer)	1.35	8.5	0.3	8–15	~1.4
Type III	1.35	8.5	0.3	(1) 8.2	0.6
				(2) 7.0	0.3
Type IV	1.32	5.5	~0.13	5.0	~0.13

[a] The value listed indicates the density in a CsCl buoyant density gradient of the visible band of virus from which the DNA was obtained.

[b] The molecular weights and sedimentation coefficients are computed from sucrose gradient profiles using the molecular-weight markers indicated in Fig. 1. The molecular weights and S values of these markers have been described previously (Carter and Khoury 1975; Carter et al. 1976). Intact DNA is the full-length infectious DNA of AAV2 which is the DNA component of the majority (90%) of the virions banding at a density of 1.4 g/cm³. The numbers in parentheses for the neutral sucrose components of type-III DNA refer to the peaks 1 and 2 indicated in Fig. 1e. The data for DNA types I, II, and III were calculated from Fig. 1. The data for type IV were from gradients to be described elsewhere (L. de la Maza, F. Jay, and B. Carter, manuscript in preparation).

1.41- or the 1.45-g/cm³ band of a CsCl gradient were sedimented in alkaline sucrose, about 90% of the DNA formed a single 15.5S peak and the remainder sedimented as a shoulder at about 12S. This 12S shoulder was pooled, neutralized, and reassociated, and is designated as incomplete DNA type II (see Table 1). The 15.5S AAV DNA component from the alkaline sucrose gradient was reassociated and sedimented in a neutral sucrose gradient to yield a major 14.5S component of intact AAV duplex. In the neutral sucrose gradient there was also a small amount of an 11.5S component, which was collected and designated as incomplete DNA type I (Table 1).

When the virions that banded in CsCl at a density of 1.35 g/cm³ were sedimented in alkaline sucrose, all of the AAV DNA sedimented at 8.5S. This DNA was pooled and reassociated, and was designated as incomplete DNA type III. All of the adenovirus DNA from the 1.35-g/cm³ virus band sedimented to the bottom of the alkaline sucrose gradient. Sedimentation of the reassociated type-III AAV DNA in neutral sucrose resulted in a broad distribution of DNA from 8S to 15.5S. This peak was divided into leading and trailing edges and designated as type-III oligomers and type III, respectively.

Alkaline sedimentation of the AAV virus that had banded at 1.32 g/cm³ in CsCl released a peak of DNA that sedimented at approximately 5S. This DNA component was reassociated and designated as DNA type IV.

It must be noted that the nomenclature of these incomplete AAV DNA preparations as listed in Table 1 is intended solely to facilitate discussion and does not indicate that these are the only size classes of incomplete molecules produced. There is in fact a continuous spectrum of AAV particles banding in a broad density range from 1.45 to 1.30 g/cm³.

Physical Properties of Incomplete AAV DNA

The physical properties and molecular weights of type-I, -II, and -III DNA molecules were examined by sedimentation in alkaline and neutral sucrose gradients (Fig. 1) and are summarized in Table 1. Type-I DNA

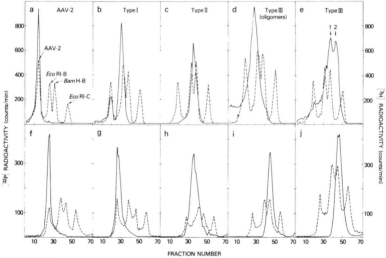

FIGURE 1

³²P-labeled DNA preparations were analyzed on neutral (a–e) or alkaline (f–j) sucrose gradients as described previously (Carter and Khoury 1975). Gradients contained ³²P-labeled AAV DNA (a, f), type-I incomplete DNA (b, g), type-II incomplete DNA (c, h), type-III oligomer fraction (d, i), or type-III incomplete DNA (e, j). All gradients contained as molecular-weight markers purified ³H-labeled duplex monomer AAV2 DNA or restriction fragments EcoRI B (38% of genome), Bam B (22%), or EcoRI C (4.5%) as indicated in (a). Sedimentation is from right to left. ————: ³²P-labeled DNA;————: ³H-labeled DNA.

sedimented in both neutral (Fig. 1*b*) and alkaline (Fig. 1*g*) sucrose gradients as a single component with a molecular weight of 1.4 × 10⁶. (However, type I is contaminated with some intact AAV DNA.) Similarly, the type-II DNA sedimented in both neutral (Fig. 1*c*) and alkaline (Fig. 1*h*) gradients as a single component with an apparent molecular weight of 0.8 × 10⁶. These properties indicate that both type-I and type-II DNA are "snapback" or "hairpin" structures.

The type-III DNA sedimented in alkaline sucrose with a mean molecular weight of 0.3 × 10⁶. In neutral sucrose the type-III DNA yielded linear duplexes of molecular weight 0.6 × 10⁶ (peak 1, Fig. 1*e*), "snapback" duplexes of molecular weight 0.3 × 10⁶ (peak 2, Fig. 1*e*), and also a variety of faster-sedimenting oligomeric structures (Fig. 1*d,e*). It is not clear whether the linear duplexes in peak 1 (Fig. 1*e*) are composed of individual strands which might form "snapback" hairpin structures if the annealing was done at lower concentration. The type-IV AAV-DNA molecules (data not shown) also reassociate to form a variety of structures analogous to those exhibited by type-III DNA.

Other experiments (L. de la Maza, F. Jay, and B. Carter, manuscript in preparation) have confirmed this interpretation of the data in Figure 1. After denaturation and rapid chilling, most DNA molecules of types I and II and about 20% of type III, but none of the complete AAV-DNA molecules, reassociate into structures resistant to digestion by S1 nuclease. The type-III oligomeric molecules appear to be joined largely by the "cohesive ends" of AAV DNA (de la Maza and Carter 1977; Berns, this volume) and can be converted to duplex molecules of molecular weight 0.6 × 10⁶ by melting at 80°C in 0.165 M NaCl (Carter and Khoury 1975). Hybridization experiments and analysis of reassociation kinetics have shown that nearly all of the incomplete AAV DNA is AAV-specific.

Restriction-Endonuclease Cleavage of Incomplete Genomes

The DNA preparations used for the sedimentation analysis (Fig. 1) were cleaved with the restriction endonuclease *Hha*I and electrophoresed in an acrylamide slab gel. The densitometer scans of the gel autoradiogram are shown in Figure 2. Additional *Hha* digests of incomplete DNA preparations are shown in the gel autoradiograms of Figure 3. The sizes of the uncleaved incomplete DNA molecules estimated by gel electrophoresis agree with those derived from neutral sucrose gradients. Cleavage with *Hha* clearly shows that as the sizes of the incomplete genomes decrease there is a progressive enrichment for the terminal regions of the AAV genome and a progressive loss of the internal regions. For example, all the *Hha* fragments from B to F are present in a low molar ratio in type-II DNA and are not seen at all in type III or IV. The fragments H, K*, and K, which map near the ends of the genome and contain the inverted repeated

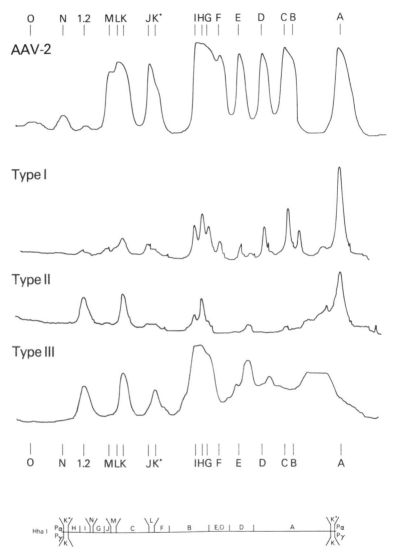

FIGURE 2

Restriction-endonuclease cleavage of incomplete AAV genomes. Samples of the ^{32}P-labeled DNA preparations that were analyzed in the sucrose gradients shown in Fig. 1 were also digested with restriction endonuclease *Hha*I. The digests were electrophoresed in an acrylamide slab gel and the gel was subsequently autoradiographed using X-ray film. The audioradiogram was then scanned in a Beckman densitometer. The peaks corresponding to individual *Hha* fragments are indicated by the letters at top and bottom. The terminal *Hha* fragment P_a, P_γ migrated off the gel. The map of the *Hha* cleavage sites on AAV DNA is shown at the bottom.

INCOMPLETE   H H A ~ I

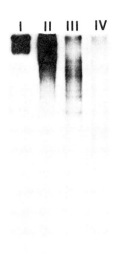

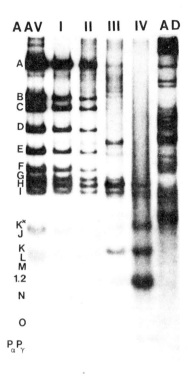

FIGURE 3

Cleavage of incomplete AAV genomes with restriction endonuclease *Hha*I. ^{32}P-labeled DNA was electrophoresed in acrylamide slab gels and auto-radiographed. The tracks on the left gel contain uncleaved, incomplete genomes of types I–IV, as indicated. The predominant component of type-IV DNA migrated at the same rate as the *Hha* J fragment (about 2.5% of the genome), and some additional, more heterogeneous material migrated at the rate of *Hha* fragments D–H (9–4.3% of the genome). The tracks in the right gel contain *Hha* digests of the ^{32}P-labeled DNA preparations indicated. The *Hha* cleavage map of AAV2 DNA is shown at the bottom. The fragment designated 1.2 in the right gel is discussed in the text.

sequences of AAV DNA, are present in all the virions. Thus, type-IV DNA contains little other than the inverted terminal repetitions which comprise 147 nucleotides at each terminus (de la Maza and Carter 1977; Berns, this volume).

The terminal Hha fragments P_α and P_γ are present in all the incomplete AAV-DNA preparations, but their enrichment is significantly less than for H, K*, and K (see, e.g., Fig. 4, track IV, right panel). Concomitant with the lesser enrichment for P_α and P_γ is the appearance in progressively larger molar amounts of a new 1.2 fragment (equivalent in size to 1.2% of the AAV genome). This is seen in both the densitometer scans (Fig. 2) and the autoradiogram (Fig. 3). The 1.2 fragment is not seen in cleavages of complete AAV2 DNA, but in the smallest molecules (such as type IV) this fragment has the highest molar concentration. This suggests that the incomplete DNA, although preferentially enriched for the ends of the AAV genome, may have a rearrangement of the region at or very close to the terminus within the inverted terminal repeat.

Cleavage with a variety of other restriction enzymes (see, e.g., Fig. 4) showed that the enrichment is for left or right terminal regions of the

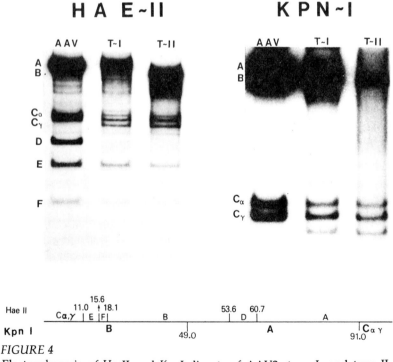

FIGURE 4

Electrophoresis of HaeII and KpnI digests of AAV2, type-I, and type-II DNA. The HaeII and KpnI cleavage maps for AAV2 DNA are shown.

genome. These digestions by enzymes such as *Hae*II or *Kpn*I that do not cleave within the inverted repetition also indicate that there is a rearrangement of the normal termini in the incomplete genomes as indicated for type-I and type-II DNA in Figure 4. In the *Hae*II digest (Fig. 4) the left terminal doublet component C$_a$ is clearly reduced in amount and a new fragment appears which is about 50–100 nucleotides shorter than C$_\gamma$. A similar phenomenon is seen in the right end for *Kpn* C$_a$ and C$_\gamma$. This appears to correspond to the relatively greater loss of the *Hha* P$_a$ and P$_\gamma$ fragments and the appearance of the *Hha* 1.2 fragment (Figs. 2 and 3).

DISCUSSION

The DNA from incomplete AAV particles that arises during the growth of this virus in the presence of helper adenovirus has been analyzed. CsCl density gradients, alkaline and neutral sucrose gradients, and restriction-endonuclease cleavage show that these molecules vary in length and genome-sequence representation. The largest molecules analyzed (type I) have a molecular weight of 1.4×10^6 in both alkaline and neutral sucrose gradients and appear to be hairpin duplexes. Restriction-endonuclease cleavage indicates that both the right and left ends of the molecule are present and that only 10–20% of the internal portion of the genome is missing from this population of molecules. Individual molecules therefore appear to be hairpin duplexes containing either the left or the right terminal 30–50% of the genome. The type-II molecules have a similar structure but contain only 20–30% of the left or the right terminal region. Type-III molecules have a more heterogeneous structural configuration. These molecules contain only the left or right 10–20% of the genome and form either hairpin molecules or non-cross-linked duplexes. Some of these molecules may form oligomeric species held by the "cohesive ends" of the AAV-DNA termini (Gerry et al. 1973; Carter and Khoury 1975; de la Maza and Carter 1977; Berns, this volume).

Analyses of the restriction cleavage maps of the incomplete DNA molecules suggest that they may have a rearrangement or a deletion involving 50–100 nucleotides within the inverted repetitious terminal sequence. For instance, cleavage with *Hae*II or *Kpn*I (Fig. 4) gave rise to a fragment that was about 50–100 nucleotides smaller than the terminal doublet obtained from the complete genome. The relative amount of this new component was inversely related to the amount of the *a* component of the regular terminal doublet, and the molar amount of this new fragment increased as the size of the incomplete particles decreased. This phenomenon was also reflected by the new 1.2% fragment present in the *Hha*I cleavage, which may arise by rearrangement or deletion of the two regular terminal fragments P$_a$ and P$_y$ and part of K or K*. It should be noted that it has not been proved rigorously that this 1.2% fragment is actually AAV DNA and the possibility remains that it might be adenovirus-specific or cell-specific.

The structural configuration of the incomplete AAV-DNA molecules suggests that they arise during replication, since the regions enriched for include the inverted repetitions which are believed to contain the replication origin (Hauswirth and Berns, this volume).

Incomplete AAV2 genomes with properties similar to the encapsidated molecules described here have been detected by pulse-labeling infected cells with [³H]thymidine and analyzing the AAV DNA in the Hirt supernatant (Hauswirth and Berns, this volume; B. Carter and L. de la Maza, unpublished). These studies show clearly that the incomplete genomes arise directly by replication rather than by subsequent degradation or recombination. The incomplete AAV genomes contained in the Hirt supernatant also yielded the 1.2 fragment when cleaved with *Hha* (B. Carter and L. de la Maza, unpublished).

We do not yet know whether the hairpin or cross-link in the incomplete DNA corresponds to an "internal" or an "external" region of the AAV genome. Several possibilities that can be suggested for the origin of these incomplete molecules are illustrated in Figure 5. The scheme for normal replication of AAV is based on those described for self-primed 3' hairpin replication (Cavalier-Smith 1974; Straus et al. 1976; Tattersall and Ward 1976). In this scheme, any nicks that occur on the template strand result in production of an incomplete molecule with an "external" hairpin or cross-link. Two models would at least give rise to "internal hairpins." In one, the strand being displaced would be nicked and the 3' end generated would fold back via some "illegitimate" base pairing and provide a primer for DNA replication. This model was proposed for adenovirus incomplete DNA molecules (Daniell 1976). Alternatively, any event that causes the DNA polymerase to turn the fork at the growing point and proceed with synthesis along the displaced strand provides the possibility to excise an internal hairpin molecule. This type of model has been proposed for inverted duplications of λdv and polyoma molecules (Chow et al. 1974; Robberson and Fried 1974).

The apparent structural alteration at the termini of the incomplete genomes as exemplified by the 1.2 fragment in the *Hha* cleavages may be explained by a decrease in length relative to the complete duplex DNA. This might arise either by failure to duplicate the terminal palindrome or by its complete loss, as illustrated in Figure 5 for the external-hairpin model. An analogous process could occur for the open end of an internal hairpin molecule. This suggests that the incomplete genomes

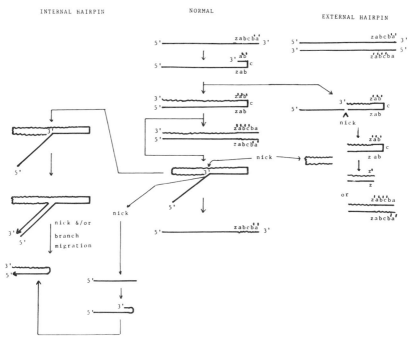

FIGURE 5
Possible models for derivation of incomplete genomes of AAV during replication.
The models are explained in the text.

may have an altered or inoperative replication origin. Consistent with this
model are preliminary experiments which indicate that type-III molecules
cannot replicate and do not interfere with either AAV- or adenovirus-
DNA replication (B. Carter and L. de la Maza, unpublished).

The density of the incomplete virus particles appears to be correlated
with the size of the incomplete DNA, which suggests that each particle
contains only one DNA molecule. This implies that packaging of AAV
DNA does not occur by a "headful" mechanism. It also implies that virus
assembly occurs by encapsidation of DNA into preformed capsids, since
any model for crystallization of viral capsids around the DNA might be
expected to yield particles with aberrant morphology with smaller DNA.
It was noted by Daniell (1976) that the DNA strands in the hairpin
molecules could have a total length up to twice that of a normal genome.
We have observed no encapsidated AAV-DNA hairpin strand longer
than a normal AAV molecule. This suggests that 1.4×10^6 daltons of DNA
is the most that can be packaged in an AAV particle.

REFERENCES

Carter, B. J. and G. Khoury. 1975. Specific cleavage of adenovirus-associated virus DNA by restriction endonuclease R.*Eco*R1—Characterization of cleavage products. *Virology* **63**:523.

Carter, B. J., K. H. Fife, L. M. de la Maza, and K. I. Berns. 1976. Genome localization of adeno-associated virus RNA. *J. Virol.* **19**:1044.

Carter, B. J., F. J. Koczot, J. Garrison, J. A. Rose, and R. Dolin. 1973. Separate helper functions provided by adenovirus for adenovirus-associated virus multiplication. *Nat. New Biol.* **244**:71.

Cavalier-Smith, T. 1974. Palindromic base sequences and replication of eukaryote chromosome ends. *Nature* **250**:467.

Chow, L. T., N. Davidson, and D. Berg. 1974. Electron microscope study of the structures of DNAs. *J. Mol. Biol.* **86**:69.

Daniell, E. 1976. Genome structure of incomplete particles of adenovirus. *J. Virol.* **19**:685.

de la Maza, L. M. and B. J. Carter. 1977. Adeno-associated virus DNA structure—Restriction endonuclease maps and arrangement of terminal sequences. *Virology* **82**:409.

Gerry, H. W., T. J. Kelly, and K. I. Berns. 1973. Arrangement of nucleotide sequences in adeno-associated virus DNA. *J. Mol. Biol.* **79**:207.

Hoggan, M. D. 1971. Small DNA viruses. In *Comparative virology* (ed. K. Maramorosch and E. Kurstak), p. 49. Academic, New York.

Johnson, F. B., C. W. Whitaker, and M. D. Hoggan. 1975. Structural polypeptides of adenovirus-associated virus top component. *Virology* **65**:196.

Robberson, D. L. and M. Fried. 1974. Sequence arrangements in clonal isolates of polyoma defective DNA. *Proc. Natl. Acad. Sci.* **71**:3497.

Straus, S. E., E. D. Sebring, and J. A. Rose. 1976. Concatemers of alternating plus and minus strands are intermediates in adenovirus-associated virus DNA synthesis. *Proc. Natl. Acad. Sci.* **73**:742.

Tattersall, P. and D. C. Ward. 1976. Rolling hairpin model for replication of parvovirus and linear chromosomal DNA. *Nature* **263**:106.

Torikai, K., M. Ito, L. E. Jordan, and H. D. Mayor. 1970. Properties of light particles produced during growth of type 4 adeno-associated satellite virus. *J. Virol.* **6**:363.

Comparison of the Terminal Nucleotide Structures in the DNA of Nondefective Parvoviruses

MARIE B. CHOW
DAVID C. WARD

*Department of Human Genetics
and Department of Molecular Biophysics and Biochemistry
Yale University School of Medicine
New Haven, Connecticut 06510*

The genomes of both defective and nondefective parvoviruses are linear, predominantly single-stranded DNA molecules of 1.4–1.7 × 10⁶ daltons (Rose 1974). Recent studies have indicated that the 3′ and 5′ terminal-nucleotide regions of the DNA from minute virus of mice (MVM) (Bourguignon et al. 1976), H-1 (Rhode 1977; Singer and Rhode 1977), Kilham rat virus (KRV) (Salzman 1977), and adeno-associated virus (AAV) (Denhardt et al. 1976; Fife et al. 1977) may exist as base-paired hairpin duplexes. Initiation and termination of DNA replication have been shown to occur close to these termini for AAV (Hauswirth and Berns 1977) and H-1 virus (Rhode 1977; Singer and Rhode 1977). In addition, models of DNA replication in which terminal hairpin structures act as primers for DNA synthesis have been proposed (Straus et al. 1976; Tattersall and Ward 1976). Since the termini of parvovirus genomes appear to be intimately involved in the DNA-replication process, it is of interest to determine whether these terminal duplex regions possess similarities in size or sequence. In this article we describe some properties of the terminal hairpin duplexes from the nondefective parvoviruses MVM, H-1, H-3, and KRV.

MATERIALS AND METHODS

Cells and Viruses

MVM, H-1, H-3, and KRV were grown on RL5E cells, a murine-sarcoma-virus-transformed rat-liver cell line (Bomford and Weinstein 1972). Initial virus inocula were kindly provided by P. Tattersall (MVM, plaque-purified strain T), L. Salzman (KRV), and H. Toolan (H-1 and H-3). Cell cultures were maintained and virus stocks were produced and purified as described previously (Tattersall et al. 1976).

DNA

Viral DNAs, radiolabeled with [^{32}P]PO$_4$ or [^{3}H]thymidine, were isolated from purified virions by sedimentation in alkaline sucrose gradients according to the method of Bourguignon et al. (1976). The specific radioactivity of these DNA samples was routinely between 5×10^4 and 2×10^5 cpm/μg. [^{3}H]thymidine-labeled polyoma DNA form II (specific activity 9×10^4 cpm/μg) was prepared and purified as described by Griffin et al. (1974). ^{32}P-labeled λ DNA was the gift of R. Young.

Enzymes

S1 nuclease was purified by the method of Vogt (1973) or was purchased from Sigma. The single-strand-specific endonuclease of vaccinia virus was the generous gift of K. Berns and K. Fife. Exonuclease I, free of exonuclease VII, was kindly provided by C. C. Richardson. Bacterial alkaline phosphatase was the gift of J. Chlebowski. The subtilisin fragment of E. coli DNA polymerase I, which lacks the 5'→3' exonuclease activity of the holoenzyme, was purchased from Boehringer-Mannheim. Exonuclease III of E. coli was purchased from New England Biolabs; T$_4$ polynucleotide kinase was obtained from P-L Biochemicals; restriction endonucleases HaeII, EcoRI, and HindIII were products of Miles Laboratories.

Enzyme-Assay Conditions

S1 NUCLEASE. Reactions were carried out essentially according to the method of Bourguignon et al. (1976). In some instances the DNA samples (at concentrations of 10–20 μg/ml) were denatured before treatment by heating for 3 min at 100°C followed by quick cooling on ice. S1-resistant products that were to be analyzed by gel electrophoresis were dialyzed against 0.5 mM EDTA and then lyophilized.

VACCINIA ENDONUCLEASE. Digestion reactions (100 μl) contained 100 mM imidazole buffer (pH 5.5), 5 mM MgCl₂, 0.5 mg/ml bovine serum albumin, 2 μg radiolabeled viral DNA, 25 μg/ml calf-thymus DNA, and 10 μl enzyme solution. Reactions were incubated at 37°C for 90 min. Reaction products were then analyzed by polyacrylamide gel electrophoresis.

EXONUCLEASE I. Assays containing 10–20 μg/ml of viral DNA were performed according to the procedure of Lehman and Nussbaum (1964) for 30 min at 37°C.

EXONUCLEASE III. Reactions were carried out as described by Richardson et al. (1964) for 30 min at 37°C. The samples were extracted twice with an equal volume of phenol [saturated with 0.01 M Tris buffer (pH 7.5), 0.001 M EDTA], ether-extracted six times, and then dialyzed against 1 mM EDTA. The dialyzed exoIII-treated DNA was then digested with S1 or vaccinia endonuclease under the conditions given above.

DNA POLYMERASE AND RESTRICTION ENDONUCLEASES. Double-stranded viral DNAs were synthesized in vitro using the subtilisin fragment of *E. coli* DNA polymerase I essentially as described by Bourguignon et al. (1976). The reactions contained 0.1 M Tris-HCl buffer (pH 7.5); 0.05 M NaCl; 0.01 M MgCl₂; 0.5 mM dATP, dCTP, dGTP, and ^{3}H- or γ-^{32}P-labeled TTP (10–20 μCi/μmole); 10–20 μg purified H-1, KRV, H-3, or MVM DNA; and 2 units DNA polymerase in a final volume of 0.5 ml. The reaction mixture was incubated for up to 90 min at 37°C, conditions which routinely yielded full-genome-length duplexes. Aliquots were removed at various times, and the polymerase was inactivated by heating at 70°C for 3 min. After cooling, part of each sample was incubated with 1 unit of restriction endonuclease (*Eco*RI, *Hin*dII, or *Hae*II) for 60 min at 37°C. The undigested and endonuclease-treated samples were then electrophoresed on 1.4% agarose slab gels. Since DNA polymerase synthesizes the complementary strand using the 3' hairpin terminus of the viral genome as a primer (Bourguignon et al. 1976), it is possible to order the restriction fragments relative to the 3' end of the viral genome by monitoring the kinetics of their appearance after digestion of the partially duplex DNA products. The physical maps obtained by this method are identical to those determined using the procedure of Smith and Birnstiel (1976).

Sequence Analysis of the 5' Terminal Nucleotides of MVM and H-1 DNA

The 5' terminal phosphate of virion DNA was removed by incubation (in 0.05 M Tris-HCl buffer, pH 8.7) with bacterial alkaline phosphatase (1–2 μg) for 15 min at 37°C followed by 15 min at 65°C. After cooling, the DNA was phenol-extracted three times and alcohol-precipitated, and the 5' terminal nucleotide was labeled with [^{32}P]PO₄ using T₄ polynucleotide kinase and 5 μM $[\gamma\text{-}^{32}P]ATP$ (New England Nuclear, specific activity 2000–3000 Ci/mmole) as described by Weiss et al. (1968). The ^{32}P-labeled DNA was then electrophoresed on a 1.4% agarose slab gel. Intact DNA was eluted from the gel and subjected to the chemical reactions described by Maxam and Gilbert (1977).

Gel Electrophoresis

Agarose (1.4%) slab gels were run as described by Sharp et al. (1973). Polyacrylamide gels were run as described by Jeppesen (1974), using a 3.0–7.5% acrylamide gradient. Gels were either dried under vacuum onto Whatman 1 paper and exposed to Kodak RP/54 X-ray film or fluorographed by the method of Bonner and Laskey (1974) and exposed to presensitized film (Laskey and Mills 1975).

RESULTS

Evidence for Hairpin Duplexes in MVM, H-1, H-3, and KRV DNA

Enzymatic and electrophoretic analyses have demonstrated the existence of hairpin duplex structures in the otherwise single-stranded genomes of these four nondefective parvoviruses. Purified, radioactively labeled DNA from each of the viruses was subjected to limit digestion with the single-strand-specific endonuclease S1. The time course of such a digestion is shown in Figure 1a. All viral DNA samples showed the presence of an S1-resistant core which contained approximately 10–12% of the input radiolabel. The double-stranded polyoma DNA control, as expected, was completely resistant to S1 nuclease. Viral DNA samples that were heat-denatured and quick-cooled immediately before S1 digestion gave the same level of resistance, indicating that the S1-resistant regions of the genome were capable of instantaneous reannealing, a characteristic of hairpin structures.

The results obtained after the viral DNAs were digested to completion with exonuclease I (exoI), a single-strand-specific 3'→5' exonuclease, are shown in Figure 1b. Whereas heat-denatured λ DNA was almost completely solubilized by exoI, 90–95% of the radioactivity in each viral DNA sample was resistant to exoI digestion. The level of exoI sensitivity did not increase when the viral DNAs were heat-denatured and quick-cooled immediately before enzyme treatment. These results indicate that at least part of the spontaneously renaturable duplex is at or near the 3' terminus of each viral genome.

Since fragmentation of an intact genome would generate an exoI-sensitive substrate, the observed degree of exoI resistance is probably an underestimate of the true resistance level. The observation that oncorna-

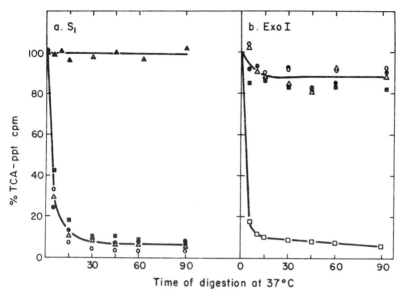

FIGURE 1

Susceptibility of parvovirus DNAs to digestion by S1 (*a*) or exonuclease I (*b*) as a function of time. Reactions, containing ^{3}H- or ^{32}P-labeled viral DNA, were incubated and assayed as described in "Materials and Methods." ●: MVM DNA; △: H-1 DNA; ■: H-3 DNA; ○: KRV DNA; ▲: polyoma DNA form II; □: heat-denatured λ DNA.

virus reverse transcriptase can utilize parvovirus DNAs efficiently as primer templates (Bourguignon et al. 1976; D. C. Ward, unpublished results) implies that the 3′ terminal nucleotide is in a duplex form. These polymerases totally lack 3′→5′ or 5′→3′ exonuclease activity and require base-paired primers in order to initiate DNA synthesis (Baltimore and Smoler 1971; Hurwitz and Leis 1972). The exoIII results presented below provide further support for the existence of a 3′ terminal hairpin duplex.

Characterization and Localization of the Duplex Structures within the Viral Genome

The S1-nuclease-resistant products were analyzed by electrophoresis in a 3.0–7.5% gradient polyacrylamide gel. An autoradiograph of a typical gel is shown in Figure 2. Two bands are generated from each viral genome upon S1 digestion. The S1-resistant fragments from all four viral DNAs appear to have the same molecular weight. The samples shown in the gel in Figure 2 were not heat-denatured before S1 digestion. However, bands of identical mobility are seen after the DNA samples are heat-denatured immediately before S1 treatment. The resistant DNA migrating near the

FIGURE 2

S1-nuclease-resistant products of [³H]TdR-labeled viral DNAs. Samples were digested with S1 for 90 min, then electrophoresed on a 3.0–7.5% polyacrylamide gel. In this experiment, samples were not denatured prior to S1 digestion. The fluorogram of the gel is illustrated. Wells *a, c, e,* and *g* contain untreated virion DNAs of H-1, MVM, H-3, and KRV, respectively. Wells *b, d, f,* and *h* contain the S1-nuclease-resistant products of H-1, MVM, H-3, and KRV DNA, respectively.

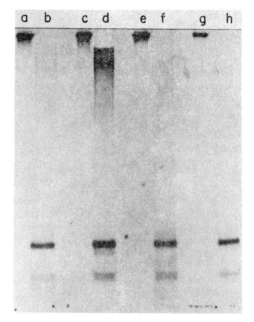

top of the gel in the MVM-DNA sample (Fig. 2, track *d*) is not spontaneously renaturable, as it is sensitive to S1 if the sample is first heat-denatured and quick-cooled. A similar species of duplex DNA had been observed in some preparations of MVM DNA (Bourguignon et al. 1976) and is observed occasionally in preparations of H-1, H-3, KRV, and BPV. The two low-molecular-weight S1-resistant fragments were found to be approximately 130 and 110 base pairs long when sized against *Hae*III restriction fragments of SV40 DNA as described by Bourguignon et al. (1976).

Since MVM DNA was shown to contain a stable 5' terminal hairpin duplex of approximately 130 base pairs (Bourguignon et al. 1976), the other three viral genomes were tested for the presence of a similar hairpin structure. After treatment with alkaline phosphatase, the 5' termini of H-1, H-3, and KRV DNAs were labeled with [³²P]PO₄ using T₄ polynucleotide kinase and [γ-³²P]ATP. The terminal-labeled DNAs were denatured and limit-digested with S1; the resultant S1-resistant products were electrophoresed on a polyacrylamide gel. The autoradiograph of that gel is shown in Figure 3. Each virion DNA contains a major S1-resistant fragment, with an electrophoretic mobility similar to that of a duplex of 130 base pairs. The secondary bands, observed to varying extents in these DNA digests, may be due to cleavage during "breathing" of the duplex, cleavage at base mismatches, or "nibbling" at the ends of

FIGURE 3
S1-nuclease-resistant products of DNA (^{32}P-labeled at the 5' terminus) from KRV, H-1, and H-3. Samples were heat-denatured and quick-cooled, digested for 90 min with S1, then electrophoresed on a 3.0–7.5% polyacrylamide gel. The auto-radiogram of the gel is shown. Wells *a*, *c*, and *e* contain undigested virion DNA from KRV, H-1, and H-3, respectively. Wells *b*, *d*, and *f* contain the S1-nuclease-resistant products of KRV, H-1, and H-3 DNA, respectively.

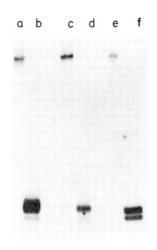

DNA duplexes by S1 (Shenk et al. 1974, 1975). These results indicate that the 5' terminal nucleotide of each viral DNA is resistant to S1 nuclease and can be isolated in a double-stranded fragment of ~130 base pairs.

Further evidence that the other S1-resistant fragment is a stable 3' terminal hairpin duplex was obtained by digesting uniformly ^{32}P-labeled MVM DNA with exoIII. This enzyme is a 3'→5' double-strand-specific exonuclease and should digest any duplex at the 3' terminus. This would therefore render any 3' terminal duplex region sensitive to a single-strand-specific nuclease. Samples of exoIII-treated MVM DNA were incubated with S1 or vaccinia endonuclease, and the resultant products were analyzed by gel electrophoresis. As shown in Figure 4, the smaller (110-base-pair) fragment is selectively lost when the

FIGURE 4
Enzymatic digestion products of MVM DNA separated by electrophoresis on a 3.0–7.5% polyacrylamide gel. Uniformly ^{32}P-labeled MVM DNA (*1*), uniformly ^{32}P-labeled MVM DNA after limit digestion with S1 nuclease (*2*) or vaccinia endonuclease (*3*), uniformly ^{32}P-labeled MVM DNA digested with exoIII prior to treatment with S1 (*4*) or vaccinia endonuclease (*5*), 5' terminal ^{32}P-labeled MVM DNA after digestion with S1 (*6*). DNA samples were not heat-denatured before digestion.

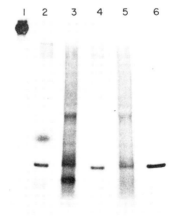

exoIII treatment precedes exposure to the second nuclease. DNA treated with S1 or vaccinia endonuclease alone shows the two fragments, as expected. MVM DNA labeled with ^{32}P at the 5' terminus again gives a single product when treated with S1 nuclease. This product comigrates with the larger (130-base-pair) fragment (Fig. 4, track 6).

It should be noted that the smaller 3' terminal hairpin duplexes are consistently found in lower yields than the larger 5' terminal hairpin duplexes (see Figs. 2 and 4). Indeed, in a previous study (Bourguignon et al. 1976) a stable 3' terminal duplex was not detected by polyacrylamide gel electrophoresis. The presence of a contaminating exoIII-type enzymatic activity in the S1-nuclease preparations employed could account for the selective loss of the 3' terminal duplex. In fact, when MVM DNA is digested with vaccinia endonuclease (Fig. 4, track 3) both terminal fragments are observed in approximately equimolar amounts.

Similarities in Nucleotide Sequences at the 5' Termini

Since both the 3' and the 5' terminal duplexes from four viruses had the same apparent molecular weight, we were interested in determining whether the nucleotide sequences in these terminal regions were identical or conserved. We have deduced sequences for approximately the first 50 nucleotides from the 5' terminus for H-1 and MVM DNA using the chemical sequencing technique of Maxam and Gilbert (1977). These sequences (shown in Fig. 5) are tentative, since we have analyzed only one strand of the duplexes. A definitive sequence assignment will require confirmation by determining the sequence of the complementary strand or by analyzing the entire region using alternative sequencing techniques. We have, on occasion, experienced difficulty in discriminating between C and T residues using the Maxam-Gilbert method. This uncertainty is reflected in the areas left blank in the MVM sequence (Fig. 5); however, we are confident that these nucleotides (designated 6–13 and 16) are pyrimidines. No assignments have been made for nucleotides 1–2 of H-1 DNA or nucleotides 1–3 of MVM DNA because these residues were not retained on the sequencing gels analyzed. There are several interesting aspects of both sequences worth noting. The fact that a defined sequence can be obtained for MVM and H-1 DNA indicates that at least the first 50 nucleotides at the 5' terminus of each genome are probably unique and nonpermuted. In addition, although the sequences for MVM and H-1 DNAs are not identical, they are highly similar. The first 20 nucleotides at the 5' end of both viral DNAs display a region rich in pyrimidines, while nucleotides 20–40 are very rich in purines. This purine-pyrimidine clustering is illustrated graphically in Figure 5.

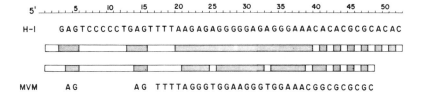

FIGURE 5

Tentative nucleotide sequences for the 5' terminus of MVM and H-1 DNAs as deduced by the method of Maxam and Gilbert (1977). The blank spaces reflect areas of present uncertainty, which is mainly due to poor C-vs.-T discrimination. The presence of pyrimidine and purine clusters is shown schematically; clear areas represent pyrimidines and shaded areas represent purines.

Similarities in Regions Outside the Terminal Duplexes

In order to investigate possible regions of homology between these viral genomes, we have prepared a series of physical maps generated from restriction-enzyme digests of double-stranded viral DNAs. These viral DNA duplexes were synthesized in ✻.tro using purified virion DNA and the subtilisin fragment of *E. coli* DNA polymerase I (Bourguignon et al. 1976). The products of the polymerase reactions comigrate in 1.4% agarose gels with the genome-length double-stranded DNA made in vivo (data not shown). A comparison of the restriction maps of H-1, MVM, and H-3 DNA for the enzymes *Hae*II, *Eco*RI, and *Hin*dIII is shown in Figure 6. All three DNAs appear to contain common restriction-enzyme cleavage sites. These occur, reading from the 3' end of the viral DNA strand, at 19.5% (*Hae*II), 20.5% (*Eco*RI), and 50.5% (*Hin*dIII) of the genome length. Although the cleavage patterns (to be published elsewhere) obtained with other restriction enzymes (e.g., *Hae*III and *Hin*f) indicate clearly that these DNAs are uniquely different, the similarities in position of many specific cleavage sites suggest that these viral genomes may share considerable regions of sequence homology.

DISCUSSION

The evidence presented here demonstrates that the parvoviruses MVM, H-1, H-3, and KRV contain hairpin duplexes at both the 3' and the 5' termini of the virion DNA. Our results indicate that the 5' terminal hairpin is ~130 base pairs and that the 3' terminal hairpin is ~110 base pairs in size. Recent experiments with BPV, a bovine parvovirus, give results identical to those described above. However, our data on the sizes of the terminal duplexes are in disagreement with the results of Lavelle and

FIGURE 6
Physical maps of MVM, H-1, and H-3 DNA for the restriction endonucleases *Eco*RI, *Hae*II, and *Hin*dIII.

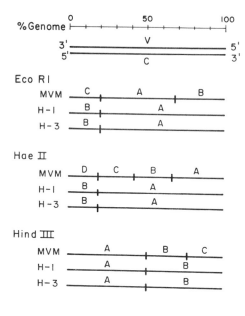

Mitra (this volume), who conclude that the 5' and 3' terminal hairpin structures in KRV DNA are 110 and 135 base pairs long, respectively. Experiments designed to resolve this apparent conflict are in progress.

Our initial studies with the 5' termini of H-1 and MVM DNAs indicate that the nucleotide sequences of the first 50 residues, though not identical, possess some interesting similarities. Nucleotides 1–20 of both viral genomes exhibit high pyrimidine content, whereas nucleotides 20–40 are extremely purine-rich. The corresponding regions of the 5' termini of KVR and H-3 DNA also exhibit a similar clustering of purine and pyrimidine bases (data not shown). H-1-virus progeny-DNA synthesis has been shown to be initiated from an origin that is close to the terminal nucleotide of the 5' hairpin duplex (Singer and Rhode 1977). The observed similarities in the mechanism of DNA replication between H-1 and the other autonomous parvoviruses (see review by Berns and Hauswirth, this volume) would suggest that they also possess a similar replication origin. Thus, the 5' terminal hairpin duplex may contain a nucleotide sequence essential for efficient polymerase recognition and binding. Since all parvoviruses depend heavily upon host-cell machinery for their replication, such a sequence(s) might reasonably be expected to be of potential significance for host-cell DNA replication as well.

Comparison of the H-1 or MVM sequences with that of the 5' terminus of the AAV genome (Fife et al. 1977) indicates no obvious similarities either in exact sequence or in the distribution of pyrimidine-rich and purine-rich tracts. The observation that the 5' terminal nucleotide sequences of MVM, H-1, H-3, and KRV are similar to each other but quite distinct from that of the AAV genome might be of biological significance. The AAVs are predominantly human and primate viruses which are incapable of complete replication without an adenovirus helper. In contrast, MVM, H-1, H-3, and KRV all replicate preferentially in rodent cells and do not require a helper virus for replication. Since the terminal regions of both parvovirus groups appear to play a central role in viral DNA replication, the observed sequence differences may be related to some aspect of AAV defectiveness or reflect the natural host range of these viruses. It is interesting to note that the nucleotide sequence around the origin of DNA replication for polyoma virus, another natural rodent virus, has clusters of purines and pyrimidine bases similar to those described above (B. Griffin, personal communication). In contrast, the nucleotide sequence around the origin of replication of SV40, a simian virus, does not exhibit this distribution of bases (Subramanean et al. 1977).

A detailed analysis of restriction-endonuclease cleavage patterns of double-stranded viral DNAs synthesized in vitro indicates that the genomes of MVM, H-1, H-3, and KRV are each unique (M. Smith, C. Astell, M. Chow, C. L. Ching, and D. C. Ward, unpublished results). However, a comparison of the *Eco*RI, *Hae*II, and *Hin*dIII maps (Fig. 6) shows that these DNAs do contain common cleavage sites, indicating possible regions of sequence similarity. Furthermore, when stable mRNA extracted from MVM-infected A-9 cells is hybridized to DNA purified from MVM, H-1, H-3, or KRV virions, the last three viral DNAs anneal up to 50% of the saturation level obtained with MVM DNA (D. Dadachanji and D. C Ward, unpublished results). These results indicate that a significant degree of sequence similarity occurs within the transcribed regions of these viral genomes in addition to that which we have observed near the 5' termini.

ACKNOWLEDGMENTS

This work was supported by Public Health Service grant GM-20124 from the National Institute of General Medical Sciences and by grant Ca-16038 from the National Cancer Institute.

REFERENCES

Baltimore, D. and D. Smoler. 1971. Primer requirement and template specificity of the DNA polymerase of RNA tumor viruses. *Proc. Natl. Acad. Sci.* **68**:1507.

Bomford, R. and I. B. Weinstein. 1972. Transformation of a rat epithelial-like cell line by murine sarcoma virus. *J. Natl. Cancer Inst.* **49**:379.

Bonner, W. A. and R. A. Laskey. 1974. A film detection method for tritium-labelled proteins and nucleic acids in polyacrylamide gels. *Eur. J. Biochem.* **46**:83.

Bourguignon, G. J., P. J. Tattersall, and D. C. Ward. 1976. DNA of minute virus of mice: Self-priming, nonpermuted, single-stranded genome with a 5'-terminal hairpin duplex. *J. Virol.* **20**:290.

Denhardt, D., S. Eisenberg, K. Bartok, and B. J. Carter. 1976. Multiple structures of adeno-associated virus DNA: Analysis of terminally labeled molecules with endonuclease R·*Hae*III. *J. Virol.* **18**:672.

Fife, K., K. I. Berns, and K. Murray. 1977. Structure and nucleotide sequences of the terminal regions of adeno-associated virus DNA. *Virology* **78**:475.

Griffin, B. E., M. Fried, and A. Cowie. 1974. Polyoma DNA: A physical map. *Proc. Natl. Acad. Sci.* **71**:2077.

Hauswirth, W. W. and K. I. Berns. 1977. Origin and termination of adeno-associated virus DNA replication. *Virology* **78**:488.

Hurwitz, J. and J. P. Leis. 1972. RNA-dependent DNA polymerase activity of RNA tumor viruses. I. Direction influence of DNA in the reaction. *J. Virol.* **9**:116.

Jeppesen, P. G. 1974. A method for separating DNA fragments by electrophoresis in polyacrylamide gradient slab gels. *Anal. Biochem.* **58**:195.

Laskey, R. A. and A. D. Mills. 1975. Quantitative film detection of ^{3}H and ^{14}C in polyacrylamide gels by fluorography. *Eur. J. Biochem.* **56**:335.

Lehman, I. R. and A. L. Nussbaum. 1964. The deoxyribonucleases of *Escherichia coli*. V. On the specificity of exonuclease I (phosphodiesterase). *J. Biol. Chem.* **239**:2628.

Maxam, A. M. and W. Gilbert. 1977. A new method for sequencing DNA. *Proc. Natl. Acad. Sci.* **74**:560.

Rhode, S. L. III. 1977. Replication process of the parvovirus H-1. VI. Characterization of a replication terminus of H-1 replicative-form DNA. *J. Virol.* **21**:694.

Richardson, C. C., I. R. Lehman, and A. Kornberg. 1964. A deoxyribonucleic acid phosphatase-exonuclease from *Escherichia coli*. II. Characterization of the exonuclease activity. *J. Biol. Chem.* **239**:251.

Rose, J. A. 1974. Parvovirus reproduction. In *Comprehensive virology* (ed. H. Fraenkel-Conrat and R. R. Wagner), vol. 3, p. 1. Plenum, New York.

Salzman, L. A. 1977. Evidence for terminal S1-nuclease-resistant regions on single-stranded linear DNA. *Virology* **76**:454.

Sharp, P. A., W. Sugden, and J. Sambrook. 1973. Detection of two restriction endonuclease activities in *Haemophilus parainfluenzae* using analytical agarose-ethidium bromide electrophoresis. *Biochemistry* **12**:3055.

Shenk, T. E., C. Rhodes, P. W. J. Rigby, and P. Berg. 1974. Mapping of mutational alterations in DNA with S1 nuclease: The location of deletions, insertions, and temperature-sensitive mutations in SV40. *Cold Spring Harbor Symp. Quant. Biol.* **39**:61.

————. 1975. Biochemical method for mapping mutational alterations in DNA with S1 nuclease: The location of deletions and temperature-sensitive mutations in simian virus 40. *Proc. Natl. Acad. Sci.* **72**:989.

Singer, I. and S. Rhode. 1977. Replication process of the parvovirus H-1. VIII. Partial denaturation mapping and localization of the replication origin of H-1 replicative form DNA with electron microscopy. *J. Virol.* **21**:724.

Smith, H. O. and M. L. Birnstiel. 1976. A simple method for DNA restriction site mapping. *Nucleic Acids Res.* **3**:2387.

Straus, S. E., E. D. Sebring, and J. A. Rose. 1976. Concatemers of alternating plus and minus strands are intermediates in adenovirus-associated virus DNA synthesis. *Proc. Natl. Acad. Sci.* **73**:742.

Subramanean, K. N., R. Dhar, and S. W. Weissman. 1977. Nucleotide sequences of a fragment of SV40 DNA that contains the origin of DNA replication and specifies the 5' ends of "early" and "late" viral RNA. III. Construction of the total sequence of *Eco*RII-G fragment of SV40. *J. Biol. Chem.* **252**:355.

Tattersall, P. and D. C. Ward. 1976. Rolling hairpin model for replication of parvovirus and linear chromosomal DNA. *Nature* **263**:5573.

Tattersall, P., P. J. Cawte, A. J. Shatkin, and D. C. Ward. 1976. Three structural polypeptides coded for by minute virus of mice, a parvovirus. *J. Virol.* **20**:273.

Vogt, V. M. 1973. Purification and further properties of single-strand-specific nuclease from *Aspergillus oryzae*. *Eur. J. Biochem.* **33**:192.

Weiss, B., T. R. Live, and C. C. Richardson. 1968. Enzymatic breakage and joining of deoxyribonucleic acid. V. End group labeling and analysis of deoxyribonucleic acid containing single strand breaks. *J. Biol. Chem.* **243**:4530.

Double-Helical Regions in Kilham-Rat-Virus DNA

GEORGE LAVELLE
SANKAR MITRA

Biology Division
Oak Ridge National Laboratory
Oak Ridge, Tennessee 37830

Kilham rat virus (KRV), a nondefective parvovirus, contains a linear single-stranded DNA with a molecular weight of about 1.5×10^6 (Robinson and Hetrick 1969; McGeoch et al. 1970; May and May 1970; Salzman et al. 1971). Details of the arrangement of nucleotide sequences and the secondary structure of the DNA of two other parvoviruses have recently been presented. The DNAs of the defective adeno-associated virus (AAV) (Gerry et al. 1973) and the nondefective minute virus of mice (MVM) (Bourguignon et al. 1976) are not randomly circularly permuted. However, AAV DNA contains a limited number of terminal-sequence permutations (Gerry et al. 1973) and an inverted terminal repetition which permits formation of single-stranded circles by self-annealing (Koczot et al. 1973; Berns and Kelly 1974). MVM DNA contains a stable 5' terminal hairpin duplex, and, in addition, the 3' terminus is self-priming for DNA synthesis (Bourguignon et al. 1976). Unlike AAV DNA, single-stranded, hydrogen-bonded circular molecules have not been seen in self-annealed preparations of MVM DNA (Bourguignon et al. 1976) or KRV DNA (G. Lavelle, D. Allison, and S. Mitra, unpublished observations).

Recent reports have indicated that the termini of KRV DNA are also double-stranded (Salzman 1977; Lavelle and Mitra 1977). In this paper the secondary structure of KRV DNA is characterized by hydroxyapatite chromatography and by reaction with single- and double-strand-specific DNases.

219

MATERIALS AND METHODS

Cell Culture and Virus

Normal rat kidney (NRK) cells (Duc-Nguyen et al. 1966) were maintained in McCoy's medium supplemented with 10% fetal-calf serum and 2 mM glutamine. Strain 171 of Kilham rat virus (Tennant et al. 1969) was propagated by infection of subconfluent monolayers of NRK cells in roller bottles. Radioactively labeled virus was grown in medium containing 10 μCi/ml [^{3}H]thymidine or in phosphate-free Eagle's minimum essential medium (MEM) containing 10 μCi/ml of [^{32}P]orthophosphate supplemented with 0.01 mM phosphate and 10% dialyzed fetal-calf serum.

Purification of Virus and DNAs

Virus was purified from lysates of infected cells by CsCl equilibrium centrifugation as described before (Lavelle and Li 1977). DNA was extracted by treatment of purified virus with 0.3 N NaOH for 20 min at 20°C and purified by velocity sedimentation in an alkaline 5–20% sucrose gradient as described by Koczot et al. (1973). Peak fractions of DNA were pooled, dialyzed, made to 0.1% in sodium dodecyl sulfate (SDS), and then extracted with an equal volume of redistilled phenol and precipitated by 2 volumes of ethanol in the presence of 0.3 M Na-acetate at −20°C overnight. Precipitates of DNA were dissolved in 0.01 M Tris-HCl (pH 8.0). The specific activity of ^{32}P-labeled DNA was 2–20 × 10^5 cpm/μg; that of ^{3}H-labeled DNA was 1–10 × 10^5 cpm/μg.

NRK cell DNA was prepared from ^{3}H- or ^{14}C-labeled cells by methods described previously (Lavelle et al. 1975). Single-stranded DNA was obtained by boiling DNA (10 min in 10 mM Tris-HCl, pH 8.0) and chilling rapidly to 0°C. ^{3}H-labeled PM2 phage DNA, purified by the method of Espejo et al. (1969), was a gift of Dr. R. Fujimura. The nicked and denatured PM2 DNA was prepared by treating 25 nmoles of DNA with 1 ng pancreatic DNase for 10 min at 37°C in 0.2 ml incubation mixture containing 50 mM Tris-HCl (pH 8.6), 3 mM MgCl$_2$, 1 mM dithiothreitol. The digestion was stopped by addition of EDTA to 5 mM followed by heating at 100°C for 3 min and rapid chilling.

Hydroxyapatite Chromatography

DNA was chromatographed on hydroxyapatite (HAP) prepared by the method of Tiselius et al. (1956) as modified by Miyazawa and Thomas (1965), using a gradient of Na-phosphate (NaP) buffer (pH 6.8). Samples in 0.02 M NaP were applied to a 1-ml column of hydroxyapatite and eluted with a 100-ml linear gradient of 0.05–0.4 M NaP at a flow rate of 1 ml/min.

Fractions of 2.5 ml were collected, and radioactivity was measured after precipitation with 10% trichloroacetic acid.

Enzymes and Assay Conditions

The purification and assay conditions of S1 endonuclease and *Escherichia coli* exonuclease I have been described (Mitra and Stallions 1976). *E. coli* exonuclease III was purified and assayed according to the method of Jovin et al. (1969). *E. coli* exonuclease VII was a gift from Dr. J. Chase and was assayed according to the published procedures (Chase and Richardson 1974).

Isolation of Viral Core DNA

Approximately 1 μg of labeled viral DNA was digested with 200 units of S1 nuclease until 70–90% of the DNA became acid-soluble. The incubation mixture was then made up to 2 mM ethylene bis(oxyethylene-nitrilo)tetraacetate (EGTA) and subsequently treated with 0.6 unit of exonuclease I (Lehman 1966) under the standard assay conditions (Mitra and Stallions 1976) at 37°C until no further DNA was acid-soluble (93–96% became acid-soluble). After removal of proteins from the mixture by extraction with an equal volume of phenol in the presence of 0.5% SDS, 0.5 M NaCl, and 10 mM EDTA, the residual "core" DNA was precipitated with alcohol and dissolved in a small volume of 10 mM Tris-HCl (pH 8.0). In order to determine which core fragment contained the 3' end, intact KRV DNA was digested with exonuclease III prior to limit digestion with exonuclease VII, as follows. One nmole of ^{3}H-labeled KRV DNA was treated with 40 units of exonuclease III in 0.06 ml incubation mixture for 30 min at 37°C (Jovin et al. 1969) along with a control without the enzyme. After heating at 80°C for 5 min, the samples were incubated with 2 units of exonuclease VII for 40 min at 37°C (Chase and Richardson 1974). After addition of Sarkosyl to 0.5%, the samples were electrophoresed in polyacrylamide gels as described below.

Polyacrylamide gel electrophoresis

DNA samples were electrophoresed in cylindrical gels (10 cm $\times$ 0.6 cm) containing 12% acrylamide, 0.07% bisacrylamide. Electrophoresis was carried out at 4 mA/tube for 4–5 hr in 40 mM Tris-acetate (pH 8.1), 20 mM Na-acetate, 1 mM EDTA, and 0.1% SDS at room temperature. The gels were then sliced into 2-mm sections and assayed for radioactivity in a toluene-Triton X-100 scintillation solvent system after digestion with 1 ml of 30% H_2O_2 overnight at 65°C. DNA was occasionally recovered from the gel slices after incubation in 2 ml of 10 mM Tris-HCl (pH 9.3), 2 mM EDTA overnight at 4°C. Eluted DNA was precipitated with 2 volumes of ethanol in the presence of 0.3 M Na-acetate.

RESULTS

Hydroxyapatite Chromatography of Intact Viral DNA

Viral DNA released from purified KRV by alkaline lysis sediments as a single major peak in sucrose gradients (Fig. 1). This DNA has been shown to have a sedimentation coefficient of 25S at neutral pH in solutions of high ionic strength (Lavelle and Li 1977). Unit-length DNA selected from peak fractions has a characteristic elution profile when chromatographed on hydroxyapatite (HAP) using a concentration gradient of NaP buffer (Fig. 2). Viral DNA elutes at a point on this gradient between the single- and double-stranded DNA markers, which suggests that the viral DNA is partly double-stranded. Viral DNA showed the same elution profile from HAP after denaturation in alkali and quick renaturation (data not shown). Thus, the partially duplex property of viral DNA is the result of intramolecular secondary structure which is contained in each intact unit-length KRV-DNA molecule. It is important to note that viral DNA that it is not purified by velocity sedimentation usually contains some DNA fragments that elute from HAP at the position of single-stranded DNA. Only intact unit-length viral DNA purified by sedimentation was used for the following enzymatic studies.

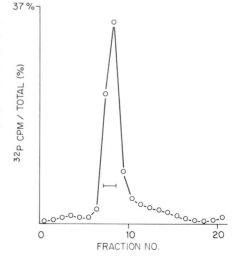

FIGURE 1

Purification of ^{32}P-labeled unit-length KRV DNA by velocity sedimentation in an alkaline sucrose gradient. Centrifugation was carried out in an SW 50.1 rotor (Beckman) at 42,000 rpm, 20°C, for 4 hr. Brackets indicate fractions that were pooled for further study. Sedimentation was from right to left.

FIGURE 2
Hydroxyapatite chromatography of unit-length KRV DNA. ●——●: ^{32}P-labeled KRV DNA. ○——○: ^{3}H-labeled denatured and native NRK cell DNA. ———: Na-phosphate molarity.

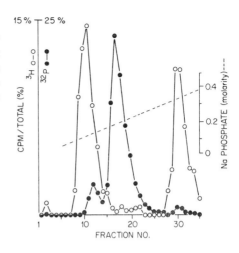

Susceptibility of Viral DNA to S1 Endonuclease and Resistance to E. coli *Exonuclease I*

As shown in Figure 3*A*, intact KRV DNA was about 95% susceptible after limit digestion with an excess of single-strand-specific S1 endonuclease. This confirms that the viral genome is essentially single-stranded. Denatured, sheared DNA from NRK cells, used as an internal marker, was similarly susceptible to S1.

FIGURE 3
(*A*) Susceptibility of ^{32}P-labeled KRV DNA and denatured ^{3}H-labeled NRK cell DNA to S1 endonuclease. A mixture of approximately 1 nmole (nucleotide phosphorus) of each DNA was treated with 80 units of S1 endonuclease in 0.25 ml, and aliquots were taken out at different times for measuring acid-insoluble radioactivity. (*B*) Digestion of ^{32}P-labeled KRV DNA and ^{3}H-labeled PM2 DNA, nicked and denatured as described in "Materials and Methods," with different amounts of exonuclease I. ●——●: ^{32}P; ○——○: ^{3}H.

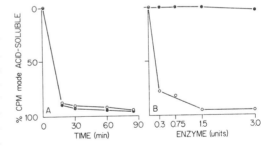

Intact viral DNA was completely resistant, however, to digestion with *E. coli* exonuclease I, a nuclease specific for the 3' OH end of single-stranded DNA (Lehman 1966). This is demonstrated in Figure 3*B* in reactions that contained nicked, denatured DNA of bacteriophage PM2 as an internal marker. PM2 DNA was up to 95% susceptible under conditions where the degradation of KRV DNA was negligible. To rule out the possibility that the resistance of KRV DNA to exonuclease I was due to phosphorylation at the 3' terminus, the DNA was treated with bacterial alkaline phosphatase prior to reaction with exonuclease I. This did not alter the resistance of viral DNA to the enzyme (results not shown). These results suggest that the 3' end of KRV DNA is double-helical. When KRV DNA was heated to 80°C prior to incubation with exonuclease I it still remained resistant to digestion, which indicated that the 3' terminal duplex reanneals instantaneously (results not shown).

Further confirmation of the 3' terminal duplex structure was obtained by treatment of intact viral DNA with *E. coli* exonuclease III (which is specific for the 3' end of double-stranded DNA) prior to reaction with exonuclease I (G. Lavelle and S. Mitra, manuscript in preparation). Under these conditions, KRV DNA was mostly (> 75%) susceptible to digestion by exonuclease I, which indicates the removal of the double-stranded structure at the 3' end of the molecule by exonuclease III. Exonuclease III alone did not make the viral DNA acid-soluble to any significant extent and did not alter the sedimentation rate of the viral DNA; thus, the absence of contaminating endonucleases was confirmed. Because the secondary structure of intact DNA renatured instantly after treatment with heat or alkali, we conclude that the double-helical DNA at the 3' end is in the form of a hairpin.

Isolation and Separation of Double-Helical DNA Segments

Two DNA fragments were resolved after sequential treatment of intact DNA with S1 endonuclease and exonuclease I and electrophoresis of the resistant DNA. The sizes of these residual cores of DNA were consistently 135 and 110 base pairs (G. Lavelle and S. Mitra, manuscript in preparation) when their mobility was compared with that of *Hind*-restriction-endonuclease fragments of bacteriophage φX174 RF I DNA (Maniatis et al. 1975).

Limit digestion of intact viral DNA with *E. coli* exonuclease VII, which is specific for both 3' and 5' ends of single-stranded DNA (and which also appeared to be contaminated with an endonuclease, as indicated by the degradation of circular single-stranded M13 phage DNA), also yielded two core fragments of sizes described above. These are shown in Figure 4. Also shown in Figure 4 is the effect of treatment of intact DNA with exonuclease III prior to digestion with exonuclease VII. This resulted in

FIGURE 4
Electrophoresis of KRV DNA core fragments produced by exonuclease VII with and without prior exonuclease-III treatment. o——o: Control without exonuclease III. ●——●: Predigested with exonuclease III.

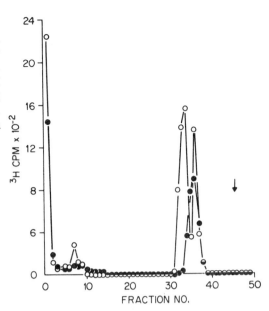

selective loss of the larger core piece (135 base pairs), indicating that it is located at the 3' end of the viral genome.

Fragments of viral core DNA were separated after S1-endonuclease and exonuclease-I digestion and chromatographed on hydroxyapatite (Fig. 5). The larger fragment (Fig. 5A) eluted at a salt concentration between single- and double-stranded DNA markers, at a position similar to that of

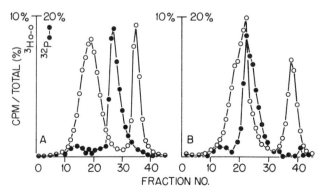

FIGURE 5
Hydroxyapatite chromatography of viral DNA core fragments. (A) The 135-base-pair fragment. (B) The 110-base-pair fragment. ●——●: ^{32}P-labeled KRV core DNA. o——o: ^{3}H-labeled denatured and native NRK cell DNA.

intact DNA (see Fig. 2). The smaller fragment (Fig. 5*B*) eluted at a mean salt concentration slightly higher than that of single-stranded DNA, but lower than that of the larger core piece.

Thermal chromatography on hydroxyapatite (Miyazawa and Thomas 1965; Niyogi and Thomas 1968) in 0.1 M NaP revealed a T_m of 88°C, almost identical to that of intact DNA (G. Lavelle and S. Mitra, manuscript in preparation). These results suggest that double-helical core regions confer on intact DNA its unusual thermal stability and elution behavior on hydroxyapatite.

DISCUSSION

The single-stranded genome of KRV has a stable and characteristic secondary structure, as evidenced by retention on hydroxyapatite, high thermal stability, and resistance to single-strand-specific exonuclease I. These properties can be attributed to two small double-helical fragments of core DNA, 135 and 110 base pairs, which can be isolated from intact viral DNA and separated by gel electrophoresis. The larger DNA fragment exhibited a chromatographic pattern on hydroxyapatite similar to that of intact DNA, which characteristically elutes at a salt concentration between those of single- and double-stranded DNA. The smaller fragment eluted at a slightly higher salt concentration than single-stranded marker DNA. The results are in agreement with those of Wilson and Thomas (1973), who demonstrated the influence of short double-helical segments on retention by hydroxyapatite of DNA containing long single chains. Furthermore, the high thermal stability of intact viral DNA (T_m = 88°C) is the same as, and can be attributed to, the high melting temperature for the larger piece of core DNA. Finally, the double-stranded nature of viral core DNA has been confirmed by base-composition analyses, which show approximate complementarity of nucleotides in core DNA, in contrast with the case for intact viral DNA (G. Lavelle and S. Mitra, manuscript in preparation).

The structure of the 3' end of KRV DNA was revealed by experiments using single- and double-strand-specific DNases. Complete resistance of intact DNA to digestion by exonuclease I indicates that the 3' terminus consists of duplex DNA. This is supported by the observation that pretreatment of intact DNA with exonuclease III resulted in susceptibility to degradation by exonuclease I (G. Lavelle and S. Mitra, manuscript in preparation). The fact that resistance to exonuclease I could not be reduced by prior denaturation supports the conclusion that the 3' end of viral DNA has the structure of a stable hairpin duplex.

Localization of the larger fragment of viral core DNA at the 3' terminus of KRV DNA was demonstrated by its selective loss after treatment of intact DNA with exonuclease III followed by exonuclease VII. One function of the 3' duplex may be to act as a primer for DNA synthesis. We have found that intact KRV DNA is a suitable template for DNA polymerase of bacteriophage T4 (S. Mitra and G. Lavelle, to be published). Similar results were obtained by Bourguignon et al. (1976) with MVM DNA. It seems probable that the 5' end of viral DNA is also a hairpin duplex, and that the small core-DNA fragment corresponds to this terminus.

The genomes of the nondefective parvoviruses KRV and MVM and the defective adeno-associated viruses share interesting structural properties of their terminal DNA sequences. The double-stranded form of AAV DNA contains self-complementary terminal sequences which are repeated at each end of the chromosome in a reverse, inverted fashion (Koczot et al. 1973). The 3' terminal hairpin of KRV DNA and the 5' terminal hairpin of MVM DNA (Bourguignon et al. 1976) consist of folded-back, self-complementary sequences. Indeed, both termini of these viral genomes exist as hairpin duplexes (see M. B. Chow and D. C. Ward, this volume). Whether these duplex termini contain inverted sequence repetitions is currently being investigated.

Because the DNAs of adenoviruses also have inverted terminal-sequence repetitions (Garon et al. 1972), and because inverted repetitions are known to occur in eukaryotic cell DNA (Wilson and Thomas 1974), it is interesting to speculate that terminal DNA structures of the type found in parvovirus genomes are important in the replication of eukaryotic DNA. Tattersall and Ward (1976) have proposed a rolling hairpin model of DNA replication which exploits such terminal duplex structures.

ACKNOWLEDGMENTS

We are grateful to Dr. R. W. Tennant for his encouragement and support of these studies. The research was sponsored by the Energy Research and Development Administration under contract with the Union Carbide Corporation.

REFERENCES

Berns, K. I. and T. J. Kelly, Jr. 1974. Visualization of the inverted terminal repetition in adeno-associated virus DNA. *J. Mol. Biol.* **82**:267.

Bourguignon, G. J., P. J. Tattersall, and D. C. Ward. 1976. DNA of minute virus of mice: Self-priming, nonpermuted, single-stranded genome with a 5'-terminal hairpin duplex. *J. Virol.* **20**:290.

Chase, J. W. and C. C. Richardson. 1974. Exonuclease VII of *Escherichia coli*. *J. Biol. Chem.* **249**:4545.

Duc-Nguyen, H., E. N. Rosenblum, and R. F. Zeigel. 1966. Persistent infection of a rat kidney cell line with Rauscher murine leukemia virus. *J. Bacteriol.* **92**:1133.

Espejo, R. T., E. S. Canelo, and R. L. Sinsheimer. 1969. DNA of bacteriophage PM2: A closed circular double-stranded molecule. *Proc. Natl. Acad. Sci.* **63**: 1164.

Garon, C. F., K. W. Berry, and J. A. Rose. 1972. A unique form of terminal redundancy in adenovirus DNA molecules. *Proc. Natl. Acad. Sci.* **69**:2391.

Gerry, H. W., T. J. Kelly, Jr., and K. I. Berns. 1973. Arrangement of nucleotide sequences in adeno-associated virus DNA. *J. Mol. Biol.* **79**:207.

Jovin, T. M., P. T. Englund, and L. L. Bertsch. 1969. Enzymatic synthesis of deoxyribonucleic acid. XXVI. Physical and chemical studies of a homogeneous deoxyribonucleic acid polymerase. *J. Biol. Chem.* **244**:2996.

Koczot, F. J., B. J. Carter, C. F. Garon, and J. A. Rose. 1973. Self-complementarity of terminal sequences within plus or minus strands of adenovirus-associated virus DNA. *Proc. Natl. Acad. Sci.* **70**:215.

Lavelle, G. and A. T. Li. 1977. Isolation of intracellular replicative forms and progeny single strands of DNA from parvovirus KRV in sucrose gradients containing gunidine hydrochloride. *Virology* **76**:464.

Lavelle, G. and S. Mitra. 1977. Secondary structure of Kilham rat virus DNA. *Am. Soc. Microbiol. Abstr.* p. 280.

Lavelle, G., C. Patch, G. Khoury, and J. Rose. 1975. Isolation and partial characterization of single-stranded adenoviral DNA produced during synthesis of adenovirus type 2 DNA. *J. Virol.* **16**:755.

Lehman, I. R. 1966. Exonuclease I (phosphodiesterase from *Escherichia coli*). In *Procedures in nucleic acid research* (ed. G. L. Cantoni and D. R. Davies), vol. 1, p. 203. Harper and Row, New York.

Maniatis, T., A. Jeffrey, and H. van de Sande. 1975. Chain length determination of small double- and single-stranded DNA molecules by polyacrylamide gel electrophoresis. *Biochemistry* **14**:3787.

May, P. and E. May. 1970. The DNA of Kilham rat virus. *J. Gen. Virol.* **6**:437.

McGeoch, D. J., L. V. Crawford, and E. A. C. Follett. 1970. The DNAs of three parvoviruses. *J. Gen. Virol.* **6**:33.

Mitra, S. and D. R. Stallions. 1976. The role of *Escherichia coli* dnaA gene and its integrative suppression in M13 coliphage DNA synthesis. *Eur. J. Biochem.* **67**:37.

Miyazawa, Y. and C. A. Thomas, Jr. 1965. Nucleotide composition of short segments of DNA molecules. *J. Mol. Biol.* **11**:223.

Niyogi, S. K. and C. A. Thomas, Jr. 1968. The stability of oligoadenylate-polyuridylate complexes as measured by thermal chromatography. *J. Biol. Chem.* **243**:1220.

Robinson, D. M. and F. M. Hetrick. 1969. Single-stranded DNA from the Kilham rat virus. *J. Gen. Virol.* **4**:269.

Salzman. L. A. 1977. Evidence for terminal S1-nuclease-resistant regions on single-stranded linear DNA. *Virology* **76**:454.

Salzman, L. A., W. L. White, and T. Kakefuda. 1971. Linear, single-stranded deoxyribonucleic acid isolated from Kilham rat virus. *J. Virol.* **7**:830.

Tattersall. P. and D. C. Ward. 1976. Rolling hairpin model for replication of parvovirus and linear chromosomal DNA. *Nature* **263**:106.

Tennant, R. W., K. R. Layman, and R. E. Hand. 1969. Effect of cell physiological state on infection by rat virus. *J. Virol.* **4**:872.

Tiselius, A., S. Hjertén, and O. Levin. 1956. Protein chromatography on calcium phosphate columns. *Arch. Biochem. Biophys.* **65**:132.

Wilson, D. A. and C. A. Thomas, Jr. 1973. Hydroxyapatite chromatography of short double-helical DNA. *Biochim. Biophys. Acta* **331**:333.

———. 1974. Palindromes in chromosomes. *J. Mol. Biol.* **84**:115.

Defective Particles of Parvovirus LuIII

HANS-PETER MÜLLER
MARKUS GAUTSCHI
GÜNTER SIEGL

Institute of Hygiene and Medical Microbiology
University of Bern
Bern, Switzerland

Virus particles containing only part of the viral genome that interfere with the replication of mature infectious virions have been described in many virus-cell systems (Huang 1973; Cole 1975; Schlesinger et al. 1975). There is also experimental evidence that particles having neither the characteristics of mature virions nor those of empty capsids can be isolated from tissue cultures infected with adeno-associated viruses (Parks et al. 1967), Kilham rat virus, H-1 virus, minute virus of mice, porcine parvovirus, feline panleukopenia virus, and LuIII virus (for references see Siegl 1976). However, no detailed studies concerning the physicochemical characteristics and the biological significance of these particles have been reported.

Propagation of parvovirus LuIII in HeLa cells frequently yields particles of this type in a quantity exceeding that of mature virions. This virus-cell system was therefore chosen as an experimental model in which to study both the characteristics of these defective or incomplete particles and the parameters influencing their synthesis in the infected cell.

METHODS

The methods used for the propagation of LuIII in HeLa and NB cells have been described (Siegl and Gautschi 1976), as have methods for the synchronization of HeLa and NB cells (Siegl and Gautschi, this volume), the extraction of virus with 0.2 M glycine buffer (pH 9), and the assay of viral

231

hemagglutinin (Hallauer et al. 1971). In contrast with previous studies, however, infectivity of virus samples was determined in a plaque assay in NB cell cultures according to the methods given by Ledinko (1971) for H-1 virus. Unless it is stated otherwise, virus particles were harvested by repeatedly freezing and thawing infected cultures and collecting the cellular debris by centrifugation at 12,000g for 30 min. Virus remaining in the supernatant was pelleted at 200,000g for 90 min and was added to the cell sediment. To achieve optimum extraction of virus, this concentrate was resuspended in 10 mM Tris-HCl (pH 7.4) and was incubated overnight at 37°C in the presence of DNase I (100 μg/ml), RNase (10 μg/ml), and 3 mM $MgCl_2$. At the end of the enzyme treatment, EDTA was added to a concentration of 2 mM and the sample was incubated at 37°C for 1 hr in the presence of 1% Sarkosyl NL-97. After centrifugation for 30 min at room temperature and 4500g, the virus released into the supernatant was analyzed either by sedimentation in linear 10–25% sucrose gradients or by equilibrium centrifugation in CsCl gradients, both made up in 10 mM Tris, 1 mM EDTA, 0.35 M NaCl, and 0.1% Sarkosyl (pH 8). Sedimentation was carried out either in a Beckman SW 27.1 rotor at 25,000 rpm and 4°C for 240 min or in an SW 50.1 rotor at 40,000 rpm for 90 min. Preparative as well as analytical density gradient centrifugations were performed in an SW 27.1 rotor at 15°C and 25,000 rpm for 60–72 hr or in a type-50 fixed-angle rotor at 42,000 rpm. Labeling and extraction of viral DNA from particles and the isolation and gradient analysis of viral replicative-form DNA from infected cells have been described in detail elsewhere (Siegl and Gautschi 1976).

RESULTS

Purification of Defective Particles

One of the main prerequisites of these studies was to establish an easy and fast technique for the isolation and quantitation of the defective particles (D particles) of LuIII virus. For this purpose, crude, concentrated virus suspensions were incubated for various times with various combinations of DNase, RNase, receptor-destroying enzyme, pronase, and EDTA, as well as Sarkosyl, and were finally centrifuged through linear 10–25% sucrose gradients made up with or without 0.1% Sarkosyl. After addition of a known quantity of ^{3}H-labeled LuIII virus particles it could be shown that up to 95% of the particles may be recovered if the cell pellets are incubated at 37°C with DNase, RNase, and 1% Sarkosyl, as outlined in "Methods." Moreover, addition of 0.1% Sarkosyl to the buffers of sucrose and CsCl gradients completely prevented the formation of virus aggregates during centrifugation. This fact is well illustrated by changes in the sedimentation spectrum of the particles. In gradients without

Sarkosyl this spectrum is always characterized by the pelleting of a significant proportion of the virus.

Characteristics of Defective Particles

Using the above methods we have demonstrated defective particles of parvovirus LuIII that sediment at 70–95S and band in CsCl gradients in the density range 1.32–1.38 g/cm³. Occasionally, distinct peaks could be detected. When D particles of various densities were isolated and recentrifuged in sucrose gradients in the presence of mature virions and empty capsids used as sedimentation markers, at least four particle species of distinct sedimentation and buoyancy behavior in CsCl could be distinguished. S values and buoyant densities of these particles were well correlated, and, as Table 1 illustrates, slower or faster sedimentation was paralleled by banding at lower or higher density, respectively.

The correlation between sedimentation and buoyancy behavior of D particles appeared to be related to the size of the DNA molecule they contained. Extraction of DNA from particles of known density and sedimentation of the nucleic acid in both neutral and alkaline gradients strongly suggested that the molecules were predominantly single-stranded. This assumption is also supported by the observation that they banded in CsCl at a density around 1.72 g/cm³, i.e., close to the

TABLE 1

Physicochemical Characteristics of Virus Particles Isolated from LuIII-Virus-Infected HeLa Cells

Particle			DNA	
density (g/cm³)	S	S (pH 13)	molecular weight (daltons)	normal genome (%)
Infectious virus				
1.41	110	16.0	1.6×10^6	100
Defective particles				
1.33–1.34	69	6.1	1.4×10^5	8.9
1.34–1.35	78	7.7	2.6×10^5	16.1
1.35–1.36	90	9.6	4.5×10^5	27.9
1.36–1.37	94	11.4	6.9×10^5	42.8

density recorded for the single-stranded DNA of mature infectious virions. The molecular weight of the DNA varied from 1.4×10^5 daltons (8.9% of the complete viral genome) in particles banding at 1.33–1.34 g/cm^3 to 6.9×10^5 daltons (42.8%) in those found at 1.36–1.37 g/cm^3 (Table 1).

Conditions Favoring the Synthesis of D Particles

Cultures of HeLa and NB cells were infected at a multiplicity of 10, and newly synthesized virus was labeled with [^{14}C]thymidine and ^{3}H-amino acids as described previously (Siegl and Gautschi 1976). Virus was harvested after complete lysis of the cells and was concentrated, purified, and analyzed by sedimentation in Sarkosyl-sucrose and Sarkosyl-CsCl density gradients. As Figure 1 shows, the virus harvests obtained from the two culture systems were quite different. In extracts obtained from HeLa cells the distribution of [^{14}C]TdR in the gradients indicated the presence of almost identical amounts of DNA in mature virions and D particles, whereas in suspensions harvested from NB cells only negligible quantities of D-particle DNA-label were found. However, since D particles contain only part of the viral genome, the real proportions of mature virions, D particles, and empty capsids are given by the fractions of the total ^{3}H-amino-acid-label in their respective density positions. Thus, virus suspensions harvested from HeLa cells contained only about 7% mature virions, 28% D particles, and 65% empty capsids. In contrast, 45% of all labeled capsids in NB cells banded like mature virions, 25% banded in the position of D particles, and only 30% behaved like empty capsids. The NB system also incorporated into particles about 5–10 times as much label as the HeLa system.

In most virus-cell systems, multiplicity of infection exerts a dramatic influence on the production of D particles. Randomly growing HeLa and NB cell cultures were therefore infected with unpurified virus pools at a multiplicity of 10^{-5} to 10^1 plaque-forming units (PFU) per cell. After labeling with [^{3}H]TdR, virus was harvested at the time of maximum cytopathogenic effect and was analyzed in Sarkosyl-sucrose gradients. Despite the 7-log difference in the concentration of virus particles in the inoculum, the amount of label incorporated into D particles relative to the quantity of label in 110S particles showed no comparable variation (see Table 2). Extreme values of only 180% and 55% were recorded in HeLa cells, and in NB cells the value of relative incorporation remained constant at about 20%.

In an extension of these dilution experiments, HeLa cells were inoculated at a multiplicity of 1 PFU/cell with 110S particles purified by repeated sedimentation in Sarkosyl-sucrose gradients as well as by banding in CsCl gradients. Even after pessimistic calculations, this virus suspen-

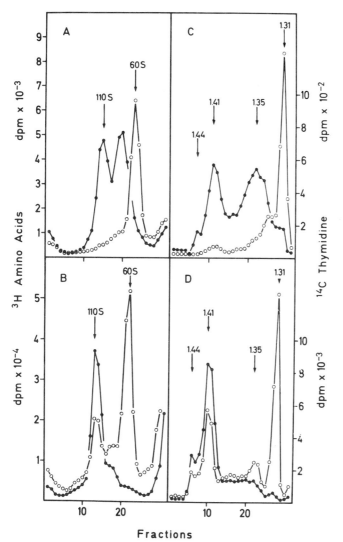

FIGURE 1

Sedimentation spectrum and buoyant-density profile of
LuIII-virus suspensions harvested from HeLa and NB cells.
Virus particles were labeled both with [¹⁴C]thymidine (●)
and with ³H-amino acids (○). (A) and (B) illustrate the
sedimentation of LuIII virus grown in HeLa and NB cells,
respectively. (C) and (D) show the density distribution of
virus isolated from HeLa and NB cells, respectively.

TABLE 2

Variation with Multiplicity of Infection (m.o.i.) in the Proportion of D Particles and 110S Virions in Harvests of HeLa and NB Cells.

Cells	m.o.i.	$\dfrac{\text{cpm 70–90S}}{\text{cpm 100S}} \times 100$
HeLa	10	180±10
	1	125±15
	10^{-1}	95±10
	10^{-2}	95±10
	10^{-3}	80±15
	10^{-4}	60±15
	10^{-5}	55±15
NB	10	20±10
	1	20±10
	10^{-1}	20±10
	10^{-2}	20±10
	10^{-3}	15±10
	10^{-4}	20±10
	10^{-5}	20±15

sion contained not more than 2–3% D particles. Infected cells nevertheless yielded virus suspensions in which D particles contained 80% as much label as found in the 110S virions. In terms of particles, this would mean that about 4–5 times as many D particles as mature virions were synthesized.

A further experiment was performed to test whether or not frequent undiluted passage of virus harvests resulted in an increase in the synthesis of D particles in HeLa cells. The passages were started with a highly purified suspension of 110S particles. As is evident from Figure 2, virus harvested at the end of the first passage already contained significant numbers of D particles. During the 30 subsequent passages the relative incorporation of label into D particles and mature virions changed several times, showing an undulating increase and decrease in the number of D particles with passage number. Waves of "high D-particle producer" or "low producer" cultures extended over about 5–8 individual passages. In contrast with the undulating relative number of D particles in virus harvests, the infectivity titer of virus suspensions declined slowly but constantly: passage 1 yielded 10^7 PFU/ml; suspensions harvested from passage number 25 contained only 10^3 PFU/ml.

FIGURE 2
Sedimentation spectra of
[³H]TdR-labeled virus particles isolated from HeLa cells
during undiluted passage of
LuIII virus. Gradients were
normalized with respect to
the number of cpm sedimenting in the 110S position

$$\left[\frac{\text{cpm (fraction)}}{\text{cpm (110S peak fraction)}} \right]$$

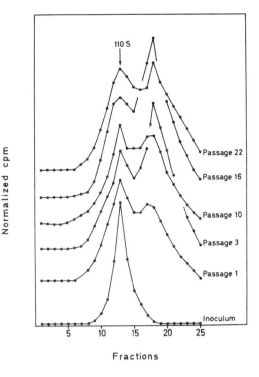

Fractions

Synthesis of D Particles Under One-Step Growth Conditions

In HeLa cells synchronized for DNA synthesis and infected at the time of
release from the synchronization block, maturation of progeny virions
could be detected as early as 9–11 hr after infection (Siegl and Gautschi
1973a,b). To study the synthesis of D particles under these conditions,
cultures were labeled at 2-hr intervals between 11 and 19 hr post-infection
(p.i.). Virus was harvested at the end of the labeling periods and was
analyzed on Sarkosyl-sucrose gradients as described. The results pic-
tured in Figure 3 indicate significant differences in the synthesis of D
particles in early and late phases of the replication cycle of LuIII virus. In
the period between 11 and 13 hr after infection only about 10% the
amount of label incorporated into 110S virions could be detected in D
particles, whereas more than 50% was found in D particles at 17–19 hr p.i.
The same experiments also showed that the rate of synthesis of mature
virions increased only up to 15 hr after infection. D particles, however,
were produced at a still increasing rate up to the end of the replication
cycle.

The D particles could be artificial products resulting from the interruption of virus maturation at the same time of harvesting. To exclude this

FIGURE 3

Synthesis of the D particles of LuIII virus in HeLa cells under one-step growth conditions. The proportion of defective particles increases almost linearly up to the end of the replication cycle.

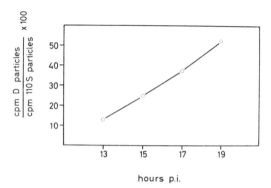

possibility, virus particles labeled in duplicate cultures were isolated either at the end of labeling or after a 2–5-hr chase period in the absence of [³H]TdR. The relative amounts of 110S and D particles detectable in these cultures at various times after infection did not change during the chase period. It may be assumed, therefore, that D particles do not represent intermediary structures in the normal process of LuIII-virus maturation.

Coinfection with Mature Virions and D Particles

Both 110S virions and D particles purified and isolated as described earlier in this paper were assayed for infectivity as well as for hemagglutinin. For the 110S virions, one hemagglutinating unit (HAU) proved to be equivalent to 225 PFU, whereas the respective value determined for D particles was 0.05 PFU/HAU. HeLa cells synchronized for DNA synthesis were infected at the time of release from the synchronization block. The inoculum contained 10 PFU/cell of 110S virion and, on the basis of hemagglutination units, 0, 50, 100, 200, and 500 times the amount of D particles. Additional cultures received 0, 100, and 500 equivalents of D particles only. The infected cultures were then labeled with [³H]TdR (10 μCi/ml), either between 9 and 20 hr after infection (for the isolation of virus particles) or between 14 and 16 hr (if virus-specific, low-molecular-weight DNA was to be extracted).

Cultures infected only with the 110S virus pool yielded newly labeled 110S virions and D particles, as well as virus-specific DNA of low molecular weight similar to that described elsewhere (see Siegl and Gautschi, this volume). Infection with purified D particles, on the other hand, did not give rise to the production of progeny particles, and no specific DNA could be extracted. Finally, mixing of mature virions and D particles did not change the relative amounts of 110S and D particles; however, both the recovery of progeny virus and the recovery of low-molecular-weight DNA decreased with increasing quantities of D particles in the inoculum.

Two hundred equivalents of D particles reduced replication of LuIII virus to an undetectable level.

DISCUSSION

As in many other virus-cell systems, the major difficulty in the study of defective particles of LuIII virus was the purification and quantitation of particles of this type. Contaminating proteins and nucleic acids can be removed readily from virus suspensions because the virion is exceptionally stable. However, purified particles tend to form stable aggregates containing mature virions, D particles, and empty capsids in unpredictable proportions. The data presented in this article now provide reliable evidence that incubation of virus suspensions at 37°C with 1% Sarkosyl and addition of 0.1% Sarkosyl to the gradients circumvents these problems. The D particles isolated from HeLa cells by means of the described technique were heterogeneous. At least four particle species of distinct sedimentation and buoyancy behavior could be separated, and it is quite possible that more sophisticated separation methods might reveal additional species. The individual particle species are also characterized by the exceptionally small size of the single-stranded DNA they contain. At present, however, there is only indirect evidence that this DNA is derived from the viral genome.

Formation of D particles was favored by several experimental conditions, such as the host-cell system, multiplicity of infection, and passage history of the inoculated virus suspension. Of these variables, the type of cells used for the propagation of LuIII virus proved to be the most important. Both the total quantity of virus particles synthesized and the composition of the virus harvests could be altered by growing LuIII either in HeLa or in NB cells. Moreover, only HeLa cells reacted to infection at low multiplicities with a reduced synthesis of D particles. In NB cells, on the other hand, the proportions of mature virions and D particles remained constant over a considerable range of input multiplicities.

It is conceivable that these D-particle structures represent intermediates in the maturation of LuIII virions comparable to those found during adenovirus assembly (Sundquist et al. 1973). In this case, they would be expected to accumulate preferentially late in the replication cycle of the virus, especially in cells supporting virus synthesis at low efficiency. The pulse-chase experiments in which the ratio of labeled 110S particles and D particles remained constant even after a prolonged chase period seem to rule out this possibility. However, this result may only reflect the reduced rate of virus maturation in HeLa cells compared to that in NB cells. An alternative mechanism for the formation of D particles is suggested by the synthesis of defective poliovirus particles (Cole 1975). Here the nucleic acid molecule of the D particles is capable of autonomous replication, but formation of particles depends on capsid proteins coded for by the RNA of standard virions. The results of experiments in which cells were infected with or without mature LuIII virus and purified D particles do not support such a hypothesis. The synthesis of virus-specific low-molecular-weight DNA could not be demonstrated in cells infected only with D particles; neither was the yield of D particles increased by simultaneous inoculation of 110S virions and defective particles. On the contrary, D particles added to the inoculum interfered with the replication of mature virions, causing a reduction in the recovery of both total virus particles and total virus-specific replicative DNA. Since a great excess of D particles was required to inhibit replication of LuIII virus completely, it is possible that this type of interference results from a competition of 110S virions and D particles for the receptors at the cell surface.

REFERENCES

Cole, C. N. 1975. Defective interfering (DI) particles of poliovirus. *Prog. Med. Virol.* **20**:180.

Hallauer, C., G. Kronauer, and G. Siegl. 1971. Parvoviruses as contaminants of permanent human cell lines. I. Virus isolations from 1960–1970. *Arch. gesamte Virusforsch.* **35**:80.

Huang, A. S. 1973. Defective interfering viruses. *Annu. Rev. Microbiol.* **27**:101.

Ledinko, N. 1971. Plaque assay of the effects of cytosine arabinoside and 5-iodo-2′-deoxyuridine on the synthesis of H-1 virus particles. *Nature* **214**:1346.

Parks, W. P., J. L. Melnick, R. Rongey, and H. D. Mayor. 1967. Physical assay and growth cycle studies of a defective adenosatellite virus. *J. Virol.* **1**:171.

Schlesinger, S., B. Weiss, and D. Dohner. 1975. Defective particles in alphavirus infections. *Med. Biol.* **53**:372.

Siegl, G. 1976. The parvoviruses. *Virol. Monogr.* **15**:1.

Siegl, G. and M. Gautschi. 1973a. The multiplication of parvovirus LuIII in a synchronized culture system. I. Optimum conditions for virus replication. *Arch. gesamte Virusforsch.* **40**:105.

———. 1973b. The multiplication of parvovirus LuIII is a synchronized culture system. II. Biochemical characteristics of virus replication. *Arch. gesamte Virusforsch.* **40**:119.

———. 1976. Multiplication of parvovirus LuIII in a synchronized culture system. III. Replication of viral DNA. *J. Virol.* **17**:841.

Sundquist, B., E. Everitt, L. Philipson, and S. Hoglund. 1973. Assembly of adenoviruses. *J. Virol.* **11**:449.

PARVOVIRUS DNA REPLICATION

Self-Primed Replication of Adeno-Associated-Virus DNA

STEPHEN E. STRAUS

Division of Infectious Disease
Department of Medicine
Washington University
St. Louis, Missouri 63110

EDWIN D. SEBRING
JAMES A. ROSE

Laboratory of Biology of Viruses
National Institute of Allergy and Infectious Diseases
National Institutes of Health
Bethesda, Maryland 20014

The growth of the defective parvovirus AAV (adeno-associated virus) is unconditionally dependent upon coinfection with a helper adenovirus. AAV contains a linear single-stranded DNA genome of approximately 1.4×10^6 daltons (Rose 1974). There are two noteworthy features of this DNA. First, the complementary plus and minus DNA strands are encapsidated separately (Rose et al. 1969). Second, the terminal base sequences of AAV DNA include inverted repetitions which may be partially or totally palindromic (Koczot et al. 1973; Berns and Kelly 1974; Spear et al. 1977).

Prior to our initial reports characterizing the replicative intermediates of AAV DNA there was little knowledge of the mechanism of DNA synthesis of this parvovirus (Sebring et al. 1975; Straus et al. 1976b). Among certain nondefective parvoviruses there was evidence for linear duplex replicative forms involved in the replication of minute virus of mice (MVM) (Dobson and Helleiner 1973; Tattersall et al. 1973), Kilham rat virus (Salzman and White 1973), and H-1 (Rhode 1974). Tattersall et al. (1973) noted that a portion of the replicating molecules of MVM DNA were capable of rapid renaturation, and this in turn suggested that MVM DNA synthesis might be mediated by a self-priming mechanism.

During the past two years the understanding of parvovirus DNA replication has advanced considerably. The present report reiterates a number of observations that had been made regarding the isolation and identification of AAV DNA replicative intermediates (Straus et al. 1976b), describes new data which further clarify the origin and nature of these intermediates, and, finally, reviews a model of self-primed DNA synthesis that was suggested by these studies and remains consistent with the more recent investigations of both the defective (Handa et al. 1976) and the nondefective parvoviruses (Gunther and May 1976; Siegl and Gautschi 1976; Tattersall and Ward 1976; Lavelle and Li 1977; Rhode 1977; Salzman 1977; Singer and Rhode 1977a,b).

MATERIALS AND METHODS

Infections were carried out in KB-cell spinner cultures at 39.5°C, unless it is stated otherwise. Procedures for the growth, purification, and assay of AAV2, and of the wild-type strain of adenovirus type 5 (Ad5WT) and the adenovirus-5 DNA-minus temperature-sensitive mutant ts125 (generous gifts of Dr. M. Ensinger and Dr. H. Ginsberg), have been described previously (Ensinger and Ginsberg 1972; Straus et al. 1976a). The conditions of infection and labeling have also been described (Straus et al. (1976a). Viral DNA was extracted by a modification (Straus et al. 1976a) of the Hirt (1967) procedure. Free-solution DNA-DNA annealing reactions, with monitoring by hydroxyapatite (HAP) chromatography, were performed as detailed by Thoren et al. (1972). Labeled viral DNA components were analyzed by sedimentation in gradients of 5–30% neutral sucrose or 10–30% alkaline sucrose (Straus et al. 1976b), by benzoylated-naphthoylated DEAE (BND)-cellulose chromatography (Sedat et al. 1967), and by digestion with the single-strand-specific (S1) nuclease of *Aspergillus oryzae* (Vogt 1973). The restriction endonuclease R·EcoRI (Yoshimori 1971) was kindly provided by Dr. P. Howley. Cleavage reactions, preparation of the 0.5% agarose-3% polyacrylamide slab gels, electrophoresis, and autoradiography were performed as described by Howley et al. (1975).

RESULTS

Purification of Replicating AAV DNA

The analysis of AAV DNA synthesis was facilitated by the preparation of an extract of replicating DNA molecules that is essentially free of contaminating host-cell and helper-adenovirus DNA. This was accomplished in two steps. First, the DNA-minus mutant ts125 at the nonpermissive temperature 39.5°C was used to support normal levels of AAV DNA synthesis while adenovirus DNA synthesis was restricted to less than 0.5% of that which is usually achieved in coinfections with the Ad5WT helper (37°C) (Straus et al. 1976a). Second, the DNA of the coinfected cells was extracted with a modified version of the Hirt procedure which permits the recovery of about 80% of the virus-specific DNA counts in the supernatant fraction while nearly all of the cellular DNA is found in the pellet. A series of control experiments revealed that this modified Hirt technique preserves the molecular form of extracted DNA molecules, i.e., it does not promote the annealing of complementary single strands, the denaturation of duplex molecules, or the introduction of single-stranded nicks.

Velocity Sedimentation of Replicating AAV DNA in Neutral Sucrose

KB cells were coinfected with Ad5WT and AAV2 (at 37°C) or with ts125 and AAV2 (at 39.5°C). Viral DNA was labeled with [³H]thymidine and extracted by the modified Hirt method. The sedimentation profiles of the Hirt supernatants are identical in neutral sucrose except for the absence of the adenovirus components (≥31S) in extracts from ts125 coinfections (Straus 1976a). Free-solution hybridization reactions demonstrated that the DNA counts in individual fractions collected from the 10–21S region of these gradients are from 91.1% to 96.4% AAV-specific. Moreover, DNA counts from this region account for an average of 95.6% of the total labeled AAV DNA in each gradient.

Pulse-Chase Kinetics of Replicating AAV DNA

KB cells were coinfected with ts125 and AAV2 (at 39.5°C) for 16 hr. The culture was pulse-labeled with [³H]thymidine for 2 min and divided into three equal portions. These were extracted either (i) immediately (Fig. 1A, D), (ii) after a 30-min chase with an excess of cold thymidine (Fig. 1B,E), or (iii) after a 6-hr chase with an excess of cold thymidine (Fig. 1C,F). The sedimentation profiles of the AAV DNA in neutral sucrose (Fig. 1A,B,C) demonstrate that the pulse-labeled-DNA counts in the 16–21S region of the gradients chase sequentially into 15S molecules and then into genome-length, single-stranded progeny DNA (20S). There are two other notable features of the gradients depicted in Figure 1. First, counts that initially appear in faster-sedimenting molecules subsequently chase into genome-length (16S in alkali) DNA. This indicates that there are concatemeric AAV DNA replicative intermediates. Second, pulse-labeled DNA that sediments slowly (<16S) does not appear to chase into single strands, which indicates that this material is not a precursor of progeny molecules.

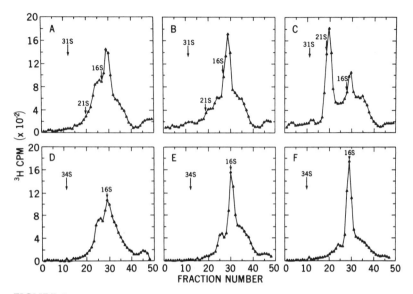

FIGURE 1

Neutral (*A–C*) and alkaline (*D–F*) sucrose gradient analyses of the pulse-chase kinetics of replicating AAV DNA. Ts125 and AAV2-infected KB cells were extracted by a modified Hirt method after a 2-min pulse of [³H]thymidine (*A,D*), a 2-min pulse and 30-min chase (*B,E*), and a 2-min pulse and 6-hr chase (*C,F*). The arrows at 31S, 21S, and 16S in the neutral gradients indicate the positions of cosedimented ¹⁴C-labeled Ad5 DNA, ¹⁴C-labeled SV40 DNA I, and ¹⁴C-labeled SV40 DNA II, respectively. The arrows at 34S and 16S in the alkaline gradients correspond to the positions of ¹⁴C-labeled Ad5 DNA and ¹⁴C-labeled SV40 DNA III, respectively.

When fractions pooled from preparative neutral sucrose gradients matching those in Figure 1 were analyzed, additional features of AAV DNA replication could be discerned (data not shown). The progeny single-stranded DNA (20S in Fig. 1C) was collected and reannealed. After resedimentation in neutral sucrose, many of the counts appeared in a peak of linear duplex molecules at 15S, which is consistent with the expectation that the single-stranded component should contain both plus and minus strands. The pulse-labeled 15S DNA (as in Fig. 1A) was denatured, neutralized, and resedimented in neutral sucrose. The dominant peak of activity appeared at 20S, which indicated that the 15S DNA includes linear duplex molecules which may be denatured to form genome-length single strands. Unexpectedly, when the same pool of 15S DNA was rerun in alkaline sucrose gradients, many of the DNA counts sedimented more rapidly than the genome-length strands. This suggested that the 15S peak might also contain hairpinlike intermediates which unfold under denaturing conditions to form continuous linear strands that exceed genome length.

Hydroxyapatite Chromatography of Replicating AAV DNA

In order to explore further the possibility that some of the intermediates contained large, self-complementary regions, we exploited the property of rapid reassociation that these molecules would be expected to possess. When pulse-labeled replicating AAV DNA was alkaline-denatured and chromatographed on HAP, 55% of labeled counts eluted as double-stranded molecules. The neutral and alkaline sucrose sedimentation profiles of these rapidly annealing molecules were then compared with those of the DNA that eluted as single strands (Fig. 2). The pool of rapidly annealing DNA included most of the counts that sedimented with forms

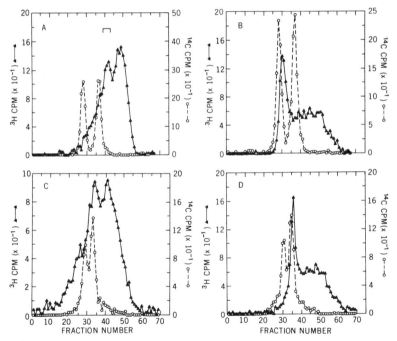

FIGURE 2

Neutral (*A*,*B*) and alkaline (*C*,*D*) sucrose gradient analyses of replicating AAV DNA after denaturation and HAP chromatography. The sedimentation profiles depicted are those of the rapidly annealing molecules that eluted as double strands (*A*,*C*) and the DNA that eluted as single strands (*B*,*D*). Cosedimented ^{14}C-labeled SV40 DNA is plotted with a dashed line. The marker peaks are at the 16S and 21S positions in neutral sucrose and at 16S and 18S in alkaline sucrose.

of less than and greater than genome length. Approximately 54% of the counts in rapidly annealing DNA that sedimented as genome-length (15S) molecules in neutral sucrose (bracketed fractions, Fig. 2A) sedimented faster than the genome-length (16S) forms in alkaline sucrose gradients (data not shown).

BND-Cellulose Chromatography of Replicating Forms

The experiments discussed above permitted us to conclude that large regions of some of the replicating molecules are self-complementary. To isolate that portion of each of these molecules that was almost completely self-complementary, AAV DNA extracted by the Hirt method was denatured, chilled, diluted, neutralized, and applied to a column of BND-cellulose. The conditions were carefully controlled to diminish markedly the probability of intermolecular annealing. The DNA that eluted as double-stranded was denatured, chilled, neutralized, and subjected to another cycle of denaturation and quick cooling followed by two cycles of BND-cellulose chromatography. Approximately 19% of the starting counts were recovered in the final double-stranded fraction. Portions of this fraction and of the starting material were compared by sedimentation in neutral and alkaline sucrose gradients. Figure 3 demonstrates that the double-stranded eluate from BND-cellulose is enriched for molecules sedimenting faster than genome-length forms in alkaline sucrose. A portion of the rapidly annealing molecules sedimented as a sharp peak of twice-genome-length (21S) DNA in alkaline sucrose (Fig. 3D). When collected and rerun in neutral sucrose, these dimers sedimented as genome-length duplexes (15S). This specific population of intermediates was selected for subsequent analysis by digestion with S1 nuclease and restriction endonuclease R·EcoRI.

The rapidly annealing DNA was also shown to include concatemers of greater than dimer length. Fractions corresponding to a tetramer (27S) were collected from alkaline sucrose. Upon neutralization and re-sedimentation in neutral sucrose gradients, the 27S DNA sedimented as duplex structures of approximately half their single-stranded length. Intermediates of greater than tetramer length were also observed but were not analyzed further.

S1-Nuclease Digestion of the Dimeric Intermediate

One reasonable model that is consistent with the data presented thus far postulates that the rapidly annealing molecules are linear forms composed of plus and minus strands which are covalently linked end to end. Under nondenaturing conditions, such molecules would form hairpin structures. Inherent in this model is the prediction that at the hairpin loop

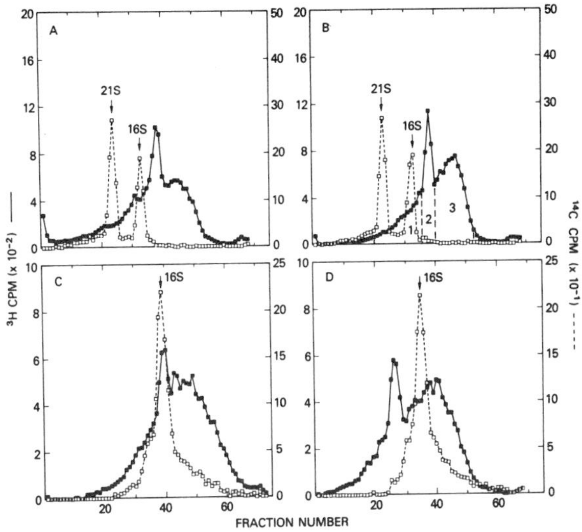

FIGURE 3
Neutral (A,B) and alkaline (C,D) sucrose gradient analyses of rapidly annealing DNA. The sedimentation profiles depicted are those of the Hirt-method-extracted DNA (A,C) and the rapidly annealing DNA isolated by BND-cellulose chromatography (B,D). Cosedimented ^{14}C-labeled SV40 DNA is plotted with a dashed line. Preparative gradients equivalent to that of (B) were divided into individual pools for subsequent examination.

there are several nucleotides that are sterically constrained and thus unable to form base pairs. That this single-stranded region is relatively short is evident from the inability of the rapidly annealing forms to be retained by BND-cellulose. To detect the presence of this single-stranded loop, we subjected the highly purified dimeric intermediates to digestion with the single-strand nuclease S1 (Fig. 4). Prior to digestion, these intermediates unfold in alkaline sucrose and sediment at 21S. Digestion of the dimer, however, generates a molecule that denatures in alkali and sediments as genome-length (16S) strands of fairly homogeneous length. Therefore, these duplex molecules appear to contain a single-stranded region at one end joining the two strands, which is consistent with a unit-length hairpin structure.

R·EcoRI Digestion of the Dimeric Intermediate

We turned next to the question of whether there is a unique polarity to the hairpin intermediate. Endonuclease R·*Eco*RI cleaves AAV DNA twice,

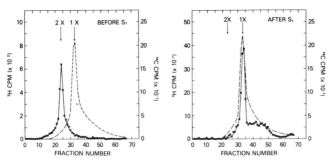

FIGURE 4
Alkaline sucrose gradients of dimer-length replicating AAV DNA (A) before and (B) after treatment with S1 nuclease. Cosedimented ^{14}C-labeled SV40 DNA III is plotted with a dashed line.

forming two long, terminal fragments, A and B, and one short, internally situated fragment, C (Carter and Khoury 1975). The dimeric intermediate was cleaved by R·EcoRI (Fig. 5). Half of the reaction mixture was applied directly to a polyacrylamide gel (column 2). The remainder of the cleaved DNA was denatured in alkali and neutralized immediately prior to electrophoresis (column 3). The cleavage patterns of the dimeric intermediate (columns 2, 3) include a slowly moving band of incompletely digested or

FIGURE 5
Autoradiogram depicting the electrophoretic pattern of an acrylamide-agarose gel containing the R·EcoRI cleavage products of (1) virion-purified, ^{32}P-labeled duplex AAV DNA, (2) dimer-length, replicating ^{32}P-labeled AAV DNA, (3) dimer-length, replicating, ^{32}P-labeled AAV DNA that was denatured and neutralized immediately prior to electrophoresis, and (4) the marker HindII and III cleavage products of ^{32}P-labeled SV40 DNA.

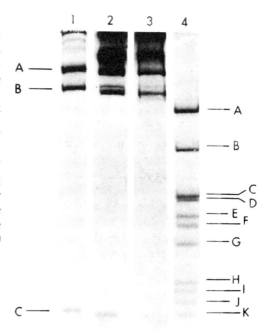

nonspecifically aggregated forms (Carter and Khoury 1975). Unlike the terminal fragments of the linear duplex AAV DNA which migrate as single bands (column 1), each of the terminal fragments of the dimeric intermediate appears to migrate as double bands. After denaturation and neutralization there is an enrichment for the faster-migrating components of each of these double bands (column 3). These observations suggest that each of the termini of the dimeric intermediate may possess either the extended linear form or the slightly shorter (and hence faster-migrating) hairpin structure.

DISCUSSION

As a model for the replication of certain viral genomes, discontinuous synthesis (Okazaki et al. 1968; Sugino et al. 1972), wherein short RNA-primed DNA segments are added sequentially and ligated to form long strands, has accumulated substantial support. However, this model fails to explain the completion of the 5′ end of a replicating linear DNA molecule unless additional mechanisms and specialized terminal structures are postulated (Watson 1972). These inherent limitations are also apparent when one considers the replication of AAV DNA. The AAV genome is single-stranded and linear. Covalently closed circular intermediates have not been observed (E. D. Sebring, unpublished results). The AAV genome possesses an inverted terminal repetition, and replication proceeds via concatemeric intermediates. AAV DNA replication generates fragments of less than genome length (10–12S). This DNA, however, does not participate in a discontinuous mode of synthesis, since, after pulse-labeling, the counts in this material do not chase into completed strands (Fig. 1C). Moreover, preliminary experiments failed to demonstrate the covalent attachment of RNA sequences to very briefly labeled AAV DNA intermediates (S. E. Straus, unpublished observations).

The data reviewed here are consistent with a model of self-primed DNA synthesis (Fig. 6). AAV DNA contains palindromic terminal sequences (Fig. 6A). The self-complementary 3′ terminus folds back on itself and serves as a primer for the synthesis of the complementary strand. AAV DNA replication may be initiated by hairpin formation at the 3′ end of either strand, in accord with the results of the R·EcoRI experiment (Fig. 5) and with the observation that virions carry either plus or minus strands.

Self-primed synthesis would generate a hairpin whose complementary strands remain covalently attached by a short, S1-nuclease-sensitive segment. A similar reinitiation of self-primed synthesis at the 3′-OH end of this unit-length hairpin would yield a twice-genome-length hairpin composed of alternating plus and minus strands. Repeated cycles of

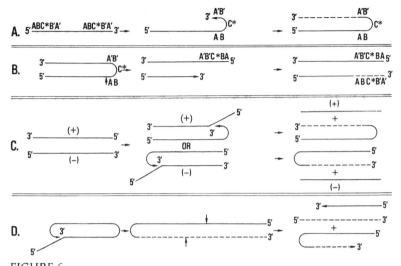

FIGURE 6

Model for the self-primed replication of AAV DNA. In (A) and (B) letters represent a palindromic terminal sequence. (A) Formation of hairpin intermediate, (B) conversion of hairpin to linear duplex form and regeneration of transferred terminal palindrome, (C) displacement synthesis of single-stranded progeny, and (D) formation and the specific cleavage of long palindromic molecules.

self-priming would result in forms several times the length of the AAV genome.

This model of self-primed replication implies the existence of a cellular or virus-specified endonuclease that might act in two ways. It may nick at a specific site near the hairpin loop, thus permitting regeneration of the terminal sequences (Fig. 6B) according to the steps outlined by Cavalier-Smith (1974) or to those recently suggested by Tattersall and Ward (1976). In addition, the endonuclease may process the long concatemers into genome-length hairpins and linear duplexes by site-specific, staggered cuts as proposed for T7 DNA (Watson 1972). After cleavage, the 3' ends are regenerated by 5'-to-3' synthesis (Fig. 6D). The genome-length linear duplexes generated by either of the putative endonuclease-dependent steps may serve as templates for the displacement synthesis of both plus and minus progeny strands (Fig. 6C).

The existence of true concatemers of AAV DNA has been questioned recently. Handa and Shimojo (1977) examined the synthesis of AAV DNA intermediates in isolated human-embryo-kidney-cell nuclei. It was reported that the concatemers may be converted to genome-length molecules by digestion with papain, trypsin, and SDS and re-extraction with phenol. We have been unable to substantiate a role for proteins in the maintenance of the concatemeric form (J. Wang and J. Rose, unpublished results). Digestion of the highly purified dimeric intermediates was performed with protease K (250 μg/ml of enzyme in 0.05 M Tris, pH 8, 0.001 M EDTA at 37°C for 30 min). In alkaline sucrose gradients the sedimentation profiles of the starting material and of the digested product were identical (21S).

Features of the self-priming model are applicable to DNA replication by other organisms as well. The data reviewed by Berns and Hauswirth (this volume) suggest that the replication of nondefective-parvovirus DNA proceeds in a similar fashion. Cavalier-Smith (1974) and Tattersall and Ward (1976) have considered in detail the utility of a self-priming mechanism for the replication of eukaryotic chromosomes and other linear DNA.

ACKNOWLEDGMENTS

We are indebted to Peter Howley for assistance in performing the endonuclease digestions, gel electrophoresis, and autoradiography. We also thank Dr. Lawrence Gelb for his critical review of this manuscript, and Patricia Garner and Cindy Cunningham for their secretarial skills.

REFERENCES

Berns, K. I. and T. J. Kelly, Jr. 1974. Visualization of the inverted terminal repetition in adeno-associated virus DNA. *J. Mol. Biol.* 82:267.

Carter, B. J. and G. Khoury. 1975. Specific cleavage of adenovirus-associated virus DNA by restriction endonuclease R·EcoRI—Characterization of cleavage products. *Virology* 63:523.

Cavalier-Smith, T. 1974. Palindromic base sequences and replication of eukaryote chromosome ends. *Nature* 250:467.

Dobson, P. R. and C. W. Helleiner. 1973. A replicative form of the DNA of minute virus of mice. *Can. J. Microbiol.* 19:35.

Ensinger, M. J. and H. S. Ginsberg. 1972. Selection and preliminary characterization of temperature-sensitive mutants of type 5 adenovirus. *J. Virol.* 10:328.

Gunther, M. and P. May. 1976. Isolation and structural characterization of monomeric and dimeric forms of replicative intermediates of Kilham rat virus DNA. *J. Virol.* 20:86.

Handa, H. and H. Shimojo. 1977. Viral DNA synthesis in vitro with nuclei isolated from adeno-associated virus type 1-infected cells. *Virology* 77:424.

Handa, H., H. Shimojo, and K. Yamaguchi. 1976. Multiplication of adeno-associated virus type 1 in cells coinfected with a temperature-sensitive mutant of human adenovirus type 31. *Virology* 74:1.

Hirt, B. 1967. Selective extraction of polyoma DNA from infected mouse cell cultures. *J. Mol. Biol.* 26:365.

Howley, P. M., M. F. Mullarkey, K. K. Takemoto, and M. A. Martin. 1975. Characterization of human papovavirus BK DNA. *J. Virol.* 15:173.

Koczot, F. J., B. J. Carter, C. F. Garon, and J. A. Rose. 1973. Self-complementarity of terminal sequences within plus or minus strands of adenovirus-associated virus DNA. *Proc. Natl. Acad. Sci.* **70**:215.

Lavelle, G. and A. T. Li. 1977. Isolation of intracellular replicative forms and progeny single strands of DNA from parvovirus KRV in sucrose gradients containing guanidine hydrochloride. *Virology* **76**:464.

Okazaki, R., T. Okazaki, K. Sakabe, K. Sugimoto, and A. Sugino. 1968. Mechanism of DNA chain growth. I. Possible discontinuity and unusual secondary structure of newly synthesized chains. *Proc. Natl. Acad. Sci.* **59**:598.

Rhode, S. L. III. 1974. Replication process of the parvovirus H-1. II. Isolation and characterization of H-1 replicative form DNA. *J. Virol.* **13**:400.

———. 1977. Replication process of the parvovirus H-1. VI. Characterization of a replication terminus of H-1 replicative-form DNA. *J. Virol.* **21**:694.

Rose, J. A. 1974. Parvovirus reproduction. In *Comprehensive virology* (ed. H. Fraenkel-Conrat and R. Wagner), vol. 3, p. 1. Plenum, New York.

Rose, J. A., K. I. Berns, M. D. Hoggan, and F. J. Koczot. 1969. Evidence for a single-stranded adenovirus-associated virus genome: Formation of a DNA density hybrid on release of viral DNA. *Proc. Natl. Acad. Sci.* **64**:863.

Salzman, L. A. 1977. Evidence for terminal S1-nuclease-resistant regions on single-stranded linear DNA. *Virology* **76**:454.

Salzman, L. A. and W. L. White. 1973. In vivo conversion of the single-stranded DNA of the Kilham rat virus to a double-stranded form. *J. Virol.* **11**:299.

Sebring, E. D., S. E. Straus, H. S. Ginsberg, and J. A. Rose. 1975. Concatemeric intermediates in adenovirus-associated virus DNA replication. *Fed. Proc.* **34**:639.

Sedat, J. W., R. B. Kelly, and R. L. Sinsheimer. 1967. Fractionation of nucleic acid on benzoylated-naphthoylated DEAE cellulose. *J. Mol. Biol.* **26**:537.

Siegl, G. and M. Gautschi. 1976. Multiplication of parvovirus LuIII in a synchronized culture system. III. Replication of viral DNA. *J. Virol.* **17**:841.

Singer, I. I. and S. L. Rhode III. 1977a. Replication process of the parvovirus H-1. VII. Electron microscopy of replicative-form DNA synthesis. *J. Virol.* **21**:713.

———. 1977b. Replication process of the parvovirus H-1. VIII. Partial denaturation mapping and localization of the replication origin of H-1 replicative-form DNA with electron microscopy. *J. Virol.* **21**:724.

Spear, I., W. Hauswirth, K. Fife, and K. Berns. 1977. Studies on the terminal structure of adeno-associated virus DNA. *Am. Soc. Microbiol. Abstr.* **S8**:280.

Straus, S. E., H. S. Ginsberg, and J. A. Rose. 1976a. DNA-minus temperature-sensitive mutants of adenovirus type 5 help adenovirus-associated virus replication. *J. Virol.* **17**:140.

Straus, S. E., E. D. Sebring, and J. A. Rose. 1976b. Concatemers of alternating plus and minus strands are intermediates in adenovirus-associated virus DNA synthesis. *Proc. Natl. Acad. Sci.* **73**:742.

Sugino, A., S. Hirose, and R. Okazaki. 1972. RNA-linked nascent DNA fragments in *Escherichia coli*. *Proc. Natl. Acad. Sci.* **69**:1863.

Tattersall, P. and D. C. Ward. 1976. Rolling hairpin model for replication of parvovirus and linear chromosomal DNA. *Nature* **263**:106.

Tattersall, P., L. V. Crawford, and A. J. Shatkin. 1973. Replication of the parvovirus MVM. II. Isolation and characterization of intermediates in the replication of the viral deoxyribonucleic acid. *J. Virol.* **12**:1446.

Thoren, M. M., E. D. Sebring, and N. P. Salzman. 1972. Specific initiation site for simian virus 40 deoxyribonucleic acid replication. *J. Virol.* **10**:462.

Vogt, V. 1973. Purification and further properties of single-strand-specific nuclease from *Aspergillus oryzae*. *Eur. J. Biochem.* **33**:192.

Watson, J. D. 1972. Origin of concatemeric T7 DNA. *Nat. New Biol.* **239**:197.

Yoshimori, R. N. 1971. "A genetic and biochemical analysis of the restriction and modification of DNA by resistance transfer factors." Ph.D. thesis, University of California at San Francisco Medical Center.

Initiation and Termination of Adeno-Associated-Virus DNA Replication

WILLIAM W. HAUSWIRTH
KENNETH I. BERNS

Department of Immunology and Medical Microbiology
College of Medicine
University of Florida
Gainesville, Florida 32610

Adeno-associated virus (AAV) is a defective parvovirus requiring coinfection with a helper adenovirus for multiplication. Plus or minus linear single strands are encapsidated separately and have molecular weights of 1.4 × 10⁶ daltons. The isolation of both complementary strands from virions in high quantity has facilitated studies of the sequence organization of the DNA which have revealed the existence of an inverted terminal repetition of 141 base pairs at both molecular ends (reviewed by Berns and Hauswirth, this volume), the sequence of which includes a terminal palindrome.

It has been suggested that terminal palindromes may play a pivotal role in AAV DNA replication (Koczot et al. 1973; Gerry et al. 1973), possibly through a palindromic sequence at the 3′ terminus which could form a hairpin and thus serve as a primer for DNA synthesis. Straus et al. (1976) have isolated covalently linked plus and minus AAV DNA strands which can "snap back" to self-anneal via hairpin structures. Similar results have been obtained by Tattersall et al. (1973) for minute virus of mice (MVM), by Gunther and May (1976) for Kilham rat virus, and by Handa and Shimojo (1977) for AAV. It was postulated that such structures originate from DNA replication involving a 3′ terminal hairpin structure as the primer. Similarly, Tattersall and Ward (1976) have postulated a 3′-hairpin primer in the replication of MVM, an autonomously

257

replicating parvovirus, on the basis of DNA-synthesis experiments, both in vivo (Tattersall et al. 1973) and in vitro (Bourguignon et al. 1976).

In order to understand AAV DNA replication better, we have used an adaptation of the pulse-labeled-DNA method described by Danna and Nathans (1972). It is possible to determine the site of termination of AAV DNA replication along the viral genome, because in full-length duplex AAV DNA that segment synthesized last is labeled most heavily. Similar analysis of the pulse-label contained in separated plus and minus single strands was done to determine the direction of DNA replication. Additionally, short DNA molecules labeled during the pulse period were analyzed in the hope that their limited genome size could serve to locate the probable origin of AAV DNA replication.

MATERIALS AND METHODS

CELLS AND VIRUSES. The growth and purification of AAV2 (strain H) and helper virus, adenovirus type 2 (Ad2), have been described by Carter et al. (1973) and Gerry et al. (1973). KB3 cells used in pulse-labeling experiments were grown in monolayers as described elsewhere (Hauswirth and Berns 1977).

INFECTION, PULSE-LABELING, AND ISOLATION OF AAV DNA. As described by Hauswirth and Berns (1977), each flask of confluent KB-cell monolayers was coinfected with AAV and Ad2, adsorbed for 2 hr, incubated for 37 hr, and [^{3}H]thymidine-pulsed. Low-molecular-weight DNA was isolated from the pooled lysates by the Hirt procedure (Hirt 1967), pronase-digested, phenol-extracted, and dialyzed. The DNA was then layered on 10–30% linear sucrose gradients containing 1 M NaCl and 5 mM EDTA in 0.05 M Tris (pH 8.0), and sedimented at 35,000 rpm in an SW 41 rotor for 18 hr using a Beckman L2-65B centrifuge. Fractions were collected by bottom puncture. Samples for alkaline sucrose gradients were made 0.1 M in NaOH and 0.1% in Sarkosyl, and were sedimented along with ^{32}P-labeled AAV DNA in 10–30% sucrose gradients (0.1 M NaOH, 1 M NaCl, and 5 mM EDTA) for 17 hr at 20°C at 30,000 rpm in an SW 41 rotor. Fractions were collected by bottom puncture.

RESTRICTION-ENZYME DIGESTION. Cleavage of AAV DNA by restriction endonucleases *Hin*dIII (Berns et al. 1975), *Hpa*II (Spear 1977), and *Hae*III (Fife 1977) are described elsewhere. Physical maps of the restriction fragments so produced are also reported in these references.

ELECTROPHORESIS OF RESTRICTION FRAGMENTS. DNA fragments were electrophoresed in 6% acrylamide or 1.4% agarose gels as described previously (Hauswirth and Berns 1977).

PULSE-LABELED GRADIENT ANALYSES. Using the method of Hauswirth and Berns (1977), DNA-containing bands were cut out, dissolved, and counted by liquid scintillation. Corrections were made for the amount of DNA present in each band and the thymidine content of each restriction fragment.

DNA STRAND SEPARATION. DNA strands were separated according to the method of Maxam and Gilbert (1977). Determinations of single-strand polarities and single-strand label gradients have been described by Hauswirth and Berns (1977).

EXONUCLEASE-III DIGESTION. Monomer-length duplex AAV DNA ($3 \mu g$, 6×10^4 cpm), isolated after a 10-min pulse label with [^{3}H]thymidine, was digested at 37°C with exonuclease III. The reaction mixture (0.3 ml) contained, in addition to ^{3}H-labeled AAV DNA, 1 μg of uniformly ^{32}P-labeled AAV DNA (5×10^5 cpm), 2 μg of sonicated calf-thymus DNA, 6 units of exonuclease III (Biolabs, 4000 U/ml) in 0.06 M Tris (pH 8.0), 0.7 mM MgCl$_2$, and 0.01 mM β-mercaptoethanol. At the indicated reaction times, the digestion was stopped and the high-molecular-weight DNA was precipitated and counted (Hauswirth and Berns 1977).

RESULTS

Pulse-Labeled DNA Species from AAV-Infected Cells

Our basic experimental plan was to isolate mature linear single strands of AAV DNA which had been labeled during a brief exposure of infected cells to [^{3}H]thymidine. That region of the DNA which had been replicated last would be labeled most heavily. Thus, infected-KB-cell monolayers were [^{3}H]thymidine pulse-labeled at 18 hr after infection and lysed, and AAV DNA was isolated and purified by Hirt (1967) extraction. In order to fractionate various mature forms of AAV DNA (Gerry et al. 1973), in addition to replicating forms of AAV DNA, from any contaminating adenovirus or cellular DNA, the AAV DNA preparation was sedimented through a neutral 10–30% sucrose gradient (Fig. 1). In addition to monomer-length AAV DNA (indicated in Fig. 1 by arrow) labeled during the pulse, other distinct size classes of DNA are apparent. One species migrating faster than monomer-length duplex AAV is identified as duplex AAV of dimer length, which chases to monomer length (W. W. Hauswirth and K. I. Berns, unpublished results). This DNA appears to be identical to the dimeric concatemers described by Straus et al. (1976). Two slower-migrating species, with apparent sizes of about 55% and 30% monomer duplex length, are also present. The possible origin and mode of synthesis of these duplexes will be discussed in a later section. The ^{3}H-pulse-labeled DNA sedimenting with linear duplex monomers was selected as most likely to represent mature AAV DNA. Sedimentation of

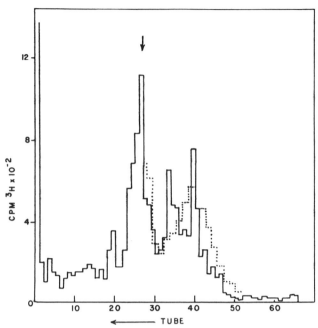

FIGURE 1

Sedimentation of pulse-labeled AAV DNA. Linear 10–30% neutral sucrose gradient of 5-min-pulse-labeled AAV DNA from a Hirt extract of infected cells. The peak corresponding to linear duplex monomer DNA, designated with an arrow, was pooled for further analysis as discussed in the text. The dotted line is the profile after a 60-min chase period.

this fraction through an alkaline sucrose gradient revealed that 94% of the pulse-label sedimented as intact linear single strands of AAV DNA indistinguishable from those single strands isolated from virions. A fraction (6%) of the label sedimented more rapidly, in a region appropriate for linear single-stranded dimers of the type observed by Straus et al. (1976). Digestion of monomer material with *Hpa*II yielded a pattern of fragments after electrophoresis and ethidium bromide staining identical to mature double-stranded AAV DNA fragments, both in relative mobility and in amount. Therefore, ^{3}H-label from this fraction is in linear duplex monomers of AAV DNA composed of two mature, linear, complementary single strands.

Gradient of Labeling in Linear Duplex Monomers

In order to determine the distribution of radioactive label incorporated into various regions of the AAV genome during the pulse, a mixture of the linear duplex monomers from the neutral sucrose gradient and uniformly ^{32}P-labeled virion AAV DNA was cleaved with *Hpa*II. After electrophoretic separation, fragments were visualized by staining with ethidium bromide, the bands were cut out from the gel, and the slices were dissolved and counted. The relative amount of label incorporated into each fragment was determined by measuring the ^{3}H/^{32}P ratio in each fragment and normalizing for the thymidine composition of the individual fragments. A physical map of the fragments has been determined (Berns and Hauswirth, this volume) and is partially presented as the abscissa in Figure 2. For the 5- and 10-min pulses, a gradient of labeling is evident that essentially has been lost by extension of the pulse duration to 20 min. The gradient of labeling is bimodal, with fragments near the ends

FIGURE 2
AAV DNA pulse-label incorporation patterns. Endo·R *Hpa*II digestion of ^{3}H-pulse-labeled duplex monomer AAV DNA. Relative [^{3}H]thymidine content of each fragment is plotted as a function of its position in the AAV genome. Each panel depicts the label pattern from cells pulsed for the indicated time periods. The genome extent of fragment A also includes fragments K and L, which were not sufficiently resolved for individual determinations. Fragment M is represented in twice-molar proportions (Spear 1977) and is therefore not reported.

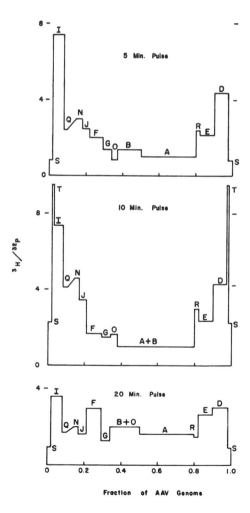

of the molecule labeled most heavily. However, the terminal fragment(s) is labeled relatively lightly. The next fragment in from the ends, fragment T, which is within the terminal-nucleotide-sequence repetition and therefore is produced from either molecular end, was resolved in the 10-min pulse and was labeled most heavily.

Because the *Hpa*II terminal fragment S occurs at both ends of the molecule it was necessary to separate the left and right ends of the molecule so that the levels of incorporation of radioactivity into terminal fragments could be determined independently for each end. *Hind*III was used to cleave the pulse-labeled linear duplex molecules once, and the two fragments were separated by agarose gel electrophoresis. Analysis after *Hpa*II cleavage of the two *Hind*III fragments confirmed that terminal fragments at both ends of the molecule had incorporated lower levels of radioactivity during the pulse than adjacent subterminal fragments.

From these data we conclude that AAV DNA replication terminates within the terminal-nucleotide-sequence repetition inside the ends of the mature DNA molecule.

Gradients of Labeling in the Complementary Strands of AAV DNA

The bimodal gradients of labeling of linear duplex monomers described in the preceding subsection demonstrated that AAV DNA replication terminates near the ends of the molecule, within the terminal repetition. Two general models of DNA replication would be consistent with these data. (1) DNA replication might begin in the middle of a double helix and proceed bidirectionally toward each end. (2) Replication might begin near either molecular terminus and proceed to the other end.

To help decide whether the origin of replication is near the middle of AAV DNA or close to the ends, the gradient of labeling established in each of the complementary single strands during the pulse was determined. Uniformly ^{32}P-labeled virion AAV DNA was cleaved using *Hpa*II. The fragments were separated electrophoretically, the gel was sliced appropriately, and the fragments were eluted. After denaturation, the strands of most fragments were separated electrophoretically.

To determine the polarity of each separated strand from an *Hpa*II fragment (i.e, to see which was from the original plus strand of intact AAV DNA and which from the minus strand), hybrid duplexes uniformly labeled with ^{32}P in either the plus or the minus strand were then cleaved using *Hpa*II and the complementary strands of the fragments were separated as described above. In each case analyzed, ^{32}P predominated in only one strand, as its polarity showed.

With these techniques, the relative incorporation of label during a pulse was determined for each strand of *Hpa*II fragments derived from linear duplex monomers (Fig. 3). For both plus and minus strands the label gradient is unimodal, with the highest activity near the 3' ends.

Unfortunately, we were unable to determine the label content of any fragment produced from within the terminal repetitions, either because individual strands could not be identified (fragment S) or because the amount of label was too low and the strand was too short for reliable determination (fragment T). Nevertheless, the labeling pattern is obviously unimodal 5'→3' and allows the unambiguous conclusion that DNA replication proceeds unidirectionally, copying each template strand from its 3' end to its 5' end.

Exonuclease-III Determination of Single-Strand Labeling Gradients

Having constructed pulse-labeling patterns for duplex DNA and both complementary strands by physical separation of restriction fragments and their composite strands, we wanted an independent method of

FIGURE 3
Single-strand AAV DNA pulse-label incorporation patterns. Relative [³H]thymidine content of Endo·R *Hpa*II fragments from each complementary strand of duplex monomer AAV DNA from 10-min-pulse-labeled cells. Top panel shows data for the plus strand (5'→3') and bottom panel depicts the minus-strand (3'→5') pattern.

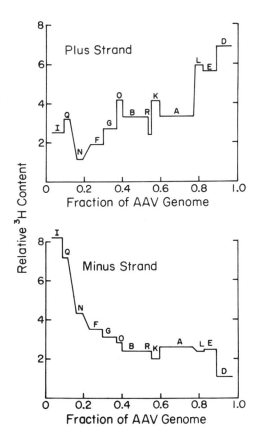

confirming these general labeling features. Exonuclease III (exoIII) from *E. coli* catalyzes the stepwise release of 5′ mononucleotides from the 3′-OH ends of duplex DNA (Richardson and Kornberg 1964) and should therefore solubilize pulse label from each 3′ end of duplex AAV DNA more quickly than total AAV DNA is digested. The course of [³H]thymidine solubilization by exoIII as a function of the fraction of the AAV genome solubilized is shown in Figure 4. If our previously deduced single-strand labeling pattern is generally accurate, pulse-label should be released more rapidly than the AAV genome is solubilized because the highest activity on each strand occurs near the 3′ end. After correction for thymine content at each extent of digestion, this is clearly the case. The experimental data are well above the 45° line at which label release equals DNA solubilization. Furthermore, the digestion is nearly complete at 50% total solubilization, which suggests little comtaminating non-exoIII activity. We conclude that exoIII digestion confirms, independently, the general shape of complementary single-strand label gradients deduced via restriction-fragment analysis.

Short DNA Molecules Labeled During the Pulse

In addition to monomer-length AAV DNA labeled during the pulse, other distinct size classes of DNA are apparent (Fig. 1). Two species sedimenting more slowly than duplex monomer are present, with apparent sizes of about 55% and 30% duplex-monomer length. After a 60-min chase period, the 55%-genome-length DNA has converted to 30% material without conversion of either to duplex monomer. The DNA in both peaks was digested with *Hpa*II and the products were electrophoresed. A densitometric scan of the resulting ethidium-bromide-stained gel shows clearly that only DNA sequences from near the genome termini are

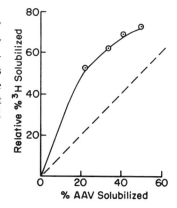

FIGURE 4
Exonuclease-III digestion of pulse-labeled AAV DNA. Percent of total [³H]thymidine activity solubilized by exoIII digestion of duplex monomer AAV DNA from 10-min-pulse-labeled cells plotted against percent of the AAV genome solubilized. The dashed line is the 45° plot that would be expected if the percent ³H solubilized equaled the percent AAV genome solubilized.

present (Fig. 5) and that almost all of the DNA can be identified as AAV-specific. Interestingly, the molar representation of *Hpa*II fragments is highest for fragment T and much lower for more internal fragments or the terminal fragments. The pulse-label content (Fig. 5) parallels this pattern; hence, after correction for the relative molarity of each fragment, there is essentially no label gradient in those molecules. This suggests that DNA synthesis is terminated nonspecifically. If this is so, then a 5'-terminal-labeled molecule should show little or no label in a normal *Hpa*II fragment, as is indeed the case (W. W. Hauswirth and K. I. Berns, unpublished results).

DISCUSSION

Features of AAV DNA replication supported by our pulse-label experiments include (a) self-primed DNA synthesis via a 3' hairpinned struc-

FIGURE 5
Relative pulse-label content and molarity of *Hpa*II fragments in slowly sedimenting AAV DNA from a 10-min pulse.

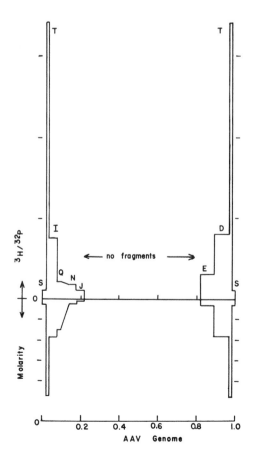

ture, (b) unidirectional synthesis from the 3' end of either template strand, and (c) termination at a 5' hairpinned structure to account for the low label incorporation we observe in terminal restriction fragments. Some or all of these processes have been incorporated into DNA-replication models recently proposed for parvoviruses (Tattersall and Ward 1976) and for AAV (Straus et al. 1976; Berns and Hauswirth, this volume).

The most consistent explanation for the structure of the two classes of slowly sedimenting DNA molecules is that they are predominantly hair-pinned duplexes containing terminal AAV sequences of random lengths. Their formation may be understood if normal initiation is followed by premature termination leading to a template-strand switch, as has been suggested for λdv (Chow et al. 1974), polyoma virus (Robberson and Fried 1974), and adenovirus (Danielle 1976). Conversion of the larger class to the smaller during a chase period may reflect digestion of a single-strand tail or possibly a processing of randomly repeated terminal regions. If these molecules are indeed products of normal initiation, one expects that the genome segment at the origin of replication will be present most frequently. Hence, the high molarity of *Hpa*II fragment T suggests that it contains or is near the AAV replication origin.

REFERENCES

Berns, K. I., J. Kort, K. Fife, E. Groggan, and I. Spear. 1975. Study of the fine structure of adeno-associated virus DNA with bacterial restriction endonucleases. *J. Virol.* **16**:712.

Bourguignon, G. J., P. Tattersall, and D. Ward. 1976. The DNA of minute virus of mice: A self-priming non-permuted single stranded genome with a 5'-terminal hairpin. *J. Virol.* **20**:290.

Carter, B. J., F. Koczot, J. Garrison, J. Rose, and J. Dolin. 1973. Separate function provided by adenovirus for adeno-associated virus multiplication. *Nat. New Biol.* **244**:71.

Chow, L. T., N. Davidson, and D. Bert. 1974. Electron microscope study of the structures of dv DNAs. *J. Mol. Biol.* **86**:69.

Danielle, E. 1976. Genome structure of incomplete particles of adenovirus. *J. Virol.* **19**:685.

Danna, K. J. and D. Nathans. 1972. Bidirectional replication of simian virus 40 DNA. *Proc. Natl. Acad. Sci.* **69**:3097.

Fife, K. H. 1977. "Structure and nucleotide sequence studies of adeno-associated virus DNA." Ph.D. thesis, Johns Hopkins University, Baltimore, Maryland.

Gerry, H. W., T. Kelly, Jr., and K. Berns. 1973. Arrangement of nucleotide sequences in adeno-associated virus DNA. *J. Mol. Biol.* **79**:207.

Gunther, M. and P. May. 1976. Isolation and structural characterization of monomeric and dimeric forms of replicative intermediates of Kilham rat virus DNA. *J. Virol.* **20**:86.

Handa, H. and H. Shimojo. 1977. Viral DNA synthesis in vitro with nuclei isolated from adeno-associated-virus-type-1-infected cells. *Virology* **77**:424.

Hauswirth, W. W. and K. Berns. 1977. Origin and termination of AAV DNA replication. *Virology* **78**:488.

Hirt, B. 1967. Selective extraction of polyoma DNA from infected mouse cell cultures. *J. Mol. Biol.* **26**:365.

Koczot, F. J., B. Carter, C. Garon, and J. Rose. 1973. Self-complementarity of terminal sequences within plus or minus strands of adenovirus-associated virus DNA. *Proc. Natl. Acad. Sci.* **70**:215.

Maxam, A. M. and W. Gilbert. 1977. A new method for sequencing DNA. *Proc. Natl. Acad. Sci.* **74**:560.

Richardson, C. C. and A. Kornberg. 1964. A deoxyribonucleic acid phosphatase-exonuclease from *Escherichia coli*. I. Purification of the enzyme and characterization of the phosphatase activity. *J. Biol. Chem.* **239**:242.

Robberson, D. L. and M. Fried. 1974. Sequence arrangements in clonal isolates of polyoma defective DNA. *Proc. Natl. Acad. Sci.* **71**:3497.

Spear, I. S. 1977. "The use of bacterial restriction endonucleases in elucidating the fine structure of AAV DNA." Ph.D. thesis, Johns Hopkins University, Baltimore, Maryland.

Straus, S. E., E. Sebring, and J. Rose. 1976. Concatemers of alternating plus and minus strands are intermediates in adenovirus-associated virus DNA synthesis. *Proc. Natl. Acad. Sci.* **73**:742.

Tattersall, P. and D. Ward. 1976. The rolling hairpin: A model for the replication of parvovirus and linear chromosomal DNA. *Nature* **263**:106.

Tattersall, P., L. V. Crawford, and A. I. Shatkin. 1973. Replication of the parvovirus MVM. II. Isolation and characterization of intermediates in the replication of the viral deoxyribonucleic acid. *J. Virol.* **12**:1446.

Effect of Phosphonoacetic Acid on the Replication of Adeno-Associated Virus

JAMES F. YOUNG
HEATHER D. MAYOR

Department of Microbiology and Immunology
Baylor College of Medicine
Houston, Texas 77030

Adeno-associated satellites are defective viruses which require the presence of a helper virus to initiate their replication cycle (Atchison et al. 1965; Hoggan et al. 1966). Coinfection of cells with adeno-associated virus (AAV) and a helper adenovirus results in the production of progeny AAV virions (Atchison et al. 1965). In contrast, herpesviruses can serve only as partial helper viruses, initiating AAV DNA synthesis (Boucher et al. 1971), RNA synthesis (Rose and Koczot 1972), and antigen synthesis (Atchison 1970; Rose and Koczot 1972), with no detectable infectious AAV progeny produced (Atchison et al. 1965; Atchison 1970; Rose and Koczot 1972).

The exact mechanism(s) by which either helper virus can promote the initiation of AAV macromolecular synthesis remains to be elucidated. This is a challenge, since many complex events are occurring simultaneously in helper-virus-infected cells. One approach to simplifying analysis of these systems is through the use of conditionally defective helper viruses. For example, these studies have shown that the HSV1-coded DNA polymerase and thymidine kinase are not necessary for complete helper function. Another approach is through the use of inhibitors of macromolecular synthesis. Such compounds have been used to dissect the herpesvirus replication cycle into discrete coordinated events. One such study determined the order and rate of viral protein synthesis and demonstrated that viral polypeptides were synthesized in three groups, designated a, β, and γ (Honess and Roizman 1974). Other investigators have shown that phosphonoacetic acid (PAA) is a potent inhibitor of herpesvirus replication (Shipkowitz et al. 1973; Overby et al. 1974; Mao et al. 1975). This compound specifically inhibits the enzymatic activity of the herpesvirus-coded DNA-dependent DNA polymerase, with little effect on cellular DNA polymerases (Mao et al. 1975). In the presence of concentrations of PAA sufficient to prevent virus growth and viral DNA synthesis, normal amounts of early virus proteins (a and β groups) are made but late virus proteins (γ groups) are reduced to less than 15% of the yields found in untreated infected cells (Honess and Watson 1977).

Because PAA can specifically shut down late herpesvirus functions, it should prove to be a useful tool in determining which protein or group of HSV proteins may be important for AAV replication.

MATERIALS AND METHODS

CELLS AND VIRUSES. A continuous line of African green monkey kidney cells (Vero) was used for growth and assay of all HSV1 stocks used in experiments. Another continuous line of African green monkey kidney cells (CV1) was used for growth and assay of simian adenovirus 15 (SV15) and AAV type 1.

HSV1 (strain KOS), which was 99.98% sensitive to PAA, and a PAA-resistant strain (HSV1Pʳ) obtained from Dr. Dorothy Purifoy, Department of Virology, Baylor College of Medicine, were used. HSV1Pʳ virus was selected by plaquing HSV1 wild-type virus in the presence of 100 μg/ml of PAA. Eleven subsequent passages with 100 μg/ml of PAA in the overlay medium resulted in a stock 99.5% resistant to the inhibiting effect of 100 μg/ml of PAA.

Growth and assay of AAV1 and SV15 were as described elsewhere (Mayor et al., this volume).

COMPLEMENTATION AND IMMUNOFLUORESCENCE. Our standard procedures have been described previously (Drake et al. 1974; Mayor et al., this volume).

CHEMICALS. Disodium phosphonoacetic acid was a generous gift of Abbot Laboratories.

RESULTS

Dose of PAA Required to Inhibit Helper Viruses

Monolayer cultures were grown on 35-mm tissue-culture dishes and infected with HSV1 multiplicities of 10 plaque-forming units (PFU) per cell. After 1 hr of adsorption at 37°C, cultures were washed, maintenance medium containing 0–500 μg of PAA was added, and incubation continued. HSV1-infected cultures were frozen at −70°C 16 hr after infection, thawed once, and assayed for virus production. SV15-infected cells were frozen and thawed three times 24 hr post-infection (p.i.) before analysis. Figure 1 illustrates the dose-response curves obtained for the effect of PAA on both HSV1 grown in Vero cells and SV15 grown in CV1 cells. These results show a drastic inhibition of HSV1 growth at PAA concentrations greater than 200 μg/ml, a finding consistent with that of Honess and Watson (1977). In contrast, only very slight inhibition of SV15 was seen with levels up to 500 μg/ml of PAA in the overlay medium. Similar findings have been obtained using isolated nuclei from cells infected with human adenovirus (Bolden et al. 1975).

Effect of PAA on AAV1-Antigen Synthesis

Monolayer cultures were grown on 12-mm cover slips and coinfected with AAV1 [1 immunofluorescence-producing unit (IFU) per cell] and

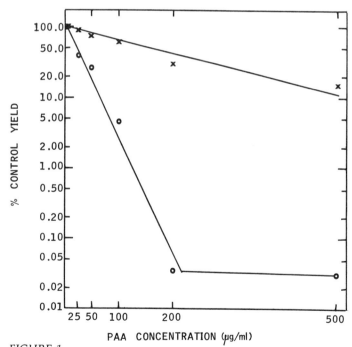

FIGURE 1
Dose-response curves of HSV1 wild type and SV15 to PAA. ○: HSV-1; x: SV15. Control values are 1.6 × 10⁸ PFU/ml for HSV1 and 7.7 × 10⁷ PFU/ml for SV15.

various multiplicities of helper viruses. After 1 hr of adsorption at 37°C, the inocula were replaced with maintenance medium containing 0–200 μg/ml of PAA and incubation continued. Cover slips were harvested at appropriate times, and the number of AAV1-antigen-positive cells was determined by indirect immunofluorescence. Figure 2 shows the effect of PAA on AAV1-antigen synthesis when HSV1 wild-type (HSV1P[s]) virus was used as a helper. There was a profound inhibition of AAV1-antigen synthesis with concentrations of PAA as low as 25 μg/ml, and complete inhibition was observed with 200 μg/ml. This effect was seen at all multiplicities of HSV1 tested. Figure 3 shows the dose-response when HSV1P[r] was used as a helper. No inhibition was seen at low concentrations of PAA, and only a slight decrease occurred in the number of AAV1-antigen-positive cells at the highest level of the drug.

When SV15 was used as the helper virus, a decrease in the number of cells producing AAV1 antigen was observed. However, as with HSV1P[r],

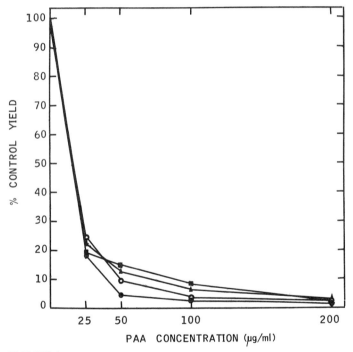

FIGURE 2
Effect of PAA on AAV-antigen synthesis with HSV1 wild type as helper. HSV1 multiplicities of infection are 10 (■), 1.0 (○), 0.1 (●), and 0.01 (▲). Control values are 24.2%, 14.4%, 4.6%, and 1.0%, respectively, for indicated multiplicities. All cover slips were harvested at 16 hr p.i.

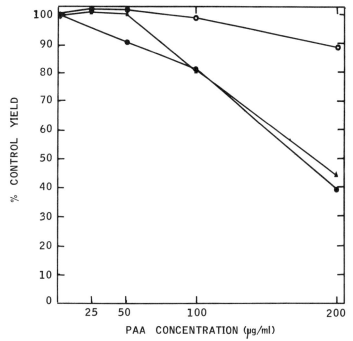

FIGURE 3

Effect of PAA on AAV-antigen synthesis with PAA-resistant strain of HSV1 as helper. Multiplicities of infection are 10 (o), 1.0 (▲), and 0.1 (●). Control values are 22.6%, 11.0%, and 2.1%, respectively, for indicated multiplicities. All cover slips were harvested at 16 hr p.i.

this decline was not nearly as dramatic as that observed with HSV1 wild-type virus (unpublished results).

Effect of Addition or Removal of PAA on AAV1-Antigen Synthesis

Monolayer cultures of Vero cells grown on cover slips were coinfected with 20 PFU/cell of HSV1P^s and 1 IFU/cell of AAV1. After 1 hr of adsorption at 37°C, maintenance medium was added to one set of cultures and incubation continued. Medium containing 200 μg/ml of PAA was then added to duplicate cultures each hour after infection. Incubation was continued until 16 hr after infection, at which time cover slips were harvested and the number of cells producing AAV1 antigen was determined. Figure 4A shows that the addition of PAA up to 5 hr after infection abolished the helper function responsible for AAV1-antigen synthesis. Addition of the drug after 5 hr resulted in less inhibition, until

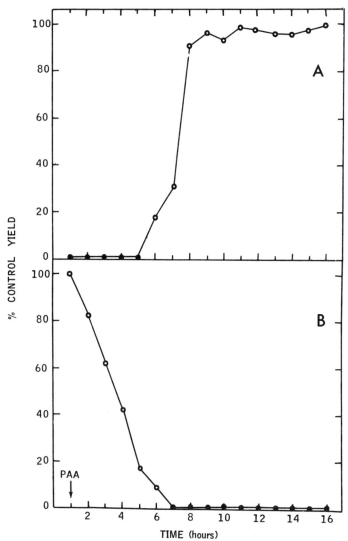

FIGURE 4

Effect of PAA on AAV-antigen synthesis. (A) Addition of 200 μg/ml of PAA at various times after infection. All cultures were harvested at 16 hr p.i. (B) Removal of PAA at various time intervals after infection. Cell cultures were harvested at 16 hr p.i.

control values were reached at 8 hr after infection. In a second set of infected cultures, the inocula were replaced after adsorption with medium containing 200 μg/ml of PAA. Duplicate cultures were changed into medium without the drug at each hour after infection and were harvested at 16 hr. Figure 4B shows the data obtained from such an experiment. A gradual decrease in the number of cells producing AAV1 antigen was seen. There were no AAV1-positive cells observed if PAA was included in the medium for more than 6 hr. The inhibition was reversible, however, since incubation in the absence of drug beyond 16 hr after infection results in AAV-positive cell numbers equal to the control (data not shown). The length of this delay is equivalent to the period during which PAA was present, and the rate of appearance of AAV1 antigen is identical to that seen in control cultures.

DISCUSSION

In this paper we have presented data showing the effect of PAA on the replication of AAV1 in tissue culture. Using helper viruses sensitive or resistant to the action of PAA, we have provided evidence of the temporal appearance of a helper function required for AAV-antigen synthesis.

PAA inhibited AAV1-antigen synthesis drastically when HSV1P[s] was used as a helper. However, when PAA-resistant viruses (HSV1P[r] or SV15) were used as helpers, this severe inhibition was not seen. The inhibition of AAV1-antigen synthesis by PAA is therefore mediated not through a direct effect on AAV1 but rather through an HSV function sensitive to PAA. Since PAA inhibits the herpesvirus-coded DNA polymerase, the inhibition of AAV-antigen synthesis could be due to a failure in AAV DNA replication (Carter et al. 1973). However, temperature-sensitive mutants of HSV1 lacking a viral DNA polymerase at the restrictive temperature have been shown to complement AAV. Therefore, a subsequent event dependent on DNA polymerase function is undoubtedly involved. Honess and Watson (1977) have shown that yields of herpesvirus proteins are greatly reduced in infected cells treated with PAA. This effect appears to be an indirect consequence of the inhibition of HSV DNA synthesis and not a direct effect on protein synthesis. Normal amounts of early virus proteins (a and β groups) are made, but late-protein (γ-group) yields are reduced. It appears, therefore, that a protein in the γ group is certainly important to the replication of AAV. This observation is supported by experiments involving the addition of PAA at various times in the HSV-AAV coinfection. Since addition of the drug 5 hr after infection is still effective in completely inhibiting AAV1-antigen synthesis, the PAA-sensitive event required by AAV must occur relatively late in the HSV cycle. These results do not mean that the a and β proteins are not also important in AAV replication, but they are obviously not sufficient.

AAV is defective in initiating its own DNA synthesis (Carter and Rose 1972). Carter et al. (1973) have also postulated a second defect at the transcriptional level on the basis of kinetic studies of AAV-adenovirus coinfected cells.

It remains to be determined whether normal amounts of AAV DNA and AAV RNA are produced in PAA-treated cells. This would indicate what role, if any, the early HSV proteins (a and β groups) play in the replication of AAV.

Thus, PAA is a useful compound capable of blocking the HSV replication cycle at the level of DNA synthesis. Analysis of the HSV system can therefore be simplified and may lead to the dissection of the helper virus function(s) utilized by AAV.

ACKNOWLEDGMENTS

This research was supported by grant Q-398 from the Robert A. Welch Foundation, Houston, Texas, and grant CA 14618 from the National Cancer Institute, U.S. Public Health Service.

REFERENCES

Atchison, R. W. 1970. The role of herpesviruses in adenovirus-associated virus replication in vitro. *Virology* **42**:155.

Atchison, R. W., B. C. Casto, and W. McD. Hammon. 1965. Adenovirus-associated defective virus particles. *Science* **149**:754.

Bolden, A., J. Aucker, and A. Weissbach. 1975. Synthesis of herpes simplex virus, vaccinia virus, and adenovirus DNA in isolated HeLa cell nuclei. *J. Virol.* **16**:1584.

Boucher, D. W., J. L. Melnick, and H. D. Mayor. 1971. Nonencapsidated infectious DNA of adeno-satellite virus in cells coinfected with herpesvirus. *Science* **173**:1243.

Carter, B. J. and J. Rose. 1972. Adenovirus-associated virus multiplication. VII. Analysis of in vivo transcription induced by complete or partial helper viruses. *J. Virol.* **10**:9.

Carter, B. J., F. Koczot, J. Garrison, J. A. Rose, and R. Dolin. 1973. Separate helper functions provided by adenovirus for adenovirus-associated virus multiplication. *Nat. New Biol.* **244**:71.

Drake, S., P. A. Schaffer, J. Esparza, and H. D. Mayor. 1974. Complementation of adeno-associated satellite virus antigens and infectious DNA by temperature sensitive mutants of herpes simplex virus. *Virology* **60**:230.

Hoggan, M. D., N. R. Blacklow, and W. P. Rowe. 1966. Studies of small DNA viruses found in various adenovirus preparations: Physical, biological and immunological characteristics. *Proc. Natl. Acad. Sci.* **55**:1467.

Honess, R. W. and B. Roizman. 1974. Regulation of herpesvirus macromolecular synthesis. I. Cascade regulation of the synthesis of three groups of viral proteins. *J. Virol.* **14**:8.

Honess, R. W. and D. H. Watson. 1977. Herpes simplex virus resistance and sensitivity to phosphonoacetic acid. *J. Virol.* **21**:584.

Mao, J. C.-H., E. E. Robishaw, and L. R. Overby. 1975. Inhibition of DNA polymerase from herpes simplex virus-infected WI-38 cells by phosphonoacetic acid. *J. Virol.* **15**:1281.

Overby, L. R., E. E. Robishaw, J. B. Schleicher, A. Rueter, N. L. Shipkowitz, and J. C.-H. Mao. 1974. Inhibition of herpes simplex virus replication by phosphonoacetic acid. *Antimicrob. Agts. Chemother.* **6**:360.

Rose, J. A. and F. Koczot. 1972. Adenovirus-associated virus multiplication. VII. Helper requirement for viral deoxyribonucleic acid and ribonucleic acid synthesis. *J. Virol.* **10**:1.

Shipkowitz, N. L., R. R. Bower, R. N. Appell, C. W. Nordeen, L. R. Overby, W. R. Roderick, J. B. Schleicher, and A. M. Von Esch. 1973. Suppression of herpes simplex virus infection by phosphonoacetic acid. *Appl. Microbiol.* **26**:264.

H-1 DNA Synthesis

SOLON L. RHODE III

Institute for Medical Research
Putnam Memorial Hospital
Bennington, Vermont 05201

The nondefective parvovirus H-1 possesses a genome of linear single-stranded DNA (Karasaki 1966; Usategui-Gomez et al. 1969) whose molecular weight is only 1.6×10^6 (McGeoch et al. 1970). Since the viral genome is a single molecule, it has the capacity to code for proteins with a total molecular weight of about 190,000 (Rhode 1977a; Singer and Rhode 1977b). Only the two capsid polypeptides VP1 and VP2' have been identified as virus-induced proteins in the infected cell (Kongsvik et al. 1974). Although the combined molecular weight of these proteins is 164,000, it is likely that VP2' is coded for by the same structural gene as VP1, since the structural polypeptides of minute virus of mice (another nondefective parvovirus) possess extensive sequence homology (Tattersall et al. 1977). However, it is certain that the number of structural genes in the H-1 genome is small and may be only one. This implies that the replication of the virus, and particularly the synthesis of its DNA, will be carried out largely by cellular enzymes and cofactors. Thus, the parvoviruses provide a small and easily characterized replicon for analysis of the mechanisms of eukaryotic cell DNA replication.

If the parvoviruses are to be used as probes of the cellular functions required for DNA replication, it is necessary to determine the functions (if there are any) of viral proteins in these processes. Isolation and characterization of mutant viruses is a powerful method for defining the role of viral proteins in the replication of the viral genome. I have used this approach in the analysis of the replication process of H-1. The purpose of this paper is to review briefly previous data and to present more recent experimental results that pertain to the mechanisms and regulation of H-1 DNA synthesis. The results are divided into sections based on the three types of DNA synthesis exhibited during the replication of H-1: (1)

synthesis of parental replicative-form (RF) DNA, (2) replication of RF DNA, and (3) synthesis of the progeny single-stranded DNA. Finally, hypothetical models detailing the mechanism of H-1 DNA synthesis and experiments testing the predictions of these models will be discussed.

MATERIALS AND METHODS

Cells and Virus

H-1, H-3, and LuIII were propagated in NB cells and infectivity was titrated by plaque assay as described by Ledinko (1967). Mutants of H-1 were isolated and characterized as described elsewhere (Rhode 1976 and manuscript in preparation). The preparation of defective interfering (DI) H-1 viruses will be described in another report (S. Rhode, manuscript in preparation).

Viral DNA

Purification of virus and intracellular viral DNA has been described (Rhode 1973, 1976). Electron microscopy of viral DNA was done as reported previously (Singer and Rhode 1977a). Digestion of H-1 DNA with restriction endonucleases and gel electrophoresis of the fragments were carried out as described previously (Rhode 1977b).

Determination of RF-DNA Synthesis

Synthesis of H-1 RF DNA was measured by determining the incorporation of [^{3}H]TdR or ^{32}P-labeled orthophosphate into RF DNA. Viral DNA was labeled with ^{32}P by the addition of carrier-free, ^{32}P-labeled orthophosphate to phosphate-free minimum essential medium (MEM) containing 10% undialyzed fetal-calf serum, or with [^{3}H]TdR (5 μCi/ml, 5 × 10^{-6} M) in the presence of FUdR (0.5 μg/ml).

The radioactivity in RF DNA of dimer or monomer length was determined by fractionating the Hirt supernatants by slab gel electrophoresis in 1% agarose gels. The Hirt supernatants were precipitated with ethanol overnight at −20°C, washed once in 70% ethanol + 30% 50 mM Tris-HCl (pH 7.5), 0.15 M NaCl, 1 mM EDTA, and redissolved in 50 mM Tris-HCl (pH 7.5), 1 mM EDTA. The samples were incubated with pancreatic RNase (50 μg/ml) for 30 min at room temperature, and 1/2 volume of 50% glycerol, 2.5% SDS, 1 mM EDTA, and 0.0125% bromophenol blue was added. Electrophoresis was for 400 V·hr at 15°C.

Viral DNA was located in wet or dry gels by autoradiography using Kodak Royal RP2 film, and the appropriate regions of the gel were excised. The ³H and/or ³²P in the gel slices were measured by liquid-scintillation spectrometry (Rhode 1977b).

RESULTS

Synthesis of Parental RF DNA

Direct analysis of the synthesis of the H-1 parental RF DNA is handicapped by a very high particle/infectivity ratio of about 10^4 under our conditions (S. Rhode and I. Singer, unpublished data; Kongsvik et al., this volume). Ordinarily, this type of experiment utilizes a virus preparation with a radiolabeled DNA whose fate can be followed after infection of the culture (Salzman and White 1973). Even with a more favorable efficiency of infection, such experiments require a method of validating that the DNA being studied represents DNA initiating infection. This is especially important because parvovirus DNA has a foldback structure at the 3' terminus and is a very active template for a variety of DNA polymerases (Bourguignon et al. 1976). Consequently, its conversion to a double-stranded parental RF DNA may not be representative of the actual infectious process.

I have obtained some information relevant to the synthesis of H-1 parental RF DNA by indirect methods that will be summarized here. The first viral synthetic event identified in H-1-infected cultures was the synthesis of DNA, upon which subsequent transcription and translation of the hemagglutinin or the virion proteins VP1 and VP2' were dependent. This event was termed HA-DNA synthesis, and it was coordinated with the cell cycle because it occurred in either the late S phase or the G2 part of the cell cycle (Rhode 1973). Incorporation of the thymidine (TdR) analog 5-bromo-2'-deoxyuridine (BUdR) into HA DNA produced an "all or none" inhibition of viral protein synthesis proportional to the BUdR/TdR ratio and inversely proportional to the multiplicity of infection (Rhode 1974b). These observations suggest that viral protein synthesis utilizes a small number of DNA templates per cell and that their number is proportional to the multiplicity of infection. A kinetic analysis of HA-DNA synthesis reveals that it occurs between 0 and 2 hr before detection of HA synthesis. I will present evidence in a subsequent section of this article that RF-DNA replication requires viral protein synthesis in NB cells. Since HA-DNA synthesis must precede viral protein synthesis, it cannot represent RF-DNA replication. By exclusion, HA-DNA synthesis is parental RF-DNA synthesis or some maturational step in parental RF-DNA synthesis that activates the RF-DNA for transcription. However, the existence of a viral protein required for RF-DNA replication (RF rep), other than VP1 or VP2', cannot be excluded. A mutant with a defective RF rep protein under restrictive conditions only would be useful for analysis of RF-DNA replication.

Replication of RF DNA

REQUIREMENT FOR VIRAL PROTEIN SYNTHESIS. Replication of RF DNA has been shown to be inhibited by treatment with the protein-synthesis inhibitor cycloheximide only if the drug is applied at or before the time of HA-DNA synthesis (Rhode 1974b). The continuation of RF-DNA replication at later times was resistant to inhibition by cycloheximide. These results suggested that the protein required for RF replication was viral, but a cellular protein synthesized concomitantly with HA could not be excluded. Similar results were obtained in parasynchronous infections using NB cultures treated with methotrexate prior to infection. These cells are not well synchronized by this treatment, but nearly all of them progress in the cell cycle to a point of competency for initiation of H-1 infection. The cells are competent for HA-DNA and HA synthesis over the relatively long period of 10 or more hours. In the experiment shown in Figure 1, advancing the time of infection advances the onset of hemagglutinin synthesis to a point somewhere within the methotrexate treatment period for H-1. The methotrexate block is obviously leaky, for addition of more potent inhibitors of DNA synthesis such as fluoro-deoxyuridine or arabinofuranosyl cytosine inhibits HA-DNA synthesis and subsequent infection (Rhode 1973; Siegl and Gautschi 1973). Optimal results are obtained in NB cells with methotrexate in the presence of approximately 10^{-7} M thymidine or undialyzed serum. In comparison with H-1, LuIII has a slower eclipse and was not induced to produce its HA during the methotrexate treatment even though the period of treatment was longer than the eclipse, as shown in Figure 1A. Immuno-fluorescent staining of these cultures for H-1 antigen showed that 95% of the cells were infected (had antigen-positive nuclei) for all three schedules at the time the HA synthesis peaked, i.e., at 16 hr in Fig. 1A, 10 hr in Fig. 1B, and 0 hr in Fig. 1C. Treatment of H-1-infected NB cultures with cycloheximide prior to HA synthesis, at times when the NB cell has been shown to be competent for initiating H-1 infection, inhibited RF-DNA replication (Table 1). Therefore, if the newly synthesized proteins required for RF-DNA replication are cellular, they must turn over rapidly. These studies suggest, but do not prove, that RF-DNA replication requires synthesis of viral protein in NB cells, and this protein is produced at the time of synthesis of the capsid proteins.

A more specific approach to this problem is to isolate mutants of H-1 defective in RF-DNA replication (RF rep⁻). One such mutant, called ts14, has been isolated. This virus is a temperature-sensitive mutant of H-1 that is thermolabile for plaque formation and production of infectious virus (S. Rhode, unpublished data). Its RF-DNA replication is defective at both permissive and restrictive temperatures, at which it incorporates [³H]TdR or [³²P]orthophosphate into RF DNA at rates of 3–7% of that of wild-type

FIGURE 1

Synthesis of H-1 and LuIII hemagglutinin in NB cells as a function of time. Replicate NB cell cultures were partially synchronized by a single 16-hr treatment with methotrexate at 0.5 µg/ml (Rhode 1976). The culture medium contained 10% undialyzed fetal-calf serum. At the conclusion of the treatment, cultures were washed twice and refed with medium containing thymidine. Virus infections were initiated by inoculating at an m.o.i. of 20 PFU/cell with 0.2 ml virus stock solution containing methotrexate for 30 min (*A*) at the termination of the block (*t*=0), (*B*) 6 hr before termination (*t*−6), and (*C*) at the inception of the block (*t*−16), as indicated by an arrow. Unadsorbed virus was removed with two washes of Hank's balanced salt solution, and the appropriate culture medium was applied. At the indicated times, duplicate cultures were harvested and the hemagglutination titer was determined with guinea-pig red blood cells. ●: H-1; ○: LuIII. Note that there are no data points between *t*−16 and *t*=0.

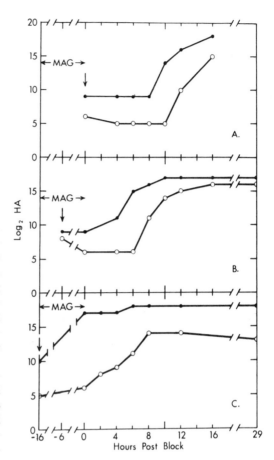

H-1. A complementation analysis of ts14 and an RF rep⁺ mutant of H-1 or wt H-3 revealed that the ts14 RF rep⁻ phenotype was *cis*-dominant in the presence of RF rep⁺ protein. In the presence of a predominance of ts14 protein and higher proportions of ts14 DNA the inhibition of RF-DNA synthesis was *trans*-acting, that is, the RF-DNA synthesis of the co-infecting RF rep⁺ virus was inhibited. The ts14 protein was found to com-

TABLE 1

Effect of Cycloheximide on H-1 RF-DNA, Hemagglutinin, and Immunofluorescent-Antigen Synthesis

	Inhibition (%)				
Experiment	Hirt pellet	dimer RF	RF	HA	FA
1. H-1 (cyclo 8 hr p.i.) [^{3}H]TdR	51	80	71	98.4	nd
Mock-infected [^{3}H]TdR	97	nd[a]	nd	nd	nd
2. H-1 (cyclo 8 hr p.i.) [^{3}H]TdR	60	88	85	98.4	87
[^{32}P]	39	82	78	nd	nd
Mock-infected [^{3}H]TdR	95	nd	nd		
[^{32}P]	71	nd	nd		
3. H-2 (cyclo 7 hr p.i.)[^{3}H]TdR	70	96	96	99.2	100
[^{32}P]	47	97	98		
Mock-infected [^{3}H]TdR	96	nd	nd		
[^{32}P]	69	nd	nd		

Replicate cultures of NB cells in 100-mm or 60-mm petri dishes were infected with a plaque-purified wt H-1 at m.o.i. of 30 or mock-infected after a 16-hr incubation with methotrexate as in Figure 1. At 8 hr p.i. (exp. 1, 2) or 7 hr p.i. (exp 3), 1/10 vol. of cycloheximide in medium (500 μg/ml) was added to the drug-treated cultures and the controls had 1/10 vol. of medium added. After 30 min of incubation the cultures were washed twice with Tris-buffered saline and then incubated with [^{3}H]TdR (5 μCi/ml, 5 × 10^{-6} M) and FUdR, 0.5 μg/ml with or without cycloheximide, for 3½ hr as before. In experiments 2 and 3 the labeling medium was a low-phosphate medium with [^{32}P]orthophosphate at 50 μCi/ml. The incorporation of radiolabel was terminated by extracting viral DNA as described in "Materials and Methods." Evidence presented elsewhere indicates that most of the radioactivity in the Hirt pellet of infected cultures is viral DNA (Singer and Rhode, this volume). Values shown are the percent inhibition caused by the cycloheximide treatment.

[a] nd: Not done.

plement RF rep⁻ defective interfering (DI) virus for RF-DNA replication as described below. From these results it was concluded that the defective RF-DNA replication of ts14 was due to an altered regulatory sequence in its DNA and not to a mutant RF rep⁻ protein.

Defective interfering viruses of H-1 have been produced at high multiplicities of infection by 20 serial propagations of virus when a helper virus was present at a multiplicity of infection of 5 (S. Rhode, manuscript in preparation). These DI viruses interfere with production of wt H-1 in a multiplicity-dependent manner, and the mechanism of interference includes inhibition of the synthesis of the virion proteins VP1 and VP2'. Different populations of DI virus are produced with different helper viruses. The most potent interfering preparation of DI virus, DI-14, was produced by propagating DI virus obtained after 14 passages with wt H-1 for 6 more passages in the presence of ts14 H-1. DI-DI-1 was produced in the same way with DI-1 as the helper virus for the last 6 passages. DI-1 is a viable defective virus that produces small plaques, which will be described elsewhere (Rhode, unpublished data). These DI viruses are defective for viral protein synthesis as determined by synthesis of hemagglutinin or synthesis of polypeptides with the electrophoretic mobility of VP1 and VP2'. An experiment was done to test the yield of RF DNA produced by DI virus and to correlate the level of DNA production with the synthesis of viral hemagglutinin (Fig. 2 and Table 2). It is clear that DI-wt (DI virus produced with wild-type helper virus) synthesized considerably less RF DNA and HA than wt H-1 at a similar multiplicity of infection. DI-14 is even more restricted in its synthesis of RF DNA at low levels of helper virus. These results establish that DI-wt and DI-14 are RF rep⁻ as well as being defective in HA synthesis. When DI-wt is supplemented with additional RF rep⁺ helper virus (either H-1 or the closely related virus H-3), the synthesis of DI RF DNA is increased (Table 3). The proportion of DI RF DNA in a mixture of DI and helper-virus RF DNA was estimated by physical mapping techniques that take advantage of the differences between the *Hin*dII restriction-endonuclease cleavage patterns of DI virus and its helper virus (Fig. 3). For DI-wt and H-3 co-infections the proportion of DI-wt RF was 80% or more. For H-1 helper viruses this value could not be determined precisely because of similarities in their marker fragments. The parvovirus LuIII, which is not as closely related to H-1 as H-3 by physical mapping techniques (S. Rhode, manuscript in preparation), showed no evidence of complementing H-1 DI virus for RF-DNA replication (Table 3). In fact, few

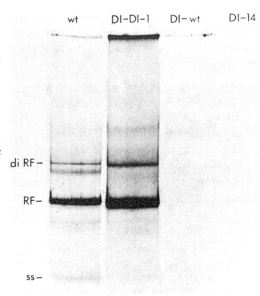

FIGURE 2
Agarose gel electrophoresis of DI RF DNA. Parasynchronous NB cultures were infected with wt H-1, DI-DI-1, DI-wt, or DI-14 as in Table 2. The Hirt supernatants were fractionated by electrophoresis in a slab gel of 1% agarose. The figure is the autoradiograph of the dried gel.

TABLE 2

Relative Yields of ^{32}P-labeled RF DNA and Single-Stranded (ss) DNA from DI Virus Compared with wt H-1

Virus	PFU/cell	RF/wt RF	ssDNA: DI/wt	HA/wt HA
wt	2.5	1.0	1.0	1.0
DI-DI-1	1.3	1.51	0.33	1.0
DI-wt	2.0	0.04	< 0.01	0.008
DI-14	0.05[a]	0.01	< 0.01	0.0001

The yields of ^{32}P-labeled RF DNA and ssDNA from NB cultures infected with wt H-1, DI-DI-1, DI-wt, and DI-14 were determined by agarose gel electrophoresis. The cultures were labeled with [^{32}P]orthophosphate at 25 μCi/ml in regular medium (MEM) from 10 to 16 hr p.i. Viral DNA was extracted and fractionated on a 1% agarose gel for 500 V·hr (Fig. 2). The areas containing RF DNA or ssDNA were excised and counted by liquid-scintillation spectrometry. Blanks were excised from the area ahead of the DNA bands. Replicate cultures were harvested for determination of the hemagglutinin titer. Values shown are the yields obtained with DI virus normalized to those obtained with wt H-1.

[a] The amount of helper virus present in DI-14 is low because ts14 is defective in its own replication, but not defective for its helper function (S. Rhode, unpublished data). The multiplicity of DI-14 particles that infect the cells under these conditions is at least 5 per cell, since this amount of DI-14 inhibits synthesis of wt H-1 and its HA by more than 95%.

DI-wt *Hind*II fragments were visible, which indicated that LuIII probably interfered with DI-wt RF-DNA synthesis (data not shown). Similar results were obtained when DI-14 was tested for complementation with these viruses (data not shown). These results suggest that H-1 RF-DNA replication requires viral protein synthesis and that the action of this protein has some specificity for H-1 DNA since LuIII does not complement H-1 DI-wt for this function whereas H-1 or H-3 wild-type viruses do.

ELECTRON MICROSCOPY OF RF-DNA REPLICATION. The geometry of RF-DNA replication has been analyzed by electron microscopy of RF DNA and its replicative intermediates (RI DNA) (Singer and Rhode 1977a). These studies used ts1 H-1, a mutant defective in progeny DNA synthesis under restrictive conditions, to avoid confusion with replicative intermediates engaged in progeny DNA synthesis. The ts1 RI DNA molecules were found to be double-stranded, branched linear molecules with the replication fork distributed randomly between the positions of 0.07 and 0.9 fractions of the genome length (Fig. 4A). Monomer-length RI DNA was considerably more common than dimer-length RI, as found for RF DNA. By a combination of physical mapping and partial-denaturation mapping it was determined that the origin of replication was within a 300-base-pair region encompassing the right end of the molecule (Singer and Rhode 1977b). This end contains the 5'-phosphoryl terminus of the

TABLE 3
Complementation of RF rep⁻ DI-wt by H-1 or H-3 for RF-DNA Synthesis

Helper virus	Total RF DNA (cpm ^{32}P)	Percentage RF DNA DI-virus-specific
None added	18,205	90–100
wt H-1	88,461	> 20[a]
ts14 H-1	115,234	> 50[a]
DI-1 H-1	112,870	> 30[a]
wt H-3	99,643	80[b]
wt LuIII	44,494	< 20[a]

Parasynchronous NB cultures were infected with DI-wt with and without addition of various helper viruses at an m.o.i. of 10–20. The viral DNAs were labeled with [^{32}P]orthophosphate, extracted, and sedimented in neutral sucrose gradients. The total RF DNA was measured by Čerenkov counting of the sucrose gradients. The proportion of the total represented by DI-wt DNA was estimated by digesting the RF DNA with HindII and analyzing the fragments by gel electrophoresis in 3% acrylamide-0.5% agarose (Fig. 3).

[a] The values shown are approximations based on the extent of exposure for certain marker bands for DI-wt.
[b] For DI-wt + H-3 the H-1 HindII D′ and D fragments and the H-3 E fragment were counted and the relative amounts of H-1 DNA and H-3 DNA were determined. The 80% value obtained may be a low estimate because some of the DI-wt RF DNA may not produce a HindII D′ or D fragment. Nevertheless, with H-3 the increase in yield of DI-wt DNA is over fourfold.

viral strand of the RF DNA (Rhode 1977b). Replicative intermediates remained bound to columns of benzolated DEAE-cellulose under conditions that eluted double-stranded DNA, so the existence of some single-stranded regions in these molecules was suggested. These regions are small, since they were not visualized by electron microscopy using the aqueous Kleinschmidt method (Fig. 4B).

The partial-denaturation studies revealed two areas in the left half of H-1 RF DNA with a lower melting temperature than the remainder of the molecule. They also determined that at the extreme right end, near the replication origin, there are sequences with a very high melting temperature, which suggests a high G-C content. This interpretation was supported by the finding of a cluster of HpaII cleavage sites, 5′-CCGG-3′, in the foldback region of the right end (Rhode 1977b).

STRUCTURE OF H-1 RF DNA. H-1 RF DNA as extracted by the Hirt method exists in two major forms, linear monomer and dimer. The dimer has

FIGURE 3

Electropherogram of the
HindII digestion products of
DI-wt RF DNA and various
helper viruses. Parasynchronous cultures of NB cells were
infected with DI-wt (2 PFU/
cell) and helper virus: wt H-1,
ts14, DI-1, and H-3 at m.o.i.
of 10–20. RF DNA was labeled
with [³²P]orthophosphate and
digested with HindII. Electrophoresis was carried out on a
3% acrylamide-0.5% agarose
gel for 540 V·hr.

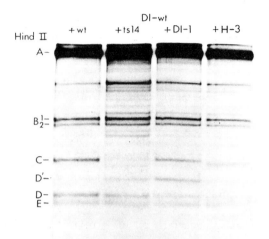

been shown to consist of two monomer molecules in "tail-to-tail" linkage, where the left end is considered the tail (Rhode 1977a). This linkage is
largely alkali-sensitive and is presumed to be hydrogen-bonded base
pairing, although some molecules contain covalently linked single-
stranded dimer-length molecules. The left end of monomer RF DNA is
present in two forms, an extended structure and a foldback structure. If
these structures are linear duplexes, then the difference in electrophoretic
mobility indicates that the foldback region is about 70 base pairs in length.
Neither structure tends to bind to benzolated DEAE-cellulose columns as
single-stranded DNA, so any single-stranded regions must be small. The
relative proportions of the two forms of the left end are about 2 to 1 in
favor of the extended structure.

The right end of RF DNA, which contains the 5' terminus of viral
DNA, also exhibits extended and foldback structures (Rhode 1977b).
In this case 90% or more of the molecules are extended, so the foldback
right end is less common than the foldback left end. The right-end
foldback region (about 105 bp) is larger than that at the left end. It also
does not appear to contain the same sequences, since it is cleaved by
HindII and by HpaII whereas the left end is not, and no circularization
of virion DNA was observed by electron microscopy after renaturation
(Singer and Rhode 1977a). The role these structures may play in the
replication of the molecule will be discussed in the last section of this
article.

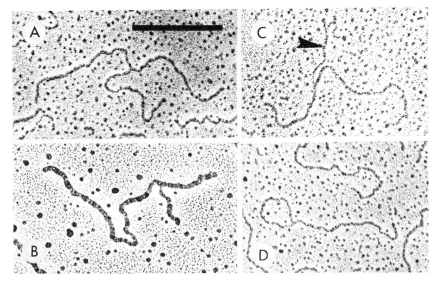

FIGURE 4

Electron micrographs of H-1 DNA, 49,000×. Bar = 0.5 μm. (A) ds RF DNA replicative intermediate (RI) isolated from NB cells infected with ts1 H-1 at T_r, replicated 52%, spread from 50% formamide onto 20% formamide. (B) RI of ds RF DNA isolated from hamster-embryo cells infected with ts1 H-1, purified by benzolated DEAE-cellulose chromatography, and prepared for EM by the aqueous Kleinschmidt method; 20% replicated. (C) Progeny single-strand RI isolated from NB cells infected with wt H-1 at 37°C. The single-stranded branch (arrowhead) is thinner and more kinked than the rest of the moleucle. (D) Unusual "lollipop" RF DNA from ts1-H-1-infected NB cells prepared as in (A). (Kleinschmidt preparations and electron microscopy were performed by I. I. Singer.)

Synthesis of H-1 Progeny DNA

ASYMMETRY OF PROGENY SINGLE-STRANDED-DNA SYNTHESIS. Until recently it was thought that the nondefective parvoviruses encapsidated only one of the two complementary strands found in their RF DNA (Rose 1974). Gel electrophoresis of H-1 virion DNA revealed a low level of DNA migrating at the position of RF DNA (Singer and Rhode 1977a). Also, labeling of the 5′ terminus of virion DNA with polynucleotide kinase resulted in a small amount of labeling at the left end of the molecule (Rhode 1977b). These studies indicated the presence of the strand (c) which is complementary to virion DNA in fewer than 1% of the total virions. That this DNA did not arise from contamination with nonencapsidated strands was demonstrated by its resistance to DNase in a preparation of purified virus (Fig. 5). The mechanism by which c strands

FIGURE 5

Resistance to DNase digestion of v- and c-strand DNA in virions. The hybridization of virion v and c strands producing a double-stranded DNA with the mobility of RF DNA was shown by electrophoresis of the DNA on a 0.7% agarose gel. Samples applied to the gel are (1) ^{32}P-labeled RF DNA, (2) ^{32}P-labeled RF DNA + ^{32}P-labeled H-1 virus, DNase treated, (3) ^{32}P-labeled virus, and (4) ^{32}P-labeled H-1 virus, DNase treated. Samples (2) and (4) were treated with 2 U pancreatic DNase in a 30-μl volume of 20 mMTris-HCl (pH 7.5), 5 mM MgCl$_2$, 1 mM dithiothreitol for 4 hr at 37°C. The reaction was stopped by addition of 1 μl of 0.5 M EDTA, 10 μl of 10% SDS, and viral DNA liberated from virions by incubation for 5 min at 100°C. The samples were incubated at 65°C for 1 hr with 0.5 M NaCl to anneal v and c strands of DNA prior to electrophoresis. DNase removed the exogenous RF DNA (sample 2) but not the virion-associated RF DNA (sample 4).

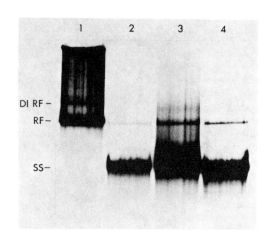

are produced and encapsidated is unknown. The transcriptional mapping of H-1 has not been done at this time, so it is not possible to describe the major virion DNA in a genetic sense as a plus or a minus strand. Therefore it will be referred to here as the v (viral) strand.

The first mutants of H-1 that we characterized proved to be thermolabile for progeny DNA synthesis (Rhode 1976; I. Singer and S. Rhode, unpublished data). Wild-type H-1 also was partially defective at 39.5°C, but the difference between wt H-1 and the mutants ts1 and ts2 was clear at 38°C. We have used two methods for detecting defective progeny DNA synthesis. One utilizes the density-label BUdR to show that the asymmetry of incorporation of BUdR into the v strand characteristic of wild-type DNA synthesis is reduced in the mutant. The second method combines an assay of extractable viral DNA with electron microscopy or virus purification to determine whether encapsidation is occurring. A mutant is considered defective in progeny DNA synthesis if it meets two criteria: (1) it must be defective in encapsidation of its DNA as determined by electron microscopy or by isolation of virus, and (2) there must be no accumulation of free viral single-stranded DNA detectable by extraction of infected cells by the Hirt method (Singer and Rhode, this volume).

All of the H-1 mutants isolated to date that are defective in progeny DNA synthesis have alterations in their capsid proteins. Their empty capsids are defective in hemagglutination of guinea-pig red cells if synthesized at 39.5°C. These results imply that one or both capsid proteins are required for progeny DNA synthesis.

The mechanism and the geometry of progeny DNA synthesis have not been studied extensively in our laboratory. Preliminary results suggest that the origin may be the same as for RF-DNA replication (I. Singer and S. Rhode, unpublished data). Density-labeling experiments with BUdR indicate that the newly synthesized v strand is in the RF DNA molecule and the pre-existing RF-DNA v strand is displaced before its encapsidation (Rhode 1974a,b). The efficiency of encapsidation as reflected by the relative size of the pool of virion DNA recovered by Hirt extraction seems to vary with both the virus and the host-cell type. The hamster cell is more efficient in producing H-1 than is the human NB cell, and the human diploid fibroblast is the least efficient cell we have studied (Singer and Rhode, this volume).

It would be desirable to isolate a mutant that is temperature-sensitive for RF-DNA replication but not for progeny DNA synthesis. The results discussed above suggest that such a mutant could be found.

Electron microscopy of RF DNA and RI DNA extracted from hamster cells infected with wt H-1 showed some molecules with thin, kinked branches suggestive of single-stranded DNA by the formamide spreading method (Fig. 4C). With the aqueous method, some double-stranded RF DNA molecules with bushlike protuberances of the sort expected for single-stranded branches were also observed. Not enough of these apparent RI DNA molecules engaged in progeny DNA synthesis were recovered for a detailed analysis. RI molecules with extensive single-stranded regions are probably subject to loss by adsorption during purification, and extra precautions need to be taken during their extraction and preparation for the Kleinschmidt procedure. A method that allows clear visualization of both single-stranded and double-stranded regions of a small DNA molecule, such as described by Wu and Davidson (1975), is also required for a reliable identification of RI molecules engaged in progeny DNA synthesis.

DISCUSSION

In this report I have discussed our studies of the three stages of H-1 DNA synthesis. The results with H-1 and those of other laboratories working with closely related parvoviruses suggest the following scheme for the DNA replication of the nondefective parvoviruses (reviewed by Rose 1974; Siegl 1976): After uncoating of the particle the first synthetic event is the synthesis of viral DNA (HA DNA), upon which subsequent transcription of viral mRNA for capsid proteins is dependent. HA-DNA synthesis appears to depend on a cell-cycle-specific event occurring between the late S phase and the early G2 portion of the cell cycle. Because the subsequent replication of RF DNA (at least to detectable levels) requires viral protein synthesis, it is likely that HA-DNA synthesis is the formation of parental RF DNA or a modification of it, such as extension of the foldback ends.

The only viral proteins detected in vivo have been the capsid proteins VP1 and VP2' (Kongsvik et al. 1974), which may be coded for by a single structural gene (Tattersall et al. 1977). Whether the parvoviruses are monocistronic or not is uncertain, but it is clear that the capsid proteins are multifunctional. They are required for progeny DNA synthesis, perhaps for RF-DNA replication, and for particle formation (Rhode 1976). They also are present in the nucleus in several different structural states, and they appear to be subject to host-cell modification (Singer and Rhode, this volume).

Complementation studies of heterotypic infections show multiplicity-dependent cross-interference for viral protein synthesis which is not "all or none." That is, the reduction in immunofluorescent-positive cells for the antigens of each virus is not proportional to the reduction in hemagglutinin of that virus as determined by neutralization assays (Rhode, manuscript in preparation). Since capsid protein synthesis is completed in an interval as short as 1–4 hr in an individual cell and in an interval of about 6–8 hr for cultures with less than perfect synchrony (Rhode 1973), it seems likely that transcription or translation of capsid proteins is inhibited by a feedback mechanism. This could explain the ability of one virus to dominate another in the mixed infections when it has a multiplicity advantage. H-1 and LuIII exhibit this cross-interference for capsid protein synthesis, yet LuIII proteins do not support H-1 RF-DNA replication. Therefore, it is unlikely that initiation of RF-DNA replication is involved in the inhibition process (S. Rhode, manuscript in preparation).

In the presence of viral proteins RF DNA undergoes semiconservative replication and produces progeny single-stranded DNA by an asymmetric strand-displacement mechanism as discussed above. The details of the mechanism and control of these synthetic processes are unknown.

The replication origin of H-1 RF DNA is located in the *Hind*II D or E fragment at the right end of the molecule (Rhode 1977b; Singer and Rhode 1977a,b), but the mechanisms of initiation of progeny DNA synthesis and RF-DNA replication are not known. There are three major contenders among the many possible mechanisms that might initiate H-1 DNA synthesis: (1) RNA-primed initiation, (2) DNA-primed initiation (which utilizes the foldback structure), and (3) protein-primed initiation. RNA-primed initiation has been demonstrated to be the source of 3'-hydroxyl sites for deoxynucleotide transferase reactions in some cases where DNA synthesis has been well studied (see, e.g., Schekman et al. 1975). H-1 RF-DNA replication was not sensitive to inhibition by α-amanitin, an inhibitor of RNA polymerase II, but there is little substantive evidence either for or against RNA-primed initiation in H-1 DNA synthesis (Rhode 1974b). The structure of parvovirus DNA with a foldback region at each end suggests the use of the DNA 3'-hydroxyl termini as primers for initiating DNA synthesis. This seems likely for parental RF synthesis. Indeed, MVM virion DNA has been shown to be an excellent primer template for various DNA polymerases (Bourguignon et al. 1976).

The foldback regions at the ends of H-1 DNA are the type of structure predicted by Cavalier-Smith in his model for replication of chromosomal ends (Cavalier-Smith 1974). He proposed that the 5' terminal region of a DNA strand exists as an inverted self-complementary or foldback region previously synthesized on the 3' terminus of the parental strand. The foldback region is transferred to the nascent strand by ligation and released from the parent strand by a site-specific endonuclease which generates the 5' terminus. The 3' terminus of the parental strand is synthesized by extension of the parent strand in the 5'→3' direction using the foldback as template.

I will present here three models for replication of H-1 DNA, with specific attention to the right end of the molecule. The first of these was inspired by our observation of "lollipop" configurations of H-1 DNA by electron microscopy (I. Singer and S. Rhode, unpublished results) (see Fig. 4D). These molecules were of uncertain significance, since end-to-side superimpositions were a possibility. They were observed only in H-1 DNA prepared from the semipermissive NB cell and not in viral DNA from hamster-embryo cultures. It is possible that in the more permissive hamster cell their existence is more transient. In this model (Fig. 6 part 1), initiation is considered RNA-primed and occurring in a replicative intermediate that has a covalently closed hairpin configuration at the right end. Depending on the rapidity with which a nickase cleaves at the foldback, a "lollipop" replicative intermediate may exist before conversion to a branched linear molecule. An important consequence of this model is that the foldback region is not transferred from a parent strand to a nascent strand. The second and third models are basically RNA- or

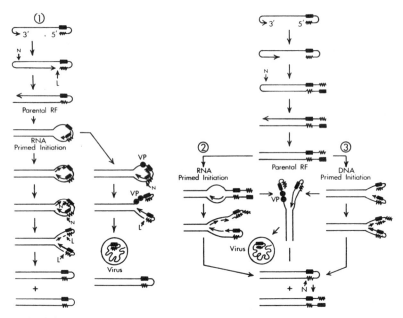

FIGURE 6

Hypothetical models of H-1 DNA synthesis. The foldback structure with its
inverted self-complementary sequences is illustrated by the solid bar paired
to the wavy line. The abbreviations used are VP for virion protein, L for
DNA ligase, and N for site-specific, single-strand endonuclease ("nickase").
The foldback structure at the left end and its replication are not included in
the figure.

DNA-primed versions of the model of Cavalier-Smith. For these models
the foldback region is transferred from the parent strand to its nascent
complement.

The question of whether the foldback region is transferred from one
strand to the other is open to experimental verification with H-1 DNA by
using the restriction endonuclease HindII to cleave the foldback region
from the body of the molecule. During H-1 DNA synthesis in the presence of a radiolabeled nucleotide such as [³H]TdR, the foldback region of
virion DNA will incorporate [³H]TdR more slowly than the remainder of
the molecule and the lag will be proportional to the average interval
between consecutive initiation events for a given molecule. These experiments are in progress. If the transfer of the foldback region does
occur, it would provide justification for an effort to isolate the site-specific
nuclease involved in this process.

It is apparent that our knowledge of the mechanisms and control
of parvovirus DNA synthesis is very incomplete. As the secrets of
these processes are revealed and virus-controlled functions are defined,

we may learn a great deal about how the eukaryotic cell replicates its genome.

ACKNOWLEDGMENTS

I gratefully acknowledge R. L. Costantino and J. Bratton for expert technical assistance, Dr. I. Singer for electron microscopy and helpful criticism of the manuscript, and V. Haas and J. Pratt for secretarial duties.

This work was supported by U. S. Public Health Service Grant CA07826-12 from the National Cancer Institute.

REFERENCES

Bourguignon, G. J., P. J. Tattersall, and D. C. Ward. 1976. DNA of minute virus of mice: Self-priming, nonpermuted, single-stranded genome with a 5'-terminal hairpin duplex. *J. Virol.* **20**:290.

Cavalier-Smith, T. 1974. Palindromic base sequences and replication of eukaryote chromosome ends. *Nature* **250**:467.

Karasaki, S. 1966. Size and ultrastructure of the H-viruses as determined with the use of specific antibodies. *J. Ultrastruct. Res.* **16**:109.

Kongsvik, J. R., J. F. Gierthy, and S. L. Rhode III. 1974. Replication process of the parvovirus H-1. IV. H-1 specific proteins synthesized in synchronized human NB kidney cells. *J. Virol.* **14**:1600.

Ledinko, N. 1967. Plaque assay of the effects of cytosine arabinoside and 5-iodo-2'-deoxyuridine on the synthesis of H-1 virus particles. *Nature* **214**:1346.

McGeoch, D. J., L. V. Crawford, and E. A. Follett. 1970. The DNA's of three parvoviruses. *J. Gen. Virol.* **6**:33.

Rhode, S. L. III. 1973. Replication process of the parvovirus H-1. I. Kinetics in a parasynchronous cell system. *J. Virol.* **11**:856.

———. 1974a. Replication process of the parvovirus H-1. II. Isolation and characterization of H-1 replicative form DNA. *J. Virol.* **13**:400.

———. 1974b. Replication process of the parvovirus H-1. III. Factors affecting H-1 RF DNA synthesis. *J. Virol.* **14**:791.

———. 1976. Replication process of the parvovirus H-1. V. Isolation and characterization of temperature-sensitive H-1 mutants defective in progeny DNA synthesis. *J. Virol.* **17**:659.

———. 1977a. Replication process of the parvovirus H-1. VI. Characterization of a replication terminus of H-1 replicative-form DNA. *J. Virol.* **21**:694.

———. 1977b. Replication process of the parvovirus H-1. IX. Physical mapping studies of the H-1 genome. *J. Virol.* (in press).

Rose, J. A. 1974. Parvovirus reproduction. In *Comprehensive virology* (ed. H. Fraenkel-Conrat and R. R. Wagner), vol. 3, p. 1. Plenum, New York.

Salzman, L. A. and W. White. 1973. In vivo conversion of the single-stranded DNA of the Kilham rat virus to a double-stranded form. *J. Virol.* **11**:299.

Schekman, R., J. H. Weiner, A. Weiner, and A. Kornberg. 1975. Ten proteins required for conversion of φX174 single-stranded DNA to duplex form in vitro. *J. Biol. Chem.* **250**:5859.

Siegl, G. 1976. The parvoviruses. *Virol. Monogr.* **15**:1.

Siegl, G. and M. Gautschi. 1973. The multiplication of parvovirus LuIII in a synchronized culture system. *Arch. gesamte Virusforsch.* **40**:119.

Singer, I. I. and S. L. Rhode III. 1977a. Replication process of the parvovirus H-1. VII. Electron microscopy of replicative-form DNA synthesis. *J. Virol.* **21**:713.

———. 1977b. Replication process of the parvovirus H-1. VIII. Partial denaturation mapping and localization of the replication origin of H-1 replicative-form DNA with electron microscopy. *J. Virol.* **21**:724.

Tattersall, P., A. J. Shatkin, and D. C. Ward. 1977. Sequence homology between the structural polypeptides of minute virus of mice. *J. Mol. Biol.* **111**:375.

Usategui-Gomez, M., H. W. Toolan, N. Ledinko, F. Al-Lami, and M. S. Hopkins. 1969. Single-stranded DNA from the parvovirus H-1. *Virology* **39**:617.

Wu, M. and N. Davidson. 1975. Use of gene 32 protein staining of single-strand polynucleotides for gene mapping by electron microscopy: Application to the $\varphi 80d_3$ ilv su$^+$ 7 system. *Proc. Natl. Acad. Sci.* **72**:4506.

Replication of Minute-Virus-of-Mice DNA

DAVID C. WARD
DINSHAW K. DADACHANJI

Department of Human Genetics
and Department of Molecular Biophysics and Biochemistry
Yale University School of Medicine
New Haven, Connecticut 06510

Replication of the single-stranded-DNA genomes of both defective (i.e., adeno-associated virus [AAV]) and autonomous parvoviruses proceeds via linear double-stranded-DNA intermediates. Monomer- and dimer-length duplex replicative forms (RFs) of viral DNA have been isolated and characterized from cells infected with AAV (Straus et al. 1976; Handa et al. 1976), minute virus of mice (MVM) (Tattersall et al. 1973; Dobson and Helleiner 1973), H-1 (Rhode 1974, 1977), LuIII (Siegl and Gautschi 1973, 1976), Kilham rat virus (KRV) (Salzman and White 1973; Gunther and May 1976), and X14 (Mayor and Jordan 1976). Higher oligomeric forms of RF DNA, such as tetramers, have also been observed in cells infected with parvoviruses [e.g., AAV (Straus et al. 1976) and MVM (Tattersall et al. 1973)], but they have not been characterized as extensively. A significant proportion of both monomer and dimer RF DNA molecules were found to be capable of instantaneous reannealing (Tattersall et al. 1973; Straus et al. 1976; Gunther and May 1976; Mayor and Jordan 1976; Rhode 1977), which suggested that the viral and complementary DNA strands were linked topologically or covalently. However, closed circular forms of viral DNA have not been detected in infected-cell lysates. These observations, coupled with recent information on viral genome structure, led to the proposal of replication models for parvovirus DNA (Straus et al. 1976; Tattersall and Ward 1976), one feature of which is the self-primed initiation of DNA synthesis. Structural studies have indicated that both the 3' and 5' terminal regions of the DNA from MVM (Bourguignon et al. 1976), H-1 (Rhode 1977; Singer and Rhode 1977; Chow and Ward, this

297

volume), KRV (Salzman 1977; Chow and Ward, this volume), H-3 (Chow and Ward, this volume), and AAV (Denhardt et al. 1976; Fife et al. 1977) exist as stable, base-paired hairpin duplexes suitable for fulfilling such a primer function. Furthermore, both initiation and termination of parvovirus DNA replication in vivo have been shown to occur close to these hairpin termini (Rhode 1977; Singer and Rhode 1977; Hauswirth and Berns 1977). In order to investigate further the possible role of self-primed synthesis in the replication of MVM DNA, we have examined the fate of ^{32}P-labeled virion DNA in parasynchronous cultures of A-9 cells. In addition, we have analyzed the EcoRI restriction-endonuclease fragments of monomeric and oligomeric RF DNA for their ability to undergo spontaneous renaturation. Finally, we have studied the conversion of monomer and dimer RF DNA to single-stranded viral DNA under pulse-chase conditions.

MATERIALS AND METHODS

Virus and Cells

The plaque-purified strain T of MVM (Tattersall 1972) was grown either in the A-9 variant of mouse L cells (Littlefield 1964) or in the RL5E line of murine-sarcoma-virus-transformed rat epithelial cells (Bomford and Weinstein 1972). Cell cultures were maintained and virus stocks produced from low-multiplicity infections (m.o.i. ~10^{-3} PFU/cell as described by Tattersall et al. 1976).

^{32}P-labeled MVM virions of high specific radioactivity were prepared by infecting exponentially growing cultures of RL5E cells at 10 PFU/cell. At 8 hr post-infection (p.i.) the medium was removed, the cells were washed twice with warm phosphate-buffered saline (PBS), and fresh Dulbecco's modified minimal essential medium containing 0.5% of the normal phosphate concentration and 100 μCi/ml of carrier-free [^{32}P]orthophosphate was added. Infected cells were harvested 48 hr p.i. and the virus was purified as described previously (Tattersall et al. 1976). Viral DNA, extracted and purified by velocity sedimentation in alkaline sucrose gradients as described by Bourguignon et al. (1976), had an initial specific activity of 1.07×10^6 cpm/μg.

Suspension cultures of A-9 cells were parasynchronized by treatment with excess thymidine at a final concentration of 2 mM for 15 hr. The cells were centrifuged gently at 37°C, the medium was removed, and the cells were washed once with PBS at 37°C and then resuspended in warm Joklik modified medium. Virus infections were initiated immediately after the cells had been resuspended in fresh medium. The level of cell synchrony, determined by mitotic-index analysis (Schindler and Schaer 1973) of parallel uninfected cultures, was generally 60–70%.

Synthesis of Double-Stranded MVM DNA in Cells Infected with [32]*P-Labeled Virus*

A 100-ml culture of parasynchronous A-9 cells (5×10^5 cells/ml) was infected with approximately 15–20 PFU/cell of ^{32}P-labeled MVM (1.5 µg viral DNA). The input virus multiplicity was estimated from the specific activity of viral DNA (1.07×10^6 cpm/µg) and the particle/infectivity ratio (300–400:1) of plaque-purified MVM, strain T (Tattersall 1972). Aliquots of 10 ml were removed at 0, 4, 6, 8, 12, and 16 hr p.i. and the cells were resuspended in 0.5 ml of 0.02 M Tris-HCl, 0.15 M NaCl, 0.01 M EDTA (pH 7.5) and added carefully to an equal volume of 1.2% sodium dodecyl sulfate in 0.02 M Tris-HCl, 0.01 M EDTA (pH 7.5). The cell lysate was incubated for 1 hr at 25°C with pronase, proteinase K, and chymotrypsin, each at a final concentration of 250 µg/ml. After digestion, intracellular DNA was fractionated by the method of Hirt (1967). The supernatant fraction was made 0.3 N in sodium acetate, 3 volumes of ethanol were added, and the DNA was precipitated after standing at $-70°C$ for 2 hr and then resuspended in 0.05 M sodium phosphate buffer (pH 6.8). The overall recovery of the input ^{32}P-label was 82–93%. Each sample was applied to a 2×0.7-cm column of hydroxyapatite (Clarkson Chemical) and washed with 5 ml of 0.05 M sodium phosphate buffer. Single- and double-stranded viral DNA was fractionated by gradient elution as described by Tattersall et al. (1973). The distribution of radioactivity was determined by Čerenkov counting. Fractions containing duplex DNA were pooled, dialyzed against 0.05 M triethylammonium bicarbonate, lyophilized, resuspended in 25–50 µl of 0.01 M phosphate buffer (pH 6.8), and digested with S1 nuclease as described by Bourguignon et al. (1976) before and immediately after heat denaturation.

Isolation of MVM RF DNA

Nonsynchronized monolayer cultures of RL5E cells ($\sim 1.5 \times 10^6$ per 100-mm dish) were infected with MVM [$\sim$10 plaque-forming units (PFU) per cell] in the standard manner. At about 9 hr p.i. the medium was removed and replaced with Dulbecco's medium containing 10% of the normal phosphate concentration and 22 µCi/ml of ^{32}P]phosphate. Cells were harvested 38 hr p.i. and washed with cold PBS, and intracellular DNA was fractionated by the method of Hirt (1967), essentially as described by Tattersall et al. (1973). The supernatant fraction was then dialyzed extensively against TNE buffer (0.01 M Tris-HCl, 0.02 M NaCl, 0.01 M EDTA, pH 7.5). One-tenth volume of 3.0 M sodium acetate (pH 6.0) and yeast tRNA (25 µg/ml final concentration) were added to the dialyzed solution and the DNA was precipitated with 3 volumes of ethanol at $-20°C$. The precipitate was redissolved in 0.5 ml of 0.01 M sodium phosphate buffer (pH 6.8) and fractionation of the ^{32}P-labeled DNA was achieved by electrophoresis on 1.4% agarose slab gels as described by Sharp et al. (1973). Half of the DNA sample was electrophoresed as isolated; the remainder was denatured by the addition of 1/10 volume of 1 N NaOH and electrophoresed after standing for 10 min at room temperature.

The major bands of ^{32}P-labeled DNA (detected by autoradiography of the wet gels) were excised and the gel slices were dissolved in a saturated solution of potassium iodide as described by Blin et al. (1975). The solutions were adjusted to a refractive index of 1.420 (a density of 1.49 g/cm^3) and centrifuged to equilibrium in an SW 50.1 rotor (46,000 rpm, 20°C, 35–40 hr). Each gradient was separated into ten drop fractions, and the distribution of ^{32}P-label was determined by Čerenkov counting. Peak fractions of ^{32}P-labeled DNA were pooled, dialyzed against TE buffer (0.01 M Tris-HCl, 0.001 M EDTA, pH 7.5), alcohol-precipitated, redissolved in a small volume of TE buffer, and stored at −20°C until used.

Digestion of RF DNA Molecules with EcoRI

Monomer, dimer, and tetramer species of RF DNA, purified as described above, were incubated for 1 hr at 37°C with 30 units of EcoRI (Boehringer-Mannheim). The reaction mixtures (30 μl) also contained 0.10 M Tris-HCl, 0.05 M NaCl, and 0.01 M MgCl$_2$, at pH 7.5. Reactions were stopped by adding 5 μl of termination buffer (Sharp et al. 1973). Where denatured products were desired, reactions were stopped by boiling for 2 min, quick-cooling on ice, and then adding termination buffer. Parallel samples of RF DNA which received no EcoRI were processed in the same way. The samples were then electrophoresed on 1.4% agarose gels. Gels were dried under vacuum onto Whatman 1 paper and exposed to Kodak XR-5 X-ray film, and the autoradiograms were quantitated by scanning with a Joyce-Loebl densitometer.

Analysis of RF DNA Using Pulse-Chase Kinetics

Twelve 100-mm dishes of nonsynchronized A-9 cells (3 × 10^6 cells/dish) were infected with MVM (10 PFU/cell) and pulse-labeled with [^{3}H]thymidine (20 μCi/ml; 45 Ci/mmole) for 20 min at 14 hr p.i. After removal of the labeling medium, cells were washed with warm PBS and Dulbecco's medium containing 0.4 mM cold thymidine, and 0.1 mM deoxycytidine was added. Two dishes were removed at 0, 30, 60, 90, 120, or 150 min after initiation of the cold chase, and the cells were washed with PBS and then immediately lysed with the SDS-EDTA solution of Hirt (1967). Intracellular DNA was fractionated after combined protease treatment as described above. DNA in each supernatant fraction was alcohol-precipitated and redissolved in 25 μl of 0.05 M sodium phosphate buffer (pH 6.8). Before electrophoresis on a 1.4 % agarose gel, 1.0 μg cold MVM DNA and 1.0 μg cold monomer duplex DNA [synthesized in vitro as described by Bourguignon et al. (1976)] was added to each sample. Bands of single-stranded DNA, monomer RF DNA, and dimer RF DNA were detected by ethidium bromide fluorescence, excised, and dissolved in Protosol (New England Nuclear), and the [³H]thymidine content was determined in a liquid scintillation counter.

RESULTS

Infection of A-9 Cells with ³²P-Labeled MVM

In an attempt to analyze directly the kinetics of MVM parental RF DNA synthesis and to quantitate the level of spontaneously renaturing DNA in this population of RF molecules, a parasynchronous culture of A-9 cells was infected with ³²P-labeled MVM at a multiplicity of ~15–20 PFU/cell. Samples were withdrawn at various times after infection and viral DNA was extracted by a modification of the Hirt procedure (Hirt 1967). This extraction technique (see "Methods") disrupts assembled virions and gives near quantitative recovery of total intracellular viral DNA in the supernatant fraction. The ³²P-labeled DNA was then separated into single- and double-stranded forms by chromatography on hydroxyapatite. As shown in Figure 1A, approximately 99% of the recovered radioactivity chromatographed as single-stranded DNA when extracted at 0 or 4 hr p.i. This is in contrast with the report by Salzman and White (1973) that approximately 50% of the input viral DNA was converted to a double-stranded form within 1 hr in KRV-infected cells. However, between 6 and 16 hr p.i. the percentage of ³²P-labeled virion DNA which chromatographed as duplex DNA increased from 2% to 10% of the total input (Fig. 1A), following a kinetic pattern similar to that observed for total virus-specific DNA synthesis in parasynchronous A-9 cell cultures (data not shown). To determine what percentage of the

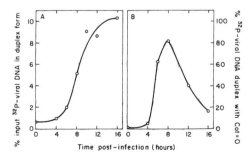

FIGURE 1
(A) Kinetics of duplex MVM DNA synthesis in a parasynchronous culture of A-9 cells after infection with ³²P-labeled virus. (B) The percentage of duplex DNA, isolated at the indicated times post-infection, which exhibits spontaneous renaturation. See "Methods" for experimental details.

Time post-infection (hours)

double-stranded DNA was capable of spontaneous renaturation, fractions containing the duplex DNA were pooled, dialyzed, heat-denatured and quick-cooled, and then immediately rechromatographed on hydroxyapatite at 4°C. As shown in Figure 1B, approximately 60% of the duplex DNA isolated 6 hr p.i. exhibited $C_0t = 0$ reassociation kinetics. The percentage of instantaneously renaturing molecules increased to 80% by 8 hr p.i. and then declined slowly to 18% by 16 hr p.i. To confirm that these results were not due to slow reannealing of separate DNA strands, samples of the DNA which re-eluted as duplex molecules after a second passage through hydroxyapatite were treated with S1 nuclease before and immediately after heat denaturation. All three DNA samples analyzed (isolated 6, 8, and 12 hr p.i.) were more than 95% resistant to S1 even after denaturation and quick-cooling. Although these observations strongly suggest that an initial event in the replication of MVM DNA requires the formation of duplex RF molecules in which parental and complementary strands are covalently attached, they do not provide definitive proof. Since MVM(T) has a particle/infectivity ratio of 300–400:1 (Tattersall 1972), one cannot be absolutely certain that studies of the fate of the total input DNA are truly representative of viral DNA which is actually initiating infection.

Digestion of In Vivo RF DNA Molecules with EcoRI

For the results described above to be consistent with a self-priming mechanism of complementary-strand DNA synthesis, spontaneously renaturing monomer RF DNA molecules should, upon digestion with an appropriate restriction enzyme, generate a spontaneously renaturing restriction fragment containing the 3' terminus of the viral DNA strand. Should initiation of progeny DNA synthesis also occur via a self-priming mechanism, using the 3'-OH terminus of the complementary strand, one might expect to isolate from RF DNA molecules a spontaneously renaturing restriction fragment containing the 5' terminus of the genome also. To determine if restriction fragments with these properties could be detected, RF DNA molecules were separated by agarose gel electrophoresis, extracted from the gel, treated with *Eco*RI, and re-electrophoresed before and after denaturation.

Figure 2 illustrates preparative 1.4% agarose gels of the ^{32}P-labeled DNA from MVM-infected RL5E cells present in the supernatant fraction of the standard Hirt extraction. Under nondenaturing conditions, bands corresponding in molecular weight to single-stranded viral DNA (SS) and duplex DNAs of monomeric (MD), dimeric (D), tetrameric (T), and oligomeric (O) genome lengths are clearly observed. Densitometric analysis indicated that the distribution of ^{32}P radioactivity among these molecular species was as follows: SS DNA 10%, MD DNA 36%, D DNA

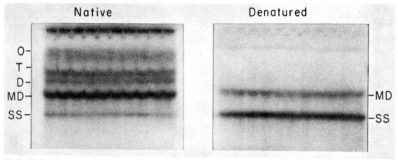

FIGURE 2

Preparative-scale fractionation of ^{32}P-labeled MVM RF DNA on 1.4% agarose gels, before and after denaturation of the DNA sample. See "Methods" for experimental details. SS: single-stranded viral DNA; MD: monomer-length DNA duplex; D: dimer-length DNA duplex; T: tetramer-length DNA duplex; O: oligomeric duplex DNA of unspecified size.

12%, T DNA 18%, and O DNA 24%. However, when the same DNA sample is electrophoresed after alkali denaturation only two prominent bands are observed. These correspond in size to SS DNA and MD DNA, constituting 64% and 32%, respectively, of the total radioactivity in the gel with a molecular weight equal to or larger than that of SS DNA. It is apparent that although 54% of the DNA molecules in this preparation are larger than monomer duplex DNA, virtually none contain single strands greater than twice genome length after denaturation.

ANALYSIS OF MONOMER RF DNA. The in-vitro-synthesized MVM DNA monomer duplex is cleaved twice by EcoRI (Bourguignon et al. 1976) at map coordinates 20.5 and 68.5 (see Chow and Ward, this volume). The 3' terminus of the viral DNA strand is in fragment C (0/20.5), the 5' terminus is in fragment B (68.5/100), and the central portion of the genome is in fragment A (20.5/68.5). When the monomer duplex of MVM DNA synthesized in vivo is digested with EcoRI and analyzed electrophoretically, both terminal fragments (B and C) appear as doublets (Fig. 3, track 3). These doublet bands differ in size by approximately 100 base pairs, a difference roughly equivalent in size to that of the hairpin duplexes shown to exist at both termini of the viral genome (see Chow and Ward, this volume). The slower-migrating member of each doublet probably represents the terminal fragment in which the hairpin duplex of the viral strand is in the extended (nonhairpin) form (▄▄▄▄▄▄); the faster-migrating band is the fragment in which the terminus remains in the foldback configuration (◖▄▄▄). The foldback fragments may exist in a true hairpin form (as illustrated above) or in a nicked form (◖ ▄▄▄). Densitometric analysis showed that 55% of the 3' terminal

FIGURE 3

Autoradiogram of a 1.4% agarose gel depicting the electrophoretic mobility of purified MVM (1) monomer RF DNA, (2) monomer RF DNA after heat denaturation, (3) monomer RF DNA after digestion with EcoRI, and (4) monomer RF DNA digested with EcoRI and then heat-denatured prior to electrophoresis. A, B, and C refer to the native forms of the restriction fragments illustrated in track 3. Both B and C fragments occur as doublets (see text). A', B', and C' designate the electrophoretic mobilities of fragments A, B, and C, respectively, after denaturation (track 4).

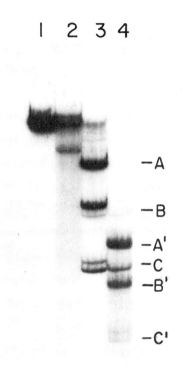

fragment was in the foldback configuration whereas only about 15% of the 5' terminal fragment was in that form.

When a parallel sample of monomer RF DNA was treated with EcoRI and then heat-denatured prior to electrophoresis, the gel pattern shown in track 4 of Figure 3 was obtained. After denaturation, fragment A migrates as a new single band (A'); the extended forms of fragments B and C migrate at the new positions, designated B' and C', respectively. In contrast, at least 70% of the faster-migrating form of each terminal fragment re-electrophoreses with its original mobility, indicating that spontaneous renaturation had occurred. The total absence of any native form of fragment A in track 4 of the gel provides an internal control which rules out the possibility that the renatured terminal fragments arise by re-annealing of separate strands.

Foldback forms of terminal restriction fragments which contained a specific single-stranded nick (⊂＿＿＿＿) would be expected to generate two sizes of single-stranded DNA upon denaturation. The denatured 3' terminal fragment (C') is indeed resolved as a doublet (Fig. 3, track 4). Although a similar doublet might also be expected upon denaturation of a nicked hairpin form of fragment B, none is observed. However, detection of the smaller strand piece would be difficult since it would contain less than 3% [<1/2 of (100 − 70) × 15% ÷ 100] of the radioactivity in all forms of the B restriction fragment.

Native monomer RF DNA (Fig. 3, track 1) is resolved into two bands upon denaturation (Fig. 3, track 2), one equivalent in size to single-stranded virion DNA and the other remigrating as intact monomer duplex. The percentage of the initial monomer RF DNA which reannealed spontaneously (59%) is similar to the value calculated from analysis of the restriction data.

ANALYSIS OF DIMER AND TETRAMER RF DNA.   Dimer and tetramer RF DNA molecules were subjected to the same series of reactions described above and their properties were compared with those of monomer RF DNA. The results are shown in Figure 4. When re-electrophoresed after purification from the preparative gel, both tetramer RF DNA (track 1) and dimer RF DNA (track 5) were found to contain a band which comigrated with the monomer DNA duplex. Several lines of evidence suggest that the latter band is not due to contaminating monomer RF DNA but is derived from the dimer and tetramer species during their isolation and purification from the preparative gel. First, tetramer RF DNA is not contaminated with dimer RF DNA, which migrated between the tetramer and monomer species on the original gel. Second, although ~58% of the native monomer RF DNA (track 9) reanneals spontaneously after heat denaturation (track 10), essentially all of the tetramer RF DNA fraction migrates as genome-length single-stranded DNA (track 2) when treated similarly. Thus, virtually all of the molecules in the tetramer RF DNA pool are completely denaturable. (It should be noted that on this gel genome-length single-stranded DNA has an electrophoretic mobility similar to that of the native *Eco*RI fragment A.) A third argument against monomer RF DNA contamination comes from the denaturation and restriction-enzyme studies of the dimer RF DNA described below.

When the dimer RF DNA fraction is denatured before electrophoresis (track 6), ~39% of the molecules migrate with a mobility equivalent to that of the monomer RF DNA and the remainder migrate as genome-length single-stranded DNA. The pattern of restriction fragments obtained after treatment of the dimer RF DNA with *Eco*RI (track 7) is similar to that observed upon digestion of monomer RF DNA (track 11), with the exception that two additional bands are observed in dimer RF DNA. One of these (indicated by the open triangle) has a molecular weight equal to the sum of fragments A + C and is presumably the result of incomplete digestion. The other (indicated by the arrow) is a fragment (designated di-Br, for dimer bridge) estimated to be 1900 base pairs in length (i.e., 39% of a genome duplex), almost exactly twice the size of the *Eco*RI fragment C. Since all partial digestion products must be larger than

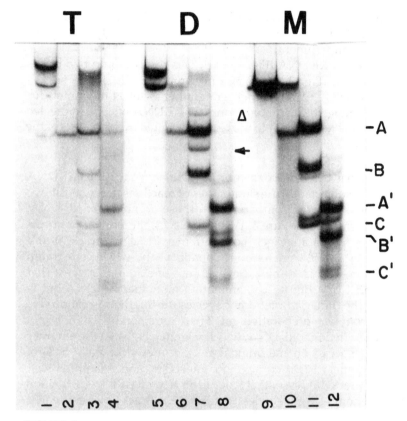

FIGURE 4

Structural analysis of monomer (M), dimer (D), and tetramer (T) forms of MVM RF DNA. The figure illustrates the autoradiogram of a 1.4% agarose gel which contains tetramer RF DNA (tracks 1–4), dimer RF DNA (tracks 5–8), and monomer RF DNA (tracks 9–12) before and after each RF DNA species was subjected to the same three reactions. Tracks 1, 5, and 9: untreated RF DNA; tracks 2, 6, and 10: RF DNA after heat denaturation; tracks 3, 7, and 11: RF DNA after digestion with *Eco*RI; tracks 4, 8, and 12: RF DNA digested with *Eco*RI and then heat-denatured prior to electrophoresis. The restriction-fragment designations (A, B, etc.) are as described in the legend to Figure 3. Open triangle and arrow refer to bands in track 7 discussed in text.

fragment A, the di-Br fragment cannot be attributed to incomplete enzyme cleavage. Such a restriction fragment would be expected, however, if the viral strands in dimer RF DNA were oriented in a "tail-to-tail" arrangement with their 3' termini overlapping, as proposed by Straus et al. (1976) and Tattersall and Ward (1976). Demonstration of a similar "bridge" fragment upon *Eco*RI digestion of the dimer RF DNA from H-1-infected cells (Rhode 1977) provides additional support for the proposed structure of dimer RF DNA.

Densitometric analysis demonstrated that only about 39% of the viral strands in the dimer RF DNA pool had their 3' termini located in the di-Br fragment; 100% of these termini would be expected in the di-Br fragment if the dimer RF DNA were totally "intact." However, both the present and previous studies (e.g., Tattersall et al. 1973; Straus et al. 1976) indicate that some multimeric RF DNA molecules break down to monomer duplex even under relatively mild conditions, presumably because they contain single-stranded nicks close to the region where the genome termini overlap. The subpopulation of the di-Br fragment which contained single-stranded nicks would be expected to be equally sensitive and thus converted to unit-length fragment C. The di-Br fragment observed might, therefore, have been derived predominantly from those dimer RF DNA molecules which contained no internal nicks. This interpretation is supported by the data shown in track 8 of Figure 4, which is an *Eco*RI digest of dimer RF DNA that was heat-denatured prior to electrophoresis. Whereas only 19% of the fragment-C doublet migrated as the faster component before denaturation (track 7), after denaturation this fragment contained 54% of the total radioactivity associated with all forms of the 3' terminal fragment (C and C'). Thus, after denaturation, at least 90% $[(54 - 19) \div 39]$ of the di-Br fragment has an electrophoretic mobility identical to that of the hairpin form of fragment C. These results indicate that the di-Br fragment contains the 3' terminus of the viral genome, that it is predominantly free of internal nicks, and that it has a structure

which, after denaturation, can reanneal spontaneously to generate two 3' hairpin terminal fragments:

(v and c represent the viral and complementary strands of the fragment).

Denaturation of the dimer RF DNA pool (track 6) showed that about 39% of the total molecules had covalently attached viral and complementary strands. The analysis of the di-Br fragment described above demonstrates that at least 90% of these molecules must have been derived from true dimers. These results strongly support the contention that the monomer RF DNA observed in the dimer RF DNA pool (track 5) are derived via disruption of nicked dimer RF DNA molecules rather than from contaminating monomer RF DNA.

One additional point worth noting about Figure 4 is that the greater the number of viral genome equivalents in the RF DNA the higher is the percentage of both terminal fragments which are observed in the extended (nonhairpin) form. In addition, spontaneous renaturation is less extensive in larger RF DNA molecules, which indicates that they contain more single-stranded nicks. Because virion-length single-stranded DNA is generated upon denaturation, these nicks are spaced one genome length apart.

Conversion of Monomer and Dimer RF DNA to Single-Stranded Viral DNA

The analyses presented above provide further evidence for the existence of concatemeric forms of viral DNA in MVM-infected cells. In an attempt to establish that these molecules are true, metabolically active, replicative intermediates, as reported previously for the concatemeric forms of AAV DNA (Straus et al. 1976), the synthesis of single-stranded progeny DNA was studied using pulse-chase kinetics. A nonsynchronous culture of MVM-infected A-9 cells was given a 20-min pulse-label with [³H]thymidine 14 hr p.i. Samples were removed immediately after the pulse-labeling period and after being chased with an excess of cold thymidine for 30, 60, 90, 120, and 150 min. Intracellular DNA was fractionated by the modified Hirt procedure and the supernatant fraction was run on a 1.4% agarose slab gel. Bands corresponding to single-stranded DNA, monomer RF DNA, and dimer RF DNA were extracted from the gel and their [³H]thymidine contents were determined. The results, shown in Figure 5, indicate that the radioactivity in both monomer and dimer RF DNA molecules is lost under the chase conditions, with a concomitant increase

FIGURE 5
Distribution of [³H]thymidine between single-stranded DNA (□), monomer RF DNA (●), and dimer RF DNA (○) in MVM-infected A-9 cells, under pulse-chase labeling conditions. See "Methods" for experimental details.

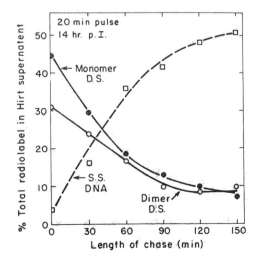

in the amount of single-stranded DNA. Although higher oligomeric forms of viral RF DNA were not detected on the gel by the ethidium bromide fluorescence, approximately 20% of the total radioactivity in the supernatant fraction of the 20-min pulse was in DNA which electrophoresed more slowly than dimer RF DNA; about 55% of these counts were lost after a 150-min chase, presumably into lower-molecular-weight forms of viral DNA. While the experimental approach did not allow precise quantitation of the distribution of radiolabel in all forms of viral DNA, it is apparent that both monomer and dimer RF DNA molecules are metabolically active and capable of generating single-stranded progeny DNA.

DISCUSSION

The data presented here are entirely compatible with the models of parvovirus DNA replication described previously (Straus et al. 1976; Tattersall and Ward 1976). Some of the salient features of the proposed replication scheme are illustrated schematically in Figure 6 in order to facilitate this discussion.

Synthesis of parental RF DNA via a self-priming mechanism would yield a monomer RF DNA molecule in which, at least initially, the viral (v) and complementary (c) strands are attached covalently through the 3' terminus of the v strand. We have observed that ^{32}P-labeled MVM virion DNA is converted into duplex DNA about 6–8 hr p.i. in a parasynchronous culture of infected cells, and that at this time the majority of the duplex DNA undergoes spontaneous renaturation. The percentage of this duplex DNA which renatures spontaneously declines as infection progresses. Thus, covalent attachment of v and c DNA strands appears to be an important initial event in viral DNA replication, even considering the particle-infectivity proviso mentioned earlier.

Analyses of the restriction fragments obtained upon digestion of monomer and dimer RF DNA with *Eco*RI demonstrate clearly that only fragments containing the 3' and 5' terminal portions of the viral genome are capable of spontaneous renaturation. However, more refined restriction-enzyme data are required to show that the strand linkage occurs precisely at the genome termini. The observation that fragments from *both* termini are found to renature spontaneously is consistent with the hypothesis that both c- and progeny v-strand syntheses are initiated via a self-priming mechanism. Nevertheless, one cannot rule out the possibility that the covalent attachment of the v and c strands in the 5' terminal fragment is due to DNA ligation

The restriction-enzyme data also provide further evidence that dimer RF DNA contains an alternating arrangement of v and c strands, as illustrated in Figure 6. Treatment of dimer RF DNA with *Eco*RI generates a

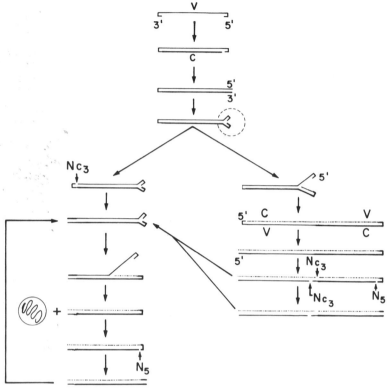

FIGURE 6

A modified rolling-hairpin model for MVM DNA replication (after Tattersall and Ward 1976). For explanation of terms see "Discussion."

new fragment not seen in monomer RF DNA. This di-Br fragment is a dimer of the *Eco*RI fragment C which contains the 3' terminus of the genome. Since the di-Br fragment generates spontaneously renaturing monomer-length fragment C upon denaturation, the v and c strands in some dimer RF DNA must be linked covalently through the 3' terminus of the v strand.

As proposed in Figure 6, progeny v DNA can be generated either from monomer RF DNA (left pathway) or from higher oligomeric RF DNA forms, of which only the dimer (right pathway) is illustrated. The demonstration that both monomer and dimer RF DNA species are metabolically active is again consistent with the proposed replication scheme. A feature common to both pathways in the self-priming mechanism is the generation of a "rabbit-eared" structure (see dashed circle in Fig. 6), via a "hairpin-transfer" process, such that the 3'-OH terminus of the c strand can prime the initiation of progeny v-strand synthesis. However, production of progeny DNA via strand-displacement synthesis from *any* form of RF DNA would require two specific types of endonucleolytic cleavages to yield the mature-virus genome.

One of these enzymatic cuts (designated Nc_3 in Fig. 6) would occur at the end of the terminal palindrome distal to the 3' end of the v strand (i.e., opposite the 3' terminus of the v strand in monomer RF DNA), leaving a 3'-OH group suitable for completion of the progeny strand by displacement synthesis. Such a displacement at the level of *monomer* RF DNA would transfer the original 3' terminus of the v strand to the c strand. Analysis of the AAV genome by Fife et al. (1977) has demonstrated terminal sequence heterogeneity consistent with such a transfer, which they designated as "flip-flop." However, transfer needs to occur only once (in the parental RF DNA) for replication to proceed from the monomer RF DNA of the autonomous parvoviruses (see Fig. 6); this is a consequence of the fact that the autonomous parvoviruses encapsidate only one strand of the RF DNA duplex. Displacement synthesis of a dimer RF DNA which is doubly nicked at Nc_3 cleavage sites (Fig. 6) would break the molecule, generating monomer RF DNA suitable for continued replication.

The second type of cleavage (designated N_5) would occur precisely at the 5' terminus of the v strand in either monomer or oligomer RF DNA, generating 5' phosphate and 3' hydroxyl ends. Displacement synthesis at this nick in either monomer or dimer RF DNA will recreate the hairpin originally present at the 5' end of the v strand and produce a new copy of this hairpin on the 3' end of the c strand which again can serve to prime further progeny v-strand synthesis. In concatemers (e.g., tetramer RF DNA) which contain overlapping 5' genome termini nicked at N_5 sites, displacement synthesis at one or both nicks will break the molecule, generating lower RF DNA species which again are fully capable of producing progeny v strands.

Virtually all studies on parvovirus DNA replication have demonstrated the existence of monomer RF DNA in infected cells. However, the amounts of oligomeric RF DNA reported have varied greatly. While in some cell-virus systems significant amounts of oligomeric RF DNA are observed (Tattersall et al. 1973; Straus et al. 1976), in other systems RF DNA larger than a dimer has not been detected (Gunther and May 1976; Mayor and Jordan 1976; Rhode 1977). Furthermore, the distribution of RF DNA molecules observed in LuIII-infected cells varies with the host cell on which the virus is propagated (G. Siegl, personal communication). The observed variability in the size distribution of the RF DNA pool may reflect the intracellular levels of the processing nuclease(s) responsible for the Nc_3 and N_5 cleavages. Cell-virus systems which exhibit high nuclease activity will replicate predominantly at the level of monomer RF DNA. A decrease in level of activity would facilitate the generation of higher oligomer forms, which would then be processed to generate progeny DNA. The isolation of the cleavage enzyme(s) and the establishment of its origin (viral or cellular) is of considerable interest.

ACKNOWLEDGMENTS

This work was supported by Public Health Service grant GM-20124 from the National Institute of General Medical Sciences and by grant CA-16038 from the National Cancer Institute.

REFERENCES

Blin, N., A. V. Gabain, and H. Bujard. 1975. Isolation of large molecular weight DNA from agarose gels for further digestion by restriction enzymes. *FEBS Lett.* **53**:84.

Bomford, R. and I. B. Weinstein. 1972. Transformation of a rat epithelial-like cell line by murine sarcoma virus. *J. Natl. Cancer Inst.* **49**:379.

Bourguignon, G. J., P. J. Tattersall, and D. C. Ward. 1976. DNA of minute virus of mice: Self-priming, nonpermuted, single-stranded genome with a 5′-terminal hairpin duplex. *J. Virol.* **20**:290.

Denhardt, D. T., S. Eisenberg, K. Bartok, and B. J. Carter. 1976. Multiple structures of adeno-associated virus DNA: Analysis of terminally labeled molecules with endonuclease R·*Hae*III. *J. Virol.* **9**:574.

Dobson, P. R. and C. W. Helleiner. 1973. A replicative form of the DNA of minute virus of mice. *Can. J. Microbiol.* **19**:35.

Fife, K., K. I. Berns, and K. Murray. 1977. Structure and nucleotide sequences of the terminal regions of adeno-associated virus DNA. *Virology* **78**:475.

Gunther, M. and P. May. 1976. Isolation and structural characterization of monomeric and dimeric forms of replicative intermediates of Kilham rat virus DNA. *J. Virol.* **20**:86.

Handa, H., H. Shimojo, and K. Yamaguchi. 1976. Multiplication of adeno-associated virus type 1 in cells coinfected with a temperature-sensitive mutant of human adenovirus type 31. *Virology* **74**:1.

Hauswirth, W. W. and K. I. Berns. 1977. Origin and termination of adeno-associated virus DNA replication. *Virology* **78**:488.

Hirt, B. 1967. Selective extraction of polyoma DNA from infected mouse cell cultures. *J. Mol. Biol.* **26**:365.

Littlefield, J. W. 1964. Three degrees of guanylic acid-inosinic acid pyrophosphorylase deficiency in mouse fibroblasts. *Nature* **203**:1142.

Mayor, H. D. and L. E. Jordan. 1976. The replication of rodent parvovirus X14. *J. Gen. Virol.* **30**:337.

Rhode, S. L. III. 1974. Replication process of the parvovirus H-1. II. Isolation and characterization of H-1 replicative form DNA. *J. Virol.* **13**:400.

————. 1977. Replicative process of the parvovirus H-1. VI. Characterization of a replication terminus of H-1 replicative form DNA. *J. Virol.* **21**:694.

Salzman, L. A. 1977. Evidence for terminal S_1-nuclease-resistant regions on single-stranded linear DNA. *Virology* **76**:454.

Salzman, L. A. and W. L. White. 1973. In vivo conversion of the single-stranded DNA of the Kilham rat virus to a double-stranded form. *J. Virol.* **11**:299.

Schindler, R. and J. C. Schaer. 1973. Preparation of synchronized cell cultures from early interphase cells obtained by sucrose gradient centrifugation. In *Methods in cell biology* (ed. D. M. Prescott), vol. VI, p. 43. Academic, New York.

Sharp, P. A., B. Sugden, and J. Sambrook. 1973. Detection of two restriction endonuclease activities in *Haemophilus parainfluenzae* using analytical agarose-ethidium bromide electrophoresis. *Biochemistry* **12**:3055.

Siegl, G. and M. Gautschi. 1973. Multiplication of parvovirus LuIII in a synchronized culture system. 2. Biochemical characteristics of virus replication. *Arch. gesamte Virusforsch.* **40**:119.

————. 1976. Multiplication of parvovirus LuIII in a synchronized culture system. III. Replication of viral DNA. *J. Virol.* **17**:841.

Singer, I. I. and S. L. Rhode III. 1977. Replication process of the parvovirus H-1. VIII. Partial denaturation mapping and localization of the replication origin of H-1 replicative-form DNA with electron microscopy. *J. Virol.* **21**:724.

Straus, S. E., E. D. Sebring, and J. A. Rose. 1976. Concatemers of alternating plus and minus strands are intermediates in adenovirus-associated virus DNA synthesis. *Proc. Natl. Acad. Sci.* **73**:742.

Tattersall, P. 1972. Replication of the parvovirus MVM. I. Dependence of virus multiplication and plaque formation on cell growth. *J. Virol.* **10**:586.

Tattersall, P. and D. C. Ward. 1976. Rolling hairpin model for replication of parvovirus and linear chromosomal DNA. *Nature* **263**:106.

Tattersall, P., L. V. Crawford, and A. J. Shatkin. 1973. Replication of the parvovirus MVM. II. Isolation and characterization of intermediates in the replication of the viral deoxyribonucleic acid. *J. Virol.* **12**:1446.

Tattersall, P., P. J. Cawte, A. J. Shatkin, and D. C. Ward. 1976. Three structural polypeptides coded for by minute virus of mice, a parvovirus. *J. Virol.* **20**:273.

Purification and Properties of Replicative-Form and Replicative-Intermediate DNA Molecules of Parvovirus LuIII

GÜNTER SIEGL
MARKUS GAUTSCHI

Institute of Hygiene and Medical Microbiology
University of Bern
CH 3010 Bern, Switzerland

Several models have been proposed recently for the replication of parvovirus DNA (Straus et al. 1976; Tattersall and Ward 1976; Singer and Rhode 1977). These have been based on studies of the secondary structure of the viral genome as well as on the isolation and characterization of double-stranded replicative-form (RF) and replicative-intermediate (RI) DNA from infected cells. So far, however, the theoretical considerations have not been confirmed experimentally; this is due in part to the lack of suitable hybridization probes for the quantitation of viral and complementary strands. Such a hybridization system would also be of advantage in the analysis of virus-specific RNA and in the characterization of the DNA found in defective parvovirus particles.

Faced with these problems in our studies of the replication of LuIII-virus DNA, we have chosen the intracellular linear double-stranded RF form of viral DNA as a source of virus-specific sequences for such a hybridization system. This paper describes the experimental conditions under which maximal amounts of RF DNA can be isolated. It also provides strong evidence that both the quantity and the type of DNA molecules extractable from infected cells vary considerably with the culture system as well as with the method and time of isolation.

MATERIALS AND METHODS

Virus and Cells

The biological properties of parvovirus LuIII and the growth characteristics of the HeLa cells have been described (Siegl and Gautschi 1973). NB cells (newborn-human kidney cells transformed by SV40) were a generous gift of Dr. J. van der Noordaa of the University of Amsterdam. These cells could be grown and synchronized by the methods used for the cultivation of HeLa cells. However, the medium had to be supplemented with 1% of a 100× concentrated solution of MEM nonessential amino acids, and the pH had to be adjusted to 7.6.

Isolation of DNA

DNA of infected cells was labeled for 2 hr in the presence of 10 μCi/ml of Amersham tritiated thymidine, [^{3}H]TdR (specific activity 47 Ci/mmole). At the end of the labeling period monolayers in 75-cm^2 plastic flasks were lysed, either with 4 ml of 0.6% sodium dodecyl sulfate, 10 mM Tris, 10 mM EDTA (pH 7.6) for 30 min at room temperature or by incubation for 15–20 hr at 37°C in 4 ml of the above solution plus 200 μg/ml of Merck proteinase K. In both cases the viscous lysate was brought to a concentration of 1 M NaCl and, after standing for 16 hr at 0°C, was centrifuged for 45 min at 15,000g at 0–4°C. The supernatant represented the starting material for the purification and concentration of low-molecular-weight DNA, whereas the pellet was dissolved in 0.1 N NaOH to allow determination of acid-insoluble radioactivity. Low-molecular-weight DNA was further purified by repeated extraction with chloroform-isoamyl alcohol (24:1, vol/vol) (Gautschi and Clarkson 1975). The aqueous phase was diluted to 0.3 M NaCl, 10 mM Tris, 1 mM EDTA (pH 8.1), and adsorbed onto Serva benzolated-naphtholated-DEAE-cellulose (BND-cellulose). DNA was sequentially eluted with 1 M NaCl, 1 M NaCl plus 0–2% caffeine, and 1 M NaCl in 0.1 N NaOH as described previously (Siegl 1973). Details of the sedimentation of DNA in neutral or alkaline CsCl, of the DNA-DNA hybridization technique, and of the digestion with the single-strand-specific nuclease S1 have been described (Siegl and Gautschi 1976). For density-gradient centrifugation 4.6 g of Cs$_2$SO$_4$ was added to 6 ml of a DNA solution in polyallomer tubes. The sample was then centrifuged to equilibrium for 40 hr at 45,000 rpm in a Beckman Ti50 rotor. Preparative linear 5–20% sucrose gradients were run in a Beckman SW27 rotor at 25,000 rpm for 22 hr. These gradients were made up in 10 mM Tris,

10 mM EDTA, 1 M NaCl, and 15 mM Na-citrate (pH 8.2). Centrifugation in both Cs_2SO_4 and sucrose gradients was carried out at 20°C. Fractions were collected from the tops of the tubes, and absorbancy at 260 nm was monitored by means of an ISCO gradient fractionator.

RESULTS

Extraction of Intracellular Viral DNA

Cultures of HeLa or NB cells synchronized for DNA synthesis were infected at a multiplicity of 10 concomitantly with release from the synchronization block. Starting 6 hr after infection, individual cultures were pulsed with [^{3}H]TdR for 2 hr at regular intervals. Intracellular DNA of low molecular weight labeled during this period was then extracted, either by the original method of Hirt (1967) or by means of the proteinase-K modification of this technique described in the "Materials and Methods" section of the present article. A comparison of the respective results showed that, in both types of cells, the proteinase-K modification released up to 10 times as much DNA into the supernatant as the original technique. The size spectra of the DNA molecules recovered in both types of extracts proved to be almost identical, and, consequently, further experiments used only the proteinase-K modification.

Maximum synthesis of low-molecular-weight DNA could be detected in HeLa cells as well as in NB cells 14–16 hr after infection (Fig. 1). However, NB cells released about 10 times as much acid-precipitable label

FIGURE 1
Low-molecular-weight DNA in protein-ase-K extracts of synchronized NB and HeLa cells at various times after infection with LuIII virus. Cpm (o) recovered in the supernatant represent (■) percent of total label incorporated by the infected cells.

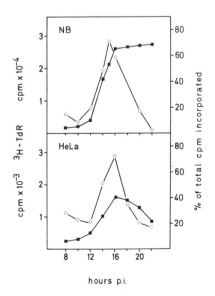

into the extract as did a comparable number of HeLa cells. Up to 70% of total [³H]TdR incorporated into NB cells could be recovered as low-molecular-weight DNA. HeLa cells, on the other hand, yielded at most 40–50% of acid-precipitable label in the supernatant.

Sedimentation Characteristics of the DNA

Sedimentation analysis of the extracted DNA in neutral and alkaline gradients confirmed our previous observation that distinct virus-specific DNA molecules can be isolated as early as 9–10 hr after infection. At that time both culture systems released almost exclusively a double-stranded DNA molecule which sedimented in neutral gradients at about 15S. With progressing time, however, the spectrum of low-molecular-weight DNA became more complex, and significant differences between HeLa and NB cells could be detected. As is evident from Figure 2, sedimentation in neutral as well as alkaline gradients revealed a broad spectrum of molecules in the DNA obtained from HeLa cells. DNA from NB cultures showed a more uniform sedimentation behavior. On the basis of the number of 16S DNA molecules in the alkaline gradients, the relative

FIGURE 2
Sedimentation profile of low-molecular-weight DNA extracted from synchronized HeLa and NB cells at 16 hr after infection. Sedimentation of DNA extracted from HeLa cells in neutral (a) and alkaline (b) gradients; NB-cell extracts sedimented at neutral (c) and alkaline (d) pH.

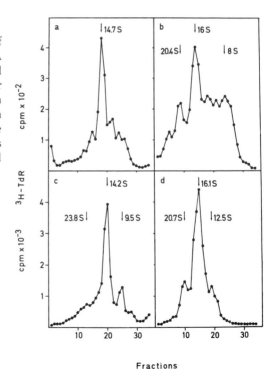

Fractions

number of molecules of twice this size (dimers, 20.4S) or even four times this size (tetramers, 26.8S), as well as the relative number of smaller molecules (6–13S), increased significantly in the HeLa extracts up to 22 hr after infection. In contrast, only monomer- and dimer-length molecules were clearly discernible throughout this period of infection in alkaline gradients loaded with NB-cell extracts. The small peak sedimenting at 9.5S in neutral gradients and at 12.5S in alkaline gradients could be detected as early as 16 hr after infection. Synchronized infected NB cells already showed signs of degeneration at this time.

Chromatography on BND-Cellulose

Low-molecular-weight DNA was freed from residual SDS and proteins by three rounds of extraction with chloroform-isoamyl alcohol before being adsorbed onto BND-cellulose. Table 1 lists the percentages of double-stranded DNA (eluted with 1 M NaCl), partially single-stranded DNA (eluted with 1 M NaCl + 2% caffeine), and largely single-stranded DNA (eluted with 0.1 N NaOH) in extracts obtained at various times after infection. The majority of the DNA molecules extracted from either cell type behaved like double-stranded DNA with more or less extensive single-stranded sequences. NB cells also showed a decrease in double-stranded DNA in parallel with an increase in firmly adsorbed single-stranded molecules between 12 and 16 hr. A similar decrease with time in the percentage of double strands was also recorded for HeLa cells. However, the relative amount of double-stranded DNA in extracts of the latter cells was almost twice as high as in the respective NB-cell samples.

To fractionate partially single-stranded DNA further, these molecules were also eluted with linear gradients of 0–1% or 0–2% caffeine. Figure 3 shows that most of the DNA was released in the presence of about 0.2% caffeine. Nevertheless, there was always a considerable fraction of labeled molecules that could only be recovered with higher caffeine concentrations. Molecules with similar elution behavior were pooled (Pools I–V in Fig. 3) and were analyzed in parallel with those in the 1 M NaCl eluate as well as the NaOH eluate. In both the NaCl eluate (Fig. 4a, b) and the 2% caffeine (Fig. 4c,d), of a DNA sample extracted from NB cells at 15 hr after infection the majority of the molecules sedimented as a relatively homogeneous band at about 15S in neutral gradients. However, at alkaline pH (Fig. 4b,d), this DNA resolved into two peaks, which, with respective S values of 16 and 20.6, would represent the monomer and dimer lengths of a single-stranded DNA of 1.5×10^6 daltons. The ratio of monomer to dimer forms was 1:1.5. In the caffeine extract, on the other hand, a ratio of 7:1 was observed.

Table 2 lists the relative amounts of molecular forms in extracts obtained at various times after infection. No significant change could be detected

TABLE 1

Fractionation by Chromatography on BND-Cellulose of Low-Molecular-Weight DNA Labeled and Extracted at Various Times After Infection

Labeling (hr p.i.)	NB cells			HeLa cells		
	1 M NaCl	2% caffeine	0.1 N NaOH	1 M NaCl	2% caffeine	0.1 N NaOH
12–14	17.9	68.8	13.3	33.8	54.4	11.8
13–15	15.8	70.0	14.2	29.0	62.0	9.0
14–16	13.4	66.2	20.4	26.8	61.9	11.1

Figures represent the percentage of cpm eluted as double-stranded (1 M NaCl), partially single-stranded (2% caffeine), and largely single-stranded (0.1 N NaOH) molecules.

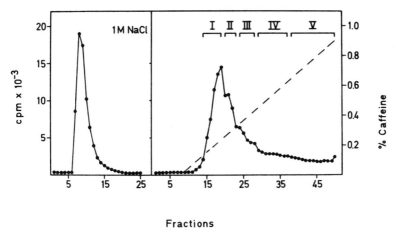

FIGURE 3
Fractionation by chromatography on BND-cellulose of low-molecular-weight DNA extracted from synchronized NB cells at 15 hr after infection. Molecules were eluted with either 1 M NaCl or a linear 0–1% caffeine (in 1 M NaCl) gradient. I–V indicate the fractions pooled for further analysis.

FIGURE 4
Sedimentation of low-molecular-weight DNA extracted from synchronized NB cells at 15 hr after infection and fractionated by chromatography on BND-cellulose. DNA eluted by 1 M NaCl and sedimented in neutral (a) and alkaline (b) gradients; DNA eluted by 2% caffeine and sedimented at neutral (c) and alkaline (d) pH.

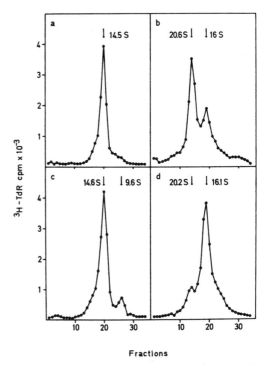

Fractions

321

TABLE 2
The Relative Numbers of 16S and 20.5S DNA Molecules in the NaCl and Caffeine
Fractions of NB-Cell extracts Labeled at Various Times After Infection

Labeling (hr p.i.)	1 M NaCl		2% Caffeine	
	16S + 20.5S[a]	16S:20.5S[b]	16S + 20.5S[a]	16S:20.5S[b]
12–14	65.4	0.6:1	64.1	11.6:1
13–15	64.6	0.7:1	64.0	7.4:1
14–16	63.3	0.4:1	69.4	3.5:1

[a] Percentage of total cpm recovered in the respective eluate.
[b] Cpm in the 16S peak vs. cpm in the 20.5S peak.

within the NaCl eluate, yet the relative number of dimers in the caffeine fraction increased with time. A further characteristic of the caffeine eluate was the presence of molecules of decreasing length trailing behind the 16S peak. Thus, molecules released at the lowest caffeine concentration sedimented rather homogeneously, but those eluting with 1–2% caffeine sedimented as a broad band between 24S and 8S, with a peak at 16S. A similar sedimentation behavior was characteristic for DNA recovered from the BND-cellulose with 0.1 N NaOH.

Distribution of Single-Stranded Regions

The extent of single-strandedness in the isolated molecules was monitored by digestion with nuclease S1. During a 60-min incubation at 37°C or 45°C in the presence of 50–300 mM NaCl, between 1% and 3.6% of the radioactivity in the NaCl-eluted DNA was rendered acid-soluble. Maximum digestion was obtained at 45°C in the presence of 50 mM NaCl. Digestion of the DNA samples pooled from the caffeine gradient (Fig. 3) yielded variable results. In one experiment, 3.5% of acid-soluble radioactivity was measured for DNA eluted with 0.2% caffeine, whereas 26% digestion was observed for molecules released with 1–2% caffeine. These extreme values could not be reproduced regularly, but the results always indicated an increase in single-strandedness in parallel with the increase in the caffeine concentration necessary for elution from BND-cellulose. When unfractionated 0–2% caffeine eluates of the DNA samples extracted at various times after infection were analyzed, the percentage of acid-soluble radioactivity decreased from 21.6% in the 12–14-hr extract to only 7.7% for DNA labeled between 14 and 16 hr.

The DNA eluted with 1 M NaCl and the molecules released with 0.2%

caffeine banded at an identical density of 1.48 g/cm^3 in Cs$_2$SO$_4$ density gradients. Fractions eluted at increased concentrations of caffeine were found at correspondingly increased densities.

Isolation of Complementary DNA Strands

The dimer-length single-stranded molecule sedimenting in alkaline gradients at 20.5S originated from a double-stranded DNA of monomer length, which sedimented at 14.5S at neutral pH. Presumably this consists of covalently linked viral and complementary strands. To cleave this hairpin structure, the double-stranded DNA was digested with S1 nuclease at 37°C for 60 min in the presence of 50, 150, and 300 mm NaCl. The samples were then sedimented in alkaline gradients. The original NB-cell extract used for these experiments contained monomer and dimer DNA in a ratio of 1:1.8. This ratio was changed to 3:1 by mere incubation at 37°C. As further experiments showed, the magnitude of this reversion increased with temperature, and after the DNA was boiled for 2 min a value of 5:1 was recorded. On the other hand, digestion with nuclease in 50 mm NaCl yielded 16S and 20.5S molecules in a ratio higher than 8.0:1. The figures recorded for digestion in 150 and 300 mm NaCl were 8.6:1 and 4.8:1, respectively.

DNA digested in 50 mm NaCl and subsequently heated to 100°C for 2 min could be digested by further addition of S1 to the same extent as a ^{14}C-labeled double-stranded HeLa DNA included in the experiments as an internal control. After S1 digestion, DNA which eluted from the 1 m NaCl wash of the BND-cellulose column was centrifuged in preparative neutral sucrose gradients. As monitored by both the distribution of radioactivity and the absorbency at 260 nm, the majority of the DNA sedimented at 14.5S. About half of the radioactivity of these double-stranded DNA molecules could be displaced during reannealing in the presence of unlabeled viral single-stranded DNA. We have to assume, therefore, that the DNA isolated from these preparative gradients contained viral genome and virus-specific complementary DNA molecules in identical amounts.

DISCUSSION

The studies described in this paper provide evidence that the spectrum of DNA molecules extracted from LuIII-infected cells may vary with the type of cell, the time elapsing between infection and extraction, and (last but not least) the method of extraction. Using randomly growing instead of synchronized cell cultures, the spectrum of replicative viral DNA will be confused further by the nucleic acid molecules produced in already degenerating cells. The latter point is well illustrated by the increase in small-size as well as high-molecular-weight DNA appearing in Hirt supernatants of synchronized HeLa cells at the end of the replication cycle of parvovirus LuIII.

The main goal of the present study, however, was the isolation and characterization of linear double-stranded RF DNA molecules, which have to be assumed to contain viral genome and complementary sequences in identical quantities. Maximal amounts of such DNA could be extracted and purified from HeLa and NB cells at 14–16 hr after infection. We have shown by electron microscopy (data not presented) that more than 95% of the molecules eluted from BND-cellulose in the presence of 1 M NaCl were linear. Less than 5% of them were branched or Y-shaped. DNA molecules isolated from infected HeLa cells had a rather broad size distribution, whereas the majority of molecules from NB cells were between 1.2 and 1.7 μm. About two-thirds of these molecules consisted of double-stranded DNA, the individual strands of which appeared to be covalently linked and formed a hairpin structure. The hairpin could be readily cleaved in its terminal single-stranded loop by S1 nuclease, and identical amounts of separable complementary DNA strands could be isolated as 14.5S double-stranded DNA molecules in preparative neutral sucrose gradients.

Sedimentation in sucrose gradients or banding of the digested extracts in Cs_2SO_4 was necessary to remove residual RNA. Lysis of cells with proteinase K and fractionation with 1 M NaCl yielded huge concentrations of transfer RNA and 5S ribosomal RNA. Because of their elevated content of double-stranded sequences, these molecules coeluted with RF DNA from BND-cellulose. Extensive digestion of extracts with RNase before adsorption onto BND-cellulose resulted in the loss of only 90–93% of the RNA. The remaining percentage proved to be enough to interfere with the spectrophotometric quantitation of LuIII RF DNA at 260 nm.

About 50% of the radioactivity of 14.5S RF DNA could be replaced by unlabeled LuIII-virion DNA in hybridization experiments. Preliminary reannealing studies also suggested that the isolated molecules are rather homogeneous and are of a genetic complexity comparable to that of polyoma DNA.

LuIII RI DNA eluted from BND-cellulose in the presence of caffeine. The majority of this partially single-stranded DNA was of monomer length and showed no covalent linkage between complementary DNA strands. However, electron microscopy revealed that about 20–30% of RI molecules were branched or Y-shaped. Such structures have been described for both LuIII and H-1 RI DNA (Siegl and Gautschi 1976; Singer and Rhode 1977). RI DNA represented the major fraction of pulse-labeled low-molecular-weight DNA in infected HeLa and NB cells. The existence of "branched" molecules and of more or less single-stranded sequences also suggested that RI DNA represents the most active species in the replication of the single-stranded genome of parvoviruses.

REFERENCES

Gautschi, J.R. and J. M. Clarkson. 1975. Discontinuous DNA replication in mouse P-815 cells. *Eur. J. Biochem.* **50**:403.

Hirt, B. 1967. Selective extraction of polyoma DNA from infected mouse cell cultures. *J. Mol. Biol.* **26**:365.

Siegl, G. 1973. Physicochemical characteristics of the DNA of parvovirus LuIII. *Arch. gesamte Virusforsch.* **43**:334.

Siegl, G. and M. Gautschi. 1973. The multiplication of parvovirus LuIII in a synchronized culture system. I. Optimum conditions for virus replication. *Arch. gesamte Virusforsch.* **40**:105

———. 1976. Multiplication of parvovirus LuIII in a synchronized culture system. III. Replication of viral DNA. *J. Virol.* **17**:841.

Singer, I. I. and S. L. Rhode III. 1977. Replication process of the parvovirus H-1. VII. Electron microscopy of replicative-form DNA synthesis. *J. Virol.* **21**:713.

Straus, S. E., E. D. Sebring, and J. A. Rose. 1976. Concatemers of alternating plus and minus strands are intermediates in adenovirus-associated virus DNA synthesis. *Proc. Natl.Acad. Sci.* **73**:742.

Tattersall, P. and D. C. Ward. 1976. Rolling hairpin model for replication of parvovirus and linear chromosomal DNA. *Nature* **263**:106.

Three Distinct Replicative Forms of Kilham-Rat-Virus DNA

GARY S. HAYWARD*

Institut für Molekulare Genetik der Universität Heidelberg
Heidelberg, Federal Republic of Germany
and Unit de Biochimie, Institut Gustave-Roussy
Villejuif, France

HERMANN BUJARD

Institut für Molekulare Genetik der Universität Heidelberg
Heidelberg, Federal Republic of Germany

MAGALI GUNTHER

Unit de Biophysique, Institut de Recherche Scientifique sur le Cancer
Villejuif, France

The DNA found in virions of all parvoviruses is single-stranded and exists in a linear rather than circular conformation. The nondefective parvoviruses, including minute virus of mice (MVM), H-1, and Kilham rat virus (KRV), package predominantly only one strand, described as the viral or v strand. In contrast, approximately equal numbers of both plus and minus strands are encapsidated by the defective adeno-associated viruses (AAV). Evidence from a number of sources indicates that one of the first steps in replication for all of these DNA molecules involves conversion to a double-stranded replicative form (Salzman and White 1973; Tattersall et al. 1973; Rhode 1974). However, as pointed out by Watson (1972), further rounds of replication as simple linear duplex DNA molecules pose certain problems related to the losses of information at 5'

* Present address: Department of Pharmacology and Experimental Therapeutics, Johns Hopkins School of Medicine, Baltimore, Maryland 21205

termini that are inherent in the nature of DNA synthesis and the action of DNA polymerases. In phage systems, these problems are largely overcome by circularization or the generation of multigenomic concatemers as intermediate structures in the replication process.

Earlier work (Gunther and May 1976) provided evidence from sedimentation, hybridization, and hydroxyapatite binding studies that some of the intracellular replicative forms of KRV DNA possessed self-renaturing properties and that multiple DNA species with differing biophysical parameters existed. Spontaneously renaturing DNA could arise from at least three known sources: (a) covalently joined twisted circular (or supercoiled) structures, (b) some form of covalent joining or cross-linking between the complementary strands, or (c) the presence of highly repetitive DNA sequences. At the time this work was performed only the first type had been observed to occur in viral DNA. Recent evidence for the existence of a duplex hairpin foldback at the 3' end of MVM viral DNA and the self-priming properties of MVM DNA synthesis in vitro (Bourguignon et al. 1976) suggested that a mechanism other than circularization or concatemer formation might be operative in the parvovirus system. In an attempt to investigate the nature of parvovirus DNA replication we set out to define further the properties and topography of the fast-sedimenting and self-renaturing forms of intracellular KRV DNA by agarose gel electrophoresis, buoyant-density studies, and electron microscopy.

MATERIALS AND METHODS

VIRUS AND RF DNA. The source of the rat virus used in this work and the procedures used for purifying virus, infecting primary rat fibroblasts and RT cells, etc. have been described in detail elsewhere (Gunther and May 1976). Intracellular KRV DNA was isolated by the selective Hirt extraction procedure from infected cells lysed 1–2 hr after addition of [^{3}H]thymidine. The supernatant DNA was deproteinized with chloroform, ethanol-precipitated, and banded to equilibrium in a CsCl density gradient. Selected ^{3}H-labeled replicative-form (RF) DNA was further purified by phenol-extraction and dialysis. Separation of the fast-sedimenting shoulder of RF DNA (18S) from the main peak of ^{3}H-labeled RF DNA (14.5S) was achieved by velocity sedimentation through 4-ml 5–20% neutral sucrose gradients at 54,000 rpm and 20°C for $2\frac{1}{2}$ hr in a swinging-bucket rotor. CsCl and sucrose gradient profiles of RF DNA extracted by the Hirt method have been described in detail (Gunther and May 1976).

REFERENCE DNA SPECIES. The ^{14}C-labeled SV40 DNA was prepared from virus passaged at high multiplicities of infection and was a gift from Dr. E.

May. The ^{32}P-labeled phage PM2 DNA was a gift from Dr. B. Revet. The single-stranded linear form of PM2 DNA was separated from single-stranded circles after denaturation of isolated form-II DNA and sedimentation through an alkaline CsCl gradient. ^{14}C-labeled single-stranded KRV DNA was extracted from purified virions grown in the presence of [^{14}C]thymidine.

AGAROSE GEL ELECTROPHORESIS.   Gels of 0.8% agarose (SeaKem) were cast in cylindrical tubes of 1 cm × 18 cm and electrophoresis of native or denatured DNA in Tris-phosphate buffer was performed as described previously (Hayward and Smith 1972). The gels were then sliced into 1.2-mm fractions, dissolved by autoclaving in 1 ml of buffer, and dried onto glass-fiber-paper squares for counting in toluene scintillation fluid.

ELECTRON MICROSCOPY.   Aqueous spreading was performed by the droplet diffusion method from a solution containing 0.15 M ammonium acetate (pH 6.7) and 0.001% cytochrome c. The DNA was picked up on carbon-coated grids and rotary-shadowed with Pt/Pd. The standard deviation for double-stranded-DNA length measurements in this laboratory has been determined to be 3.5%. Denaturation and unfolding of single-stranded DNA was carried out by heating to 60°C in 26% dimethylsulfoxide and 5% formaldehyde (method B of Bujard 1970). Artificial cross-links were introduced with higher concentrations of formaldehyde. The length of single-stranded DNA under these conditions is reduced to approximately two-thirds the length it would have in a double-stranded conformation.

BUOYANT DENSITY IN CESIUM CHLORIDE GRADIENTS.   DNA samples were mixed with CsCl in 0.01 M Tris, 0.001 M EDTA (pH 8) to an initial density of 1.70 g/cm³. Centrifugation was carried out at 35,000 rpm for 20 hr at 25°C in an SW 50.1 rotor.

SEDIMENTATION IN ALKALINE CESIUM CHLORIDE.   Gradients were prepared from CsCl solutions of 1.25 and 1.45 g/cm³ in 0.5 M KOH. DNA samples were centrifuged through 12-ml gradients in polyallomer tubes at 37,000 rpm and 20°C for 6 hr in an SW 40 rotor.

RESULTS

Characteristics of Viral DNA in Hirt Extracts

The DNA used in these studies was obtained from rat fibroblasts labeled with [^{3}H]thymidine for 1–2 hr at 12 hr after KRV infection. A modified Hirt extraction procedure was used to isolate low-molecular-weight DNA, which was then banded to equilibrium in a CsCl buoyant-density gradient. Approximately 60% of the incorporated ^{3}H radioactivity was recovered in the Hirt supernatant, and only that portion (~40–70%) with the buoyant density of double-stranded KRV DNA (1.705 g/cm³) was selected for further study. Note that this represents a deliberate selection for replicative forms (RF DNA), and that replicative-intermediate structures containing, for example, single-stranded regions or growing polynucleotide chains together with single-stranded mature viral DNA are mostly discarded. The RF DNA gives rise to a major DNA species sedimenting at 14.5S in neutral sucrose gradients and a smaller shoulder of faster-sedimenting material of approximately 18S (Gunther and May 1976). Previous hybridization experiments have indicated that at least 80–90% of the ³H-labeled DNA in both of these RF DNA fractions is of viral origin.

Agarose Gel Electrophoresis of Native RF DNA

Electrophoresis through agarose gels resolves double-stranded DNA species with different molecular weights and also species of the same size but with different topography (Aaij and Borst 1972; G. S. Hayward, H. Bujard, and B. Revet, manuscript in preparation). Figure 1a shows the separation of total KRV ³H-labeled RF DNA into two distinct species by electrophoresis through 0.8%-agarose gels. Comigrating reference species in these gels included ³²P-labeled PM2 viral DNA (a mixture of form-I supercoils, form-II nicked circles, and form-III linear DNA species, all of molecular weight 6.4 × 10⁶) and also forms I and II of ¹⁴C-labeled

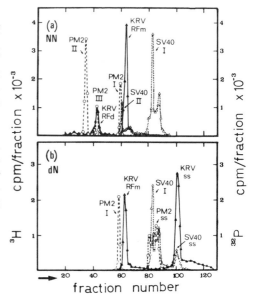

FIGURE 1

Agarose gel electrophoresis of unfractionated KRV RF DNA. A mixture of ³H-labeled KRV RF DNA (▲——▲), ³²P-labeled PM2 viral DNA (o- - - -o), and ¹⁴C-labeled SV40 viral DNA (△·····△) was divided into two equal parts. One sample was applied to gel a as native DNA (top panel) and the other to gel b after denaturaturation in 0.1 M NaOH (lower panel). Electrophoresis was carried out for 17 hr at 3 V/cm and 4°C through 0.8% agarose.

SV40 viral DNA (molecular weight 3.2×10^6). The SV40 DNA also included at least 50% defective molecules containing deletions that give rise to multiple bands migrating slightly ahead of the major form-I and form-II species.

Isolation of the native 14.5S and shoulder 18S RF DNA on sucrose gradients prior to gel electrophoresis confirmed that the 14.5S peak corresponds to the faster-migrating species (Fig. 2*a*) and that the fast-sedimenting shoulder resolves into a single discrete peak that comigrates with linear PM2 DNA (Fig. 2*c*). From previous work with KRV and other parvovirus DNAs, we expected that the major RF species would be a monomer duplex form of KRV DNA with a molecular weight twice that of the viral strand. The migration of the 14.5S species close to the position for linear SV40 DNA (3.2×10^6) is consistent with this interpretation if one assumes that it too has a linear conformation. However, the migration of the fast-sedimenting form is far too slow for either a nicked circular form or a supercoiled form with this same molecular weight. Therefore, a likely

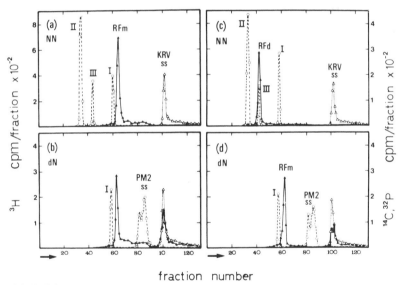

fraction number

FIGURE 2

Agarose gel electrophoresis of native and denatured forms of the isolated 14.5S and 18S KRV DNA. The two different sedimentation components of RF DNA were separated on neutral sucrose gradients. Mixtures of each ^{3}H-labeled KRV RF DNA (▲——▲) together with ^{32}P-labeled PM2 viral DNA (○- - - - -○) and ^{14}C-labeled KRV single-stranded viral DNA (△·····△) were subjected to electrophoresis through 0.8% agarose gels as described in Figure 1. Left panels: the major monomer RF peak (14.5S DNA) in the native state (gel *a*) and after denaturation (gel *b*). Right panels: the fast-sedimenting RF species (18S) in the native state (gel *c*) and after denaturation (gel *d*).

alternative possibility, considering its comigration with linear PM2 DNA, is that of a linear duplex dimer of approximate molecular weight 6.4×10^6.

Electron Microscopy of Isolated 14.5S and 18S RF DNA

The most convincing evidence for the linear nature of both types of RF KRV DNA came from electron microscopy. However, this was not easily accomplished directly from Hirt extracts: our initial studies clearly indicated the presence of a great excess of unlabeled cellular DNA fragments in such DNA preparations. A scheme to assess the extent of cell DNA contamination using a DNA preparation labeled with low levels of [^{14}C]thymidine prior to infection and with the regular brief pulse of high-specific-activity [^{3}H]thymidine 12 hr after KRV infection showed that as much as 26% of the isolated DNA in the 14.5S peak and 48% of the 18S DNA peak was still of cellular origin even after our regular purification by isopycnic banding of Hirt-method-extracted DNA once in CsCl followed by sedimentation through neutral sucrose gradients. To avoid this problem, all electron-microscope studies were performed with ^{3}H-labeled viral RF DNA that had been purified by three successive cycles of isopyncnic centrifugation in CsCl density gradients before separation of the 14.5S and 18S forms on neutral sucrose gradients.

Examples of typical photomicrograph fields of highly purified native RF DNA molecules are presented in Figure 3 *A* and *B*. No significant number of circular forms were ever observed in either the 14.5S or the 18S DNA samples. Histograms of the contour lengths of both types of RF DNA measured relative to standard replica grating lines are given in Figure 4. Admixed form-II PM2 circles were also measured as an internal size reference in the case of the 14.5S DNA sample. Most of the 14.5S DNA consisted of a unimodal population of linear molecules of average contour length 49.9% that of PM2 DNA, corresponding to a molecular weight of 3.2×10^6. Similarly, the bulk of the 18S DNA fell within a somewhat broader but largely unimodal peak of linear molecules with contour lengths 93.5–94.9% that of PM2 (depending on subjective interpretations of the size range to include in the peak). These measurements correspond to molecular weights of 6.0–6.1×10^6, or 5–7% less than the expected length for a full "dimer" of the 14.5S DNA.

Agarose Gel Electrophoresis of Denatured RF DNA

Agarose gel electrophoresis also resolves single-stranded DNA species of different molecular weights or conformations and in many instances even separates the two complementary strands from denatured duplex DNA (Hayward 1972). Figure 1*b* shows the gel electrophoresis profile of the mixture of total ^{3}H-labeled KRV RF DNA plus reference PM2 and

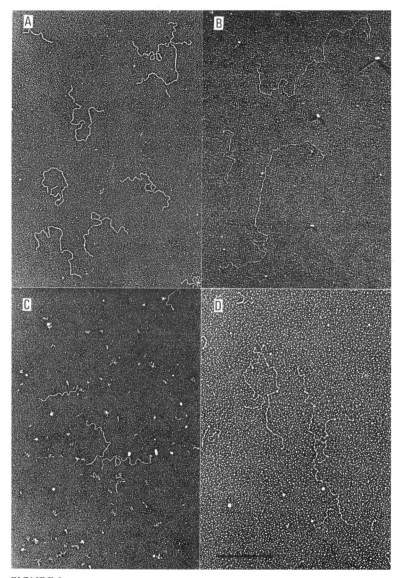

FIGURE 3

Electron microscopy of native and denatured forms of isolated 14.5S and 18S KRV RF DNA. (*A*) Native 14.5S DNA under aqueous spreading conditions. (*B*) Native 18S DNA under aqueous spreading conditions. (*C*) Alkali-denatured 14.5S DNA under aqueous conditions. (*D*) Isolated 18S DNA after mild cross-linking, denaturation, and unfolding in formaldehyde and dimethylsulfoxide. Bar indicates 1 μm.

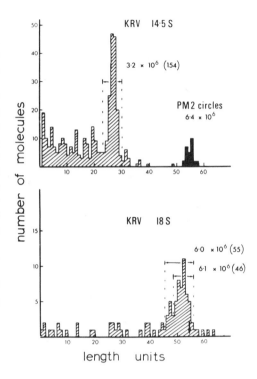

FIGURE 4
Histograms of the measured lengths of DNA molecules observed in the isolated 14.5S and 18S DNA preparations. Hatched areas represent linear molecules; solid area represents admixed form-II PM2 circular molecules. Horizontal arrows and dashed lines indicate size ranges of molecules included in the peak areas for average molecular weight estimates relative to that of PM2 DNA. The number of molecules included in each peak area is given in parentheses. Vertical arrow in the lower panel denotes the expected average length for molecules of exactly twice the size of the monomer RF (i.e., 6.4 × 10⁶).

SV40 DNA species after denaturation with 0.1 M NaOH. Under these conditions the DNA moves out of the alkaline pH into a gel of neutral pH, and consequently the supercoiled form-I PM2 and SV40 DNAs spontaneously renature and migrate at the same rate as before denaturation (compare with Fig 1a). However, denatured form-II and form-III DNAs behave as the single-stranded forms, and in the case of PM2 the separated linear and circular forms of both the H and L strands give a complex pattern of four bands, all with molecular weight 3.2 × 10⁶. The ³H-labeled KRV DNA again yields two distinct peaks. Most of the ³H-labeled DNA behaves as a single-stranded species of molecular weight 1.6 × 10⁶ (comigrating with SV40 single strands) or less, but approximately 35% apparently still migrates at the monomer duplex position. Significantly, although no ³H-labeled DNA remains at the dimer duplex position, neither does any comigrate with the PM2 single strands, which would be the expected position for double-length single strands of KRV DNA (molecular weight 3.4 × 10⁶). When the isolated 14.5S and 18S KRV DNA species were denatured separately (Fig. 2b and d), each gave rise to approximately equal amounts of ³H-labeled DNA at both the monomer duplex position and the single-stranded position (comigrating with ¹⁴C-labeled KRV viral strand added as a reference). The single-stranded denatured 18S DNA (Fig. 2d) gives a bimodal peak, which suggests the possibility of separation between the viral and complementary strands in these gels.

Buoyant Density of Denatured RF DNA

To confirm that the denatured RF DNA which comigrates with the original native 14.5S species actually has a double-stranded structure rather than having an exceptionally slow mobility caused by an unusual single-stranded conformation, we have studied the buoyant-density properties of denatured RF DNA in CsCl gradients. The results with unfractionated RF DNA are shown in Figure 5 a and b. As expected, the native form of RF DNA gave a unimodal peak at 1.705 g/cm^3, but the denatured form split into two fractions: one banding close to the position of reference KRV viral strand DNA (1.725 g/cm^3) and the other remaining at the density of duplex RF DNA (1.705 g/cm^3). Therefore, a substantial portion of the KRV RF DNA appears to be capable of spontaneous renaturation to re-form duplex structures.

Covalent Joint Between Viral and Complementary Strands in Monomer RF DNA

Since there are no circular forms present in the KRV RF DNA, one explanation for the spontaneous renaturing of some of these molecules could be a covalent joint or cross-link(s) between the two complementary strands. If this is the case, a portion of the DNA should sediment as single strands of twice the molecular weight of the viral strand under denaturing conditions. The experiment shown in Figure 6 indicates that monomer ^{3}H-labeled RF DNA (14.5S), when subjected to rate zonal sedimentation through alkaline CsCl density gradients, does yield a discrete minor peak that cosediments with isolated linear PM2 single strands, corresponding to a molecular weight of 3.2 × 10^6. Since this material renatures spontaneously upon removal of the denaturing conditions, we conclude that it must contain a viral DNA strand joined covalently to a complementary strand.

Electron-microscopic examination of the denatured 14.5S RF DNA under aqueous spreading conditions shows spontaneous renaturation of a portion of the DNA to linear duplex structures; the remainder stays in the typical single-stranded "bush" conformation (Fig. 3C).

Nature of the Monomer RF

The experiments described above indicate that the 14.5S DNA, although it migrates as a single band in agarose gels, exists as two species in approximately equal amounts: one that renatures rapidly and one that

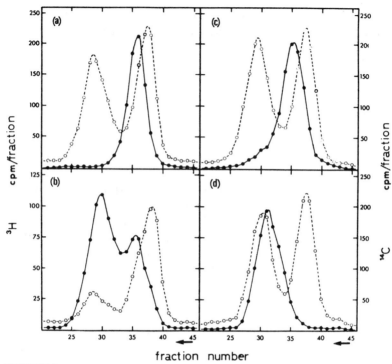

FIGURE 5

Neutral CsCl buoyant-density profiles of native and denatured RF DNA. (*a*) Unfractionated native RF DNA. (*b*) Unfractionated RF DNA after alkali denaturation and neutralization. (*c*) The renatured 14.5S DNA fraction recovered from a neutral sucrose gradient after denaturation of the monomer RF DNA. (*d*) The approximately 24S DNA fraction recovered from neutral sucrose gradients after denaturation of isolated monomer RF DNA. Reference buoyant-density species consisting of a mixture of ^{14}C-labeled KRV single-stranded viral DNA (1.725 g/cm[3]) and ^{14}C-labeled rat-cell DNA (1.700 g/cm[3]) were added to all samples. ●: ^{3}H-labeled KRV DNA; ○: ^{14}C-labeled DNA. Buoyant density increases in the direction shown by arrows.

does not. Taking advantage of this fact, one can separate DNA from these two species by sedimentation in neutral sucrose gradients after denaturation (not shown). One fraction sediments at 14.5S and consists entirely of spontaneously renatured duplex material, as shown by buoyant-density procedures (Fig. 5*c*). The second sediments at 24S and contains only single-stranded DNA (Fig. 5*d*). Because this single-stranded ^{3}H-labeled DNA has a somewhat lower average buoyant density (1.723 g/cm[3]) than that of the reference ^{14}C-labeled KRV viral strand (1.725

FIGURE 6

Zone sedimentation of monomer RF DNA through an alkaline CsCl gradient. ●: ³H-labeled KRV 14.5S DNA; ○: ³²P-labeled linear single strands of PM2 DNA. Monomer- and dimer-length KRV single strands are denoted by m and d. Arrow indicates direction of sedimentation.

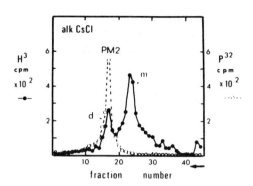

g/cm^3), we presume that the denaturable species of monomer RF DNA contains a mixture of both viral and complementary strands and that the complementary strand must have a buoyant density of approximately $1.721 \ g/cm^3$. Therefore, the difference between the two forms of monomer RF is probably simply the presence or absence of cross-linking. Experiments designed to distinguish by electron microscopy between a terminal covalent linkage and natural internal cross-links by denaturation in formaldehyde were not entirely successful; the technique itself introduced random cross-links. However, we did observe that at least some molecules contained apparently natural linkages across one end of the duplex but that none had linkages at both ends (not shown).

Nature of the Dimer RF

Denaturation of the 18S dimer RF DNA yielded a mixture of monomer-length single strands and monomer duplex DNA. The latter represents the renatured form of self-complementary dimer-length single strands consisting of one full viral strand bound covalently to a complementary strand. We interpret this evidence to indicate that the linear duplex dimers consist of two monomer duplex units joined not in tandem but in a "back-to-back" orientation. To account for the large numbers of monomer-length single strands released upon denaturation, we conclude that on average one strand in every two in the dimer RF contains a relatively specifically located interruption between the viral-strand and the complementary-strand units.

Alternative dimer structures, such as a noncovalent binding between free complementary "sticky" ends of the two monomer RF forms, would also be possible according to the data available at present. We have attempted to discriminate between these two possibilities in a number of ways: (1) We were unable to demonstrate any dimer formation by the monomer DNA under conditions expected to promote associations through possible complementary end sequences. (2) We have been unable to dissociate any such putative "sticky" ends of dimer molecules by heat treatment of purified dimer DNA. (3) Electron-microscopic examination of denatured dimer DNA after deliberate random cross-linking with formaldehyde demonstrated the existence of at least some molecules with two intact dimer-length single strands but without co-valent joints across their termini (Fig. 3D).

DISCUSSION

The experiments described here and by Gunther and May (1976) provide definitive evidence for the existence of at least the following three distinct RF forms of KRV DNA:

(1) regular double-stranded linear monomers of molecular weight 3.2×10^6 containing only monomer-length single strands,

(2) double-stranded rapidly renaturing linear monomers, also of molecular weight 3.2×10^6, consisting of double-length covalently joined viral and complementary strands, and

(3) linear duplex dimers of molecular weight $6.1–6.4 \times 10^6$ consisting of approximately equal proportions of monomer-length single strands and covalently joined dimer strands that renature spontaneously to form monomer duplexes (we believe that the monomer units in these dimers are joined back to back rather than in tandem, and that they represent immediate products of the replication of the covalently joined mono-mers).

Our experiments do not provide any information about precursor-product relationships among these three forms of newly synthesized viral DNA or any proof that they are all necessarily involved in the pathway to synthesis of progeny single-stranded DNA. The proportions of the dif-ferent forms remained relatively constant at approximately 2:2:1 when labeled at various different times after infection.

Since our studies were completed, a number of laboratories have pub-lished information bearing on the DNA replication mechanism for other parvoviruses (Siegl and Gautschi 1976; Straus et al. 1976; Tattersall and Ward 1976; Rhode 1977). The process can be conveniently divided into three stages: (1) synthesis of the complementary strand, (2) replication of the duplex RF forms, and (3) synthesis of the progeny single strands. There appears to be general agreement for all parvoviruses studied that the first step proceeds by a 3'-end self-priming mechanism on the viral or parent single strand to form covalently joined monomer duplexes. A further hairpin foldback initiation at the 3' end of the complementary strand followed by continuous elongation using both viral and com-plementary strands in turn as the template leads to back-to-back dimers and higher oligomers by a mechanism now generally referred to as "rolling-hairpin" synthesis. This model was formulated independently by Straus et al. (1976) for AAV and by Tattersall and Ward (1976) for MVM. Virtually nothing is known about the third stage of replication; the scheme proposed for MVM assumed that staggered single-strand nicks would be introduced into the oligomeric duplex DNA from which chain elongation at the 3' end of the complementary strand would be sufficient to both displace the adjacent viral strands and regenerate the terminal hairpin sequences.

The evidence for the "rolling-hairpin" scheme proposed for AAV replication was the existence of fast-sedimenting oligomeric RFs that contained self-renaturing multi-genome-length single strands. In contrast, we find little evidence for tetrameric or larger RFs of KRV DNA (see Fig. 1a) and no evidence at all for single strands of greater than dimer length. Similarly, the existence of the non-covalently-joined monomer RFs in KRV-infected cells, together with the internal nicking and apparent lack of covalent joints at the ends of our dimer RF, is somewhat inconsistent with the published replication schemes. Although it is true that our particular DNA extraction procedure may not have released higher oligomers into the Hirt supernatant, it seems more likely that there may be fundamental differences in the replication schemes for the defective and nondefective parvoviruses necessitated by different requirements for packaging both, versus only one, of the progeny single strands. In addition, the rate of nicking between viral and complementary strands may be much greater for KRV, H-1, LuIII, and possibly also MVM than for AAV. Even if higher oligomeric forms are generated by the nondefective parvoviruses, it may be necessary to break them down again into the non-covalently-joined monomer RFs before progeny single-strand synthesis can occur (Gunther and Revet, this volume).

ACKNOWLEDGMENTS

One of us (G.S.H.) was a Postdoctoral Fellow of the European Molecular Biology Organization in 1972 and 1973, when this work was performed.

REFERENCES

Aaij, C. and P. Borst. 1972. The gel electrophoresis of DNA. *Biochim. Biophys. Acta* **269**:192.

Bourguignon, G. J., P. J. Tattersall, and D. C. Ward. 1976. DNA of minute virus of mice: Self-priming, nonpermuted, single-stranded genome with a 5'-terminal hairpin duplex. *J. Virol.* **20**:290.

Bujard, H. 1970. Electron microscopy of single-stranded DNA. *J. Mol. Biol.* **49**:125.

Gunther, M. and P. May. 1976. Isolation and structural characterization of monomeric and dimeric forms of replicative intermediates of Kilham rat virus DNA. *J. Virol.* **20**:86.

Hayward, G. S. 1972. Gel electrophoretic separation of the complementary strands of bacteriophage DNA. *Virology* **49**:342.

Hayward, G. S. and M. G. Smith. 1972. The chromosome of bacteriophage T5. I. Analysis of the single-stranded DNA fragments by agarose gel electrophoresis. *J. Mol. Biol.* **63**:383.

Rhode, S. L. III. 1974. Replication process of the parvovirus H-1: II. Isolation and characterization of H-1 replicative form DNA. *J. Virol.* **13**:400.

————. 1977. Replication process of the parvovirus H-1: VI. Characterization of a replication terminus of H-1 replicative form DNA. *J. Virol.* **21**:694.

Salzman, L. A. and W. White. 1973. In vivo conversion of the single-stranded DNA of the Kilham rat virus to a double-stranded form. *J. Virol.* **11**:299.

Siegl, G. and M. Gautschi. 1976. Multiplication of parvovirus LuIII in a synchronized culture system. III. Replication of viral DNA. *J. Virol.* **17**:841.

Straus, S. E., E. D. Sebring, and J. A. Rose. 1976. Concatemers of alternating plus and minus strands are intermediates in adeno-associated virus DNA synthesis. *Proc. Natl. Acad. Sci.* **73**:742.

Tattersall, P. and D. C. Ward. 1976. Rolling hairpin model for replication of parvovirus and linear chromosomal DNA. *Nature* **263**:106.

Tattersall, P., L. V. Crawford, and A. J. Shatkin. 1973. Replication of the parvovirus MVM: II. Isolation and characterization of intermediates in the replication of the viral DNA. *J. Virol.* **12**:1446.

Watson, J. D. 1972. Origin of concatemeric T7 DNA. *Nat. New Biol.* **239**:197.

DNA Replication of Kilham Rat Virus: Characterization of Intracellular Forms of Viral DNA Extracted by Guanidine Hydrochloride

ANNA TAI LI
GEORGE C. LAVELLE
RAYMOND W. TENNANT

University of Tennessee–Oak Ridge Graduate School of Biomedical Sciences and Biology Division, Oak Ridge National Laboratory Oak Ridge, Tennessee 37830

Kilham rat virus (KRV) is a member of the subgroup of nondefective parvoviruses which are able to replicate without helper virus in most host cells (Toolan 1968; Rose 1974). The DNA extracted from purified KRV has been shown to be linear and single-stranded (Robinson and Hetrick 1969; Salzman et al. 1971). Double-stranded replicating intermediates have been identified in KRV-infected cells by several groups (Salzman and White 1973; Gunther and May 1976; Li et al. 1977). The production of single-stranded DNA from a linear replicative intermediate has been demonstrated by the isolation of such molecules from adenovirus-infected cells (Sussenbach et al. 1972; van der Eb 1973; Lavelle et al. 1975) and from parvovirus-infected cells (Tattersall et al. 1973; Rhode, 1974; Gunther and May 1976). These studies suggest that progeny DNA synthesis occurs by an asymmetric strand displacement similar to that proposed by Sussenbach et al. (1972).

In this study we have utilized a new procedure for the separation of intracellular viral DNA from host DNA (Lavelle and Li 1977) to examine the properties of viral replicative intermediates and the characteristics of progeny strand production.

MATERIALS AND METHODS

Cells and Viruses

Infections with KRV were carried out in normal rat kidney (NRK) monolayer cell cultures (Duc-Nguyen et al. 1966). Strain 171 of KRV was obtained originally from Dr. L. Kilham (Tennant et al. 1969). Procedures for the growth and purification of KRV and viral DNA have been described (Tennant and Hand 1970; Lavelle and Li 1977). Infections with adenovirus type 2 (Ad2) and adeno-associated virus type 2 (AAV2) were carried out in spinner cultures of KB cells. DNAs of Ad2 and AAV2 were prepared from purified viruses as described previously (Rose et al. 1975; Lavelle et al. 1975).

Labeling and Extraction of DNA

KRV-infected NRK cells were pulse-labeled with [^{3}H]thymidine (TdR) (Schwarz/Mann, 50–60 Ci/mmole) added to the cultures (10 μCi/ml or 100 μCi/ml) at either 20 or 23 hr after infection for intervals specified in the individual experiments. Chasing with unlabeled TdR was performed by washing pulse-labeled infected cells three times with cold phosphate-buffered saline (PBS) and refeeding with warmed medium supplemented with 4×10^{-4} M unlabeled thymidine to minimize further incorporation of [^{3}H]TdR. The selective recovery of KRV viral DNA was accomplished by sedimentation of infected cells through a linear 5–20% sucrose gradient containing 4 M guanidine hydrochloride (Gu-HCl) as described previously (Lavelle and Li 1977). Aliquots of all fractions were precipitated and washed with trichloroacetic acid (TCA) and collected on nitrocellulose filters. The radioactivity of the precipitates was determined by counting the filters in a toluene-based scintillation fluid. Fractions were pooled and precipitated for 10 hr or more with 2.5 volumes of cold 95% ethanol and 0.05 volume of 3 M Na-acetate. Ethanol precipitates were spun down at $-20°C$ in a Sorvall centrifuge, resuspended in $0.1 \times SSC$, and digested with 100 μg/ml RNase A (Worthington, 4800 units/mg) for 1 hr at 37°C. SDS was added to all samples to 1% final concentration and samples were extracted three times with equal volumes of phenol saturated with 0.001 M Tris-HCl buffer (pH 8.00). Samples were dialyzed against $0.1 \times SSC$ for 24 hr, with two changes of solution. Dialyzed samples were ethanol-precipitated and subjected to neutral sucrose sedimentation or CsCl equilibrium sedimentation.

Uninfected log-phase NKR cells were labeled for 24 hr with 2 μCi/ml

[³H]TdR, and the cell DNA was extracted by the method of Hirt (1967). DNA from the Hirt pellet was used as a source of cellular DNA.

Sucrose Density Gradient Sedimentation

Velocity sedimentation was carried out either in 5–20% neutral sucrose gradients containing 1 M NaCl, 10 mM Tris-HCl (pH 8.0), 1 mM EDTA, and 0.15% Sarkosyl or in 5–20% alkaline sucrose gradients containing 0.3 N NaOH, 0.7 M NaCl, 1 mM EDTA, 0.15% Sarkosyl. Neutral gradients were centrifuged in a Beckman SW 50.1 rotor at 42,000 rpm for $3\frac{1}{2}$ hr at 20°C. Alkaline gradients were run under the same conditions for 4 hr. Sucrose gradients containing 4 M Gu-HCl also contained 10 mM Tris-HCl (pH 8.0) and 1 mM EDTA.

Digestion by S1 Nuclease

[³H]TdR-labeled DNA was treated with the single-strand-specific S1 endonuclease (Sigma) in reaction mixtures containing 0.3 M NaCl, 0.1 mM $ZnSO_4$, 0.02 M Na-acetate (pH 4.5), and 10 μg/ml of native calf-thymus DNA. Incubation was carried out at 45°C for 2 hr with 6–15 units of enzyme per ml of reaction mixture. Resistant DNA was assayed by TCA precipitation and counting as described above. Control experiments showed that limit digestion of added ³H-labeled single-stranded DNA was reached within 80 min in the digestion reactions described in this article.

Chromatography of DNA on Benzolated-Naptholated-DEAE (BND)-Cellulose Columns

Chromatography of DNA on BND-cellulose columns followed a modified version of the procedure of Levine and Kang (1970) and Horwitz (1971). BND-cellulose (Gallard-Schlesinger) was suspended in 0.3 N NaCl, 0.001 M EDTA, 0.01 M Tris (pH 8.1), and 20% ethanol. The buffer solution was removed by centrifugation and the wet cellulose was packed into a 2.5-ml disposable syringe to a height of 1–2 cm. The column was then washed with 0.3 N NaCl, 0.001 M EDTA, 0.01 M Tris-HCl buffer (pH 8.1) until the OD_{260} fell below 0.5. Viral DNA samples were phenol-extracted, ethanol-precipitated, resuspended, and dialyzed against 0.3 M NaCl, 0.001 M EDTA, 0.01 M Tris-HCl buffer (pH 8.1), and then adsorbed onto the column at a flow rate of 0.5–1.0 ml/min at room temperature. The column was washed with 20 ml of the same buffer. Completely double-stranded (ds) DNA was eluted in 6 ml of 1 M NaCl, 0.001 M EDTA, 0.01 M Tris-HCl (pH 8.1). After equilibration with 1 M NaCl-containing buffer, a 24-ml gradient of 0–1.5% caffeine in 1 M NaCl, 0.001 M EDTA, 0.01 M Tris-HCl buffer (pH 8.1) was passed through the column and 2-ml fractions were collected. Carrier RNA was added and the entire fraction was precipitated and counted as above.

RESULTS

Isolation and Purification of Intracellular Forms of KRV DNA

The one-step extraction of intracellular DNA by guanidine hydrochloride (Gu-HCl) results in a good separation of high- from low-molecular-weight DNA, as shown in Figure 1. When uninfected cells which had been labeled for 1 hr with 100 μCi/ml [³H]TdR were sedimented through the Gu-HCl gradients, 90% of the radioactivity incorporated into the high-molecular-weight cell DNA was present in the bottom fractions, and little radioactivity above background was detectable throughout the rest of the gradient. On the other hand, when KRV-infected cells were labeled 23–24 hr after infection with 100 μCi/ml [³H]TdR and extracted in the same manner, the high-molecular-weight cell DNA was again detected in the bottom fractions but three other peaks of DNA were also detected along the gradient. Three main pools were collected from gradients, as indicated in Figure 1; Pool I was the fastest-sedimenting species. These samples were further characterized by sedimentation through a neutral 5–20% sucrose gradient. Relative to double-stranded AAV DNA (14.5S), Pool-III DNA (in neutral sucrose) had a sedimentation coefficient of 17S, which is similar to the S value reported for the double-stranded form of H-1 DNA (Rhode 1974). The DNA in Pool II appeared to be more heterogeneous and had S values from 17S to 23S. Pool-I DNA had the same sedimentation coefficient as

FIGURE 1
Sedimentation of KRV-infected and mock-infected cells in neutral sucrose gradients containing 4 M Gu-HCl. Subconfluent monolayers of NRK cells were infected (●——●) or mock-infected (○——○) and pulse-labeled for 1 hr at 23 hr p.i. with 100 μCI/ml [³H]TdR. Cells were layered onto separate gradients as described in the "Methods" section. Sedimentation was from right to left. The position where ¹⁴C- labeled KRV virion DNA sedimented in a parallel gradient is indicated by the arrow. Pools of DNA were collected as indicated.

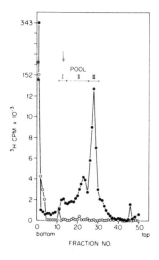

purified single-stranded KRV virion DNA, i.e., 25S (A. T. Li, G. Lavelle, and R. W. Tennant, manuscript in preparation).

Velocity Sedimentation in Alkaline Sucrose Gradients

The behavior of these three pools of DNA in alkaline sucrose gradients was examined. The cosedimentation of ^{32}P-labeled AAV DNA with ^{14}C-labeled KRV virion DNA is shown in Figure 2A. Denatured AAV DNA appears to be slightly smaller than the KRV DNA. DNA samples analyzed in Figure 2B–D were obtained by Gu-HCl extraction of KRV-infected cells which were labeled with 100 μCi/ml [^{3}H]TdR from 20 to 28 hr post-infection (p. i.). Figure 2B shows that purified Pool-I DNA cosedimented with ^{14}C-labeled KRV virion DNA. No DNA significantly larger than genome length is detectable in the gradient. The results of the sedimentation of Pool-III DNA with KRV DNA marker are shown in Figure 2C. The majority of Pool-III DNA was of genome length, but some DNA of up to twice genome length (arrow) was also detectable. Little or no DNA shorter than one genome length was observed in the Pool-III material. Pool-II material was more heterogeneous. As shown in Figure

FIGURE 2
Alkaline sucrose sedimentation of intracellular KRV DNA with KRV marker DNA. ^{14}C-labeled KRV virion DNA (o——o in each graph) was sedimented with (A) ^{32}P-labeled AAV DNA, (B) ^{3}H-labeled Pool-I DNA, (C) ^{3}H-labeled Pool-III DNA, (D) ^{3}H-labeled Pool-II DNA, (E) 5-min-pulse ^{3}H-labeled Pool-III DNA and (F) 5-min-pulse ^{3}H-labeled Pool-II DNA. The position in the gradient calculated for DNA of twice the KRV genome length is indicated by the arrow in each panel.

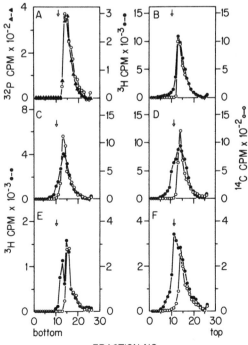

2D, a sizable fraction of the DNA (~35% of total) was longer than genome length. DNA of genome length (~ 40% of total) and DNA of less than genome length (~ 25% of total) (Fig. 2D) were also observed. In order to further characterize viral DNA molecules actively undergoing replication, viral DNA was extracted from infected cells which were pulsed for 5 min at 23 hr p.i. with 100 μCi/ml [³H]TdR. Alkaline sucrose sedimentation profiles of purified Pool-II and -III material from this experiment are shown in Figure 2E and F. Pulse-labeled Pool-III DNA (Fig. 2E) contained two discrete species of molecules, one of genome length (~55%) and the other of twice the genome length (~45%). On the other hand, pulse-labeled Pool-II DNA (Fig. 2F) showed one major peak of DNA of twice the genome length (~50%) with a prominent shoulder trailing up to the position of genome-length DNA (~ 41%). Very little DNA of less than genome length was found in either Pool III or Pool II.

S1-Nuclease Digestion of Intracellular Viral DNA

The secondary structure of the DNA in each Gu-HCl-gradient pool was examined by digestion with the single-strand-specific endonuclease S1 (Sutton 1971). To estimate the extent of single-strandedness in the different viral DNA forms, DNA was subjected to S1-nuclease digestion before and after denaturation. Resistance to digestion by S1 nuclease of pulse-labeled and uniformly labeled DNA is shown in Table 1. In the uniformly labeled DNA pools, native Pool-I DNA exhibited the lowest S1 resistance (18%), Pool-III DNA exhibited the highest resistance (93%), and Pool-II DNA was intermediate (87%), which indicated that Pool-I DNA is mostly single-stranded, Pool-III DNA is mostly double-stranded, and Pool-II DNA contains some single-stranded-DNA regions. Under repeated testing, Pool-II DNA showed greater variations in its S1 sensitivity than either Pool-I or Pool-III DNA. When the same DNA species were alkali-denatured and subjected to S1 digestion, Pool-I DNA showed no significant changes, Pool-II material was 44% resistant, and Pool-III material was 33% resistant. Denaturation by boiling and quick cooling gave similar results (not shown). Pulse-labeled DNA from Pools II and III were similar in S1 resistance to their uniformly labeled counterparts before denaturation. However, significantly greater resistance to S1 was observed for short-pulsed Pool-II and -III DNA compared with uniformly labeled DNA after alkali denaturation. The nondenaturability of significant fractions of pulse-labeled Pool-II and -III DNA suggests the presence of a covalent linkage between the two strands of replicating DNA duplexes and agrees with the results of the alkaline sucrose gradients (Fig. 2C–F), in which molecules of twice the genome length were observed in both Pool-II

TABLE 1
S1-Endonuclease Digestion

| DNA source | Treatment[a] | Average percentage resistance to S1 nuclease | |
		pulse-labeled[b]	uniformly labeled[c]
Pool I	native	nd	18
Pool I	alkali-denatured	nd	11
Pool II	native	88	87
Pool II	alkali-denatured	58	44
Pool III	native	91	93
Pool III	alkali-denatured	50	33
NRK cell DNA	native	nd	95
NRK cell DNA	alkali-denatured	nd	7
KRV virion DNA	native	nd	8
KRV virion DNA	alkali-denatured	nd	6

[a] Denaturation was achieved by incubation of DNA sample with an equal volume of 1 N NaOH for 15 min, followed by quick cooling on ice and neutralization with an equivalent amount of 1 N HCl.

[b] Pulse-labeled DNAs were obtained from Gu-HCl extraction of KRV-infected NRK cells which had been pulsed with 100 μCi/ml [³H]TdR for 5 min at 23 hr p.i.

[c] Uniformly labeled DNAs were obtained from the extraction of infected cells which had been labeled from 20 to 28 hr p.i. with 100 μCi/ml [³H]TdR.

nd: Not determined

and Pool-III DNA populations. Enrichment of this spontaneously renaturing fraction in short-pulse-labeled DNA indicates a major role for such covalent interstrand linkages within replicating molecules.

BND-Cellulose Chromatography

The DNA species in Pools I–III were analyzed futher by chromatography on BND-cellulose. The elution patterns of ¹⁴C-labeled Ad2 virion DNA and of ³H-labeled KRV virion DNA are shown in Figure 3A. Essentially no radioactivity was eluted in the 0.3 M NaCl-containing buffer. All of the input Ad2 DNA was eluted with the 1 M NaCl-containing buffer. On the other hand, purified KRV virion DNA was released from the resin only by the application of the caffeine gradient. Figure 3B shows that uniformly labeled Pool-I DNA was fractionated into three components, one eluting in 1 M NaCl-containing buffer and the other two eluting at different concentrations of caffeine. Of the two subpopulations resolved in the caffeine gradient, one eluted at a low caffeine concentration and the other (60% of the total recovered radioactivity) eluted at the same caffeine

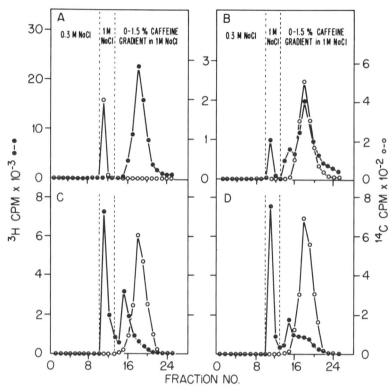

FIGURE 3

BND-cellulose chromatography of intracellular KRV DNA pools. (A) ³H-labeled KRV virion DNA (●——●) was chromatographed with ¹⁴C-labeled Ad2 DNA (○——○). (B–D) ¹⁴C-labeled KRV DNA (○——○ in each graph) was chromatographed with (B) ³H-labeled Pool-I DNA, (C) ³H-labeled Pool-III DNA, and (D) ³H-labeled Pool-II DNA.

concentration as marker KRV DNA (Fig. 3B). The linear relationship between the proportion of single-strandedness in a DNA molecule and the concentration of caffeine at which it elutes from BND-cellulose (Iyer and Rupp 1971) suggests that the material which elutes at low caffeine concentration is presumably double-stranded DNA which contained some short single-stranded regions. The DNA which coeluted with marker virion DNA at high caffeine concentrations is probably single-stranded progeny DNA.

When Pool-III DNA was examined in the same manner (Fig. 3C), 60% was eluted by high salt alone and 37% by low caffeine in high salt. However, when Pool-III DNA was digested with S1 nuclease before chromatography on the BND-cellulose, all radioactivity was eluted by 1 м

NaCl (A. T. Li, G. Lavelle, and R. W. Tennant, manuscript in preparation). Therefore, Pool III probably consists of two main populations of DNA, one which is fully double-stranded and another which contains small single-stranded regions. These small regions may comprise single-stranded loops covalently joining viral and complementary strands, as discussed above. BND-cellulose chromatography of Pool-II material gave the profile shown in Figure 3D. Approximately 55% of the label eluted with the 1 M NaCl wash, and 45% was eluted by the caffeine gradient. In contrast with caffeine-eluted material from Pool-III DNA (Fig. 3C), Pool-II DNA was eluted by the caffeine gradient over a broad range of caffeine concentration, which suggested that single-stranded regions of various lengths up to unit viral length are present on these double-stranded replicative intermediates. The DNA eluted by 1 M NaCl may represent completely double-stranded intermediates in replicative-form replication. However, because DNA in Pool-II is always detected as a leading shoulder of the Pool-III DNA, it is possible that Pool-II DNA is contaminated with double-stranded molecules from Pool-III. When short (5-min) pulses were given to infected cells such that only Pool-II material was labeled significantly, the amount of label eluted by 1 M NaCl was much reduced (A. T. Li, G. Lavelle, and R. W. Tennant, manuscript in preparation). Thus, extensive single-stranded regions are characteristic of replicating intermediates in Pool-II DNA.

Pulse-Chase Experiment for Identification of Replicative Intermediates

The kinetic relationships of the three forms of viral DNA extracted by Gu-HCl were examined by pulse-chase experiments. Figure 4 shows the Gu-HCl sucrose sedimentation patterns of intracellular DNA after a 5 min pulse with [³H]TdR followed by various intervals of chase incubation with excess unlabeled TdR. After the 5-min pulse (Fig. 4A) the majority of radioactivity was found in the 17–23S (Pool-II) region of gradients, which indicated that Pool-II DNA is the replicating DNA. Some radioactivity was also detected in the 17S (Pool-III) and 25S (Pool-I) positions, which suggested that some molecules had been completed within the 5-min period. After a 15-min chase period, a redistribution of label from Pool II to Pool III and Pool I was observed (Fig. 4B), which suggested that Pool III and Pool I are derived from Pool II. With a 1-hr chase, more label was observed in the progeny-DNA (Pool-I) region and a corresponding decrease in label was observed in the 17S region (Fig. 4C). A further increase in proportion of label in the progeny region was observed when the chase time was increased to 10 hr (Fig. 4D). The amount of label in the 17–23S region decreased progressively during the different chase periods employed, while the label in the 17S material did not completely chase into single-stranded progeny DNA. The persistence of up to 50%

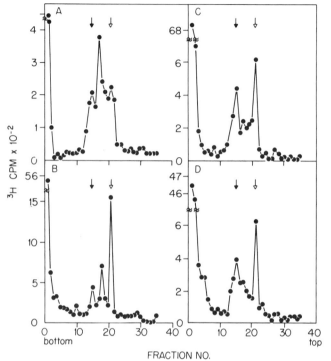

FIGURE 4

Pulse-chase kinetics of KRV replicating DNA molecules. KRV-infected NRK cells were pulsed for 5 min with 100 μCi/ml [³H]TdR at 23 hr p.i. and (A) loaded onto gradients immediately, (B) chased for 15 min with excess unlabeled TdR, (C) chased for 1 hr, or (D) chased for 10 hr. Sedimentation was from right to left in all cases. The positions of ¹⁴C-labeled KRV virion DNA are indicated by solid arrows. The approximate positions of 17S RF are indicated by the open arrows.

total viral DNA in the 17S region may be explained in at least two ways: (a) that label is incorporated in both viral and complementary strands of the 17S linear replicative form (RF), but only the viral strand is displaced into single-stranded progeny DNA, or (b) that label is incorporated into both strands of the 17S RF but only part of this RF population acts as precursor to replicative intermediates and progeny DNA. Though further experiments are required to distinguish between the two possibilities, the fact that Pool-I DNA from the Gu-HCl does not self-anneal to form double-stranded molecules indicates that single-stranded progeny DNA of one polarity only is made, and this tends to support the first possibility.

DISCUSSION

Our results indicate that within the infected cell and at a time of active viral DNA replication three populations of virus-specific molecules can be obtained by extraction with guanidine hydrochloride. The molecules in Pool I have characteristics of mature virion DNA: they are of genome length in alkaline sucrose, have the same density (1.7285 g/cm^3 in CsCl) and sedimentation coefficient as DNA purified by phenol-SDS procedures from purified virions (A. T. Li, G. Lavelle, and R. W. Tennant, manuscript in preparation), and are highly susceptible to digestion by S1 endonuclease. Pool-III DNA possesses characteristics of the linear replicative form, as reported by Tattersall et al. (1973) for MVM, by Salzman and White (1973) and Gunther and May (1976) for KRV, and by Rhode (1974) for H-1 virus. The presence of a covalent linkage between the viral and complementary strands in 33% of the uniformly labeled Pool-III DNA and in 50% of short-pulse-labeled Pool-III DNA was demonstrated by the ability of these molecules to reanneal in a unimolecular fashion following alkali denaturation. The enrichment of viral DNA molecules which reanneal instantly after denaturation in short-pulse-labeled material suggests a dynamic role of the covalent linkage in the replication of these molecules.

The resistance of Pool-II DNA to digestion by S1 nuclease and its chromatographic behavior on BND are consistent with those of double-stranded molecules containing varying amounts of single-strandedness.

During a short pulse, Pool II was the major species labeled and presumably contained the actively replicating intermediates (RI). The results of the pulse-chase experiments show that label incorporated into these RI molecules was chased first into RF DNA and single-stranded progeny, and suggest that these RIs are the immediate precursors to progeny single-stranded DNA and to the stable RF. The primary shift of label from RI to RF within 15 min indicates a rapid turnover for RI and suggests that RI consists of two subpopulations, one involved in RF production via semiconservative replication, such as that described by Rhode (1977), and the other involved in the production of single-stranded progeny DNA.

The results of these experiments are consistent with a strand-displacement mechanism for the production of single-stranded progeny DNA. Two alternatives postulated for progeny production are shown in Figure 5 A and B. In Figure 5A, a replicative form is converted into a replicative intermediate by a non-self-primed displacement of the viral strand leading to the formation of a single-stranded viral strand and a new RF molecule. Such a model would predict the presence of newly synthesized DNA strands of various lengths up to genome length, as has been found for adenoviruses (Horwitz 1971; van der Eb 1973). The model shown in Figure 5B is very similar to that proposed by Straus et al. (1976) for AAV

A STRAND-DISPLACEMENT WITHOUT SELF-PRIMED DNA SYNTHESIS

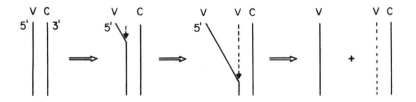

B STRAND-DISPLACEMENT WITH SELF-PRIMED DNA SYNTHESIS

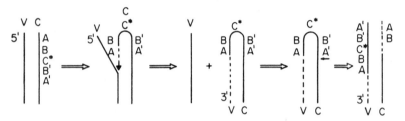

FIGURE 5

Two models for the production of single-stranded progeny DNA. V and C represent viral and complementary strands, respectively. A and A' and B and B' represent complementary sequences. C* represents unpaired bases in the terminal single-stranded loop. Broken lines indicate newly synthesized DNA.

except that in the case of the KRV asymmetric displacement of viral strand in the RI is expected. In this model the initiation of displacement depends on the 3' end of the complementary strand of the RF to serve as a primer. The existence of hairpin duplex regions on the 5' and 3' ends of the virion DNA have been reported by various groups (Bourguignon et al. 1976; Salzman 1977; Lavelle and Mitra, this volume; Chow and Ward, this volume). Tattersall and Ward (1976) have proposed a rolling-hairpin model for parvovirus DNA replication which utilizes such terminal palindromes to form multimeric-length RF molecules. The existence of palindromic sequences at the 5' end of the viral strand would entail the same phenomenon on the 3' end of the complementary strand in the RF. This model also predicts the occurrence of DNA strands of different lengths, with DNA of twice the genome length the dominant species. Our results of alkaline sucrose sedimentation, S1 digestion, and BND-cellulose chromatography are consistent with the expectations of the model shown in Figure 5B.

ACKNOWLEDGMENTS

We thank S. K. Niyogi and S. Mitra for advice and discussion and for critically reading the manuscript. A. T. Li held a graduate research assistantship from the University of Tennessee–Oak Ridge Graduate School of Biomedical Sciences. This research was carried out at the Biology Division of Oak Ridge National Laboratory, Oak Ridge, Tennessee, and was sponsored by the Energy Research and Development Administration under contract with Union Carbide Corporation.

REFERENCES

Bourguignon, G. J., P. J. Tattersall, and D. C. Ward. 1976. DNA of minute virus of mice: Self-priming, nonpermuted, single-stranded genome with a 5′-terminal hairpin duplex. *J. Virol.* **20**:290.

Duc-Nguyen, H., E. N. Rosenblum, and R. F. Zeigel. 1966. Persistent infection of a rat kidney cell line with Rauscher murine leukemia virus. *J. Bacteriol.* **92**:1133.

Gunther, M. and P. May. 1976. Isolation and structural characterization of monomeric and dimeric forms of replicative intermediates of Kilham rat virus DNA. *J. Virol.* **20**:86.

Hirt, B. 1967. Selective extraction of polyoma DNA from infected mouse cells. *J. Mol. Biol.* **26**:365.

Horwitz, M. S. 1971. Intermediates in the synthesis of type 2 adenovirus deoxyribonucleic acid. *J. Virol.* **8**:675.

Iyer, V. N. and W. D. Rupp. 1971. Usefulness of benzoylated-naphthoylated DEAE cellulose to distinguish and fractionate double-stranded DNA bearing different extents of single-stranded regions. *Biochim. Biophys. Acta* **228**:117.

Lavelle, G. C. and A. T. Li. 1977. Isolation of intracellular replicative forms and progeny single-strands of KRV DNA in sucrose gradients containing guanidine hydrochloride. *Virology* **76**:464.

Lavelle, G. C., C. Patch, G. Khoury, and J. Rose. 1975. Isolation and partial characterization of single-stranded adenoviral DNA produced during synthesis of adenovirus type 2 DNA. *J. Virol.* **16**:775.

Levine, A. J. and H. S. Kang. 1970. DNA replication in SV40 infected cells. I. Analysis of replicating SV40 DNA. *J. Mol. Biol.* **50**:549.

Li, A. T., G. C. Lavelle, and R. W. Tennant. 1977. Isolation and characterization of the replicative intermediates and progeny DNA of the parvovirus Kilham rat virus (KRV). *Fed. Proc.* **36**:1085.

Rhode, S. L. III. 1974. Replication process of the parvovirus H-1. II. Isolation and characterization of H-1 replicative form DNA. *J. Virol.* **13**:400.

————. 1977. Replication process of the parvovirus H-1. Characterization of a replicative terminus of H-1 replicative form DNA. *J. Virol.* **21**:694.

Robinson, D. M. and F. M. Hetrick. 1969. Single-stranded DNA from the Kilham rat virus. *J. Gen. Virol.* **4**:269.

Rose, J. A. 1974. Parvovirus reproduction. In *Comprehensive virology* (ed. H. Fraenkel-Conrat and R. W. Wagner), vol. 3, p. 1. Plenum, New York.

Salzman, L. A. 1977. Evidence for terminal S1-nuclease-resistant regions on single-stranded linear DNA. *Virology* **76**:454.

Salzman, L. A. and W. White. 1973. In vivo conversion of the single-stranded DNA of Kilham rat virus to a double-stranded form. *J. Virol.* **11**:299.

Salzman, L. A., W. L. White, and T. Kakefuda. 1971. Linear, single-stranded deoxyribonucleic acid isolated from Kilham rat virus. *J. Virol.* **7**:830.

Straus, S. E., E. D. Sebring, and J. A. Rose. 1976. Concatemers of alternating plus and minus strands are intermediates in adenovirus-associated virus DNA synthesis. *Proc. Natl. Acad. Sci.* **73**:742.

Sussenbach, J. S., D. J. Ellens, and H. S. Jansz. 1973. Studies on the mechanism of replication of adenovirus DNA. II. The nature of single-stranded DNA in replicative intermediates. *J. Virol.* **12**:1131.

Sutton, W. D. 1971. A crude nuclease preparation suitable for use in DNA reassociation experiments. *Biochim. Biophys. Acta* **240**:522.

Tattersall, P. and D. C. Ward. 1976. Rolling hairpin model for the replication of parvovirus and linear chromosomal DNA. *Nature* **263**:106.

Tattersall, P., L. V. Crawford, and A. J. Shatkin. 1973. Replication of the parvovirus MVM. II. Isolation and characterization of intermediates in the replication of the viral deoxyribonucleic acid. *J. Virol.* **12**:1446.

Tennant, R. W. and R. E. Hand. 1970. Requirement of cellular synthesis for Kilham rat virus replication. *Virology* **42**:1054.

Tennant, R. W., K. R. Layman, and R. W. Hand. 1969. Effect of cell physiological state on infection by rat virus. *J. Virol.* **4**:872.

Toolan, H. W. 1968. The picodnaviruses: H-1, RV and AAV. *Int. Rev. Exp. Pathol.* **6**:135.

van der Eb, A. J. 1973. Intermediates in type 5 adenovirus DNA replication. *Virology* **51**:11.

DNA-Polymerase Activity Associated with Virions of Kilham Rat Virus

LOIS ANN SALZMAN
LOUISE McKERLIE
PHYLLIS FABISCH
FRANK KOCZOT

Laboratory of Biology of Viruses
National Institute of Allergy and Infectious Diseases
National Institutes of Health
Bethesda, Maryland 20014

The genome of Kilham rat virus (KRV) is a linear single-stranded DNA molecule with a molecular weight of approximately 1.6×10^6 (Robinson and Hetrick 1969; Salzman et al. 1971) which contains terminal palindromic sequences (Salzman 1977). Until recently, little information has been available concerning the mechanism of KRV DNA replication in infected cells. It has been shown that 40–50% of the viral DNA is converted into a double-stranded linear form within 1 hr after infection of rat nephroma cells (Salzman and White 1970). Two forms of KRV replicative DNA have been isolated from rat fibroblast cells 12 hr after infection and shown to be monomer- and dimer-length DNA duplexes (Gunther and May 1976). However, the enzyme(s) responsible for the synthesis of replicative-form or progeny DNA has yet to be identified.

It has been reported that a protein with DNA-polymerase activity is associated with KRV virions, even after isopycnic centrifugation in cesium chloride and velocity sedimentation in sucrose (Salzman 1971). The KRV-associated enzyme has been purified and shown to differ in a number of characteristics from the DNA polymerases found in uninfected cells (Salzman and McKerlie 1975a,b). Nevertheless, it is still unclear whether the virion-associated enzyme is viral or cellular in origin. In this paper we review some of the properties of this enzyme, describe the isolation of a protein-DNA complex from a purified virion which exhibits DNA-polymerase activity, and discuss (where appropriate) the virion-associated enzyme's possible implications for KRV DNA replication.

MATERIALS AND METHODS

PREPARATION OF PURIFIED VIRIONS. KRV (strain 308) was grown in monolayer cultures of rat nephroma cells for 5–7 days and was radiolabeled with [³H]thymidine or [¹⁴C]thymidine as described previously (Salzman and Jori 1970). The virus was extracted from cell debris at pH 9.0 for 5 hr at 37°C and sedimented to equilibrium in two successive isopycnic centrifugations in cesium chloride at an average density of 1.41 g/cm³ (Salzman and Jori 1970). The KRV virions were pooled (CsCl density 1.39–1.41 g/cm³), dialyzed, and sedimented in a 5–30% sucrose gradient (pH 9.0) (Salzman 1971). Sucrose-gradient-purified virions were used in all experiments described.

Purified adeno-associated virus type 1 (AAV1) was obtained from Dr. J. Rose. The virus was purified by isopycnic centrifugation in cesium chloride (Rose et al. 1966).

PURIFICATION OF KRV-ASSOCIATED DNA POLYMERASE. Nonidet P-40 (NP-40) at a final concentration of 4% and dithiothreitol at a final concentration of 5 mM were added to the purified virus preparation. The solution was sonicated vigorously three times, each time for 15 sec at 4°C, and then incubated at 37°C for 20 min (Salzman and McKerlie 1975a). Deoxycholate was added to a final concentration of 0.5% and the solution was kept on ice for an additional 20 min. No intact virions were detected when the preparation was viewed under the electron microscope. The disrupted virus was dialyzed and successively eluted from DEAE-cellulose, DNA-cellulose, and phosphocellulose columns (Salzman and McKerlie 1975a). The phosphocellulose-purified protein was used for the enzyme studies described. We have not determined whether the enzyme copurified with a viral capsid protein.

ASSAY FOR KRV-ASSOCIATED DNA POLYMERASE. Incubation mixtures contained in 200 μl the following: 20 mM Tris-HCl (pH 8.9), 2 mM β-mercaptoethanol, 5 μg bovine serum albumin, 10 mM MgCl₂, 0.02 M KCl, 60 μg activated salmon-sperm DNA, 10 μM each of dCTP, dGTP, TTP, and 10 μM dATP labeled with 0.5 μCi ³H (specific activity 0.5 mCi/μmole). The DNA was precipitated in 5% trichloroacetic acid at 4°C, collected on a Whatman GF/C filter, washed with 5 aliquots of 5 ml of 5% trichloroacetic acid, dried, and counted in 10 ml of Spectrafluor in a Beckman scintillation counter.

**DETERMINATION OF NUCLEASE ACTIVITY OF KRV-ASSOCIATED DNA POLYM-
ERASE.**   DNA endonuclease activity was determined by incubation of
the enzyme (0.6 μg protein) with 1 μg of the closed circular form (form I)
of SV40 DNA labeled with [^{3}H]thymidine (62,000 cpm/μg). Incubation
was at 37°C for 1 hr in the absence of deoxynucleoside triphosphates. The
incubation mixture was then sedimented in a neutral sucrose gradient
and the conversion of form I (21S) to form II (16S) was followed (Salzman
and McKerlie 1975a).

PREPARATION OF 16S AND 8–10S KRV DNA.   A 0.2-ml (4 × 10^8 PFU/ml) aliquot
of purified KRV was incubated at pH 12.2 for approximately 5 min at room
temperature. The sample was then layered onto an 11-ml gradient of
5–20% sucrose in 0.3 M Tris, 0.7 N NaCl, 0.2 M NaOH, 0.15% Sarkosyl,
0.001 M EDTA (pH 10.0). The gradients were centrifuged for 6 hr at 10°C at
41,000 rpm in a Beckman SW 41 Ti rotor. Fractions of 0.2 or 0.4 ml were
collected from the bottoms of the tubes and immediately neutralized by
the addition of 0.1 or 0.2 ml of 1 M Tris (pH 7.4).

ELECTRON MICROSCOPY.   DNA from the alkaline sucrose gradient was
mounted for electron microscopy by aqueous or formamide techniques
(Koczot et al. 1973).

RESULTS

*DNA-Polymerase Activity Associated with KRV After Sedimentation in a Sucrose
Gradient*

KRV was purified by two successive isopycnic centrifugations in cesium
chloride. The dialyzed virus was then studied by velocity sedimentation
in a 5–30% linear sucrose gradient (pH 9). As seen in Figure 1, the

FIGURE 1
Association of enzymatic activity
with KRV. KRV purified by two
CsCl₂ isopycnic centrifugations
was added to a 5–30% linear sucrose gradient (pH 9.0) and centrifuged as reported previously
(Salzman 1971). Fractions were
collected from the bottom of the
centrifuge tube and tested for
DNA-polymerase activity.   ●:
Hemagglutination units; ○: acid-
precipitable radioactivity in
DNA-polymerase assay. (Reprinted, with permission, from
Salzman 1971.)

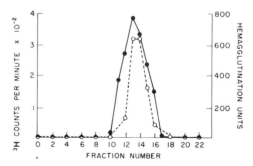

hemagglutinating activity of the KRV proteins and the acid-precipitable radioactivity from the assay for DNA polymerase parallel each other closely. This indicates a close association of the enzyme activity with the virion.

Isolation of the Associated DNA Polymerase

In order to separate the enzymatic activity from the intact virion, we sonicated the virions, treated them with a reducing agent and NP-40, and then incubated them with deoxycholate. The protein with enzymatic activity recovered after this procedure eluted in a single sharp peak from DEAE-cellulose, DNA-cellulose, and phosphocellulose columns. About 20% of the initial enzyme activity was recovered after the three chromatography steps. The enzyme was purified 63-fold and had a specific activity of 1125 units (nmoles/60 min) per mg of protein (Salzman and McKerlie 1975b). The molecular weight of the enzyme (76,000 ± 3000) was determined by sedimentation in a 5–20% sucrose gradient (pH 8.4) with markers of known sedimentation constants (Salzman and Mc-Kerlie 1975b).

Maximal activity of the reaction mixture is dependent on the presence of enzyme, activated DNA, a divalent cation, and all four nucleotide triphosphates (Table 1). In the presence of Mg^{++} as a cation, the addition of 0.02 M KCl stimulated the activity two to threefold. When three nucleotides were absent the enzyme activity was reduced to about 20%. As shown in Table 1, the KRV-associated DNA polymerase can utilize dADP in place of dATP as a substrate for polymerization, which suggests the presence of a nucleoside diphosphokinase activity similar to that found associated with both prokaryotic and eukaryotic DNA polymerases (Miller and Wells 1971; Sedwick et al. 1972).

The activity of the KRV-associated DNA polymerase is dependent on added primer template (Table 1). The enzyme prefers activated to native or single-stranded DNA. It requires a 3' OH terminus for activity, because micrococcal nuclease-treated DNA is inactive, and can copy a variety of synthetic DNA molecules. Oligo-homopolymers that have a polyribo strand in place of the polydeoxyribo strand do not serve as templates. For example, $d(T)_9$–poly-r(A) exhibited no template activity when [³H]ATP, [³H]dTTP, or [³H]UTP was used as a substrate. Under identical conditions, activity of the rat-nephroma cellular R-DNA polymerase or C11 could be demonstrated (Salzman and McKerlie 1975b).

Nuclease Activity Associated with DNA Polymerase

The phosphocellulose-purified KRV-associated DNA polymerase was also examined for endonuclease activity by determining its ability to

TABLE 1
Requirements of the KRV-Associated DNA Polymerase

Components of enzyme incubation	Activity (%)
Complete	100
− enzyme	0
− MgCl	14
− bovine serum albumin	90
− β-mercaptoethanol	100
− dCTP, − dGTP, − TTP	20
− dATP + dADP	71
− activated DNA (DNase-I-treated dsDNA)	0
+ dsDNA	30
+ ssDNA	5
+ micrococcal nuclease-treated dsDNA	16
+ d(T)$_9$r(A)$_n$ + [³H]dATP	100
+ d(T)$_9$r(A)$_n$ − [³H]dATP + [³H]ATP or +[³H]dTTP or + [³H]UTP	0

In the complete system 2.7 nmoles of deoxynucleotide were incorporated by 1.2 μg of enzyme protein in the presence of 20 μM Tris-HCl (pH 8.9); 2 mM β-mercaptoethanol; 5 μg bovine serum albumin; 10 mM MgCl$_2$; 0.02 M KCl; 60 μg activated salmon-sperm DNA; 10 nM each of dCTP, dGTP, TTP; and dATP or ATP or dTTP or UTP labeled with 0.5 μCi ³H (specific activity 0.5 mCi/μmole). ds: double-stranded; ss: single-stranded salmon-sperm DNA.

"nick" the double-stranded circular form of SV40 (component I). After incubation with the enzyme and sedimentation in a neutral gradient, 49% of the double-stranded circular component I is converted to a form which migrates in the area of component II (one linear strand, one double strand) and component III (two linear DNA strands). Further analysis of the enzyme-treated SV40 DNA in an alkaline sucrose gradient showed that DNA-polymerase endonuclease activity appears to have cleaved both strands of SV40 component I in several places. Utilizing an assay which involved the release of acid-soluble radioactivity from radioactive native and denatured adenovirus DNA, we could detect no exonuclease activity associated with the enzyme (data not shown).

Isolation of a Nucleoprotein Complex from KRV

Disruption of the defective parvovirus AAV at pH 12.2 followed by its sedimentation in a sucrose gradient at pH 12.2 results in the release of AAV DNA from its capsomeres. The DNA sediments in a single symmetrical peak at 16S (Koczot et al. 1973). If KRV is disrupted and sedimented in a sucrose gradient, as above, the KRV DNA is found in two discrete peaks: one at 16S and a second at 8–10S. The ratio of the optical density readings at 260 and 280 nm in the 8–10S peak indicated that protein was associated with the DNA. However, examination of both peaks of KRV DNA, using either aqueous or formamide mounting techniques, revealed only linear single-stranded DNA molecules.

Because exposure to pH 12.2 for several hours would probably have an adverse effect on any enzyme activity present, we disrupted the KRV (which contained [³H]thymidine in the DNA) at pH 12.2 for 2–5 min and then sedimented the preparation in a 5–20% sucrose gradient at pH 10. AAV virions were treated in an identical manner and sedimented in a separate gradient as a sedimentation marker. Fractions from the gradient were then assayed for DNA-polymerase activity. A peak of DNA-polymerase activity was found in the KRV gradient at the position of the 8–10S DNA peak (Fig. 2). Virtually 100% of the polymerase activity applied was recovered in the 8–10S region of the gradient. In addition, the specific activity of the 8–10S enzyme fraction was approximately 200–400 units per mg of protein whereas that of the input sample was 25 units/mg.

These results suggested that DNA-polymerase activity could be isolated as a stable complex of protein and viral DNA under the disruption conditions employed. However, since the profile of polymerase activity was not perfectly coincident with the peak of 8–10S DNA, it was possible that the protein and the DNA had cosedimented independently with the same S value. To distinguish between these possibilities, fractions from the 16S and 8–10S regions of the gradient were centrifuged to equilibrium in cesium chloride (Fig. 3). The [³H]thymidine-labeled DNA from the 16S fraction banded as a sharp peak with a density of 1.72–1.73 g/cm³, the

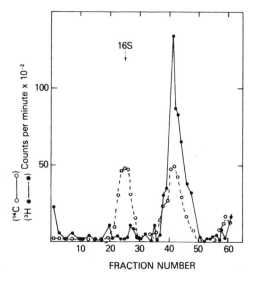

FIGURE 2
KRV labeled with [¹⁴C]thymidine disrupted in alkali (pH 12.2) and sedimented in a 5–20% sucrose gradient at pH 10. Sedimentation is from right to left. Fractions (0.2 ml) were collected from the bottom of the tube. ●——●: [¹⁴C]thymidine KRV DNA; o- - - -o acid-precipitable ³H radioactivity in DNA-polymerase assays. The position of ³H-labeled AAV DNA (16S) sedimented in a parallel gradient is indicated by an arrow.

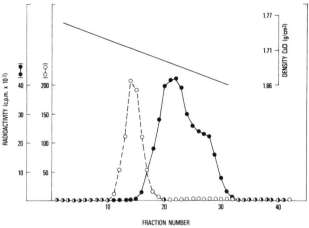

FIGURE 3
Isopycnic centrifugation in cesium chloride (average density
1.72 g/cm³) of KRV DNA isolated from a pH 10 sucrose
gradient. o-----o: [³H]thymidine KRV DNA from 16S peak;
●——●: [¹⁴C]thymidine KRV DNA from 8–10S peak.

reported density of virion DNA (Salzman et al. 1971). In contrast, DNA
from the 8–10S region banded as a broad peak with a density range of
1.66–1.69 g/cm³. The significant decrease in buoyant density suggested
that the DNA was tightly bound to protein.

Fractions containing the 8–10S KRV nucleoprotein complex were combined, condensed, disrupted at 100°C in mercaptoethanol and sodium
dodecyl sulfate, and electrophoresed with molecular-weight markers.
The gel of the 8–10S fraction revealed only one band of protein, in
addition to the ovalbumin marker, when stained with Coomassie blue
(Fig. 4). The molecular weight of this protein, determined by the procedure of Shapiro et al. (1967), was 75,000 ± 3000. This is virtually the
same molecular weight as that determined by velocity sedimentation for
the virion-associated DNA polymerase purified chromatographically
(Salzman and McKerlie 1975a).

Electron Microscopy of DNA in the 16S and 8–10S Peaks

Only linear single-stranded DNA could be detected in either the 16S or
the 8–10S peak with the electron microscope. The mean contour lengths
and size distributions of the DNA in both peaks were determined using
the circular single-stranded DNA of φX174 as an internal standard. The
DNA of the 16S peak was quite uniform in size and had a mean contour

FIGURE 4
Electrophoresis of the protein from the nucleoprotein complex in a 7.5%
acrylamide gel in 0.1 M sodium phosphate buffer (pH 7.2), 0.1% SDS.
Migration is from left to right. Ovalbumin, used as a marker in this gel, is
indicated by the arrow.

length of 1.51 ± 0.21 μm (Table 2). This value agrees with the mean length
reported by Salzman et al. (1971) for intact KRV DNA extracted from
virions, 1.53 ± 0.19, and is shorter than the mean length (1.83 ± 0.07 μm)
of φX174 DNA (Davidson and Szybalski 1971). Though some shorter
DNA fragments were observed in the 16S fraction, 75% of the molecules
had a contour length between 1.45 and 1.80 μm, with a mean value of 1.62
± 0.08 μm. DNA molecules from the 8–10S peak had a trimodal length
distribution (Table 2); only 20% of the molecules had a mean length
equivalent to that of intact KRV DNA. The reason for the observed DNA
fragmentation is unclear, but it may result from the endonuclease activity
of the KRV-associated DNA polymerase.

TABLE 2
Electron-Microscope Measurement of KRV-DNA Contour Lengths
After Alkaline Sucrose (pH 10) Sedimentation

Sedimentation coefficient $S_{20,w}$	Contour-length distribution (μm)		Mean contour length (μm)
16	1.45–1.8	(75%)	1.51±0.21
	1.0–1.45	(25%)	
8–10	0.4–0.75	(38%)	0.6±0.09
	0.75–1.3	(42%)	1.02±0.14
	1.4–1.8	(20%)	1.58±0.07

The contour lengths of 50 molecules in the $16S_{20,w}$ group and 42 molecules in the $8–10S_{20,w}$
group were measured (Salzman and Koczot 1977).

DISCUSSION

Purified virions of KRV have three structural proteins, with molecular weights of 76,000 ± 7000 (VP1), 62,000 ± 6000 (VP2), and 55,000 ± 6000 (VP3) (Salzman and White 1970). Each virion contains approximately 60 copies of VP2 and 8–10 copies of VP1 and VP3 (Salzman and White 1970). The DNA polymerase associated with the KRV virion has a molecular weight of 75,000 ± 3000 (Salzman and McKerlie 1975a), which is very close to that of VP1. However, our results thus far do not allow a definitive statement as to whether the enzyme is an integral part of the virion itself or a cellular enzyme which adventitiously associates with the particles.

The KRV-associated enzyme shares some properties with, but also has characteristic differences from, the DNA polymerases of rat nephroma cells. In its properties of stimulation of KCl, pH optimum (8.0), K_m of dNTP, and Mg optimum, the purified KRV-associated DNA polymerase resembles most closely the cytoplasmic enzyme C11, or R-DNA polymerase (Salzman and McKerlie 1975a, b). However, the virus-associated enzyme and the rat-nephroma cellular enzyme C11 vary in template specificity and in chromatographic properties. For example, the C11 enzyme can utilize the ribo strand of $d(T)_9 \cdot poly\text{-}r(A)$ to polymerize thymidylic acid under the conditions described (Table 1), whereas the virus-associated enzyme cannot. However, the experimental differences observed do not rule out the possibility that the KRV-associated enzyme may represent a modification of a cellular enzyme, such as C11, or a host-specified DNA polymerase, not normally detected in the cell, that is induced by virus infection. Nevertheless, the specific affinity of this polymerase for the KRV virion suggests that it may play an important role in KRV-DNA replication.

The association of enzyme activity with KRV may not be a general property of all parvoviruses, but may depend on the virus and the cell line in which it is grown. We have not detected DNA-polymerase activity in association with AAV, H-1, or minute virus of mice. However, detection of polymerase activity may depend on the procedure used for purification of the virus. For example, calcium ions, which are used in the purification of some parvoviruses, could lead to inactivation of DNA-polymerase activity. The inclusion of calcium ion in the extraction of the major cytoplasmic DNA polymerase a from cultured human cells leads to an irreversible inactivation of the enzyme (Eichler et al. 1977).

Our most recent study of the protein associated with KRV involves disrupting the virion in alkali and then sedimenting in an alkaline (pH 12.2 or 10.0) sucrose gradient. Whereas disruption of AAV particles results in a single peak of DNA sedimenting at 16S, two DNA-containing peaks are seen upon disruption of KRV virions. The first sedimenting at

16S, contains DNA with a mean contour length (1.53 ± 0.19 μm) close to that of the intact KRV genome (Salzman et al. 1971), but it exhibits little, if any, DNA-polymerase activity. The second peak, sedimenting at 8–10S, contains both DNA and protein. The DNA of the 8–10S peak is heterogeneous in size, containing 20% full-length KRV DNA and two classes of DNA fragments with mean contour lengths of 0.6 ± 0.09 and 1.02 ± 0.14 μm. The 8–10S peak contains virtually all of the DNA-polymerase activity in the gradient and reveals only a single polypeptide species of approximately 75,000 daltons when analyzed by gel electrophoresis. The pronounced decrease in the buoyant density of the DNA in the 8–10S peak after centrifugation to equilibrium in cesium chloride suggests that this polypeptide is tightly associated with the viral DNA. It is possible that this association alters the configuration of the DNA, thereby causing a change in its sedimentation properties. None of the DNA from the 8–10S peak banded at the density of free virion DNA, although at least 20% of the DNA in this pool was of unit length as determined by electron microscopy.

Preliminary results indicate that the 8–10S peak of DNA and protein can incorporate radioactive thymidine into acid-precipitable radioactivity in the absence of added template. If the enzyme utilizes the KRV DNA efficiently as a template, we may gain insight into the DNA replicative intermediates involved in the replication of single-stranded viral DNA. Further studies in this area are in progress.

REFERENCES

Davidson, H. and W. Szybalski. 1971. In *The bacteriophage lambda* (ed. A. D. Hershey). Cold Spring Harbor Laboratory, Cold Spring Harbor, New York.

Eichler, D. C., P. A. Fisher, and D. Korn. 1977. Effect of calcium on the recovery and distribution of DNA polymerase a from cultured human cells. *J. Biol. Chem.* **252**:4011.

Gunther, M. and P. May. 1976. Isolation and structural characterization of monomeric and dimeric forms of replicative intermediate of Kilham rat virus DNA. *J. Virol.* **20**:86.

Koczot, F. J., B. J. Carter, C. F. Garon, and J. A. Rose. 1973. Self complementarity of terminal sequences within plus or minus strands of adenovirus-associated virus DNA. *Proc. Natl. Acad. Sci.* **70**:215.

Miller, L. K. and R. D. Wells. 1971. Nucleoside diphosphokinase activity associated with DNA polymerases. *Proc. Natl. Acad. Sci.* **68**:2298.

Robinson, D. M. and R. M. Hetrick. 1969. Single-stranded DNA from the Kilham rat virus. *J. Gen. Virol.* **4**:269.

Rose, J. A., M. D. Hoggan, and A. J. Shatkin. 1966. Nucleic acid from an adeno-associated virus: Chemical and physical studies. *Proc. Natl. Acad. Sci.* **56**:86.

Salzman, L. A. 1971. DNA polymerase activity associated with purified Kilham rat virus. *Nat. New Biol.* **231**:174.

————. 1977. Evidence for terminal S₁ nuclease-resistant regions on single-stranded linear DNA. *Virology* **76**:454.

Salzman, L. A. and L. A. Jori. 1970. Characterization of the Kilham rat virus. *J. Virol.* **5**:114.

Salzman, L. A. and F. Koczot. 1977. Isolation of nucleoprotein from the parvovirus KRV. *Virology* (in press).

Salzman, L. A. and L. McKerlie. 1975a. Characterization of the deoxyribonucleic acid polymerase associated with Kilham rat virus. *J. Biol. Chem.* **250**:5583.

———— 1975b. Nuclear and cytoplasmic deoxyribonucleic acid polymerases from rat nephroma cells. *J. Biol. Chem.* **250**:5589.

Salzman, L. A. and W. L. White. 1970. Structural proteins of Kilham rat virus. *Biochem. Biophys. Res. Commun.* **41**:1551.

Salzman, L. A., W. L. White, and T. Kakefuda. 1971. Linear single-stranded deoxyribonucleic acid isolated from Kilham rat virus. *J. Virol.* **7**:830.

Sedwick, W. D., T. S. F. Wong, and K. Korn. 1972. Purification and properties of nuclear and cytoplasmic deoxyribonucleic acid polymerases from human KB cells. *J. Biol. Chem.* **247**:5026.

Shapiro, A. L., E. Vinuela, and J. V. Maizel, Jr. 1967. Molecular weight estimation of polypeptide chains by electrophoresis in SDS polyacrylamide gels. *Biochem. Biophys. Res. Commun.* **28**:815.

DNA-Polymerase Activity in Parvovirus-Infected Cells

ROBERT C. BATES
CYNTHIA P. KUCHENBUCH
JOHN T. PATTON
ERNEST R. STOUT

Department of Biology
Virginia Polytechnic Institute and State University
Blacksburg, Virginia 24061

As parvoviruses contain a DNA genome, it is certain that a DNA polymerase(s) of viral or cellular origin plays a role in the replication of viral DNA. It was reported by Salzman (1971) that Kilham rat virus (KRV) contains a virion-associated DNA polymerase which corresponds in molecular weight to the capsid protein, VP1 (72,000 daltons) (Salzman and McKerlie 1975). However, Rhode (1973) was not able to repeat this finding for KRV or H-1 virus. Because of the potential importance of such an enzyme in the replication process of parvoviruses, we examined three different parvoviruses, bovine parvovirus (BPV), LuIII, and H-1, for a virion-associated DNA-polymerase activity. Our results show that although at least one cellular DNA polymerase tended to copurify with the viruses through several purification steps, highly purified virus free of contaminating cellular material did not have DNA-polymerase activity.

It is well established that the nondefective parvoviruses replicate optimally in actively dividing cells requiring one or more cellular functions expressed during the late S or G2 phase of the cell cycle (Tennant and Hand 1970; Salzman et al. 1972; Rhode 1973; Siegl and Gautschi 1973; Tattersall et al. 1973; Parris and Bates 1976). The levels of cellular DNA-polymerase activities in synchronized HeLa cells fluctuate with the cell cycle (Spadari and Weissbach 1974). Activity of DNA polymerase γ increases prior to and in parallel with cellular DNA synthesis; that of

DNA polymerase α increases during S phase and reaches a peak after cellular DNA synthesis has reached maximal levels. DNA-polymerase-β activity remains essentially constant throughout the cell cycle. This observation, combined with our previous findings (Parris and Bates 1976) that cellular DNA synthesis and BPV DNA synthesis were temporally separated in synchronized cells, suggested that a correlation could be drawn between the expression of DNA polymerases in the cell cycle and virus replication. Therefore, we report here the levels of cellular DNA polymerases, viral DNA synthesis, and production of progeny virus in infected synchronized cells.

MATERIALS AND METHODS

Cell Cultures and Virus Propagation

Primary bovine fetal spleen (BFS) cell cultures were prepared and maintained as described previously (Parris and Bates 1976). Chimpanzee liver (CL) cell line was obtained from Flow Labs and maintained by standard methods. BFS cells used for cell-cycle studies were synchronized with hydroxyurea (HU) as described by Parris et al. (1975). H-1 virus was purchased from the American Type Culture Collection, and LuIII was supplied by Dr. K. Soike. Virus for purification studies was prepared by infecting parasynchronous cell cultures of BFS (with BPV) or CL (with H-1 or LuIII) in roller bottles with 1–5 PFU/cell of virus. Hemagglutination and infectivity assays were performed as described previously (Bates et al. 1972; Bates and Storz 1973; Parris and Bates 1976).

Virus Purification

Procedure 1, consisting of neuraminidase treatment of infected-cell lysates for 16 hr at 37°C followed by centrifugation in CsCl, was performed as described by Salzman and Jori (1970). Procedure 2 was used as reported by Tattersall et al. (1976) except that a 3-ml sucrose layer was used in the sucrose-CsCl step gradient and the gradient was centrifuged for 24–36 hr. Virions banding between 1.39 and 1.41 g/cm³, the position of infectious "full" virus, were collected for further study.

Cell Fractionation

For cell-cycle experiments, 1×10^6 BFS cells were seeded in petri dishes (60 × 15 mm) in synchronization medium (Parris et al. 1975) and incubated for 32 hr at 37°C. Mock-infected and BPV-infected cells were collected by scraping the cells into isotonic buffer (0.14 M NaCl, 1.5 mM MgCl$_2$, 0.1 M Tris-HCl, pH 7.4) containing 5 mM dithiothreitol. When advanced cytopathic effect was evident in infected cultures, detached cells were centrifuged from the medium and combined with cells scraped from the plates. The cells were broken in a Potter-Elvehjem homogenizer, with all steps performed at 4°C. Cell breakage was monitored by light microscopy, and stained preparations revealed unbroken nuclei free of cytoplasm. Nuclei were pelleted by centrifugation at 1000g. The supernatant fraction was centrifuged at 37,000g to sediment organelles and membranes, and the resultant cytoplasmic supernatant was combined with the nuclei. The nuclei-cytoplasmic fraction was sonicated in three 15-sec bursts and clarified by centrifugation at 20,000g for 20 min. The supernatant of this centrifugation was stored at −196°C until used in DNA-polymerase assays.

DNA-Polymerase Assays

DNA-polymerase activity was measured as the incorporation of radioactively labeled substrate into acid-precipitable product. Reaction mixtures in a total volume of 100 μl contained the following components. Polymerase-a assay: 20 mM KPO_4 (pH 7.2); 0.2 mM each dATP, dCTP, dGTP, and [^{3}H]dTTP (80 cpm/pmole); 7.5 mM $MgCl_2$; 50 μg DNase-activated calf-thymus DNA (Schlabach et al. 1971); and 0.5 mM dithiothreitol. Polymerase-β assay: 50 mM Tris-HCl (pH 8.5); 0.2 mM each dATP, dCTP, dGTP, and [^{3}H]dTTP (80 cpm/pmole); 7.5 mM $MgCl_2$; 50 μg activated DNA; 0.5 mM dithiothreitol; 0.2 M NaCl; and 10 mM N-ethylmaleimide. Polymerase-γ assay: 50 mM Tris-HCl (pH 7.5), 0.2 mM [^{3}H]dTTP (80 cpm/pmole), 0.5 mM $MnCl_2$, 2.5 μg poly(rA)·oligo(dT)$_{\overline{10}}$ (Weissbach et al. 1975), 0.5 mM dithiothreitol, and 0.1 M KCl. Purified virus was also disrupted and assayed for viral DNA polymerase by the method of Salzman and McKerlie (1975).

DNA polymerase a was also measured using a poly(dT)·oligo(rA) template primer (Spadari and Weissbach 1975); the results were the same as observed with activated DNA.

DNA-Polymerase-a Antibody

IgG (20 μg/ml) prepared against bovine DNA polymerase a was a generous gift from Dr. F. J. Bollum. Enzyme preparations were incubated with antibody in 20 mM Tris-HCl (pH 7.5) at 20°C for 20 min and then assayed in the standard DNA-polymerase-a reaction mixture.

Polyacrylamide Gel Electrophoresis

Slab gels were prepared using the SDS-Tris-glycine system of Laemmli (1970) as modified by Anderson et al. (1973). A 9-cm resolving gel (7.5%

acrylamide, 0.2% bis-acrylamide) was overlaid with 1 cm of a 4% stacking gel. Virus preparations were prepared for electrophoresis by boiling for 2 min in a solution containing 1 mM Tris-HCl (pH 6.8), 2% SDS, 0.6 M 2-mercaptoethanol, 10% glycerol, and 0.01% bromophenol blue.

Virus samples (3–20 μl) and marker proteins were applied to different gel wells. Electrophoresis was performed at room temperature at 1 mA/well until the tracking dye entered the resolving gel and then at 5.0 mA/well until the dye front reached the bottom of the slab.

Cylindrical gels were prepared using the neutral SDS-phosphate system of Maizel (1969). Virus samples for electrophoresis were disrupted in 0.01 M NaPO$_4$ (pH 7.2), 1.5% SDS, and 1.5% 2-mercaptoethanol at 100°C for 2 min. The solution was then adjusted to 10% glycerol and 0.01% bromophenol blue. The resolving gel was 7.5% acrylamide, overlaid with a 2.5%-acrylamide stacking gel. Electrophoresis was at room temperature at 5 mA/tube.

After electrophoresis, both slab and cylindrical gels were fixed in 20% TCA for 1 hr followed by staining in 0.2% Coomassie blue R in 50% methanol and 7% acetic acid for 3 hr. Destaining was in 10% methanol and 7% acetic acid overnight. The destained gels were scanned at 590 nm. Molecular weights of viral proteins were determined by the method of Chrambach and Radbard (1971).

Cellular and Viral DNA Synthesis

The rate of cellular DNA synthesis was measured by [³H]thymidine (TdR) (15 Ci/mmole) incorporation into acid-precipitable material as described previously (Parris et al. 1975). BPV DNA from infected cells was analyzed on sucrose gradients containing guanidine-HCl by the method of Lavelle and Li (1977). Infected cells (1 × 10⁶) were pulse-labeled with [³H]TdR 50 Ci/mmole) for 30 min, collected, and lysed by layering on 17-ml 5–20% sucrose gradients containing 4 M guanidine-HCl (10 mM Tris-HCl, pH 8.0). The gradients were centrifuged at 25,000 rpm for 16 hr in an SW 27.1 rotor and then fractionated, and radioactivity in regions corresponding to viral DNA was determined.

RESULTS

DNA-Polymerase Activities Associated with BPV Preparations

In preliminary experiments, BPV preparations obtained by purification procedure 1 were found to contain DNA-polymerase activity. In order to determine the identity of this DNA-polymerase activity (copurifying cellular enzyme or virion-associated enzyme), activities of DNA polymerases α, β, and γ were measured at each step of the purification procedure

(Table 1). BPV was purified from 5×10^7 to 1×10^8 infected cells by procedure 1, following a single-cycle infection in parasynchronous BFS-cell cultures. As shown in Table 1, the activities of polymerases α and γ were decreased 40-fold or greater by purification step 5 (second CsCl gradient), when compared to the starting infected-cell lysate. In contrast, polymerase β increased in activity during intermediate purification steps and in step 5 remained at levels comparable to those found in the starting cell lysate. It is apparent from the data presented in Table 1 that BPV purified from BFS cells by this method was not free of contaminating cellular DNA-polymerase activity. Further, polymerase β remains closely associated with the virus and copurifies with it. Disruption of the virus purified to step 5 (Table 1) by detergent treatment (Salzman and McKerlie 1975) resulted in a twofold increase in polymerase activity. This may reflect release of viral enzyme or disaggregation of contaminating cellular enzymes.

The virus band from step 5 was electrophoresed on a 7.5% gel in a cylindrical tube using a neutral SDS-phosphate buffer system. Three protein bands with molecular weights corresponding to reported values for BPV (Johnson and Hoggan 1973) were visualized. However, when the same sample was electrophoresed on a 7.5% SDS-polyacrylamide slab gel using a Tris-glycine buffer system, several additional bands were detected (Fig. 1). Although not directly determined, these bands may represent contaminating cellular DNA polymerases. The prominent 70,000–72,000-dalton protein located between the BPV structural proteins (Fig. 1) was observed at each purification step. This protein can be removed by further purification of the virus, either on sucrose gradients or by procedure 2 (described below), and apparently is not a BPV struc-

TABLE 1

DNA-Polymerase Activities During Purification of BPV by Procedure 1

Purification step	Enzyme units (pmoles/hr)		
	α	β	γ
1. Infected-cell lysate[a]	2175	50	389
2. Lysate plus neuraminidase (16 hr at 37°C)	611	78	434
3. 1,000g supernatant	218	86	301
4. Virus band (d = 1.39–1.41 g/cm³), first CsCl gradient	26	59	7
5. Virus band (d = 1.39–1.41 g/cm³), second CsCl gradient	17	51	10

Data from Salzman and Jori 1970.

[a] 5×10^7 to 1×10^8 infected cells.

FIGURE 1

Densitometric tracing of Coomassie-blue-stained polyacrylamide slab gel electrophoretogram of BPV (step 5) purified by procedure 1. Arrows indicate the position of virion proteins as determined for BPV purified by procedure 2 (see Fig. 2).

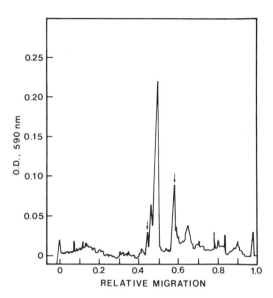

tural protein. In contrast with the KRV-associated enzyme activity (Salzman 1971), when BPV from step 5 was analyzed on a sucrose gradient DNA-polymerase activity did not cosediment with hemagglutinating activity (data not shown).

Since purification procedure 1 did not provide virus free of contaminating cellular DNA polymerases, it was impossible to determine whether the virus had an essential associated DNA polymerase. Therefore, purification procedure 2, which incorporates a sucrose-CsCl step gradient, was adopted (Table 2). By step 4 of this purification procedure, a substantial reduction in levels of a and γ polymerases was achieved, while the level of β remained high. Virus purified beyond step 6 had no detectable activity of DNA polymerase a, β, or γ. Virus from step 8 was treated with detergent and assayed as described by Salzman and McKerlie (1975). No DNA-polymerase activity was detected.

The addition of phenylmethylsulfonyl fluoride (PMSF) in step 2 (Table 2) reduced the level of a-polymerase activity threefold. Since this protease inhibitor might also affect the activity of a viral DNA polymerase, PMSF was omitted from one of the purification runs, but again no DNA-polymerase activity was detected.

BPV from step 8 of purification procedure 2 was electrophoresed on a 7.5% SDS-polyacrylamide slab gel (Fig. 2). Two distinct protein bands with molecular weights corresponding to 81,200 and 60,400 daltons were evident. The 70,000–72,000-dalton protein seen in virus purified by procedure 1 was present in only the first four steps of purification procedure 2.

TABLE 2

DNA-Polymerase Activities During Purification of BPV by Procedure 2

Purification step	Enzyme units (pmoles/hr)		
	α	β	γ
1. Infected-cell lysate[a]	87,135	9728	648
2. Lysate plus PMSF[b]	27,917	4005	628
3. Homogenate	13,562	8035	656
4. 27,000g Supernatant	17,611	7882	405
5. Supernatant plus $CaCl_2$[c]	347	396	412
6. Suspended $CaCl_2$ pellet	0	189	0
7. 12,000g Supernatant	0	0	0
8. Virus band (d = 1.39–1.41 g/cm³)[d], sucrose-CsCl step gradient	0	0	0

Data from Tattersall et al. 1976.

[a] 1×10^8 infected cells.

[b] Final concentration: 1 mM.

[c] Final concentration: 25 mM.

[d] Protein concentration: 250–500 μg/ml.

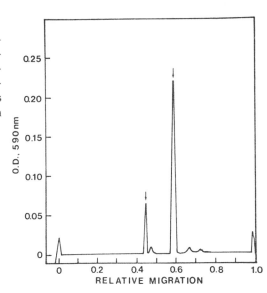

FIGURE 2

Densitometric tracing of Coomassie-blue-stained polyacrylamide gel electrophoretogram of BPV (step 8) purified by procedure 2. Arrows indicate the positions of virion proteins.

Search for Virion-Associated DNA-Polymerase Activity in Three Parvoviruses

Three parvoviruses, BPV (as described above), H-1, and LuIII, were purified by procedure 2 and assayed for DNA-polymerase activity at each step of the purification process. The data clearly show that this purification procedure was effective in removing contaminating DNA-polymerase activity from these viruses regardless of the host cell in which the virus was propagated (Table 2, Fig. 3). However, it can be seen that contaminating polymerase activity was not removed as rapidly from BPV preparations during the initial purification steps as from H-1 and LuIII (Fig. 3). This may be due to inherent differences of the viruses or to differences in the host cells in which they were propagated.

Each of the virus preparations purified to step 8 was examined for purity by electron microscopy and electrophoresis on polyacrylamide gels (Fig. 4) and tested for hemagglutinating activity and infectivity (Table 3). Then each virus was assayed in two ways for polymerase activity. First, undisrupted virus was assayed in *a*-, *β*-, *γ*-polymerase reaction mixtures. Second, virus disrupted by detergent treatment was assayed for viral polymerase by the method described by Salzman and McKerlie (1975). DNA-polymerase activity was not detected by either procedure. Therefore, we conclude that BPV, H-1, and LuIII do not contain an essential virion-associated DNA polymerase required for productive infection.

Pattern of DNA-Polymerase Activities in Synchronized Mock-Infected and BPV-Infected BFS Cells

Cells synchronized by hydroxyurea (HU) were mock-infected with MEM or infected with BPV (multiplicity of infection 10 PFU/cell) immediately upon removal of HU at the beginning of the S phase. Samples were prepared from both cultures and assayed for DNA polymerases a, β, and γ as described in "Materials and Methods." In parallel cultures within the same experiment, cellular DNA synthesis was measured as incorporation of [^{3}H]TdR into acid-precipitable material. As shown in Figure 5, DNA-polymerase-γ activity increased prior to and in parallel with cellular DNA synthesis in S phase, reaching maximal levels 2 hr before the peak in cellular DNA synthesis. In contrast, DNA-polymerase-a activity increased during S phase, reaching maximal levels at 8 hr, 2 hr after the peak rate of DNA synthesis (Fig. 5). DNA polymerase β remained at an essentially constant level throughout the period of the experiment (Fig. 5A). These patterns of enzyme activity and DNA synthesis were consistent and reproducible.

In BPV-infected cells, the activity profile of polymerase a differed markedly from that in uninfected cells (Figs. 5A and 6A). Polymerase-a

activity increased more slowly during the period corresponding to S phase, but exceeded the activity seen in mock-infected cells by 12 hr and continued to increase through 20 hr p.i. At this time, the levels of polymerase *a* were approximately threefold higher than in uninfected cells.

FIGURE 3
DNA-polymerase activities at each step during purification of BPV, LuIII, and H-1 by procedure 2. T = sum of a, β, and γ activities.

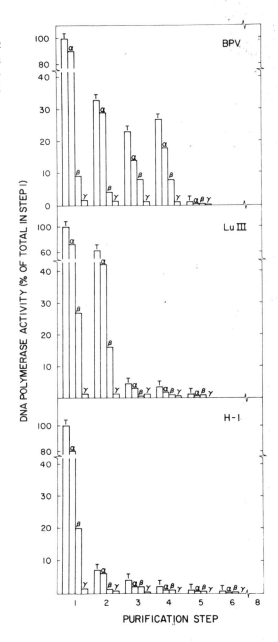

FIGURE 4

Polyacrylamide gel electrophoretogram of parvovirus proteins. BPV (well 4), H-1 (well 3), and LuIII (well 2) were purified to step 8 in procedure 2. Molecular-weight marker proteins (wells 1 and 5) were phosphorylase a (100,000), transferin (human) (76,600), bovine serum albumin (68,000), catalase (55,000), and ovalbumin (45,000).

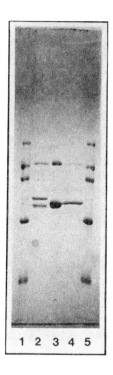

TABLE 3
Properties of Purified Parvoviruses Tested for DNA-Polymerase Activity

Virus[a]	Host cell[b]	Virus band (g/cm³)	HAU (10^{-3})[c]	PFU (10^{-8})[d]	Virion proteins (m.w. $\times 10^{-3}$)	
					mean	n
BPV	BFS	1.39–1.41	51.2	1.0	81.2 ± 0.5	4
					60.4 ± 1.0	4
H-1	CL	1.39–1.41	102.4	0.12	80.9 ± 0.8	4
					59.8 ± 1.3	4
LuIII	CL	1.39–1.41	12.8	0.5	80.9 ± 0.8	4
					61.2 ± 0.8	4
					58.3 ± 0.8	4

[a] Purified by procedure 2.
[b] BFS: bovine fetal spleen; CL: chimpanzee liver.
[c] Total hemagglutinin units (HAU) in 50 μl of virus preparation added to DNA-polymerase assays.
Total plaque-forming units (PFU) in 50 μl of virus preparation added to DNA polymerase assays.

FIGURE 5

Cyclic expression of DNA-polymerase activities and cellular DNA synthesis in mock-infected synchronized BFS cells. (*A*) DNA polymerase α (●——●), β (△——△), and γ (○——○) activities were determined immediately before and at 4-hr intervals after removal of HU. (*B*) Cellular DNA synthesis was measured by [³H]TdR incorporation into acid-precipitable material (■——■). Cells were pulse-labeled for 30 min with [³H]TdR (0.5 μCi/ml) at the indicated intervals after removal of HU.

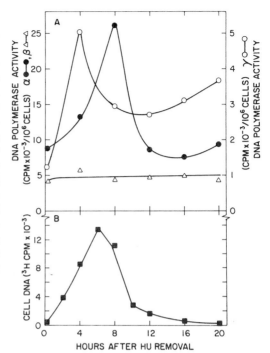

Polymerase-γ activity closely paralleled that in mock-infected cells, but remained slightly elevated between 4 and 20 hr p.i. (Fig. 6*A*). Levels of DNA polymerase β did not differ from those described for mock-infected cells (Figs. 5*A* and 6*A*).

Temporal Correlation of DNA-Polymerase Activity and Viral Replication

We have reported that cellular and BPV DNA synthesis are separated temporally in S and late S/G2 phases of the cell cycle (Parris and Bates 1976). From this observation and the results described above concerning cyclic fluctuations of DNA-polymerase activities in synchronized cells, it appeared possible to correlate DNA-polymerase activity and viral replication temporally. Parallel infected cultures within the same experiment were analyzed for DNA-polymerase activity, synthesis of viral DNA, and production of infectious virus (Fig. 6*A,B*). BPV DNA synthesis was first detected at 8 hr p.i. in cultures which were pulse-labeled with [³H]TdR and centrifuged into sucrose gradients containing guanidine-HCl. Maximal levels of viral DNA synthesis were reached at 16–20 hr p.i. (Fig. 6*B*). As seen in Figure 6*B*, production of infectious progeny virus paralleled viral DNA synthesis closely. The pattern of polymerase-α activity (Fig.

FIGURE 6
DNA-polymerase activities and virus replication in synchronized BFS cells. Cells were infected with BPV (multiplicity of infection 10 PFU/cell) immediately after removal of HU. (A) Activities of DNA polymerases a (●——●), β (△——△), and γ (○——○) were assayed as described in "Materials and Methods." (B) Rate of BPV DNA synthesis (■——■) was determined by pulse-labeling infected cells with [³H]TdR (50 μCi/ml) for 30 min. Viral DNA was isolated on 5–20% sucrose gradients containing 4 M guanidine-HCl. Infectivity (□——□) was determined for parallel cultures by the plaque assay on BFS cells.

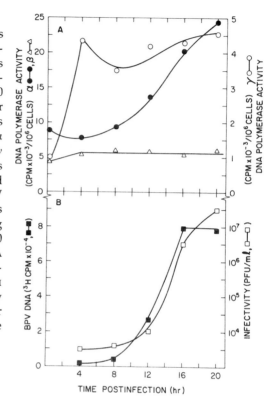

6A) was strikingly similar to those described for viral DNA synthesis and virus production (Fig. 6B). The close temporal correlation between these activities suggests that polymerase a may play a role in replication of parvovirus DNA.

Inhibition of DNA Polymerase a by Anti-a Antibody

Because of the distinctively different patterns of DNA polymerase a observed in mock-infected and virus-infected cells, it was important that we independently verify that this activity was indeed polymerase a. Enzyme preparations from the cell cycle which had maximal polymerase-a activity (8 hr mock-infected and 20 hr BPV-infected) were incubated with anti-bovine DNA polymerase a prior to DNA-polymerase assay. The results in Table 4 show that 0.83 μg/ml of antibody inhibited enzyme activity 65% in preparations from mock-infected cells and 72% in preparations from infected cells. At 1.67 μg/ml of antibody, inhibition increased to 73% and 83%, respectively.

TABLE 4
Inhibition of DNA-Polymerase Activity of Mock-Infected and BPV-Infected
BFS Cells by Anti-Bovine DNA Polymerase a

Anti-bovine DNA polymerase a (μg/ml)	DNA-polymerase activity[a] (cpm/10^6 cells)	
	8 hr mock-infected	20 hr BPV-infected
0	22,021	26,411
0.83	7,753 (65%)[b]	7,393 (72%)
1.67	5,854 (73%)	5,182 (80%)

[a] Enzyme preparations preincubated with antibody at 20°C for 10 min, then assayed in DNA polymerase a reaction mixture.
[b] Percent inhibition.

DISCUSSION

We have reported here that BPV purified by procedure 1 contained DNA-polymerase activity (Table 1). Although this finding was consistent with that for KRV purified by the same method (Salzman 1971; Salzman and McKerlie 1975), we found that the DNA-polymerase activity was separated from the virion either by sedimentation through sucrose or by an alternate purification procedure. Virus purified to step 8 in purification procedure 2 retained infectivity but was free of detectable DNA-polymerase activity (Table 2). Attempts by other labs to repeat the finding by Salzman (1971) of a virion-associated polymerase have not been successful. Rhode (1973) reported that H-1 and KRV purified by the method of Salzman and by a detergent method were free of DNA-polymerase activity. Similarly, highly purified MVM had no DNA-polymerase activity (D. C. Ward, personal communication). We have tested three different parvoviruses (BPV, H-1, and LuIII) for DNA-polymerase activity following purification by procedure 2 and have found no activity. Therefore, it appears that the presence of a virion-associated DNA polymerase is not a general property of the nondefective Parvoviridae.

Analysis of each purification step for activity of contaminating cellular DNA polymerases a, β, and γ has provided information concerning the tendency for a DNA polymerase to copurify with the virus. In purification procedure 1, the major contaminating protein as detected by gel electrophoresis had a molecular weight of 70,000–72,000 daltons (Fig. 1). In addition to this protein being detected in the final virus band from procedure 1, it was also seen in electrophoretograms of samples from the first four steps of procedure 2 and occasionally as a minor protein in virus purified to step 8. It cannot be directly determined from the experiments described here whether this protein is a contaminating cellular DNA polymerase. Our data do show, however, that DNA polymerase β copurifies more extensively with the virus than does polymerase a or γ (Tables 1 and 2). Copurification of polymerase β with the virus might be explained by the basic nature of the protein allowing for ionic interactions with the negatively charged virus during purification. Since the reported molecular weight for polymerase β is 45,000 daltons (Chang 1973), it is unlikely that the contaminating protein (70,000–72,000 daltons) described above is polymerase β.

Cellular and BPV DNA synthesis were temporally separated during the cell cycle in synchronized BFS cells (Parris and Bates 1976). DNA synthesis was maximal in BFS cells 4–6 hr after entry into S phase, whereas synthesis of low-molecular-weight viral DNA was first detected at 8 hr p.i. and reached a peak at 16 hr p.i., when cellular DNA synthesis had decreased to background levels. It has been reported that DNA-polymerase activities in synchronized HeLa cells fluctate during the cell cycle (Spadari and Weissbach 1974). In these cells DNA polymerase γ increases in parallel with but prior to cellular DNA synthesis, and DNA polymerase a increases during late S phase, reaching peak activity after completion of cellular DNA synthesis. DNA-polymerase-β activity remains constant. In a similar manner, the patterns of activity of the three DNA polymerases in synchronized BFS cells (Figs. 5A and 6A) resemble those found by Spadari and Weissbach (1974) in synchronized HeLa cells. The cyclic patterns of these DNA polymerases in synchronized cells may provide insight into their function and possible role in viral replication. Of particular interest is our finding that polymerase-a activity closely parallels that of viral DNA synthesis and production of infectious virus in BPV-infected synchronized cells (Fig. 6).

We believe that the elevated level of polymerase activity late in the cell cycle of BPV-infected cells is cellular DNA polymerase a for the following reasons: The DNA-polymerase assay procedure which we have described is highly selective. Using purified DNA polymerases from KB cells (kindly furnished by Dr. M. Q. Arens), we have determined that the contribution of polymerases β and γ to the a assay is less than 10%. The synthetic polymer poly(dT)·oligo(rA) is a template primer for only polymerase a (Spadari and Weissbach 1975). With this template, we observed the same activity and variation throughout the cell cycle as with activated calf-thymus DNA (Fig. 6A). Additionally, antibody prepared against bovine DNA polymerase a inhibited the activity 80% at the highest level of antibody used.

The specific function of each eukaryotic DNA polymerase in DNA replication remains obscure. Although one or more of the cellular DNA polymerases may play a role in parvovirus DNA replication, until now no data have been available implicating any such involvement. We have provided evidence which suggests that DNA polymerase a may be involved in viral DNA replication. However, we recognize that a temporal correlation of these events does not provide direct evidence for the role of polymerase a in the synthesis of parvoviral DNA. Further, since the levels of polymerase γ were slightly elevated in infected cells (Fig. 6A), we cannot rule out a role for this enzyme in the replication process. We are now examining DNA-polymerase activity and viral DNA synthesis in isolated nuclei to provide direct evidence for DNA-polymerase function in the synthesis of parvovirus-specific DNA.

ACKNOWLEDGMENTS

This work was supported by American Cancer Society Grant IN-117, and in part by Grant 1892740 of the College of Arts and Sciences of Virginia Polytechnic Institute and State University. We wish to thank Asim Esen for assistance with the gel electrophoresis procedures used in this study.

REFERENCES

Anderson, C. W., P. R. Baum, and R. F. Gesteland. 1973. Processing of adenovirus 2-induced proteins. *J. Virol.* **12**:241.

Bates, R. C. and J. Storz. 1973. Host cell range and growth characteristics of bovine parvoviruses. *Infect. Immun.* **7**:398.

Bates, R. C., J. Storz, and D. E. Reed. 1972. Isolation and comparison of bovine parvoviruses. *J. Infect. Dis.* **126**:531.

Chang, L. M. S. 1973. Low molecular weight DNA polymerase from calf thymus chromatin. *J. Biol. Chem.* **248**:3789.

Chrambach, A. and D. Radbard. 1971. Polyacrylamide gel electrophoresis. *Science* **172**:440.

Johnson, F. B. and M. D. Hoggan. 1973. Structural proteins of Haden virus. *Virology* **51**:129.

Laemmli, U. K. 1970. Cleavage of structural proteins during the assembly of the head of bacteriophage T4. *Nature* **227**:680.

Lavelle, G. and A. T. Li. 1977. Isolation of intracellular forms and progeny single strands of DNA from parvovirus KRV in sucrose gradients containing guanidine hydrochloride. *Virology* **76**:464.

Maizel, D. V. 1969. Acrylamide gel electrophoresis of proteins and nucleic acids. In *Fundamental techniques in virology* (ed. K. Habel and N. P. Salzman), p. 334. Academic, New York.

Parris, D. S. and R. C. Bates. 1976. Effect of bovine parvovirus replication on DNA, RNA, and protein synthesis in S phase cells. *Virology* **73**:72.

Parris, D. S., R. C. Bates, and E. R. Stout. 1975. Hydroxyurea synchronization of bovine fetal spleen cells. *Exp. Cell Res.* **96**:422

Rhode, S. L. 1973. Replication of the parvovirus H-1. I. Kinetics in a parasynchronous cell system. *J. Virol.* **11**:856.

Salzman, L. A. 1971. DNA polymerase activity associated with purified Kilham rat virus. *Nat. New Biol.* **231**:174.

Salzman, L. A. and L. E. Jori. 1970. Characterization of the Kilham rat virus. *J. Virol.* **5**:114.

Salzman, L. A. and L. McKerlie. 1975. Characterization of the DNA polymerase associated with Kilham rat virus. *J. Biol. Chem.* **250**:5583.

Salzman, L. A., W. L. White, and L. McKerlie. 1972. Growth characteristics of Kilham rat virus and its effect on cellular macromolecular synthesis. *J. Virol.* **10**:573.

Schlabach, A., B. Fridlender, A. Bolden, and A. Weissbach. 1971. DNA-dependent DNA polymerases from HeLa nuclei. II. Template and substrate utilization. *Biochem. Biophys. Res. Commun.* **44**:879.

Siegl, G. and M. Gautschi. 1973. The multiplication of parvovirus LuIII in a synchronized culture system. I. Optimum conditions for virus replication. *Arch. gesamte Virusforsch.* **40**:105.

Spadari, S. and A. Weissbach. 1974. The interrelationship between DNA synthesis and various DNA polymerase activities in synchronized HeLa cells. *J. Mol. Biol.* **86**:11.

―――. 1975. RNA-primed DNA synthesis: Specific catalysis by HeLa cell DNA polymerase *a*. *Proc. Natl. Acad. Sci.* **72**:503.

Tattersall, P., L. V. Crawford, and A. J. Shatkin. 1973. Replication of the parvovirus MVM. II. Isolation and characterization of intermediates in the replication of the viral DNA. *J. Virol.* **12**:1446.

Tattersall, P., P. J. Cawte, A. J. Shatkin, and D. C. Ward. 1976. Three structural polypeptides coded for by minute virus of mice, a parvovirus. *J. Virol.* **20**:273.

Tennant, R. W. and R. E. Hand. 1970. Requirement for cellular synthesis for Kilham rat virus replication. *Virology* **42**:1054.

Weissbach, A., D. Baltimore, F. Bollum, R. Gallo, and D. Korn. 1975. Nomenclature of eukaryotic DNA polymerases. *Eur. J. Biochem.* **59**:1.

TRANSCRIPTION, TRANSLATION, AND VIRUS MATURATION

Adeno-Associated-Virus RNA Synthesis In Vivo and In Vitro

FRANCIS T. JAY
CATHERINE A. LAUGHLIN
LUIS M. DE LA MAZA
BARRIE J. CARTER
Laboratory of Experimental Pathology
National Institute of Arthritis, Metabolism, and Digestive Diseases
National Institutes of Health
Bethesda, Maryland 20014

WILLIAM J. COOK
Laboratory of Pathology
National Cancer Institute
National Institutes of Health
Bethesda, Maryland 20014

Previous analyses of RNA transcription of the adeno-associated-virus (AAV) genome have established the general characteristics and physical map for the stable mRNA synthesized in infected cells. These results are discussed more extensively in the review by Carter in this volume. Briefly, a single, stable, 20S polysomal mRNA species is transcribed from the AAV-DNA minus strand. The map position of this mRNA extends from about 0.17 to 0.88 map units (Fig. 2). The 20S AAV RNA is poly-adenylated, at the 3' terminus (Carter 1976), and methylated, both in the 5' cap position and internally (B. Moss, A. Gershowitz, and B. J. Carter, unpublished), by post-transcriptional modification. Previous experiments provided no firm evidence for a primary transcript as a precursor to the 20S RNA. However, labeling studies did reveal, in both the nucleus and the nonpolysomal region of the cytoplasm, a heterogeneous population of 4–18S AAV RNA which was less metabolically stable than the 20S species (Carter and Rose 1974). Moreover, there was some suggestion from hybridization experiments that the nuclear and nonpolysomal cytoplasmic AAV RNA may contain sequences complementary to a slightly larger region of the DNA minus strand than did the polysomal 20S mRNA (Carter 1976). We have investigated this further using an altered procedure for isolation of AAV RNA from infected cells and found evidence which suggests that the 20S AAV mRNA is derived by posttranscriptional cleavage of a larger primary transcript.

We have begun analysis of AAV RNA transcription in several in-vitro systems to study the molecular mechanism of this process in more detail. We describe a preliminary characterization of AAV transcription complexes isolated using the Sarkosyl procedure of Gariglio and Mousset (1975). We also describe experiments which show that in-vitro transcription of AAV DNA with purified KB-cell RNA polymerase II was symmetric, in contrast with the apparent asymmetric transcription in vivo.

METHODS

Adeno-associated virus type 2 of the H strain [AAV2(H)] was grown in KB spinner cells at 37°C using a wild-type adenovirus-2 helper as before (Carter and Rose 1974) or at 39.5°C using a DNA-negative, temperature-sensitive mutant of adenovirus 5, H5ts125 (Ginsberg et al. 1974; Straus et al. 1976a).

Radioisotope labeling and purification of AAV DNA and separation of the complementary strands of bromodeoxyuridine (BUdR)-substituted DNA were as described by Carter and Khoury (1975).

Preparation of specific fragments of duplex AAV2 DNA by cleavage with restriction endonuclease *Hin*cII or *Bam*HI was as described by Carter et al. (1976).

RNA-DNA hybridization reactions were performed either on nitrocellulose filters (Carter and Rose 1974) or in solution, and were analyzed with S1 nuclease (Carter et al. 1976).

RNA was isolated from KB cells infected with AAV2 and Ad2 using a procedure, based on that of Hirt (1967), to be described in detail elsewhere (F. T. Jay, L. M. de la Maza, and B. J. Carter, manuscript in preparation). Briefly, cells were lysed in SDS, and chromatin and high-molecular-weight DNA were precipitated with 1.0 M NaCl at 0°C overnight, followed by centrifuging at 10,000 rpm in a Sorvall RC2-B centrifuge. RNA from the pellet and supernatant fractions was recovered and purified by extensive digestion with proteinase K and extraction with phenol-chloroform. After two additional cycles of digestion with DNase and extraction with phenol-chloroform the purified RNA was used for hybridization experiments.

Viral transcription complexes were isolated from cells infected at 39.5°C with AAV2 and H5ts125, essentially according to the method of Gariglio and Mousset (1975). Briefly, infected-cell nuclei, prepared by Dounce homogenization, were lysed in Sarkosyl (0.25%) and NaCl (0.1 M) and the chromatin was pelleted at 257,000g for 15 min in the Beckman SW 50 rotor. The supernatant was taken for assay of transcription activity and characterization of the AAV viral transcription complex.

DNA-dependent RNA polymerase II was assayed and purified from uninfected or adenovirus-infected KB cells according to the procedure of Schwartz and Roeder (1975).

RESULTS

Mapping of AAV RNA Synthesized In Vivo

KB cells were harvested 16 hr after infection with AAV2 and Ad2 and fractionated by the Hirt procedure as described in the "Methods" section. The RNA was then purified from both the pellet (chromatin-associated) and supernatant fractions and annealed with the separated plus or minus strands of ^{32}P-labeled AAV DNA. The proportion of ^{32}P-labeled DNA forming an S1-resistant hybrid was determined as a function of RNA concentration. As shown in Figure 1, the AAV RNA in the supernatant and pellet fractions was complementary to 81% and 91%, respectively, of the minus-strand DNA sequence. Mixing of excess amounts of pellet and supernatant RNA resulted in saturation of approximately 90% of the minus strand (data not shown). These data show that the supernatant AAV RNA sequences are a subset of those in the pellet, and that the extent of transcription on the minus strand as revealed by the reaction with the pellet RNA is significantly greater than that observed in previous studies.

Both pellet and supernatant RNA appear to show a reaction of 14–15% with the plus strand (Fig. 1 and Table 1). This is probably due to DNA-DNA annealing resulting from cross-contamination with some minus strands, since, in the absence of any RNA, 10–11% of the plus-strand preparation became nuclease-resistant. This is consistent with previous studies which did not detect any AAV RNA complementary to the plus strand (Carter et al. 1972; Carter 1976). The cross-contamination is the consequence of separating AAV strands in CsCl, a method which gives very little contamination of minus with plus strands but significant (5–10%) contamination of plus with minus strands.

The genome location of the AAV RNA sequences present in the pellet or supernatant fractions was mapped by annealing the RNA with denatured DNA fragments obtained by cleavage of AAV ^{32}P-labeled-DNA duplexes with restriction endonuclease *Bam*HI or *Hin*cII. The fragments

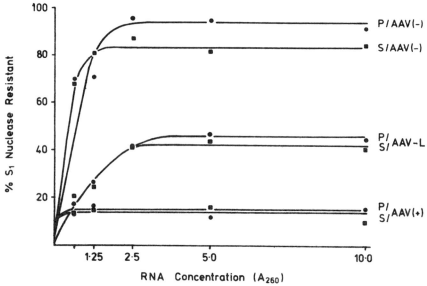

FIGURE 1
Annealing of AAV DNA with RNA from infected cells. Denatured, ^{32}P-labeled
AAV DNA was annealed with RNA isolated from the pellet or supernatant
fraction of KB cells 20 hr after infection with AAV2 and Ad2. The proportion of
DNA resistant to S1 nuclease was then determined (Carter et al. 1976). The DNA
preparations used were denatured intact AAV duplex DNA (AAV-L), purified
minus-strand [AAV(−)], or purified plus-strand [AAV(+)]. RNA preparations
were pellet (●) or supernatant (■) fractions as indicated.

used have been mapped previously (Berns et al. 1975; Carter et al. 1976) as
shown in Figure 2. The annealing reactions were performed under conditions in which little or no DNA-DNA annealing occurred (usually less
than 1–4%) so that the proportion of DNA which became resistant to
nuclease S1 was a measure of RNA-DNA hybrid formation. If there is no
RNA annealing to the plus strand, then for a DNA fragment completely
within the transcribed region the maximum protection against nuclease
digestion is 50%. Annealing experiments for each DNA fragment were
performed with increasing concentrations of RNA, and the plateau levels
of hybridization were determined as for the experiment in Figure 1.
   The *Hinc* C fragment (Table 1) showed approximately 50% annealing
with either pellet or supernatant RNA, and therefore this region of the
genome (0.31–0.51 map units) is completely within the transcribed
region. *Bam* B, *Bam* A, and *Hinc* D (Table 1) all showed less than 50%
annealing with either pellet or supernatant RNA. Furthermore, for each
of these three fragments the extent of annealing with the supernatant
RNA was less than that obtained with the pellet RNA. The poly(A)-

TABLE 1
Mapping of AAV RNA

[32]P-labeled AAV DNA	RNA	Proportion of DNA resistant to S1 nuclease (%)	RNA location on genome map start	stop
Intact	pellet	46.1		
	supernatant	42.0		
Plus strand	pellet	15.0[a]		
	supernatant	14.0		
Minus strand	pellet	91.0[b]		
	supernatant	81.0		
Bam B	pellet	37.1	0.06	
	supernatant	30.5	0.09	
	supernatant (pA+)	18.8	0.14	
Bam A	pellet	46.8		0.96
	supernatant	43.1		0.89
	supernatant (pA+)	39.5		0.84
Hinc D	pellet	34.1		0.96
	supernatant	12.8		0.90
	supernatant (pA+)	39.5		0.92
Hinc C	pellet	51.4		
	supernatant	49.7		

[a,b] Hybridization reactions contained DNA and RNA as indicated and were incubated and subsequently digested with S1 nuclease as described by Carter et al. (1976). Before annealing, the mixture was denatured by heating at 100°C for 5 min. The "zero-time" values for control reactions that were denatured but not subsequently incubated were not more than 1–2% and have been subtracted. For reactions with intact DNA or restriction fragments the annealing reaction was for 2 hr. In the absence of any added RNA there was no additional annealing of DNA above the zero-time value. For reactions with purified plus or minus strands annealing was for 48 hr. In the absence of added RNA, 11% of the plus-strand preparation (a) and 4.5% of the minus-strand preparation (b) became S1-resistant. This is taken to represent cross-contamination and has not been subtracted.

containing molecules from the supernatant RNA were fractionated by chromatography on poly(U)-Sepharose and annealed also with restriction fragments *Bam* B, *Bam* A, and *Hinc* D. For fragments *Bam* A and *Hinc* D, at the right genome terminus, this poly(A) RNA (Table 1) yielded a result similar to the bulk supernatant RNA. At the left end of the genome, the reaction of the poly(A) RNA with *Bam* B was significantly less than that of the supernatant RNA.

The simplest transcription map which can be drawn from the data in Table 1 is shown in Figure 2. On this map, the poly(A) RNA extends from approximately 0.14 to 0.88 map units, which corresponds to the previous estimates of approximately 0.17 to 0.88 using RNA isolated by the SDS-hot phenol method (Carter et al. 1976). The total supernatant map

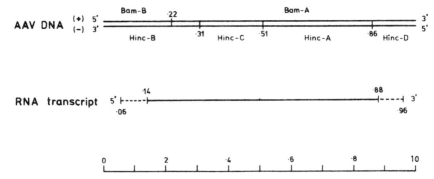

FIGURE 2

AAV transcription map. The indicated restriction fragments have been mapped (Berns et al. 1975; Carter et al. 1976). The region of the AAV genome contained in supernatant, poly(A+) RNA is represented by the solid line. The presumptive map of the additional AAV RNA precursor sequences present in the pellet RNA is indicated by the broken line.

extends from approximately 0.09 to 0.89 map units. This coincides with the poly(A) RNA map at the 3' terminus but is 0.05 map units longer at the 5' end. The pellet RNA maps from 0.06 to 0.96 map units, which is significantly longer at both termini than either supernatant or poly(A)-containing RNA.

The sizes of the transcribed regions for pellet and supernatant RNA estimated by summing the fragment data agree well with the values from separated plus or minus strands or denatured duplex molecules, which indicates the internal consistency of the data (Table 1). These results imply, but do not provide absolute proof, that the pellet RNA represents a precursor of the poly(A)-containing AAV mRNA which is cleaved at both the 3' and 5' termini.

Isolation of AAV Transcription Complexes

The above results suggest that AAV transcription occurs in a structure that precipitates in high salt. As a first step towards determining the actual template for AAV transcription we have isolated viral transcription complexes (VTC) using the Sarkosyl procedure of Gariglio and Mousset (1975). For these experiments, cells were infected at the nonpermissive temperature with Ad5ts125 and AAV2. Nuclei prepared from infected cells labeled for a prolonged period with [³H]thymidine were lysed with Sarkosyl and the supernatant was sedimented in neutral sucrose. As Figure 3A shows, in the preparations from ts125-infected cells there was little or no labeled DNA present, whereas in those from cells coinfected

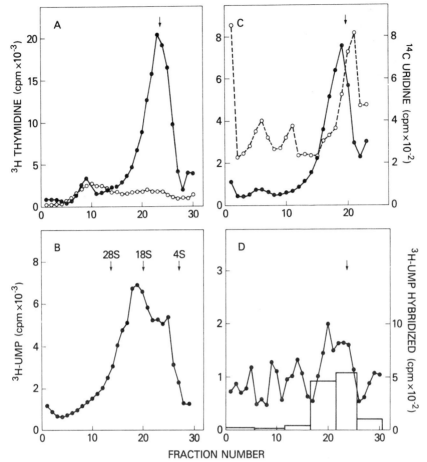

FIGURE 3

Sucrose-gradient analysis of viral transcription complexes. All gradients contained 5–20% sucrose and were centrifuged for 4 hr at 39,000g and 5°C in an SW 41 rotor. (A) Infected KB cells were labeled with [³H]thymidine 6–20 hr after infection. Sarkosyl supernatants were prepared at 20 hr, and centrifuged. The position of ³²P-labeled AAV DNA duplex is shown by the arrow. ●: Ad5ts125 + AAV2-infected cells; ○: Ad5ts125-infected cells. (B) The Sarkosyl supernatant from infected cells was assayed for endogenous RNA-polymerase activity (Table 1). The reaction mixture was then centrifuged in a 5–20% gradient. Arrows indicate the position of ³²P-labeled ribosomal RNA. (C) Infected cells were labeled with [³H]thymidine and [¹⁴C]uridine from 19.5 to 20 hr after infection. Sarkosyl supernatants were prepared at 20 hr and sedimented in a 5–20% gradient. The arrow marks the position of the ³²P-labeled AAV DNA duplex. ○: ¹⁴C radioactivity; ●: [³H] radioactivity. (D) The Sarkosyl supernatant was prepared from infected cells at 20 hr after infection and sedimented in a sucrose gradient. Gradient fractions were then assayed for RNA polymerase activity (●). The RNA synthesized in vitro was isolated from the pooled fractions and annealed to AAV DNA immobilized on nitrocellulose filters as indicated by the histogram. The amount of RNA that annealed to Ad DNA was negligible.

with AAV2 and ts125 there was a large peak of AAV DNA sedimenting at 14–15S.

The supernatant fraction of the Sarkosyl lysates from uninfected or ts125-infected cells showed little or no transcription activity, but there was a great stimulation of activity if cells were also infected with AAV (Table 2). Hybridization assays (data not shown) indicate that most of this transcription activity represents AAV RNA synthesis and that at least 85–90% of the AAV transcription activity was present in the Sarkosyl supernatant. The AAV transcription activity in the pelleted chromatin has not been characterized further. The characteristics of the AAV VTC are summarized in Table 2. RNA synthesis was inhibited by actinomycin D and also α-amanitin at 1–2 μg/ml. This indicates that AAV RNA is probably synthesized by an RNA polymerase II, as are other eukaryotic mRNAs. Heparin showed no inhibition of RNA synthesis, which implies that extensive reinitiation did not occur. Incorporation of ^{3}H-labeled UTP by the VTC was linear for 20 min and continued for at least 1 hr.

The physical characteristics of the AAV VTC were analyzed in sedimentation experiments as shown in Figure 3. When cells were labeled for

TABLE 2
Properties of the Viral Transcription Complex

		UMP incorporated	
Exp.	Incubation conditions	pmoles	percent of control
1	ts125-infected cells	0.09	2.8
	ts125 + AAV2-infected cells	3.26	100
2	uninfected KB cells	0.07	3.0
	ts125 + AAV2-infected cells	2.44	100
	omit ATP, CTP, GTP	0	0
	omit MnCl$_2$	0	0
	+ 25 U heparin	3.48	143
3	ts125 + AAV2-infected cells	12.4	100
	+ α-amanitin (1 μg/ml)	0	0
	+ actinomycin D (5 μg/ml)	0.74	5.9

Sarkosyl supernatants were prepared as described and assayed for endogenous RNA-polymerase activity. Each reaction mixture (250 μl) contained 11.5 μmoles ammonium sulfate; 7 μmoles KCl; 1.5 μmoles NaF; 0.5 μmole dithiothreitol; 0.5 μmole MnCl$_2$; 41 μmoles NaCl; 0.1 μmole each of ATP, GTP, and CTP; 0.001 μmole UTP; 19 μmoles Tris (pH 7.9); 3.34 μmoles ^{3}H-labeled UTP; and 0.21% Sarkosyl. Reactions were incubated for 60 min at 23°C and terminated by dilution in sterile water and addition of an equal volume of 25% TCA containing 25 mM sodium pyrophosphate. The precipitates were collected on Millipore filters, dried, and counted.

30 min with [^{14}C]uridine or [^{3}H]thymidine immediately prior to preparation of VTC (Fig. 3C) most of the labeled DNA sedimented in a broad
peak at 16–20S, which is characteristic of AAV replicative intermediates
(Straus et al. 1976b). The RNA was heterogeneous, with about half
sedimenting more slowly than 15S and the remainder sedimenting much
faster. There was a small shoulder of labeled RNA at about 20S.

When the VTC were prepared from unlabeled cells and then incubated
in vitro with ^{3}H-labeled UTP prior to sedimentation (Fig. 3B), most of the
^{3}H-labeled complex sedimented in a broad zone with a peak at 20S and a
shoulder at 8–12S. When the VTC were first sedimented through the
neutral sucrose gradient and then individual fractions were assayed for
incorporation of ^{3}H-labeled UTP, a similar radioactivity profile was
observed (Fig. 3D). There was a major peak at 20S and a lesser amount of
8–12S material. There was, in addition, some faster-sedimenting material
analogous to that seen when the RNA was labeled in vivo (Fig. 3C).
However, as shown by hybridization (Fig. 3D), this faster-sedimenting
species is not AAV RNA, whereas most of the RNA in the 10–20S region is
AAV-specific.

These experiments demonstrate that more than 85% of the AAV transcription activity can be solubilized as a nucleoprotein complex [as shown
by filter binding experiments (C. A. Laughlin, unpublished)] which sediments with a major peak at 20S and coincides with the leading edge of
replicating molecules. In addition, there is a slower-sedimenting component which may represent VTC in which the templates are AAV
incomplete molecules (see de la Maza et al., this volume).

Transcription of AAV2 DNA with Purified RNA Polymerase II

It has been shown that in vitro the *E. coli* RNA polymerase transcribes
AAV DNA symmetrically (Rose and Koczot 1971), in contrast with the
asymmetric in vivo transcription. Since AAV RNA appears to be synthesized by the eukaryotic RNA polymerase II, we studied in-vitro transcription using this enzyme and purified AAV DNA. The RNA polymerase II isolated from either uninfected KB cells or Ad2-infected KB cells
yielded similar results. As shown in Table 3, denatured AAV DNA was
about twice as efficient as a template as was native AAV DNA. Separated
plus or minus strands of AAV DNA were equally efficient templates,
which indicated a lack of strand-specific transcription (Table 3). This was
confirmed by additional experiments (data not shown) which indicated
that the RNA synthesized on either denatured or native AAV DNA
template was at least 30% resistant to digestion with ribonucleases A and
T1 in high salt and also hybridized equally well to either plus or minus
strands of AAV DNA. Also, the RNA products synthesized with either
template were heterogeneous in size, ranging from about 6S to 26S,

TABLE 3
In Vitro Transcription of AAV DNA by RNA Polymerase II

	pMoles ³H-labeled UMP incorporated with enzyme from	
Template (AAV DNA)	uninfected cells	infected cells
Minus strand	1.34	2.87
Plus strand	1.84	2.95
Denatured	1.44	2.76
Native	0.60	1.27

Enzyme reactions (in a final volume of 50 μl) contained 0.3 μmole each of ATP, GTP, and CTP; 0.003 μmole UTP; 4 μCi ³H-labeled UTP; 0.1 μmole MnCl$_2$; 2.5 μmoles Tris-HCl (pH 7.9); 4 μmoles (NH$_4$)$_2$SO$_4$; and 50 ng of purified plus-strand or minus-strand DNA or 100 ng of native or denatured DNA. Reactions with uninfected KB-cell enzyme contained 1.7 units of RNA polymerase II (specific activity = 450 units/μg protein), and reactions with the Ad2-infected KB-cell enzyme contained 3.2 units (specific activity = 1780 units/μg protein).

which is equivalent to nearly a full-length transcript. These results suggest that the purified RNA polymerase II initiates at random on either strand of AAV DNA. Thus, the apparent strand-specific AAV transcription observed in vivo required other factors in addition to RNA polymerase II. It is possible, though less likely, that the apparent strand-specific transcription in vivo results from post-transcriptional control by rapid degradation of the plus-strand transcript.

DISCUSSION

The mapping experiments reported here using AAV RNA from cells fractionated by the Hirt procedure indicate that there may be post-transcriptional cleavage of AAV RNA. If the RNA in the Hirt pellet is indeed a precursor of the AAV 20S mRNA, then post-transcriptional processing involves cleavage of at least 320–400 nucleotides at the 5' end and 160–200 at the 3' end. This model, involving post-transcriptional cleavage of AAV RNA at both the 5' and 3' termini, is of interest in view of current concepts about cellular mRNA processing (Perry 1976). It was originally argued that mRNA arose from the 3' end of hnRNA because poly(A) tails were found in high-molecular-weight hnRNA. More recently it has been suggested that mRNA arises from the 5' end because of the presence of methyl caps in the hnRNA. It is, of course, likely that post-transcriptional cleavage schemes may be different in detail for different mRNAs. Thus, specific cleavage patterns must be determined for individual messages.

The additional "precursor" AAV RNA sequences were not observed before (Carter et al. 1976). This RNA may have been lost into phenol or interface regions during extraction because of attachment to template, chromatin, or other proteins. This fraction was not specifically enriched in the hot-phenol isolation procedure. Alternatively, processing of these sequences may be very rapid. We have no kinetic measurements as yet to determine the relative concentration of these "precursor" sequences and of stable AAV mRNA.

The current work does not rigorously identify a promoter for AAV transcription, but it is presumably at or to the left of 0.06 map unit. This is significantly to the left of the previous estimate of 0.17–0.18, which was based on mapping of stable RNA (Carter et al. 1976). It is not clear whether the mapping experiments reported here have detected the left extremity of AAV transcription. For instance, if processing begins before transcription is completed, the terminal region of the nascent transcript may be present at only very low concentration. Thus, precise location of a putative promoter requires detection and mapping of a transcript containing a 5' terminal di- or triphosphate. Similarly, it is not clear whether the transcription terminus is actually at 0.96 map unit or futher downstream to the right. In any case, since the inverted terminal repetition in AAV DNA is about 100 nucleotides and this region can form a hairpin (see Berns et al., this volume), the transcription promoter and the terminator may each be within 100 nucleotides of this hairpin structure. It is therefore tempting to speculate that these sequences might have some influence in transcription control as well as replication. It may even be possible that these regions act in a fashion analogous to bacterial insertion sequences in which transcription initiation and termination are controlled by the directional polarity of the inverted sequence.

It is not yet clear why the "nascent" or "precursor" AAV RNA sequences are found in the Hirt pellet. Several explanations can be suggested:

(1) The transcription template may be integrated into the host-cell genome. It appears that AAV DNA can integrate into human-cell DNA (see Handa et al., this volume). However, isolation of a soluble viral transcription complex by Sarkosyl lysis suggests that not all the transcription templates can be integrated. There is, however, some AAV transcription activity which is not readily solubilized by this procedure.

(2) The transcription template may be a high-molecular-weight concatemer of AAV DNA. Although it is reported that AAV replicative intermediates (RI) include concatamers up to four genomes in length (Handa et al. 1976; Straus et al. 1976b), they were released into the Hirt supernatant when a proteolytic digestion was performed on the cell lysate prior to the Hirt precipitation. Also, Handa and Shimojo (1977) found most of the pulse-labeled AAV DNA RI molecules in the Hirt

supernatant, even in the absence of prior proteolytic digestion. This suggests that the AAV transcription templates are a different population than most of the AAV DNA RI molecules. Martin et al. (1976) have reported that oligomeric forms of SV40 DNA appeared in the Hirt pellet. Also, LuIII replicative-form DNA is reported to appear predominantly in the Hirt pellet, although it is not clear whether this is due entirely to its size or also to its close association with cellular chromatin (Siegl and Gautschi 1976; Gautschi et al. 1976).

(3) The viral transcription template or the RNA-processing apparatus may be closely associated with host-cell chromatin. It is possible that the AAV transcription template is closely associated with the cellular chromatin, as noted above for LuIII RF DNA. Alternatively, the nascent RNA may be chromatin-associated, perhaps because of the location of the RNA-processing apparatus. For instance, in the RNA isolation procedure described here, the 32S ribosomal RNA precursor is located in the Hirt pellet (F. T. Jay, L. M. de la Maza, and B. J. Carter, manuscript in preparation), and this is known to be not a nascent transcript but an intermediate stage in the processing of rRNA.

Further characterization of the isolated VTC may provide information on the nature of the AAV transcription template. It is not yet clear how the AAV RNA transcripts in the soluble VTC are related to those in the Hirt pellet, but experiments to determine this are now in progress.

In summary: The experiments reported here point to a close association of AAV transcription activity with the host-cell chromatin. This may provide a rationale for lack of strand specificity in in-vitro transcription with purified RNA polymerase II. Thus, AAV, as well as nondefective parvoviruses, may be a useful probe for analyzing the arrangement and function of cellular chromatin.

REFERENCES

Berns, K. I., J. Kort, K. J. Fife, W. Grogan, and I. Spear. 1975. Study of the fine structure of adeno-associated virus DNA with bacterial restriction endonucleases. *J. Virol.* **16:**712.

Carter, B. J. 1976. Intracellular distribution and polyadenylate content of adeno-associated virus RNA sequences. *Virology* **73:**273.

Carter, B. J. and G. Khoury. 1975. Specific cleavage of adenovirus-associated virus DNA by restriction endonuclease R.*Eco*R1—Characterization of cleavage products. *Virology* **63:**523.

Carter, B. J. and J. A. Rose. 1974. Transcription in vivo of a defective parvovirus: Sedimentation and electrophoretic analysis of RNA synthesized by adenovirus-associated virus and its helper adenovirus. *Virology* **61:**182.

Carter, B. J., G. Khoury, and J. A. Rose. 1972. Adenovirus-associated virus multiplication. Extent of transcription of the viral genome in vivo. *J. Virol.* **10:**1118.

Carter, B. J., K. H. Fife, L. M. de la Maza, and K. I. Berns. 1976. Genome localization of adeno-associated virus RNA. *J. Virol.* **19**:1044.

Gariglio, P. and S. Mousset. 1975. Isolation and partial characterization of a nuclear RNA polymerase–SV40 DNA complex. *FEBS Lett.* **56**:149.

Gautschi, M., G. Siegl, and G. Kronauer. 1976. Multiplication of parvovirus LuIII in a synchronized culture system. IV. Association of viral structural polypeptides with the host cell chromatin. *J. Virol.* **20**:29.

Ginsberg, H. S., M. J. Ensinger, R. S. Kaufman, A. J. Mayer, and U. Lindholm. 1974. Cell transformation: A study of regulation with type 5 and 12 adenovirus temperature-sensitive mutants. *Cold Spring Harbor Symp. Quant. Biol.* **39**:419.

Handa, H. and H. Shimojo. 1977. Viral DNA synthesis in vitro with nuclei isolated from adenovirus-associated virus type 1 infected cells. *Virology* **77**:424.

Handa, H., H. Shimojo, and K. Yamaguchi. 1976. Multiplication of adeno-associated virus type 1 in cells coinfected with a temperature-sensitive mutant of human adenovirus type 31. *Virology* **74**:1.

Hirt, B. 1967. Selective extraction of polyoma DNA from infected mouse cell cultures. *J. Mol. Biol.* **26**:385.

Martin, M. A., P. M. Howley, J. C. Byrne, and C. F. Garon. 1976. Characterization of supercoiled oligomeric SV40 DNA molecules in productively infected cells. *Virology* **71**:28.

Perry, R. P. 1976. Processing of RNA. *Annu. Rev. Biochem.* **45**:605.

Rose, J. A. and F. J. Koczot. 1971. Adenovirus-associated virus multiplication. Base composition of the deoxyribonucleic acid strand species and strand specific in vitro transcription. *J. Virol.* **8**:771.

Schwartz, L. B. and R. G. Roeder. 1975. Purification and subunit structure of deoxyribonucleic acid-dependent ribonucleic acid polymerase II from the mouse plasmocytoma MOPC 315. *J. Biol. Chem.* **250**:3221.

Siegl, G. and M. Gautschi. 1976. Multiplication of parvovirus LuIII in a synchronized culture system. III. Replication of viral DNA. *J. Virol.* **17**:841.

Straus, S. E., H. S. Ginsberg, and J. A. Rose. 1976a. DNA-minus temperature-sensitive mutants of adenovirus type 5 help adenovirus-associated virus replication. *J. Virol.* **17**:140.

Straus, S. E., E. D. Sebring, and J. A. Rose. 1976b. Concatemers of alternating plus and minus strands are intermediates in adenovirus-associated virus DNA synthesis. *Proc. Natl. Acad. Sci.* **73**:742.

Characterization of Adeno-Associated-Virus Polypeptides Synthesized In Vivo and In Vitro

ROBERT M. L. BULLER
JAMES A. ROSE

Laboratory of Biology of Viruses
National Institute of Allergy and Infectious Diseases
National Institutes of Health
Bethesda, Maryland 20014

AAV virions have been shown to be assembled from three polypeptides whose molecular weights were previously estimated to be 87,000, 73,000, and 62,000 (Rose et al. 1971). At present, little is known concerning the actual means by which the AAV mRNA, whose theoretical coding capacity does not exceed 95,000 daltons, is translated into capsid polypeptides totaling 222,000 daltons (Rose et al. 1971). This discrepancy could be understood in at least two ways: (1) virus-induced host polypeptides might be incorporated into the virion, as in the case of papovaviruses (Frearson and Crawford 1972; Lake et al. 1973), or (2) the three structural polypeptides may be coded for by overlapping sequences in the viral message, and thus may share extensive regions of identical amino acid sequence. The objective of this study, therefore, was to characterize AAV-specific polypeptides, both in vivo and in vitro, and to determine a probable mechanism for their synthesis.

MATERIALS AND METHODS

Cells and Virus

KB cells (3×10^5 cells/ml) in suspension culture were infected with adenovirus type 5 alone (denoted Ad) or with Ad5 and AAV2 simultaneously (denoted Ad/AAV) at multiplicities of 10 $TCID_{50}$ units per cell each. Confluent monolayers of African green monkey kidney (AGMK) cells in 25-cm^2 flasks were preinfected for 24 hr with SV40 (10 plaque-forming units per cell) prior to infection with Ad5 and AAV2 (10 $TCID_{50}$ units each per cell).

Isotopic Labeling of Intracellular Protein

Infected cultures (1.5 × 10^7 cells) suspended in modified Eagle's medium (MEM) containing reduced concentrations of either L-methionine or total amino acids were marked with ^{35}S-labeled L-methionine (>300 Ci/mmole) or U-^{14}C-labeled protein hydrolysate (54 mCi/matom of C), respectively. Labeling periods were terminated either by a chase in MEM supplemented with 1000 times the normal concentration of the labeled amino acid(s) or by harvesting into Tris-saline buffer as will be described elsewhere (R. M. L. Buller and J. A. Rose, manuscript in preparation).

For the production of virions uniformly labeled in proteins, Ad/AAV-infected KB cells (1.5 × 10^7 cells) were incubated at 18 hr post-infection (p.i.) for 4 hr with 10 μCi/ml of U-^{14}C-labeled protein hydrolysate in MEM reduced tenfold in total amino acid concentration. At 22 hr p.i. the concentration of deficient amino acids was normalized, and the incubation was continued until 30 hr p.i., at which time virions were purified from infected cells as described previously but with the trypsin treatment omitted (Rose et al. 1966, 1969).

Polyacrylamide Gel Electrophoresis (PAGE)

Total cell polypeptides were analyzed by electrophoresis through a 6–17% linear gradient of total acrylamide (5% in N,N'-methylene-bisacrylamide) employing a discontinuous buffer system (Laemmli 1970; Studier 1973; Marsden et al. 1976). Virion polypeptide molecular weights were calculated from coelectrophoresis with the following protein standards: L-glutamate dehydrogenase (53,000 daltons), hemiglobin (a chain = 14,000 and β chain = 14,600 daltons), catalase (60,000 daltons), carbonic anhydrase (29,000 daltons), and RNA polymerase (β' = 157,500, β = 150,000, σ = 90,000, and a = 40,000 daltons).

Purification of AAV mRNA

Phenol-chloroform-extracted total cytoplasmic RNA (4.3 mg/ml) was hybridized at 68°C for 8 min with 85 μg of fragmented AAV DNA in 0.1% SDS, 0.18 M $NaPO_4$ buffer (pH 6.8) (Lewis et al. 1975). AAV-specific RNA-DNA hybrids were enriched by urea-hydroxyapatite chromatography as described by Lewis et al. (1975). The hybrids were denatured, and the mRNA was separated from DNA fragments by sedimentation in 0–15% wt/wt DMSO-sucrose gradients (Moss and Koczot 1975).

In Vitro Protein Synthesis

Each reaction mixture contained 40% vol/vol cell-free S-30 from KB cells prepared as described by McDowell et al. (1972), 16 mM HEPES (pH 7.5), 60 mM KCl, 1 mM Mg-acetate, 2.4 mM dithiothreitol, 1.4 mM ATP, 0.28 mM GTP, 8 mM creatine phosphate, 0.4 mg/ml creatine kinase, 0.6 mM spermidine, 300 μg/ml crude initiation factors from rabbit reticulocytes (Schreier and Staehelin 1973), 0.12 mM (each) of 20 unlabeled amino acids (L-methionine omitted), 245 μCi/ml ^{35}S-labeled L-methionine (>300 Ci/mmole), and approximately 0.05–0.1 μg hybridization-selected RNA. The reaction mixture was incubated at 30°C for 30 min, stopped by the addition of 2% SDS, 5% 2-mercaptoethanol, 10% glycerol, and 0.002% bromophenol blue, and then held at 100°C for 5 min. Cold trichloroacetic acid (TCA) precipitable counts were assayed and the mixtures were then stored above liquid nitrogen until electrophoresis.

RESULTS

AAV-Specified Polypeptides in KB Cells

Although other investigators have identified at least three AAV structural polypeptides (Rose et al. 1971; Johnson et al. 1971), the possibility that one or more of these polypeptides was specified by the KB-cell genome was not ruled out. Three criteria can be considered when determining whether a polypeptide is virus-coded: (1) the appearance of a polypeptide of unique electrophoretic mobility in an infected-cell lysate, (2) the presence of the same polypeptide after infection of different cell lines, and (3) virus-specific mRNA-directed synthesis of the polypeptide in vitro.

To detect AAV-induced polypeptides, an Ad/AAV-infected KB culture and an Ad-infected KB culture were pulsed with 50 μCi/ml of U-^{14}C-labeled protein hydrolysate for 90 min at 25.5 and 16 hr p.i., respectively. Track *b* of Figure 1 reveals five bands (A, B, C, D, and E) which were specific to the Ad/AAV-infected culture. Of the five AAV-induced polypeptides, only A, B, and C were found in virions (Fig. 1, track *a*). Molecular-weight estimates of these polypeptides are summarized in Table 1.

Although the five AAV-specific polypeptides observed in a dual infection of KB cells had unique electrophoretic mobilities, it could be argued that one or more of these polypeptides was simply a host polypeptide induced by virus infection. To test this possibility further, AGMK cells

FIGURE 1
AAV-induced polypeptides in KB cells. Total cell protein from
Ad/AAV (track *b*) and Ad-infected KB cells (track *c*) analyzed by
PAGE. Track *a*: polypeptides from purified virions.

were used as an alternative host for the virus. Ad/AAV infection of
primary AGMK cells is abortive for both AAV and Ad5 unless cells are
additionally infected with SV40 (Rabson et al. 1964; Blacklow et al. 1967;
Straus et al. 1977). Therefore, AGMK cells were preinfected for 24 hr with
SV40 before the addition of Ad/AAV or Ad alone. At 52 hr after SV40
infection, infected cells were marked with 25 μCi/ml of ^{35}S-labeled
L-methionine for 2 hr. As illustrated in Figure 2 (track *a*, squares), AAV-

TABLE 1
Estimated Molecular Weights of AAV2-Induced Polypeptides

Gel component[a]	Molecular weight × 10⁻³ daltons	No. of independent analyses
A	90.7 ± 2.3[b]	11
B	71.6 ± 1.6	12
C	60.0 ± 2.0	12
D	24.9 ± 1.7	8
E	15.8 ± 0.9	7

[a] 263,000 Daltons of polypeptides are specified by AAV2 in KB-3 cells.
[b] Maximum coding capacity of the AAV mRNA with a molecular weight of
0.9×10^6 daltons is 95,000 daltons of protein [as calculated from average
nucleotide molecular weight (322), average amino acid molecular weight
(110), and 3' terminal poly(A) sequence (200) (Carter and Rose 1974; Carter
1976)].

FIGURE 2
AAV-induced polypeptides in AGMK cells. Ad/AAV/SV40-infected AGMK cells (track *a*) and Ad/SV40-infected AGMK cells (track *b*) analyzed by PAGE. Track *c*: polypeptides from purified virions.

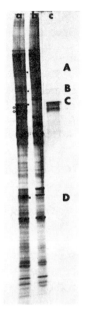

specific polypeptides A, B, C, and D were also detected in this second cell line. The apparent absence of polypeptide E is due to its deficiency in methionine residues (R. M. L. Buller and J. A. Rose, manuscript in preparation). We have occasionally observed some degradation of structural polypeptides upon repeated freezing and thawing of purified virion preparations that were stored above liquid nitrogen for several weeks. Interestingly, other polypeptides (track *a*, circles) were detected in Ad/AAV-infected cells which comigrated with such breakdown products from authentic virion polypeptides.

The third criterion for determining the genetic origin of AAV-induced polypeptides is whether or not they can be produced by translation of AAV-specific RNA in vitro. AAV-specific RNA from Ad/AAV-infected KB cells was selected by liquid hybridization to fragmented AAV DNA, enriched by hydroxyapatite chromatography, and sedimented through a DMSO-sucrose gradient. The specificity of RNA from Ad/AAV-infected and Ad-infected KB cells selected in this manner was examined further by filter hybridization to AAV and Ad DNA (Table 2). It should be noted that when hybridization-selected AAV RNA is released through denaturation and then rehybridized to fresh filter-bound AAV DNA, no more than 67% of the input counts could be reannealed (Rose and Koczot 1971). As shown in Figure 3, purified AAV mRNA directed the in vitro synthesis of polypeptides A, B, C, and D (track *c*, squares), as well as other novel nonstructural polypeptides whose molecular weights were 75,000, 61,400, 44,500, 30,200, 22,400, and 13,200 (track *c*, circles). Although

TABLE 2
Specificity of Hybridization-Selected RNA

Source of total cytoplasmic RNA	[³H]uridine-labeled RNA bound to AAV2 DNA[a] (%)	[³H]uridine-labeled RNA bound to Ad5 DNA[a] (%)
Ad5/AAV2-infected KB cells	51.2	2.3
Ad5-infected KB cells	1.8	10.4

[a] Results based on average of two filters, each of which contained 5 μg of DNA. The hybridization procedure was essentially that of Rose and Koczot (1971).

polypeptide E is present, it remained undetected in this experiment because the in vitro system employed ³⁵S-labeled L-methionine as the radiolabeled amino acid (R. M. L. Buller, unpublished results). Even though all of the L-[³⁵S]methionine-labeled AAV-specified polypeptides detected in vivo were also synthesized in vitro, the relative quantities of each made in vitro were different from the in vivo situation (compare Fig.

FIGURE 3
In vitro protein synthesis programmed with purified AAV-specified mRNA. Polypeptide synthesis utilizing KB-cell extracts and mRNA selected by hybridization to AAV DNA, from an Ad-infected culture (track *b*) or an Ad/AAV-infected culture (track *c*) analyzed on 6–17% polyacrylamide gels. Virion polypeptides are shown in track *a*.

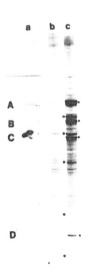

1, track *b* and Fig. 3, track *c*). Because polypeptides A, B, C, D, and E are synthesized in both human and monkey cells, and because they are produced in an in vitro protein-synthesizing system programmed with purified AAV mRNA, we conclude that these five polypeptides are all virus-coded.

Mode of Synthesis of AAV-Specified Polypeptides

As already noted, the sum of the individual molecular weights of the virus-specified polypeptides exceeds by a factor of 2.7 the theoretical coding capacity of the single virus message (Table 1; Carter and Rose 1974). Although the synthesis of overlapping polypeptides from a single mRNA could, in theory, be accomplished by control of initiation or termination or by proteolytic processing of the nascent or completed polypeptides (or some combination of these processes), only the proteolytic mechanism has been observed in virus-infected eukaryotic cells (Jacobson and Baltimore 1968). In an attempt to elucidate the mechanism of synthesis of the AAV-induced polypeptides, two kinds of experiments were carried out. In the first type of experimental approach, a short pulse of radioactive amino acids followed by a chase with excess nonradioactive amino acids was utilized to define whether the generation of AAV-induced polypeptides was directly coupled to protein synthesis or resulted from post-translational cleavage of a primary product (Schlesinger and Schlesinger 1972; Anderson et al. 1973). In the second type of experiment, amino acid analogs, which may modify the folding configuration of the substituted polypeptide, were employed to block presumptive cleavage of polypeptides (Jacobson and Baltimore 1968). Interference with production of one or more AAV polypeptides would be evidence for a proteolytic processing mechanism.

To detect rapid post-translational processing an Ad/AAV-infected culture was harvested at 18.5 hr p.i. and concentrated 50-fold in amino-acid-free medium. After incubation for 5 min, U-^{14}C-labeled protein hydrolysate (200 μCi/ml) was added and the incubation was continued for 10 min. Labeling was then terminated by addition of chase medium as described in "Materials and Methods," and samples were withdrawn for analysis by polyacrylamide gel electrophoresis (PAGE) at 0, $\frac{1}{4}$, $\frac{1}{2}$, 1, and 2 hr. A control Ad-infected culture was labeled between 18 and 19$\frac{1}{2}$ hr p.i. with 50 μCi/ml of U-^{14}C-labeled protein hydrolysate. As indicated in Table 3, the net incorporation of radioactive amino acids into TCA-precipitable radioactivity ceased within the first 15 min of the chase. Inspection of Figure 4 reveals no obvious precursor-product relationship among the known AAV-specified polypeptides. Additional experiments employing pulse times as short as 1 min also failed to reveal a redistribution of radioactivity among the virus-specified polypeptides during

TABLE 3

Incorporation of U-¹⁴C-Labeled Protein Hydrolysate into TCA-Precipitable Radioactivity in the Presence of Unlabeled Amino Acids

Chase (hr)	Cpm per 5 μl[a]	Percentage of zero-time chase counts[b]
0	128,170	100
$\frac{1}{4}$	91,184	71
$\frac{1}{2}$	78,477	61
1	81,163	63
2	96,871	76

[a] TCA-precipitable radioactivity.
[b] Chase media contained 1000 times the normal concentration of unlabeled amino acids.

a chase, nor did any polypeptide larger than polypeptide A appear (R. M. L. Buller and J. A. Rose, manuscript in preparation). This finding argues strongly that proteolytic processing, if it occurs, operates at the level of the nascent polypeptide as opposed to cleavage of a mature precursor polypeptide.

FIGURE 4

The stability of pulse-labeled AAV-induced polypeptides in KB cells. PAGE analyses of pulse-labeled Ad/AAV-infected KB cells chased for 0 (track *b*), $\frac{1}{4}$ (track *c*), $\frac{1}{2}$ (track *d*), 1 (track *e*), and 2 hr (track *f*). Pulse-labeled proteins (no chase) from Ad-infected KB cells are analyzed in track *g*. Virion polypeptides are shown in track *a*.

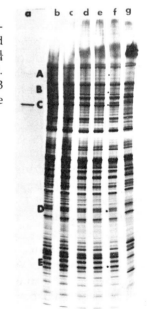

In a further attempt to detect an AAV-specified polypeptide larger than 90,200 daltons and a concomitant reduction of AAV-induced structural and nonstructural polypeptides, several amino acid analogs were tested for their effect on polypeptide synthesis. Utilizing D,L-*p*-fluorophenylalanine (20 mM), L-canavanine (13 mM), L-azetidine-2-carboxylic acid (10 mM), and L-ethionine (10 mM) (both alone and in combination), it was found that only L-canavanine interfered with the production of AAV-induced polypeptides. At 18 hr p.i., 3-ml subcultures of Ad/AAV- or Ad-infected KB cells were incubated in the presence of either no analog or 13 mM L-canavanine. The radioactive amino acid L-[35S]methionine (50 μCi/ml) was then added for a 1-hr labeling period and cell pellets were collected and analyzed by PAGE (Fig. 5). This analysis was characterized by the absence of polypeptides C and D (track *b*) and by a failure to detect any new polypeptide that could be considered a precursor of polypeptide A. However, two novel polypeptides (75,900 and 67,600 daltons) specific to L-canavanine treatment of Ad/AAV-infected cells were produced (track *b*, arrows). It is of interest to note the many differences between the untreated (track *e*) and treated (track *d*) polypeptide patterns from Ad-infected KB cells. This is especially notable in the hexon region (track *e*, arrow) of the gel, where a novel polypeptide was detected (track *d*, arrow), and suggests the existence of a hexon precursor protein. Finally, it is apparent that AAV infection depresses the synthesis of hexon and other Ad-specific polypeptides during a dual infection (compare arrows in tracks *c* and *e*).

FIGURE 5
Effect of L-canavanine on the synthesis of AAV-induced polypeptides. Ad/AAV-infected (tracks *b* and *c*) and Ad-infected (tracks *d* and *e*) KB cells were pulse-labeled in the presence (tracks *b* and *d*) or absence (tracks *c* and *e*) of 13 mM L-canavanine. Virion polypeptides are displayed in track *a*.

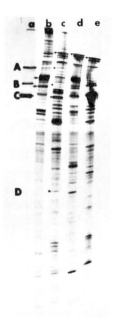

DISCUSSION

Examination of Ad/AAV-infected KB and AGMK cells by SDS-PAGE has revealed five polypeptides specific to AAV2 infection. Further evidence that these polypeptides are virus-coded was their production by an in vitro protein-synthesizing system programmed with highly purified AAV-specific mRNA. Of these five polypeptides, the three largest (A, B, and C) comigrated with the structural polypeptides of highly purified AAV virions.

Pulse-chase experiments utilizing short pulses of radioactivity suggest strongly that all of the viral polypeptides detected arose during nascent polypeptide chain synthesis rather than by processing of a completed translation product. Excluding post-translational cleavage, there are at least three basic mechanisms by which the AAV polypeptides might be generated: there could be more than one functional initiation or termination site, or proteolytic sizing might occur while synthesis of the polypeptide chain is in progress. At present, there is no experimental evidence in mammalian systems for more than one functional initiation site in each mRNA. Although studies of the initiation of polypeptide synthesis with tobacco mosaic virus (Hunter et al. 1976) and Sindbis virus (Simmons and Strauss 1974; Cancedda et al. 1975) have shown that intact virion RNA (which acts as mRNA in vivo) contains at least two sites for the initiation of protein synthesis, only one of these sites appears to be functional in vitro (Simmons and Strauss 1974; Cancedda et al. 1975; Hunter et al. 1976). The detection of AAV-specified nonstructural polypeptides D and E suggests a mechanism in which, in the simplest case, one of two alternative cleavages of the nascent 90,200-dalton polypeptide produces either the B polypeptide (71,600 daltons) plus the E polypeptide (15,800 daltons) or the C polypeptide (60,000 daltons) plus the D polypeptide (24,900 daltons). This scheme requires that the capsid polypeptides contain related sequences, which is supported by the detection of antigenic homologies between the B and C polypeptides (Johnson et al. 1972). In addition, during L-canavanine treatment of Ad/AAV-infected cells (which blocks the synthesis of polypeptides C and D) two polypeptides of 75,900 and 67,600 daltons were preferentially synthesized. These AAV-specific L-canavanine-substituted polypeptides could represent either the buildup of normal cleavage intermediates, which might only exist transiently in the absence of the analog, or the product of an aberrant processing event due to exposure of a previously cryptic cleavage site(s). The former explanation is supported by the large number of discrete nonstructural polypeptides synthesized in vitro by the AAV-mRNA-programmed system.

ACKNOWLEDGMENTS

We would like to thank Dr. M. Bloom for providing the highly purified AAV2 stock and Dr. C. F. Garon for assistance with photography. Technical help by Mrs. D. C. Ortt was greatly appreciated.

REFERENCES

Anderson, C. W., P. R. Baum, and R. F. Gesteland. 1973. Processing of adenovirus-2-induced proteins. *J. Virol.* **12**:241.

Blacklow, N. R., M. D. Hoggan, and W. P. Rowe. 1967. Immunofluorescent studies of the potentiation of an adenovirus-associated virus by adenovirus 7. *J. Exp. Med.* **125**:755.

Cancedda, R., L. Villa Komaroff, H. F. Lodish, and M. Schlesinger. 1975. Initiation sites for translation of Sindbis virus 42S and 26S messenger RNAs. *Cell* **6**:215.

Carter, B. J. 1976. Intracellular distribution and polyadenylate content of adeno-associated virus RNA sequences. *Virology* **73**:273.

Carter, B. J. and J. A. Rose. 1974. Transcription in vivo of a defective parvovirus: Sedimentation and electrophoretic analysis of RNA synthesized by adenovirus-associated virus and its helper adenovirus. *Virology* **61**:182.

Frearson, P. M. and L. V. Crawford. 1972. Polyoma virus basic proteins. *J. Gen. Virol.* **14**:141.

Hunter, T. R., T. Hunt, J. Knowland, and D. Zimmern. 1976. Messenger RNA for the coat protein of tobacco mosaic virus. *Nature* **260**:759.

Jacobson, M. F. and D. Baltimore. 1968. Polypeptide cleavages in the formation of poliovirus proteins. *Proc. Natl. Acad. Sci.* **61**:77.

Johnson, F. B., N. R. Blacklow, and M. D. Hoggan. 1972. Immunological reactivity of antisera prepared against the sodium dodecyl sulfate-treated structural polypeptides of adenovirus-associated virus. *J. Virol.* **9**:1017.

Johnson, F. B., H. L. Ozer, and M. D. Hoggan. 1971. Structural proteins of adenovirus-associated virus type 3. *J. Virol.* **8**:860.

Laemmli, U. K. 1970. Cleavage of structural proteins during the assembly of the head of bacteriophage T4. *Nature* **227**:680.

Lake, R. S., S. Barban, and N. P. Salzman. 1973. Resolutions and identification of the core deoxynucleoproteins of the simian virus 40. *Biochem. Biophys. Res. Commun.* **54**:640.

Lewis, J. B., J. F. Atkins, C. W. Anderson, P. R. Baum, and R. F. Gesteland. 1975. Mapping of late adenovirus genes by cell-free translation of RNA selected by hybridization to specific DNA fragments. *Proc. Natl. Acad. Sci.* **72**:1344.

Marsden, H. S., I. K. Crombie, and J. H. Subak-Sharpe. 1976. Control of protein synthesis in herpesvirus-infected cells: Analysis of the polypeptides induced by wild type and sixteen temperature-sensitive mutants of HSV strain 17. *J. Gen. Virol.* **31**:347.

McDowell, M. J., W. K. Joklik, L. Villa Komaroff, and H. F. Lodish. 1972.

Translation of reovirus messenger RNA synthesized in vitro into reovirus polypeptides by several mammalian cell extracts. *Proc. Natl. Acad. Sci.* **69**:2649.

Moss, B. and F. Koczot. 1976. Sequence of methylated nucleotides at the 5'-terminus of adenovirus-specific RNA. *J. Virol.* **17**:385.

Rabson, A. S., G. T. O'Connor, I. K. Berezesky, and F. J. Paul. 1964. Enhancement of adenovirus growth in African green monkey kidney cell cultures by SV40. *Proc. Soc. Exp. Biol. Med.* **116**:187.

Rose, J. A. and F. Koczot. 1971. Adeno-associated virus multiplication. VI. Base composition of the deoxyribonucleic acid strand species and strand-specific in vivo transcription. *J. Virol.* **8**:771.

Rose, J. A., M. D. Hoggan, and A. J. Shatkin. 1966. Nucleic acid from an adeno-associated virus: Chemical and physical studies. *Proc. Natl. Acad. Sci.* **56**:86.

Rose, J. A., K. I. Berns, M. D. Hoggan, and F. J. Koczot. 1969. Evidence for a single-stranded adenovirus-associated virus genome: Formation of a DNA density-hybrid on release of viral DNA. *Proc. Natl. Acad. Sci.* **64**:863.

Rose, J. A., J. V. Maizel, J. K. Inman, and A. J. Shatkin. 1971. Structural proteins of adenovirus-associated viruses. *J. Virol.* **8**:766.

Schlesinger, S. and J. M. Schlesinger. 1972. Formation of Sindbis virus proteins: Identification of a precursor for one of the envelope proteins. *J. Virol.* **10**:925.

Schreier, M. H. and T. Staehelin. 1973. Initiation of mammalian protein synthesis: The importance of ribosome and initiation factor quality for the efficiency of the in vitro system. *J. Mol. Biol.* **73**:329.

Simmons, D. T. and J. H. Strauss. 1974. Tanslation of Sindbis virus 26S RNA and 49S RNA in lysates of rabbit reticulocytes. *J. Mol. Biol.* **86**:397.

Straus, S. E., R. M. L. Buller, and J. A. Rose. 1977. Host cell restriction of adenovirus-associated virus (AAV) multiplication in African green monkey kidney (AGMK) cells. In *Proceedings of the 77th Annual Meeting of the American Society of Microbiology*, p. 302.

Studier, F. W. 1973. Analysis of bacteriophage T7 early RNAs and proteins on slab gels. *J. Mol. Biol.* **79**:237.

Adeno-Associated-Virus Polypeptides—Molecular Similarities

F. BRENT JOHNSON
DONALD A. VLAZNY
TERRELL A. THOMSON
PHYLLIS A. TAYLOR
MICHAEL D. LUBECK

Department of Microbiology
Brigham Young University
Provo, Utah

There are at least three structural polypeptides in purified adeno-associated virus (AAV) particles, and their combined molecular weights exceed the coding capacity of the viral genome (Johnson et al. 1971, 1975; Rose et al. 1971). The major polypeptide (VP1) has an estimated molecular weight of about 66,000; the two minor polypeptides have molecular weights of about 80,000 (VP2) and 92,000 (VP3). It has been suggested that these polypeptides may arise from a polypeptide precursor molecule which becomes processed post-translationally, generating the polypeptides observed in the mature virion (Johnson et al. 1971, 1975; Rose et al. 1971). Such a mechanism would predict that the precursor polypeptide would have a molecular weight compatible with the single, stable RNA species transcribed in AAV replication (Carter 1974; Carter and Rose 1974), about 120,000 daltons, and that the structural polypeptides would have overlapping amino acid sequences. These peptide similarities have been examined by a variety of methods. In this article we describe some molecular and immunochemical relationships among these peptides.

MATERIALS AND METHODS

Virus Production and Protein Purification

The AAV strains (Hoggan et al. 1966; Blacklow et al. 1967), the adenovirus helper (Rowe et al. 1955), and the methods of producing and purifying

411

AAV from KB-cell suspension cultures (Johnson et al. 1971) have been described. The capsid proteins were analyzed and purified using SDS-polyacrylamide gel electrophoresis (SDS-PAGE) as described previously (Johnson et al. 1971, 1977). Purification of the iodinated proteins for the peptide mapping studies was performed by SDS-PAGE using gels cross-linked with diallyltartardiamide (DATD) (Anker 1970).

Iodination of Virion Proteins

The methods and conditions for radioiodination and fragmentation of the proteins used in the radioimmunoprecipitation assays have been described elsewhere (Johnson et al. 1977). A heterogeneous population of polypeptide fragments was obtained, the smallest of which was approximately 10,000 daltons. Proteins used in the peptide mapping studies were labeled as follows: Purified virions were suspended in 50 μl of 0.01 M phosphate buffer (pH 7.2) and were disrupted by treatment in 1.0% SDS at 100°C for 2 min. The proteins were then iodinated by the chloramine-T method (McConahey et al. 1966). A 10-μl quantity of ^{125}I (50 μCi) was added to the virus suspension, followed by 10 μl of chloramine T, resulting in a final chloramine-T concentration of 0.1 mg/ml. The mixture was agitated for 5 min at room temperature, whereupon the reaction was terminated by the addition of 10 μl of sodium metabisulfite (2.5 mg/ml in 0.01 M phosphate buffer, pH 7.2). Unreacted iodide was removed by dialysis against the phosphate buffer.

Isoelectrofocusing

Viral polypeptides were focused in 5-cm polyacrylamide gel columns. Duplicate 5% polyacrylamide gels containing a 7 M concentration of urea and a 5% volume of 40% (wt/vol) pH 3.5–10 ampholine solution (LKB 1809-101) were cast, and a pH gradient was established in them by electrophoresis at a constant voltage of 130 V for 3 hr immediately before sample application. Anode and cathode solutions consisted of 0.86% phosphoric acid and 2% (vol/vol) ethanolamine, respectively. A sample of purified, ^{14}C-labeled AAV1 whole virions was boiled for 30 min in the presence of 0.01 M dithiothreitol (DTT) and 7 M urea. After the addition of a one-third volume of 50% sucrose, equal amounts of the sample were applied at the cathode end of each of the two gels. Isoelectrofocusing was continued for 5 hr at room temperature using 130 V. After the focusing, one gel was sliced on a template at intervals of 2.5 mm. Each fraction was allowed to elute for 8 hr in 1 ml of freshly boiled distilled water at room temperature, after which the pH of each was measured with a Beckman pHasar-I pH meter with a combination microelectrode. All slices and eluents were then prepared for liquid scintillation counting in Bray's solution. The second gel was sliced at 1.25-mm intervals and each fraction was allowed to elute for 30–60 min in 1 ml of freshly boiled distilled water. Each slice was then removed and cut along the diameter; one half was prepared for immediate scintillation counting and the other was soaked for 16 hr at room temperature in 100 μl of a solution containing 0.03% phosphate buffer (pH 7.2), 1% SDS, 0.006 M DTT, 12% sucrose, and 0.04 mg/ml Pyronin Y tracking dye. Fractions of interest (both gel slice and 100 μl preparatory solution) were overlayed onto 11-cm 1% SDS-7.5% polyacrylamide gels (pH 7.2) for comparison with a parallel-electrophoresed AAV1 sample.

Antisera

Antisera prepared in guinea pigs against each of the three SDS-treated structural polypeptides and against adenovirus mixed polypeptides (SDS-treated) were used to demonstrate cross-reactivity and AAV specificity of the viral proteins. Their production has been reported (Johnson et al. 1972). Rabbit anti-guinea-pig globulin (Grand Island Biological) was used as a precipitating agent for the guinea-pig globulin.

Radioimmunoprecipitation Assay

The radioimmunoprecipitation (RIP) test was modified from that reported by Ozer et al. (1969). Details of the assay are described elsewhere (Johnson et al. 1977). The optimal proportions of the globulin-antiglobulin reaction were determined in RIP by precipitating various concentrations of ^{125}I-labeled antisera with various concentrations of antiglobulin. It was found that a 1:10 dilution of antiglobulin gave maximum precipitation with a 1:10,000 dilution of each antiserum. These proportions of reagents were held constant in all subsequent RIP assays.

Competition Assays with Unlabeled Antigen

Unlabeled viral proteins and nonhomologous control proteins were employed as competing antigens in each labeled antigen–homologous antibody system. The antisera were mixed with various concentrations of unlabeled antigen and incubated at 37°C for 4 hr and then overnight at 4°C. ^{125}I-labeled antigen was then added at a concentration previously shown to give maximum precipitation, and the preparation was incubated as above. Addition of antiglobulin, incubation, centrifugation, and calculation of precipitate were carried out as in the standard RIP assays. The competition by the unlabeled antigen in each system was quantitated by comparing the percentage of precipitation with that found for the controls without competitor.

Peptide Mapping

Stained purified proteins were carefully excised from the DATD gels, pooled (4–5 gel slices) in polyallomer centrifuge tubes (Beckman), and digested with 2 ml of 2% periodic acid at room temperature for about 20 min. A carrier protein, bovine serum albumin (Armour, fraction V), was then added (200 μg), and the proteins were precipitated with 2 ml of cold 50% trichloroacetic acid (TCA). After incubation on ice for 4 hr, the precipitates were pelleted at 25,000g for 2 hr. The precipitates were then solubilized in cold 1 N NaOH and reprecipitated with cold 20% TCA. This procedure was repeated once more to remove residual SDS. The proteins were then washed sequentially in cold 20% TCA, acetone-1 M HCl (40:1 vol/vol; $-20°$C), and acetone ($-20°$C), whereupon the washed precipitates were dried under a stream of nitrogen. The purified proteins were then suspended in 0.05 M NH$_4$HCO$_3$ (pH 8.6) and digested with diphenyl-carbamoyl chloride-treated trypsin (Cal Biochem). The ratio of enzyme to substrate was 1:100. The enzymatic hydrolysis was allowed to proceed for 6 hr at room temperature with intermittant agitation, after which the digests were lyophilized. The peptides were suspended in electrophoresis buffer and applied to cellulose thin-layer chromatography sheets by repeated spotting under a current of warm air. Electrophoresis was conducted at pH 2.1 (acetic acid:formic acid:H$_2$O, 80:20:900 vol/vol) for about 1 hr at 1000 V. The plates were then dried under a current of warm air and allowed to equilibrate at room temperature before development in the second dimension by chromatography. The chromatography buffer contained butanol:pyridine:acetic acid:H$_2$O in the ratio of 97:75:15:60 vol/vol, pH 5.3. Radioactive peptides were located by autoradiography on GAF HR 1000 high-resolution medical X-ray film.

RESULTS

Isoelectrofocusing of VP1, VP2, and VP3

Purified ^{14}C-labeled AAV1 virions were dissociated in 7 M urea and the proteins were isoelectrofocused. The results are show in Figure 1A. Most of the AAV protein focused in a single peak with an isoelectric point between 5.0 and 5.3; the peak fraction was at pH 5.1. The gel was cut into 1.25-mm slices and each slice was cut in half, one half to be counted for radioactivity (Fig. 1A) and the other half to be overlaid on SDS-polyacrylamide gels and electrophoresed. A purified virus standard was included in the run to allow for identification of the polypeptides. Viral proteins could be detected only in fractions 27–30 from the isoelectric-focusing gel (arrows in Fig. 1A). The material that failed to enter the

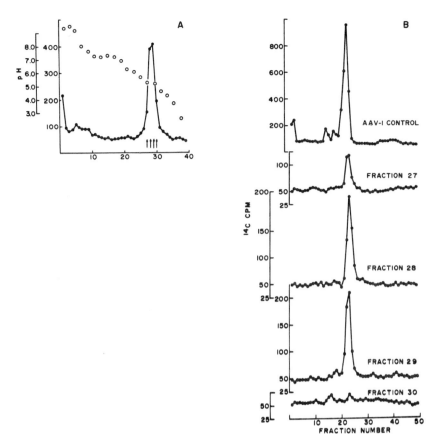

FIGURE 1
Determination of the isoelectric point of the AAV capsid proteins. (*A*) Analytical isoelectrofocusing of AAV1 polypeptides in 5% polyacrylamide-urea gel. The polypeptides focused in a single peak at an isoelectric point of pH 5.0–5.3. (*B*) Identification of the polypeptide components of this peak (the fractions indicated by the arrows in *A*).

focusing gel (fraction 1) likewise did not migrate under SDS-PAGE analysis. This material may be the same as that which does not enter standard SDS-polyacrylamide gels of whole virions. The identification of the protein in fractions 27–30 is shown in Figure 1*B*. Fractions 27 and 28 contained only VP1. The greatest amount of VP1 was in fractions 28 and 29. VP2 was found in fraction 29 and VP3 in fraction 30. Although the amount of radioactivity migrating as VP2 and VP1 (fractions 29 and 30) is low, the counts are statistically significant and reproducibly obtained. These data show that VP1, VP2, and VP3 have very similar isoelectric points (between pH 5.0 and 5.3) but that VP3 is slightly more acidic than either VP2 or VP1. Attempts to resolve the polypeptides further by focusing on shallower pH gradients were unsuccessful; a single peak of protein was always observed.

Precipitation of the Viral Proteins

The results of immunoprecipitation assays involving the viral proteins and antibody prepared against either VP1, VP2, or VP3 are shown in Figure 2. Each of the viral proteins was tested with all three antisera.

The precipitation of the polypeptides in the three antigen pools by anti-VP1 serum is shown in Figure 2A. The relatively lower level of

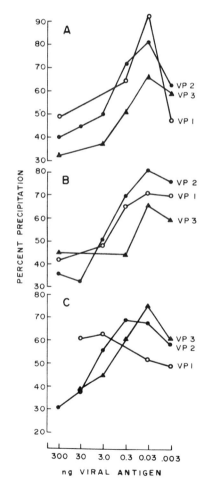

FIGURE 2
Immunoprecipitation of ^{125}I-labeled, separated AAV structural proteins by antipolypeptide sera. Varying concentrations of the three proteins were mixed with constant amounts of antiserum (1:10,000) and antiglobulin (1:10). Levels of precipitation were then determined. Significant amounts of cross-reactivity were noted. (A) Precipitation of VP1, VP2, and VP3 by anti-VP1 serum. (B) Precipitation of the proteins by anti-VP2 serum. (C) Precipitation of the proteins by anti-VP3 serum.

precipitation at the higher antigen concentrations was probably due to the antiserum being saturated with viral protein and the excess protein remaining in the supernate. At higher dilutions. the amount of unbound antigen diminished and the peak indicated that most of the labeled antigen was bound by available antibody. This marked the point where antigen-antibody affinity was best measured. The protein concentration of each antigen preparation was equalized before the reactants were mixed, so that the peaks occurred at the same concentration of each protein. The optimal reaction occurred with 0.03 ng of antigen in most cases. That 100% precipitation was not reached was probably because the proteins fragmented during iodination (Johnson et al. 1977), so that fragments without antigenic sites were possibly labeled but not pre-cipitable. It is also possible that some fragments had no homologous antibody molecules in the reagent antiserum. Figure 2A shows that VP1 was precipitated most effectively by anti-VP1, while VP2 and VP3 reacted to lesser degrees. Controls containing no antibody routinely showed background precipitation of only 10–15% of the radioactivity.

Anti-VP2 and anti-VP3 showed similar patterns for each of the poly-peptides (Fig. 2 B and C). In each case, the homologous reaction gave the best precipitation; the other proteins were bound at lesser levels.

The relative precipitation of the proteins by each antiserum suggests that they are structurally related to each other. As seen from Figure 2B, anti-VP2 precipitated VP1 more effectively than it did VP3; this indicates that, antigenically, VP1 resembles VP2 more closely. Figure 2C shows that VP2 has a greater affinity for anti-VP3 than does VP1; this suggests that, immunologically, the VP2 molecule is more similar to VP3.

Competition with Unlabeled Antigens

Unlabeled VP1, VP2, VP3, ovalbumin, SDS-treated ovalbumin, and whole AAV3 virions were tested for their blocking ability against each of the three antisera and their homologous labeled antigens. Reaction mix-tures containing optimal concentrations of labeled antigen, antibody, and antiglobulin were mixed with varying concentrations of unlabeled com-peting antigens. The amount of competition offered by the unlabeled antigen under conditions of saturation gave an estimate of the relatedness between the labeled antigen and the unlabeled competitor.

The results of the VP1–anti-VP1 system inhibited by the viral poly-peptides are shown in Figure 3A. As expected, the VP1 reaction was almost completely inhibited by VP1. VP2 and VP3 inhibited the reaction to levels of approximately 64% and 54%, respectively. These data are consistent with the observation in Figure 2 that VP2 is more closely related to VP1 than is VP3.

When the VP2–anti-VP2 system was inhibited by the viral poly-

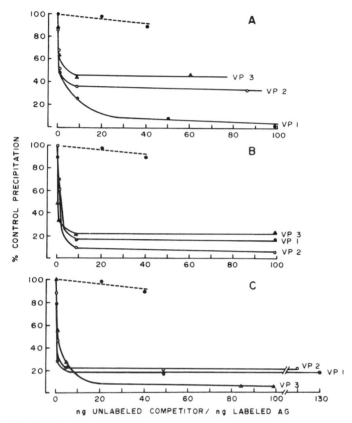

FIGURE 3

Immunocompetition assays employing unlabeled VP1, VP2, and VP3 antigen pools as competitors in the three antiserum systems. (*A*) Competition in the VP1–anti-VP1 system. (*B*) Competition in the VP2–anti-VP2 system. (*C*) Competition in the VP3–anti-VP3 system. Dashed lines depict control levels of competition by SDS-treated ovalbumin.

peptides, the results shown in Figure 3*B* were obtained. At saturation, VP2 almost completely inhibited its homologous reaction; VP1 and VP3 inhibited the reaction to approximately 84% and 78%, respectively. This substantiates the previous findings (Fig.2) that VP1 is more closely related to VP2 than is VP3. Similar results were found in the VP3–anti-VP3 system in that VP3, the homologous competitor, inhibited the reaction almost completely. However, VP1 and VP2 were not adequately discriminated, as both inhibited the homologous reaction to 80%. These results were found to be consistent in repeated experiments.

Control competition assays demonstrated the inhibitory effects shown above to be due to specific antigen-antibody combinations. Ovalbumin was shown to lack inhibitory effects in the VP1, VP2, and VP3 systems (Fig. 3). SDS-treated ovalbumin was tested in the mixed-polypeptide system (antiserum to all three polypeptides tested against labeled VP1, VP2, and VP3). The SDS-ovalbumin complexes did not inhibit these reactions, which showed that SDS was not detected serologically in these reactions, but that the reactions were due to immunologic determinants on the protein molecules themselves. Moreover, antisera to SDS-treated Ad2 polypeptides did not react with SDS-treated AAV polypeptides in RIP assays (not shown). Similarly, no reaction occurred when the three AAV VP antisera were tested against SDS-treated Ad2 polypeptides (Johnson et al 1977); this confirmed again the specificity of the reactions for AAV polypeptides.

The specificity of the antisera for the SDS-treated polypeptides and not for the native configuration of the proteins was demonstrated by competition with nondissociated AAV3 virions. Even at high concentrations of competing virus, little blocking of the precipitation was observed (data not shown).

Peptide Maps

Tryptic digests of the purified radioiodinated capsid polypeptides were separated in two dimensions on cellulose thin-layer plates. The peptide locations were determined by autoradiography. Figure 4 shows the comparative fingerprints of VP1, VP2, and VP3. There were at least 30 peptides in common, demonstrating extensive areas of peptide homology. At the same time, VP1 appeared to possess one unique peptide (1) and to share another peptide (2) with VP2 that was absent in the VP3 map. The

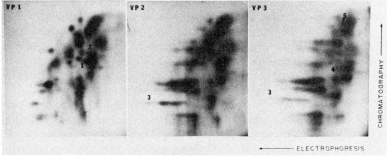

FIGURE 4
Tryptic peptide maps of VP1, VP2, and VP3 showing numerous common peptides and a small number of unique peptides.

fingerprints also indicated that VP2 and VP3 shared one peptide (3) that was absent in the VP1 map, and that VP3 had two peptides (4 and 5) that were not found in the VP1 or VP2 maps. All of the peptides found in VP2 appear to be present in maps of VP1 or VP3.

DISCUSSION

This investigation provides further insight into the relationships among the structural polypeptides of AAV. Evidence presented elsewhere (Johnson et al. 1977) shows similarities in amino acid incorporation among the three polypeptides and a kinetic relationship between the putative high-molecular-weight precursor, VP0, and the three capsid proteins. We demonstrate here that the AAV polypeptides possess similar isoelectric points, a high degree of antigenic cross-reactivity, and similar polypeptide fingerprints. Taken together, the results indicate that these viral proteins share extensive sequences of amino acid homology. Similar amino acid sequences imply a common origin of the proteins. It is likely that the AAV structural polypeptides arise by cleavage of VP0 resulting in polypeptides with overlapping segments. Alternatively, the peptides could arise from multiple initiation sites on the AAV mRNA molecule. Additional experiments are required to resolve this question.

REFERENCES

Anker, H. S. 1970. A solubilizable acrylamide gel for electrophoresis. *FEBS Lett.* 7:293.

Blacklow, N. R., M. D. Hoggan, and W. P. Rowe. 1967. Isolation of adenovirus-associated viruses from man. *Proc. Natl. Acad. Sci.* 58:1410.

Carter, B. J. 1974. Analysis of parvovirus mRNA by sedimentation and electrophoresis in aqueous and nonaqueous solution. *J. Virol.* 14:834.

Carter, B. J. and J. A. Rose. 1974. Transcription in vivo of a defective parvovirus: Sedimentation and electrophoretic analysis of RNA synthesized by adenovirus-associated virus and its helper adenovirus. *Virology* 61:182.

Hoggan, M. D., N. R. Blacklow, and W. P. Rowe. 1966. Studies of small DNA viruses found in various adenovirus preparations: Physical, biological and immunological characteristics. *Proc. Natl. Acad. Sci.* 55:1467.

Johnson, F. B. 1975. Biosynthetic and structural relationships among AAV polypeptides. In *Abstracts of Third International Congress for Virology*, p. 182.

Johnson, F. B., N. R. Blacklow, and M. D. Hoggan. 1972. Immunological reactivity of antisera prepared against the sodium dodecyl sulfate treated structural polypeptides of adenovirus-associated virus. *J. Virol* 9:1017.

Johnson, F. B., H. L. Ozer, and M. D. Hoggan. 1971. Structural proteins of adenovirus-associated virus type 3. *J. Virol.* 8:860.

Johnson, F. B., C. W. Whitaker, and M. D. Hoggan. 1975. Structural polypeptides of adenovirus-associated virus top component. *Virology* 65:196.

Johnson, F. B., T. A. Thomson, P. A. Taylor, and D. A. Vlazny. 1977. Molecular similarities among the adenovirus-associated virus polypeptides and evidence for a precursor protein. *Virology* (in press).

McConahey, P. J. and F. J. Dixon. 1966. A method of trace iodination of proteins for immunologic studies. *Int. Arch. Allergy* **29**:185.

Ozer, H. L., K. K. Takemoto, R. L. Kirschstein, and D. Axelrod. 1969. Immunochemical characterization of plaque mutants of simian virus 40. *J. Virol.* **3**:17.

Rose, J. A., J. V. Maizel, Jr., J. K. Inman, and A. J. Shatkin. 1971. Structural proteins of adenovirus-associated viruses. *J. Virol.* **8**:766.

Rowe, W. P., R. J. Huebner, J. W. Hartley, T. G. Ward, and R. H. Parrot. 1955. Studies of the adenoidal-pharyngeal-conjunctival (APC) group of viruses. *Am. J. Hyg.* **61**:197.

Antigenic Differences Among AAV Species of Different Densities

AUROBINDO ROY
GUNTER F. THOMAS
JOHNNA F. SEARS
M. DAVID HOGGAN

National Institute of Allergy and Infectious Diseases
National Institutes of Health
Bethesda, Maryland 20014

When purified virions of adeno-associated virus (AAV) are centrifuged to equilibrium in CsCl gradients four bands are usually observed (Hoggan et al. 1966; Hoggan 1971; Johnson et al. 1971; Johnson and Hoggan 1973). The lightest band, with a density between 1.32 and 1.34 g/cm³, contains noninfectious particles which are devoid of DNA. Additional noninfectious virions which contain DNA molecules of less than genome length are found in the density range of 1.34–1.38 g/cm³. The major band of infectious AAV exhibits a density of 1.39–1.42 g/cm³. However, particles which contain a complete viral genome are found also at a density of 1.46–1.51 g/cm³. The resolution of infectious virions into two classes of particles in CsCl has been shown to occur with other parvoviruses, for example, Haden virus (Johnson and Hoggan 1973), minute virus of mice (MVM) (Clinton and Hayashi 1975, 1976), LuIII (Gautschi and Siegl 1973), and H-1 (Kongsvik et al., this volume). The polypeptide composition and the biological properties of these two species of virions are different in the cases of MVM (Clinton and Hayashi 1976) and H-1 (Kongsvik et al., this volume). In this report we have employed immune electron microscopy and infectivity assays to detect physical and biological differences between the "heavy" ($\rho = 1.46$–1.51 g/cm³) and "light" ($\rho = 1.39$–1.42 g/cm³) virions of AAV.

MATERIALS AND METHODS

GROWTH AND PURIFICATION OF VIRUS. AAV1 was grown in KB spinner culture with adenovirus type 2 (Ad2) (Johnson et al. 1971), and the virions of different densities were separated by banding three times in isopycnic CsCl gradients (Hoggan et al. 1966).

INFECTIVITY TITRATION. AAV1 was titrated in human-embryo-kidney tube cultures preinfected with Ad2 helper. Infectious titer is reported as the highest dilution of virus capable of inducing AAV1 complement-fixing antigen in 50% of the tubes ($TCID_{50}$) (Hoggan et al. 1966).

PREPARATION OF ANTISERA. After separation by isopycnic centrifugation, the two bands containing the complete AAV1 genome were used for the production of hyperimmune sera in guinea pigs as described previously (Hoggan et al. 1966). Hereafter, the major infectious band (1.39–1.42 g/cm^3) will be designated as light (L) AAV, and serum produced against it as anti-L serum. Similarly, the minor band with higher density (1.46 g/cm^3) is designated as heavy (H) AAV and its corresponding antiserum as anti-H serum.

SEROLOGY. Complement-fixation tests were carried out in microtiter plates as described previously (Hoggan et al. 1966). Immune electron microscopy (IEM) was done according to the method of Thomas and Hoggan (1974).

RESULTS

Immune Electron Microscopy of Heavy and Light AAV1 Particles

The results of experiments in which H and L AAV1 particles were reacted with anti-H and anti-L sera and examined by immune electron microscopy are demonstrated in Figure 1. When H particles of AAV were reacted with anti-H serum, medium-size to large antibody-virus aggregates were formed (Fig. 1A). When the same H antiserum was reacted with L particles, a few small aggregates could be found by searching a number of fields; however, the majority of the particles were single and naked (Fig. 1B). Conversely, when anti-L serum was reacted with H virus as shown in Figure 1C, the majority of particles were found to be free of antibody. Nonetheless, a few particles heavily coated with antibody were observed. We feel that these represent contaminating L-band particles in the H preparation. When L serum was used in a homologous reaction (anti-L serum vs. anti-L particles), large aggregates containing several thousand particles associated with 4+ antibody could be seen (Fig. 1D).

We have noticed repeatedly that antibody-virus aggregates formed with H and L particles and their homologous antisera each have their own

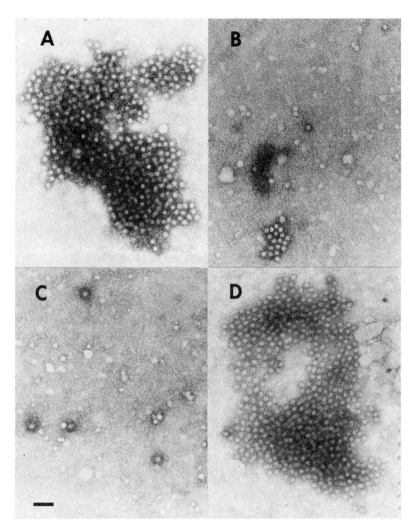

FIGURE 1
Demonstration of antigenic differences between heavy (H) and light (L) AAV1 particles using immune electron microscopy. (*A*) AAV H vs. anti-AAV-H serum. (*B*) AAV L vs. anti-AAV-H serum. (*C*) AAV H vs. anti-AAV-L serum. (*D*) AAV L vs. anti-AAV-L serum. (Bar represents 100 nm.)

characteristics and can be recognized under the electron microscope. Aggregates formed by H virus and its antiserum appear loosely clumped, with the center-to-center spacing between particles greater than the center-to-center spacing seen with L virus and its homologous serum. The higher-magnification electron micrographs shown in Figure 2 *A* and *B* demonstrate the results of mixing both H and L virus with both

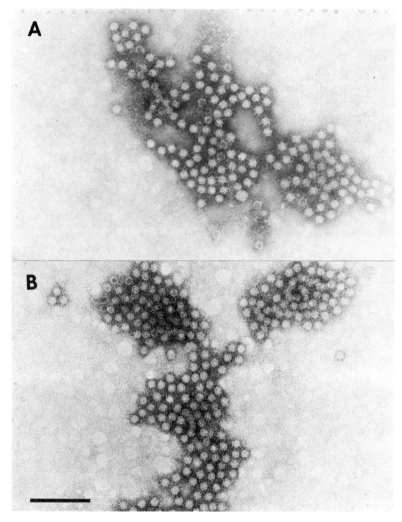

FIGURE 2
Demonstration of homogeneous heavy-particle aggregates and homogeneous light-particle aggregates in a mixed reaction. (A) Characteristic AAV-H-particle aggregate. (B) Characteristic AAV-L-particle aggregate. (Bar represents 100 nm.)

antisera. In such a combination, aggregate configurations are formed. An aggregation of loosely packed particles with the characteristics of H particles is illustrated in Figure 2A. Figure 2B shows a representative L-particle aggregate. The two aggregates shown were found on the same grid. In such a mixture, individual aggregates were always homogeneous with respect to particle type. In addition, we rarely (if ever) saw unreacted particles, although they were fairly common in reactions involving one virus type and its homologous antiserum.

Infectivity of H and L Particles

We reported that the H band contained fewer particles than the L band and that these particles were always smaller (Hoggan 1971). In general, no grossly visible band can be seen in the H density region, although particles can be detected using the electron microscope. Occasionally, however, we find preparations in which the widths of H and L bands appear almost equal. With one such preparation used in our immune-electron-microscopy studies we have also been able to compare directly the H and L particles in terms of physical particle counts and infectivity. The results are shown in Table 1. It should be noted that even though the L band contained only 18% more particles than the H band it had over 300 times the infectivity.

DISCUSSION

Our present observations clearly suggest that there is a difference in the antigenicity of H and L particles of AAV, even though we are unable to completely separate the two populations. Whereas our earlier studies indicated that the same three polypeptides are present in both H and L virus bands, the relative amounts of the two minor polypeptides (92,000 and 80,000 daltons) compared to the major polypeptide (66,000 daltons) were less in H particles than in L particles (Johnson et al. 1971). We further reported that the average diameter of H particles was 12% less than the diameter of the L particles (Hoggan 1971) and suggested that these findings might reflect a lack of, or configurational alteration of, a protein at the capsid surface. Recent support for this theory comes from an

TABLE 1
Relative Infectivity of Heavy and Light AAV Particles

	Physical particles per ml	$TCID_{50}$ per ml
Heavy band (> 1.46 g/cm^3)	3.2×10^{12}	3×10^6
Light band (1.39 g/cm$_3$)	3.9×10^{12}	1×10^9

experiment in which we measured the ratio of nucleic acid to protein in the two bands by double-labeling the virions. We found that H particles have a higher DNA/protein ratio than L particles. In one experiment the ratio of ^{3}H-labeled DNA to ^{14}C-labeled protein was 0.575 for H particles and 0.424 for L particles.

How our data on AAV relate to the findings of Clinton and Hayashi (1975) on minute virus of mice is not clear. They found a proteolytic activity associated with infected cell culture fluid which converted one of the MVM polypeptides into a smaller one and at the same time converted H MVM particles into L MVM particles. Of course, if this change represented a simple loss of protein, the particles would become denser, not lighter. Clinton and Hayashi have postulated that the capsid protein may undergo a conformational change which might alter the L particles' ability to be penetrated or solvated by CsCl. Clinton and Hayashi (1975, 1976) have also reported that H MVM particles were as infectious as L MVM particles although they adsorbed very poorly to the host cells. If the cell monolayers were washed after adsorption, most of the H-particle infectivity was lost.

We have found that the H particles of AAV are much less stable in CsCl than the L particles. After two to three bandings they tend to lose all structural integrity, and the proteins migrate to the top of the gradient. In contrast, L particles have been rebanded as many as five times without appreciable loss. Whether this relative instability or the fact that they adsorb very poorly to KB cells (K. I. Berns, personal communication) accounts for the apparent lower infectivity of H particles is currently under study.

REFERENCES

Clinton, G. M. and M. Hayashi. 1975. The parvovirus MVM: Particles with altered structural proteins. *Virology* **66**:261.
——. 1976. The parvovirus MVM: A comparison of heavy and light particle infectivity and their density conversion in vitro. *Virology* **74**:57.
Gautschi, M. and G. Siegl. 1973. Structural proteins of parvovirus LuIII. *Arch. gesamte Virusforsch.* **43**:326.
Hoggan, M. D. 1971. Small DNA viruses. In *Comparative virology* (ed. K. Maramorosch and E. Kurstak), p. 43. Academic, New York.
Hoggan, M. D., N. R. Blacklow, and W. P. Rowe. 1966. Studies of small DNA viruses found in various adenovirus preparations: Physical, biological and immunological characteristics. *Proc. Natl. Acad. Sci.* **55**:1467.
Johnson, F. B. and M. D. Hoggan. 1973. Structural proteins of Haden virus. *Virology* **51**:129.
Johnson, F. B., H. L. Ozer, and M. D. Hoggan. 1971. Structural proteins of adeno-associated virus type 3. *J. Virol.* **8**:860.

Siegl, G. 1973. Physicochemical characteristics of the DNA of parvovirus LuIII. *Arch. gesamte Virusforsch.* **43**:334.

Thomas, G. F. and M. D. Hoggan. 1974. Immune electron microscopy (IEM) of canine adeno-associated virus. In *Proceedings of the 32nd Annual Meeting of the Electron Microscopy Society of America* (ed. C. J. Arcineaux), p. 256. Claitor's, Baton Rouge, Louisiana.

Comparison of Parvovirus Structural Proteins: Evidence for Post-Translational Modification

JANE L. PETERSON
RODERIC M. K. DALE
ROGER KARESS
DANIEL LEONARD
DAVID C. WARD

Department of Human Genetics
and Department of Molecular Biophysics and Biochemistry
Yale University School of Medicine
New Haven, Connecticut 06510

Minute virus of mice (MVM), a nondefective parvovirus, contains three structural polypeptides, designated A, B, and C, with approximate molecular weights of 83,000, 64,000, and 61,000, respectively (Clinton and Hayashi 1975, 1976; Tattersall et al. 1976; Richards et al. 1977). Full virus particles contain the DNA genome and all three polypeptides, whereas empty virions contain only the A and B proteins. Tattersall et al. (1977) have shown, by analysis of tryptic and chymotryptic peptides, that the sequence of polypeptide B is almost identical to that of C and is contained within the sequence of polypeptide A. In addition, the B protein in the full virus could be converted by proteolysis in vitro to a polypeptide which has the same molecular weight as the C protein synthesized in vivo (Clinton and Hayashi 1976; Tattersall et al. 1977). However, the fingerprint analysis suggested that the in vivo cleavage did not occur at either a trypsin-sensitive or a chymotrypsin-sensitive site. The B protein of empty virions, in contrast with that of the full virus, was resistant to proteolysis.

431

On the basis of these observations, Tattersall et al. (1977) proposed a scheme for MVM maturation in which a specific proteolysis of the B polypeptide in the full virion is required for optimal viral infectivity. Kongsvik et al. (1974) reported that proteolytic treatment of intact H-1 virions converted capsid protein VP2′ to a polypeptide that comigrates with the virion protein VP2 (see also Kongsvik et al., this volume). We have examined the structural polypeptides of MVM, H-1, H-3, and KRV in order to explore the possibility that nondefective parvovirus may undergo a common scheme of virion maturation involving specific proteolysis or post-translation modification.

MATERIALS AND METHODS

Virus and Cells

Kilham rat virus (KRV), MVM, H-1, and H-3 were propagated in the transformed rat-liver cell line RL5E isolated by Bomford and Weinstein (1972). LuIII was grown on HeLa-cell monolayers. Cell cultures were maintained and virus stocks produced as described by Tattersall et al. (1976). Initial virus inocula were kindly provided by P. Tattersall (MVM, plaque-purified strain T), H. Toolan (H-1 and H-3), L. Salzman (KRV), and G. Siegl (LuIII).

All virus preparations were purified essentially according to the procedure of Tattersall et al. (1976). ^{32}P-labeled virions were purified from virus-infected RL5E cells grown from 8 to 48 hr post-infection in minimal essential medium containing 2% of the normal phosphate concentration and 40 μCi/ml of [^{32}P]orthophosphate (New England Nuclear). Virus labeled with [^{3}H]adenosine was prepared from infected cells exposed to 3 μCi/ml of [^{3}H]adenosine from 8 to 48 hr post-infection.

SDS-Polyacrylamide Gel Electrophoresis

Discontinuous polyacrylamide gel electrophoresis in the presence of sodium dodecyl sulfate (SDS) was performed according to the method of Laemmli (1970). Samples were prepared and run on 10% polyacrylamide gels as described by Tattersall et al. (1976). Fluorography was carried out by the method of Bonner and Laskey (1974), and the gels were exposed to presensitized film (Laskey and Mills 1975). Coomassie-blue staining and autoradiographic exposures were quantitated by scanning with the Joyce-Loebl densitometer.

Trypsin Treatment

The full and empty virions were digested with bovine trypsin as follows. Approximately 5 μg of viral protein in 50 mM Tris, 0.5 mM EDTA (pH 8.7) was treated with trypsin at a final concentration of 40 μg/ml for 30 min at 25°C. More trypsin was then added to give a final concentration of 80 μg/ml, and the digestion was continued for 30 min. The reaction was terminated by the addition of 0.002 M PMSF in *n*-propanol.

Two-Dimensional Gel Electrophoresis

Capsid proteins were analyzed using the two-dimensional gel system of O'Farrell (1975). Samples were dialyzed against 50 mM triethylammonium bicarbonate (pH 8.0), lyophilized, and resuspended in sample buffer (9 M urea, 2% NP-40, 5% 2-mercaptoethanol, 0.1% SDS, 5% ampholytes, pH 5–7). The virus particles were then disrupted by heating for 5 min at 60°C. Lyophilized cytoplasmic protein samples were treated with S1 nuclease as described by Peterson and McConkey (1976). The proteins were separated on the basis of their isoelectric point in the first dimension (isoelectric-focusing gel) and by molecular weight in the second dimension (SDS-polyacrylamide gel).

The pH gradient of the isoelectric-focusing gel was measured with a Bio Rad Gel Pro-pHiler; 10% polyacrylamide gels were routinely used in the second dimension. Proteins will be referred to by apparent isoelectric point and by molecular weight in thousands of daltons. For example, a protein with an isoelectric point of pH 7.0 and a molecular weight of 45,000 daltons will be designated (7.0-45).

Basic-pH Two-Dimensional Gels

The components of the first-dimension isoelectric-focusing gel were the same as those described by O'Farrell (1975), except that the ampholyte mixture was composed of 1% pH 3–10, 2% pH 7–9, and 2% pH 9–11 ampholytes. Three times the standard amounts of catalysts were used, since gels containing basic ampholytes polymerize very slowly and are quite unstable during electrophoresis. The gels were allowed to sit for 30 min after polymerization and then prerun for 30 min at 0.5 mA per gel with 0.05 M sodium hydroxide in the lower (negative electrode) tank and 0.0017% phosphoric acid in the top (positive electrode) tank. After the prerun, both tank buffers were discarded. The samples, prepared as above with 5% ampholytes of pH 7–9 (instead of pH 5–7), were loaded onto the acidic end of the gel and overlaid with 8 M urea containing 5% ampholytes of pH 3–10. Tank solutions were 0.3 M sodium hydroxide and 0.0017% phosphoric acid. The gels were run for 4 hr at 400 V or until the marker cytochrome c was focused sharply. This procedure was necessary because the pH gradient of basic gels decays quite rapidly, causing the proteins to move continuously through the gel during electrophoresis. The gels were extruded and electrophoresed in the second dimension on a 10% SDS-polyacrylamide gel in the standard manner (O'Farrell 1975).

Isolation of B Protein from MVM

The viral B protein was isolated by continual elution of a 6%-SDS discontinuous tube gel as described by Lee and Sinsheimer (1974). The purified B protein was dialyzed extensively against 50 mM triethyl-ammonium bicarbonate and lyophilized.

Detection of o-Phosphoserine and o-Phosphothreonine

^{32}P-labeled B proteins were hydrolyzed in 6 N HCl for 7 hr at 105°C. The hydrolysate was lyophilized and resuspended in 2.5% formic acid, 7.8% acetic acid, and the phosphoamino acids were separated on cellulose plates (Eastman Kodak) by the method of Bitte and Kabat (1974).

Acid and Base Hydrolysis

^{32}P-labeled B protein was hydrolyzed in 0.25 N HCl or 0.25 N NaOH for 120 min at 66°C (Goff 1974). The protein was precipitated with 5% TCA, collected on GF/C filters (Whatman), dried, and counted in liquid-scintillation fluid.

Enzymatic Digestions

^{32}P-labeled and [^{3}H]adenosine-labeled MVM B proteins (in 50 mM Tris-HCl, 0.5 mM EDTA, pH 8.7) were treated as follows: (a) with 25 μg/ml micrococcal nuclease (Sigma) and 20 mM $CaCl_2$ for 30 min at 37°C, (b) with 25 μg/ml DNase (Worthington) and 20 mM $MgCl_2$ for 30 min at 37°C, (c) with 25 μg/ml RNase (Worthington), after adjustment of pH to 6.0 with HCl, for 30 min at 37°C, and (d) with 30 μg/ml alkaline phosphatase for 30 min at 22°C. TCA-precipitable protein was processed as above.

RESULTS

Molecular-Weight Comparison of Viral Proteins

The capsid proteins of four nondefective parvoviruses, MVM, H-1, H-3, and KRV, were compared by SDS gel electrophoresis (Fig. 1). The empty virions of H-1, H-3, and KRV contain only two types of polypeptide, similar in size to the A and B polypeptides of empty MVM particles. In contrast, all full-virus preparations contain a third protein species, similar

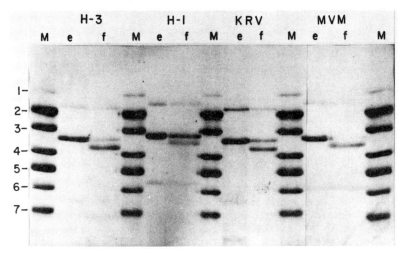

FIGURE 1
10% SDS-polyacrylamide gel of the empty- and full-virus proteins from four different parvoviruses. Molecular-weight standards (M) were (1) phosphorylase A, m.w. 100,000, (2) transferrin, m.w. 76,000, (3) bovine serum albumin, m.w. 68,000, (4) pyruvate kinase, m.w. 57,000, (5) glutamate dehydrogenase, m.w. 53,000, (6) fumarase, m.w. 48,000, and (7) aldolase, m.w. 40,000.

to the C polypeptide of MVM. The two polypeptides found in the empty virions are present in approximately the same molar ratio for each virus (approximately 15% A, 85% B). While the A polypeptide of the full virions again constitutes about 15% of the total particle protein, the amounts of B and C polypeptides vary significantly in the different viruses. Similar results have also been obtained with LuIII virus (data not shown). The molecular-weight values calculated from such gels are summarized in Table 1. To demonstrate that the slight differences in observed molecular weights were significant, full virions of MVM and H-3 were mixed and coelectrophoresed. Six polypeptide bands were obtained, as expected (Fig. 2).

Although a small amount of a fourth polypeptide species can be seen in some of the virus preparations (Fig. 1), they are probably nonviral contaminants. A fourth polypeptide (m.w. ~50,000) was found associated with MVM virions purified from rat cell lines (Tattersall et al. 1976). This polypeptide, absent in virus grown on murine cell lines, was shown to be a nonessential protein because virus particles lacking it are fully infectious. A protein of similar molecular weight has also been observed in preparations of KRV (Salzman and White 1970) and in H-1 (Kongsvik and Toolan 1972).

TABLE 1
Molecular-Weight Determination of Viral Polypeptides

	Polypeptide		
Virus	A	B	C
MVM	84,500	65,000	62,000
H-3	80,000	63,500	60,000
H-1	86,000	66,000	62,000
KRV	81,000	63,500	59,500
LuIII	85,000	65,500	62,000

Data are based on two separate determinations for H-3, H-1, and KRV, one determination for the LuIII polypeptides, and numerous determinations for MVM polypeptides (Tattersall et al. 1976).

Tryptic Digests of Full and Empty Virus Particles

Intact full and empty virions of KRV and H-3 were subjected to tryptic digestion and then analyzed by SDS-gel electrophoresis. The results (Fig. 3 and Table 2) show that these viruses are similar to MVM and H-1 in their trypsin sensitivity. The B polypeptide of full virions is readily converted to a C-like protein, whereas the B polypeptide of the empty particles is mostly resistant to cleavage. Tattersall et al. (1977) suggested that the availability of the tryptic cleavage site in the B protein of full MVM virions results from a conformational change induced by the presence of viral DNA. The small amount of B-protein cleavage seen in the empty particles may reflect the presence of defective virions which contain only a small portion of the viral genome (E. Faust, P. Tattersall, and D. C. Ward,

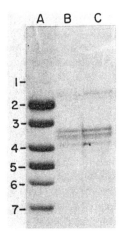

FIGURE 2
10% SDS-polyacrylamide gel of a mixture of H-3 and MVM full viruses. Well *B* contains 2 μg of each virus and well *C* contains 4 μg of each virus. The molecular-weight standards (well *A*) were the same as those in Figure 1.

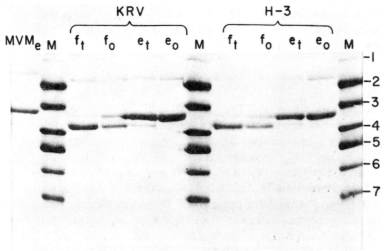

FIGURE 3
Comparison of H-3 and KRV proteins by SDS-polyacrylamide gel electrophoresis before (e_0 and f_0) and after (e_t and f_t) treatment with trypsin. The molecular-weight standards were the same as those in Figure 1.

TABLE 2
Conversion of B to C Protein in the Presence of Trypsin

Virus	Trypsin	Percent of total viral protein[a]			
		A	B	C	B + C
H-3 empty	−	12.7	87.3	0	87.3
	+	8.3	91.7	0	91.7
H-3 full	−	8.5	26.5	64.9	92.4
	+	8.5	7.5	83.9	91.4
KRV empty	−	9.7	87.1	3.2	90.3
	+	9.1	80.9	10.0	90.9
KRV full	−	10.2	30.3	59.5	89.8
	+	11.9	7.5	80.6	88.1
MVM empty	−	13.0	87.0	0	
	+	13.0	85.0	2.0	
MVM full	−	13.0	72.5	14.5	87.0
	+	12.0	17.2	70.8	88.0

[a] Values determined from densitometry of Coomassie-blue-stained proteins. It has been observed that the estimation of percent of total protein is less consistent when scanning Coomassie-blue-stained protein than autoradiographic exposures of labeled protein. This may account for the variability in the percent of A protein in the different viruses.

unpublished observations). These results suggest that the conformational properties of the full- and empty-capsid proteins of these four parvoviruses are quite similar.

Two-Dimensional Gel Electrophoresis

O'Farrell and Goodman (1976) reported that the VP1 capsid protein of SV40, a known phosphoprotein (Tan and Sokol 1972), exhibited microheterogeneity when examined by two-dimensional gel electrophoresis. The VP1 heterogeneity was observed both in purified virus preparations and in infected-cell extracts that were frozen immediately after harvesting. Two-dimensional gel electrophoresis of full and empty virions of MVM shows that all three viral proteins exhibit microheterogeneity in the isoelectric-focusing dimension (Fig. 4). The A protein, which has been reported to contain a region rich in basic amino acid residues not found in the B or C polypeptides (Tattersall et al. 1977), does not enter the standard O'Farrell isoelectric-focusing gels. However, it can be focused using a more basic gel system (see "Methods" and Fig. 4c). Each viral protein can

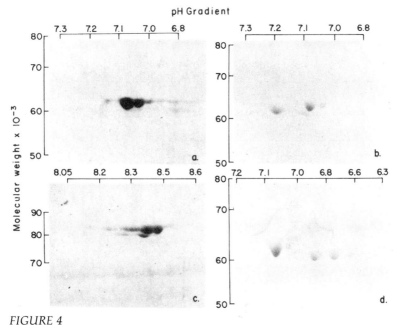

FIGURE 4

Two-dimensional gels of (*a*) the B protein from MVM empty virions, (*b*) the B and C proteins from MVM full virus, (*c*) the H-3 and MVM A proteins resolved on a basic-pH two-dimensional gel, and (*d*) a mixture of the B proteins from H-3 and MVM empty virions.

be resolved into 2–4 species which differ in apparent isoelectric point by ~0.05 pH units. Considerable caution must be exercised in preparing samples for these gels. Six to ten different isoelectric species of each viral protein are observed when virions are disrupted by boiling for 2 min in sample buffer. This high degree of microheterogeneity may reflect the thermal or chemical instability of the substituent responsible for the different isoelectric forms (see below). For example, all the proteins were electrophoresed in the presence of the reducing agent β-mercaptoethanol. We have not ruled out the possibility that extensive heterogeneity might be due to the lability of the protein substituent to reducing agents. Nevertheless, we believe that the observed heterogeneity of the MVM proteins in virions disrupted at 60°C is not artifactual, since viral proteins in infected-cell lysates show a similar pattern of microheterogeneity (Fig. 5) but most of the other cellular proteins are resolved as single protein species. However, the MVM B protein is not the only protein in the infected-cell cytoplasm to exhibit heterogeneity—cytoplasmic actin, for example, also exhibits micro-heterogeneity (Rubinstein and Spudich 1977).

The B and C protein species have pI values in this system between 7.0 and 7.2 (Fig. 4b), whereas the isoelectric forms of the A protein exhibit pI values between pH 8.2 and 8.5 (Fig. 4c). The capsid proteins of the other viruses studied exhibited a similar microheterogeneity in the isoelectric dimension. For example, Figure 4d shows the heterogeneity observed in the H-3 B protein compared to that of the MVM B protein. The H-3 protein is more acidic than the MVM B protein, with a pI range between 6.6 and 6.9.

FIGURE 5
A portion of a two-dimensional-gel fluorograph of ^{35}S-labeled cytoplasmic proteins from MVM-infected cells. The viral B protein is noted by an arrow.

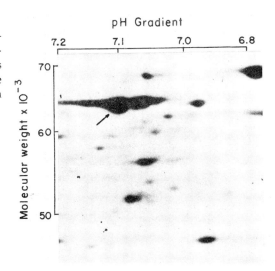

Phosphorylation of Viral Proteins

To determine whether the viral proteins were phosphorylated, and, if so, whether the phosphorylation could account for the observed isoelectric heterogeneity, MVM and H-3 virions were purified after growth in the presence of [32P]orthophosphate. An autoradiograph of a one-dimensional SDS-polyacrylamide gel clearly showed that all three MVM proteins contained 32P radioactivity. The distribution of radioactivity between the viral polypeptides was determined from the densitometer tracing of the gel autoradiograph (Fig. 6). The A protein of empty MVM particles contained 14% of the radioactivity and the B protein contained 86%; this is similar to the known molar ratio of these polypeptides within the virion. The B and C proteins of the full virus contained approximately 35% and 55% of the 32P-label, respectively. However, it was difficult to quantitate accurately the 32P content of the A protein because a large amount of 32P from the viral DNA streaked into the top of the gel. Nevertheless, a 32P-labeled A-protein band could be seen on the auto-radiograph, and its 32P content was estimated to be approximately 10% of the total radioactivity.

Analysis of 32P-labeled MVM empty virions by two-dimensional gel electrophoresis revealed that the specific radioactivity of the viral proteins increased with decreasing isoelectric point. As shown in Figure 7, the largest and most basic of the Coomassie-blue-stained B-protein spots (#1, 7.1-65) contained little or no 32P. Polypeptide #2 (7.0-65) was 32P-

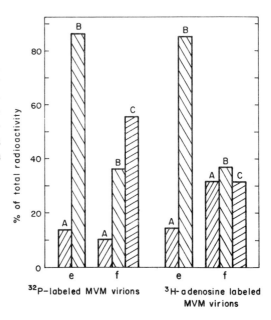

FIGURE 6
Distribution of [32P]ortho-phosphate and [3H]adenosine labels between MVM proteins (A, B, and C) in full and empty virions. The relative radio-activity content of each pro-tein was determined by densi-tometry of an autoradiograph (32P) or a fluorograph (3H) of a 10% polyacrylamide gel.

FIGURE 7
Coomassie-blue-stained, ^{32}P-labeled MVM empty proteins on a two-dimensional polyacrylamide gel compared with the corresponding autoradiogram. The Coomassie-blue stain (CB) and the autoradiogram (AR) were lined up exactly in the isoelectric dimension by a mark of radioactive ink on the gel. The gel and the autoradiogram are offset in the molecular-weight dimension to demonstrate the observed differences in the distributions of Coomassie-blue-stained and ^{32}P-labeled proteins.

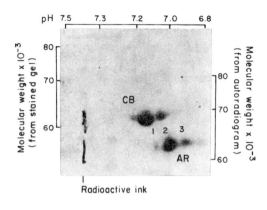

labeled, but its specific radioactivity was half that of polypeptide #3 (6.9-65). This pattern of radiolabeling would be expected if the microheterogeneity was a result of protein phosphorylation. A similar distribution of ^{32}P-label was seen with the different isoelectric forms of the MVM A protein and the H-3 B protein (data not shown). In the case of the MVM A protein, two of the ^{32}P-labeled bands (8.2-85 and 8.25-85) had very little Coomassie-blue stain, suggesting that there are minor species of A that are highly phosphorylated. We have experienced considerable difficulty in obtaining exact superimpositions of the autoradiographs and the Coomassie-blue-stained gel. In virtually all cases, the ^{32}P-autoradiograph spot appears on the acidic side of the corresponding Coomassie-blue-stained spot. Although the cause and the significance of this apparent displacement are unclear, it may reflect the lability of the phosphorylated substituent during isoelectric focusing.

To further characterize the nature of the protein phosphorylation, the B polypeptides of ^{32}P-labeled MVM and H-3 empty virions were purified. The isolated polypeptides contained only 10% of the virion-associated ^{32}P-label, which indicated either a labile linkage or the presence of ^{32}P-labeled DNA fragments in the empty virion pools. The purified B proteins were subjected to complete acid hydrolysis, and the hydrolysates were analyzed by thin-layer electrophoresis. None of the radioactivity from either viral protein comigrated with phosphoserine or phosphothreonine markers. Indeed, all of the ^{32}P-label from both viral protein hydrolysates coelectrophoresed with inorganic phosphate. The ^{32}P-labeled MVM B protein was then subjected to a variety of chemical and enzymatic treatments. These results, given in Table 3, eliminate the highly acid-labile amino acids, phosphohistidine and phosphoarginine, as likely phosphorylation candidates. The results are also inconsistent with the phosphorylated moiety being RNA or DNA alone.

Labeling of MVM Proteins with [³H]Adenosine

In order to determine whether the ³²P-labeled moiety might be in an ADP-ribose complex (Nishizuka et al., 1968; Goff, 1974; Zillig et al. 1975), MVM virions were purified from cells grown in the presence of [³H]adenosine and the polypeptides were analyzed by SDS-polyacrylamide gel electrophoresis. The relative radioactivity content of each viral protein was determined from densitometry tracings of the bands in the fluorograph of the stained gel (Fig. 6). All three protein species contained the [³H]adenosine-label. The distribution of radioactivity in the A and B proteins of the empty virion was similar to that seen with the ³²P-label. In contrast, the A protein of the full virion contained over 30% of the total radioactivity, which indicated preferential labeling of this polypeptide. When analyzed on two-dimensional gels, each isoelectric form of the MVM B protein was found to have approximately the same specific radioactivity (data not shown), in sharp contrast with the observed distribution of ³²P-label.

[³H]adenosine-labeled MVM B protein was purified from empty virions, and the chemical and enzymatic stability of the [³H]adenosine substituent investigated. Only about 10% of the radiolabel of intact particles was recovered in the purified B protein; this recovery level was similar to that obtained with ³²P-labeled virus. As seen in Table 3, both

TABLE 3
Stability of ³²P-Label and [³H]Adenosine Label in Isolated MVM B Protein After Chemical and Enzymatic Treatments

Treatment	n^a	³²P	[³H]adenosine	n^a
0.25 N HCl, 66°C, 120 min	2	70	61	2
0.25 N NaOH, 66°C, 120 min	2	86	68	2
Alkaline phosphatase, 30 μg/ml, 22°C, 30 min	2	100	100	2
Micrococcal nuclease, 25 μg/ml, 37°C, 30 min	2	76	90	3
DNase, 25 μg/ml, 37°C, 30 min	4	58	51	3
RNase, 25 μg/ml, 37°C, 30 min	3	86	93	3

TCA precipitable cpm (%)

[a] n: Number of separate determinations.

[32]P- and ^{3}H-labels were totally resistant to alkaline phosphatase and only partially sensitive to treatment with either DNase, RNase, 0.25 N HCl, or 0.25 N NaOH. Since the acid and base conditions employed were sufficient to totally hydrolyse ADP-ribose–protein complexes (Goff 1974), we conclude that the viral protein is not extensively substituted with ADP-ribose.

DISCUSSION

Our analyses of the number, molecular weights, and molar ratios of the capsid proteins from four nondefective parvoviruses suggest that the protein compositions of these virions are quite similar. Further, the selective susceptibility of the B protein in the full virions of KRV, H-3, H-1, and MVM to proteolytic cleavage suggests that these viral capsids are similar in their conformational properties and therefore may undergo a maturation scheme similar to that proposed for MVM by Tattersall et al. (1977).

The results presented here demonstrate that all three species of viral proteins can exist as phosphoproteins. Two to four distinct isoelectric forms of each protein can be resolved by two-dimensional gel electrophorsis. This microheterogeneity appears to reflect subpopulations of the viral protein which possess different levels of phosphorylation. Although viral polypeptides can also be radiolabeled with [3H]adenosine, the different isoelectric forms all possess similar specific radioactivities.

Our preliminary attempts to characterize the nature of the protein modification(s) have been unsuccessful. The ^{32}P-label is not found as phosphoserine, phosphothreonine, phosphohistidine, or phosphoarginine, nor is it in a form sensitive to alkaline phosphatase. Since only a small amount of ^{32}P is solubilized after treatment with nucleases, the substituent is unlikely to be purely DNA or RNA. In addition, the resistance of both [32P]orthophosphate and [3H]adenosine labels to mild acid hydrolysis argues against the presence of an ADP-ribose substituent. While we have not been able as yet to identify the structure of the phosphorylated and/or adenylated substituent(s), it is apparent that these viral proteins are subject to post-translational modification. Indeed, the difference in the distributions of [32P]orthophosphate and [3H]adenosine labels in the stained protein, observed in both one- and two-dimensional gels, suggests that the two radiolabels may reside in different chemical moieties and that neither modification may be directly responsible for the charge heterogeneity.

The biological significance of the different isoelectric species of viral proteins is unclear at present. The microheterogeneity may reflect viral polypeptides at different stages of post-translational modification during virion maturation. We have observed that intact full and empty virions of

MVM can each be resolved by isoelectric focusing into several distinct subpopulations (J. L. Peterson and D. C. Ward, unpublished results). The possibility that post-translational modification of viral proteins may regulate or modulate viral maturation events, such as DNA encapsidation or capsid-protein proteolysis, merits further investigation.

The finding that the B protein (pI 7.0–7.2) of MVM has a pI at least one full pH unit lower than the A protein (pI 8.2–8.5) is in agreement with the results of two-dimensional chymotryptic peptide maps of the viral proteins reported by Tattersall et al. (1977). They showed that the A protein contained several basic chymotryptic peptides not found in the B or C proteins. The observation that the A protein of H-3 virus also possesses a high isoelectric point indicates that the basic region of this polypeptide may be conserved between the parvoviruses and may interact with the DNA in full particles in a similar fashion to the histones incorporated into polyoma (Frearson and Crawford 1972) and SV40 (Lake et al. 1973).

ACKNOWLEDGMENTS

This work was supported by Public Health Service grant GM-20124 from the National Institute of General Medical Sciences and by grant CA-16038 from the National Cancer Institute.

REFERENCES

Bitte, L. and D. Kabat. 1974. Isotopic labeling and analysis of phosphoproteins from mammalian ribosomes. *Methods Enzymol.* **30**:5163.

Bomford, R. and I. B. Weinstein. 1972. Transformation of a rat epithelium-like cell line by murine sarcoma virus. *J. Natl. Cancer Inst.* **49**:379.

Bonner, W. M. and R. A. Laskey. 1974. Fluorographic detection of ³H, ¹⁴C and ³⁵S in acrylamide gels. *Eur. J. Biochem.* **46**:83.

Clinton, G. M. and M. Hayashi. 1975. The parvovirus MVM. Particles with altered structural proteins. *Virology* **66**:261.

———. 1976. The parvovirus MVM; A comparison of heavy and light particle infectivity and their conversion in vitro. *Virology* **74**:57

Frearson, P. M. and L. V. Crawford. 1972. Polyoma virus basic protein. *J. Gen. Virol.* **14**:141.

Goff, C. 1974. Chemical structure of a modification of the *Escherichia coli* ribonucleic acid polymerase polypeptides induced by bacteriophage T4 infection. *J. Biol. Chem.* **249**:6181.

Kongsvik, J. R. and H. W. Toolan. 1972. Capsid components of parvovirus H-1. *Proc. Soc. Exp. Biol. Med.* **139**:1202.

Kongsvik, J. R., J. F. Gierthy, and S. L. Rhode. 1974. Replication process of the parvovirus H-1. IV. H-1-specific proteins synthesized in synchronized human NB kidney cells. *J. Virol.* **14**:1600.

Laemmli, U. K. 1970. Cleavage of structural proteins during the assembly of the head of bacteriophage T4. *Nature* **227**:680.

Lake, R. S., S. Bartan, and N. P. Salzman. 1973. Resolution and identification of the core deoxynucleoproteins of the simian virus 40. *Biochem. Biophys. Res. Commun.* **54**:640.

Laskey, R. A. and A. D. Mills. 1975. Quantitative film detection of ^{3}H and ^{14}C in polyacrylamide gels by fluorography. *Eur. J. Biochem.* **56**:335.

Lee, A. S. and R. L. Sinsheimer. 1974. A continuous electroelution method for the recovery of DNA restriction enzyme fragments. *Anal. Biochem.* **60**:640.

Nishizuka, Y., K. Ueda, T. Honjo, and O. Hayaishi. 1968. Enzymic adenosine diphosphate ribosylation of histone and polyadenosine disphosphate ribose synthesis in rat liver nuclei. *J. Biol. Chem.* **243**:3765.

O'Farrell, P. H. 1975. High resolution two dimensional gel electrophoresis. *J. Biol. Chem.* **246**:6159.

O'Farrell, P. H. and H. M. Goodman. 1976. Resolution of simian virus 40 proteins in whole cell extracts by two dimensional electrophoresis. Heterogeneity of the major capsid protein. *Cell* **9**:289.

Peterson, J. L. and E. H. McConkey. 1976. Non-histone chromosomal proteins from HeLa cells. *J. Biol. Chem.* **251**:548.

Richards, R., P. Linser, and R. W. Armentrout. 1977. Kinetics of assembly of a parvovirus, minute virus of mice, in synchronized rat brain cells. *J. Virol.* **22**:778.

Rubenstein, P. A. and J. A. Spudich. 1977. Actin microheterogeneity in chick embryo fibroblasts. *Proc. Natl. Acad. Sci.* **74**:120.

Salzman, L. A. and W. L. White. 1970. Structural proteins of Kilham rat virus. *Biochem. Biophys. Res. Commun.* **41**:1551.

Tan, K. B. and F. Sokol. 1972. Structural proteins of simian virus 40: Phosphoproteins. *J. Virol.* **10**:985.

Tattersall, P., A. Shatkin, and D. C. Ward. 1977. Sequence overlap between the structural polypeptides of parvovirus MVM. *J. Mol. Biol.* **111**:375.

Tattersall, P., P. Cawte, A. Shatkin, and D. Ward. 1976. The structural polypeptides coded for by minute virus of mice, a parvovirus. *J. Virol.* **20**:273.

Zillig, W., H. Fujiki, and R. Mailhammer. 1975. Modification of DNA-dependent RNA polymerase. *J. Biochem.* **77**:7p.

Maturation of Minute-Virus-of-Mice Particles in Synchronized Rat-Brain Cells

RANDY RICHARDS
PAUL LINSER
RICHARD W. ARMENTROUT

Department of Biological Chemistry
University of Cincinnati College of Medicine
Cincinnati, Ohio 45267

Virus particles produced in cells infected with minute virus of mice (MVM) can be separated into at least three distinct types, which differ by their density in cesium chloride. The least dense of these, the empty particle, contains polypeptide B as its major component in addition to polypeptide A. Two density classes of full particles (1.42 and 1.46 g/cm³) are observed; both contain polypeptide C in addition to A and B. The lighter of these has polypeptide C as its major protein, whereas the denser one resembles more closely the empty capsid in its polypeptide composition. There is a progressive loss of the denser full particle and accumulation of the other full particle during the course of infection in randomly growing cells (Clinton and Hayashi 1976). Tryptic peptide mapping of the capsid proteins suggests that polypeptide B may be converted to polypeptide C by proteolytic cleavage (Tattersall et al. 1977). However, the light full particle has been observed rather late in the course of infection, after the time required for a single cycle of virus growth. The late appearance of this particle type raises the possibility that the putative processing event is a nonspecific degradation occurring in disintegrating

447

cells, rather than a specific step in virus assembly and maturation. We have therefore examined the kinetics of the appearance and turnover of these two particle types in synchronized cell cultures infected with MVM.

MATERIALS AND METHODS

Cells and Virus

RT-7 cells, isolated from a rat brain tumor by Dr. W. Au, were maintained in antibiotic-free minimal essential medium (F-11, Gibco) supplemented with 5% heat-inactivated fetal-calf serum (FCS, Gibco).

Plaque-purified MVM was kindly supplied by Dr. Peter Tattersall and was grown in RT-7 cells essentially as described by Richards et al. (1977). Hemagglutinin was measured using a 25-μl sample plus 25 μl of 5% mouse red blood cells in phosphate-buffered saline (PBS). Using this assay, the empty capsid and the 1.42-g/cm^3 virus particles have similar specific hemagglutinin titers, 470–479 hemagglutinating units (HAU) per μg protein, whereas the 1.46-g/cm^3 particle has a somewhat lower titer (150 HAU/μg protein).

Virus to be used to infect cells was first purified by sedimentation in a sucrose gradient. The particles sedimenting at 110S [2 × 10^6 plaque-forming units (PFU) per HAU] were filter-sterilized and diluted in F-11 medium.

Synchronization

Monolayers of RT-7 cells grown to 50% confluency in 150-cm^2 Corning T flasks in F-11 medium with 5% heat-inactivated FCS were refed with fresh F-11 medium containing 10% FCS. Twenty-four hours later the flasks were shaken vigorously to remove mitotic cells and loosely adhering cells. The monolayers were then refed with media prepared by mixing equal amounts of F-11 medium containing 10% FCS that had been preincubated with cells for 16 hours and fresh F-11 medium supplemented with 10% FCS (preconditioned medium). Two hours after the initial detachment, the monolayers were shaken gently to remove mitotic cells. Detached mitotic cells from several T flasks were pooled, and samples for cell count and mitotic index determination were processed immediately. Aliquots of cells were then plated into either 75-cm^2 Falcon T flasks (for analysis of virus production) or 25-cm^2 Falcon T flasks (as control samples for monitoring the degree of synchrony). Cells were infected by exposure to 5–50 PFU/cell for 1 hr, 3 hr after detachment of mitotic cells.

The degree of synchronization of the control cultures was monitored by following the mitotic index and by determining the fraction of labeled nuclei by autoradiography. At intervals (see "Results") during the experiment, 25-cm² T flasks were exposed to 1 μCi/ml methyl-[³H]thymidine for 30 min. The monolayer was then washed with ice-cold PBS. The cells were swollen and lysed by exposure to 0.5% Na-citrate solution for 10 min at 37°C. The isolated nuclei were then pelleted at low speed and resuspended in ice-cold Carnoy's fixative for 30 min. The sample was spread on a microscope slide and air-dried. The slides were washed extensively in 5% trichloroacetic acid and then in distilled H₂O to remove excess label, and were subsequently dipped in liquified Kodak NTB2 nuclear track emulsion. After one week of exposure at 4°C the slides were developed in Kodak D-19 chemical developer and stained with Giemsa.

Isolation of Nuclei

Nuclei were isolated by the method of Wray (1975). Phase-contrast microscopy showed that more than 95% of the nuclei were free of cytoplasmic tags. Only 4% or less of the cellular DNA was found in the cytoplasmic sample.

Gel Electrophoresis

SDS-polyacrylamide gel electrophoresis was performed as described by Laemmli (1970). Approximate molecular weights were determined from standard proteins run in the same gel with the virus samples.

RESULTS

Proteins of Virus Particles

In sucrose gradients, virus particles could be readily separated by sedimentation into two classes, full virus (110S) and empty capsids (70S) (Rose 1974). [³H]thymidine is associated only with the 110S particles; in these preparations the 70S particles do not contain detectable levels of DNA (Richards et al. 1977).

The proteins of the 70S and 110S particles were analyzed on SDS-acrylamide gels (Fig. 1). In accord with the results of Tattersall et al. (1976), the empty capsid (70S) contains two proteins, one of 85,000 daltons and the other of 68,000 daltons. When the 110S material is centrifuged to equilibrium in CsCl, two rather broad bands of virus are obtained, with peak densities at 1.46 and 1.42 g/cm³. The 1.42-g/cm³ virus particles contain the 85,000-dalton protein and a 65,000-dalton protein, whereas the virus of density 1.46 contains principally proteins of 85,000 and 68,000 daltons. These molecular weights for the viral proteins in the 110S particles are consistent with those found in the extensive studies of

FIGURE 1
SDS-polyacrylamide gel electrophoresis performed on the proteins of MVM particles. Virus particles were denatured in SDS (1%) and β-mercaptoethanol (1%), as described by Laemmli (1970), and subjected to electrophoresis in 8.75% gels at 20 mA/gel and stained with Coomassie blue. (a) 1.42-g/cm³ and (b) 1.46-g/cm³ particles isolated by CsCl equilibrium centrifugation; (c) empty virus purified by sucrose gradient centrifugation; (d) 1.46-g/cm³ virus mixed with an equal amount of empty virus; (e) 1.42-g/cm³ virus mixed with an equal amount of empty virus. The protein standards, run in the same gel as the viral polypeptides (A), (B), and (C), are (D) phosphorylase A (m.w. 92,500), (E) ovalbumin (m.w. 45,000), and (F) ¹²⁵I-labeled BSA (m.w. 68,000). The position indicated for protein standard F was determined by autoradiography of this gel after photography.

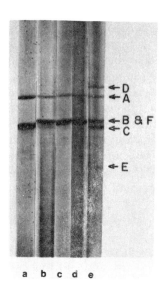

Tattersall et al. (1976). In addition, these results confirm the observation of Clinton and Hayashi (1975) that full MVM occurs as two particles of separate densities, each with a different major capsid protein.

To clarify the sequence of events in the assembly of MVM we have analyzed virus replication in a single-step infection of highly synchronized cells. As measured by the fraction-of-labeled-nuclei method, between 75% and 90% of the synchronized cells go through the S phase together, and fewer than 1% of the cells are synthesizing DNA during the G1 phase.

Kinetics of Virus Assembly

The kinetics of virus assembly was examined both by electron miroscopy and by biochemical analysis of synchronized infected RT-7 cultures. In order to correlate the two methods closely, samples examined by electron microscopy were taken from experiments analyzed by biochemical techniques. Observations obtained by electron microscopy are presented elsewhere (Richards et al. 1977); they resemble closely the results reported by Singer et al. from their studies of H-1 infections (Singer and Toolan 1975; Singer 1975).

At 14, 20, 28, and 37 hr after mitotic detachment, samples of the synchronized, infected cultures were exposed to [³H]thymidine (150 µCi/ml) for 3 hr. The virus particles were then extracted from each sample and the full virus was separated from empty capsids on sucrose gradients. The rate of synthesis of the full virus can be measured by the amount of [³H]thymidine label incorporated into the 110S region of these gradients. The total amount of virus accumulated was determined from the hemagglutination titers of the 110S (full virus) and 70S (empty capsid) regions.

As early as 17 hr after mitotic detachment slight amounts of both full and empty virus can be measured. The rates of accumulation of the full and empty particles are similar, but the empty capsid may begin to be synthesized slightly before the full virus (Fig. 2). There is a high rate of assembly of both full and empty particles in the interval of 28–31 hr after cell synchronization. Following this burst of full-virus assembly, virus synthesis continues, but at a greatly diminished rate.

Conversion of 1.46-g/cm³ Particles to 1.42-g/cm³ MVM

In order to determine if the 1.46-g/cm³ particle is processed to form the 1.42-g/cm³ particle in vivo, we have performed a "pulse-chase" experi-

FIGURE 2
Rates of virus assembly and accumulation in synchronized cells. RT-7 cells were mitotically detached and infected 3 hr later. [³H]thymidine (150 μCi/ml) was added for 3 hr at 14, 21, 28, and 37 hr after detachment. The labeled full virus was isolated from a sucrose gradient (o---o), and total full (●——●) and empty (x——x) viruses were measured by hemagglutination.

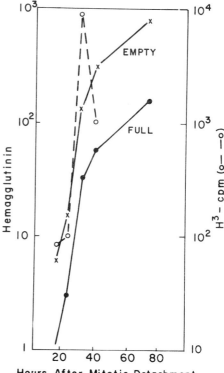

ment. Infected, synchronized RT-7 cultures were exposed to [³H]thymidine (150 μCi/ml) for 2 hr during the period of the highest rate of virus assembly (28–30 hr after detachment). The medium was then removed and the synthesis of DNA was inhibited by the addition of medium containing 25 μg/ml hydroxyurea (HU), and excess unlabeled thymidine (2.5 mmoles) was added to the medium to dilute any labeled thymidine released to internal nucleotide pools by the breakdown of DNA.

Preliminary experiments have shown that, under these "chase" conditions, net [³H]thymidine incorporation ceases within 20 min.

Samples were taken at intervals after the addition [³H]thymidine and after "chase" conditions were instituted. The full-virus material (110S) was isolated from each sample on sucrose gradients. Throughout the experiment, the total incorporation of [³H]thymidine into the cultures was also followed, and the amount of labeled virus released into the culture medium was determined.

As shown in Figure 3, [³H]thymidine is incorporated into the cultures at a linear rate from the time of addition until "chase" conditions begin at 120 min. Net incorporation into the infected cultures then ceases abruptly. Prior to the "chase," label is incorporated into 110S virus particles at a rate that is similar to the rate of incorporation into the whole cultures. Net incorporation of [³H]thymidine into the full virus particles stops immediately upon addition of HU and unlabeled thymidine. There is no significant decrease in the total amount of label in virus particles (cellular virus plus virus in the medium) during the 120 min of the "chase" period.

FIGURE 3
Incorporation of [³H]thymidine into cells and into 110S virus before and after addition of hydroxyurea and unlabeled thymidine. Total cellular [³H]thymidine incorporation was determined on an aliquot of the infected cultures after sonication (o---o). The total amount of label in 110S virus was determined by TCA precipitation across sucrose gradients run on both tissue-culture fluid and virus extracted from the cultures at each time point (●——●). At 240 min, 11% of the total viral 110S label is found in the medium.

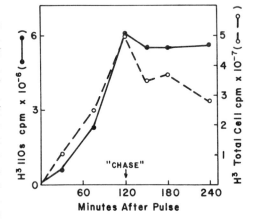

The 110S labeled MVM from each cell sample was analyzed on CsCl density gradients; the results are shown in Figure 4. At the end of a 30-min labeling period, over 96% of the [³H]thymidine in full virus particles is found at the density of 1.46 g/cm³ (Fig. 4A). However, even in this brief interval detectable amounts of 1.42-g/cm³ material are observed. By 120 min, 72% of the labeled virus is of the 1.46-g/cm³ class while 28% is found at 1.42 g/cm³ (Fig. 4C). After 120 min under "chase" conditions, only 26% of the virus-associated label is found at 1.46 g/cm³, and 74% is present in the 1.42-g/cm³ particle (Fig. 4F). These results are summarized in Figure 5. In these experiments, we have shown that [³H]thymidine incorporation into DNA stops abruptly upon addition of HU and cold thymidine to the cultures. Net [³H]thymidine incorporation into 110S virus stops at the same time. Label is lost from the 1.46-g/cm³ particle after the "chase" begins, but the accumulation of the 1.42-g/cm³ particle continues at an undiminished rate. The total amount of label lost from the 1.46-g/cm³ particle is comparable to the total amount of label gained by the 1.42-g/cm³ particle. In addition, the rate of loss of label from the 1.46-g/cm³ virus approximates the rate of label accumulation by the 1.42-g/cm³ virus. The simplest explanation for these observations is that the 1.46-g/cm³ particles are a direct precursor of the 1.42-g/cm³ particles.

FIGURE 4
CsCl gradients of full MVM isolated from infected synchronized RT-7 cells. The full virus isolated by sucrose gradients in the experiment described in the legend to Figure 3 was centrifuged to equilibrium in CsCl (type-40 rotor, 48 hr, 35,000 rpm). Lambda phage was used as a density marker, and the viral peaks were confirmed by hemagglutinin titer. (*A*) 30 min, (*B*) 75 min, (*C*) 120 min, (*D*) 150 min, (*E*) 180 min, (*F*) 240 min.

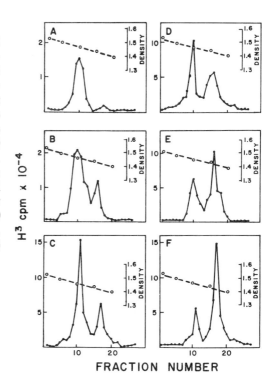

FIGURE 5
Kinetics of processing of 1.46-g/cm³ MVM and accumulation of 1.42-g/cm³ MVM. The total radioactivity of the infected cultures for each density species of full virus was determined from Figure 4 and is summarized in this figure. ●——●: 1.46-g/cm³ particles. o---o: 1.42-g/cm³ particles.

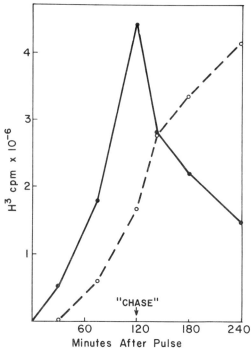

Virus Released to the Medium

During the course of these experiments about 11% of the total 110S labeled virus can be found in the culture medium. In conjunction with cell samples, the culture fluid was analyzed on sucrose gradients. The amount of virus released into the medium was determined by quantitating the amount of [³H]thymidine label in the 110S region of these gradients. The 110S material isolated from the culture fluid of the 240-min sample was analyzed on a CsCl density gradient: 27% of the [³H]thymidine label was of 1.46 g/cm³ density and 73% was of 1.42 g/cm³ density. Thus, full viruses of both density classes, as well as empty capsids, are found free in the medium. These results are consistent with the idea that disintegration of the infected cells, as observed in electron micrographs, is the primary mechanism of virus release.

Cellular Site of Virus Processing

During the course of these experiments a proportion of 1.46-g/cm³ virus particles were not converted to 1.42-g/cm³ particles. We have attempted to explain the persistence of unprocessed particles by determining

TABLE 1
Compartmentalization of Newly Synthesized Full MVM from
Infected Synchronized RT-7 Cells

Hours after labeling	Nuclear ³H-labeled MVM	Cytoplasmic ³H-labeled MVM
6	99,400 (73%)	36,300 (27%)
13	17,300 (21%)	64,300 (79%)

Cells were synchronized and infected as described in the text. At 26 hr after mitotic detachment the cells were labeled, and label was "chased" as described in the text. The separation of nuclear and cytoplasmic fractions and the determination of [³H]thymidine in 110S virus are described in "Methods."

whether virus processing is restricted to a particular cellular site. RT-7 cells were synchronized and infected as before. The cultures were exposed to [³H]thymidine (150 μCi/ml) for 6 hr beginning at 26 hr after mitotic detachment. The medium was then removed and the cultures were exposed to "chase" conditions for 7 hr. Samples were taken at the end of the 6-hr labeling period and at the end of the 7-hr chase. For each sample, the cytoplasm and nuclear fractions were separated. Full virus was isolated from the nuclear and cytoplasmic fractions of each sample by sucrose-gradient centrifugation. The 110S-virus peak was then analyzed on CsCl density gradients. Labeled 110S virus can be found in both the nucleus and the cytoplasm at the end of the 6-hr pulse, but by the end of the 7-hr chase most of the [³H]thymidine-labeled virus has left the nucleus, and the labeled virus in the cytoplasm has increased by 77% (Table 1).

Table 2 shows the results of the CsCl centrifugation of the 110S virus isolated from the nuclear and cytoplasmic fractions in this experiment.

TABLE 2
Distribution of Newly Synthesized 1.46-g/cm³ and 1.42-g/cm³ MVM in the Nucleus and Cytoplasm of Infected Synchronized RT-7 Cells

Hours after pulse	Nuclear ³H-labeled MVM		Cytoplasmic ³H-labeled MVM	
	1.46 g/cm³	1.42 g/cm³	1.46 g/cm³	1.42 g/cm³
6	27,900 (28%)	72,000 (72%)	12,000 (36%)	21,600 (64%)
13	1,200 (11%)	15,400 (89%)	26,000 (41%)	37,200 (59%)

[³H]thymidine-labeled MVM isolated from synchronized RT-7 cells (see legend to Table 1) was centrifuged to equilibrium in CsCl.

The proportion of 1.46-g/cm³ particles in the nucleus drops from 28% of the total at the end of the 6-hr labeling period to 11% of the total at the end of the 7-hr chase. In the cytoplasm the relative proportion of 1.46-g/cm³ particles to 1.42-g/cm³ particles remains essentially the same.

From these data it appears that processing of the 1.46-g/cm³-class particles occurs rapidly in the nucleus. Particles are transported after assembly to the cytoplasm, and this transport appears to occur regardless of whether the particles have been processed. Once the 1.46-g/cm³ particles have reached the cytoplasm, processing occurs at a greatly reduced rate or not at all.

DISCUSSION

It has been shown that synthesis of MVM appears to require some function of the host-cell S phase (Tattersall 1972). We have shown that assembly of the virus particles occurs abruptly some 8 hr after the end of the S phase. These infected cells do not undergo subsequent mitosis. Thus, virus assembly and processing do not appear to require the host cell to be in S phase.

In these experiments we have presented evidence that the 1.46-g/cm³ class of MVM particles are processed to form virus of density 1.42 g/cm³. Processing of the 1.46-g/cm³ particles is observed at the time of maximal virus assembly in synchronized cells. Detectable amounts of the processed 1.42-g/cm³ virus are formed within 30 min. The processing event occurs most rapidly in the infected-cell nucleus, the site of virus assembly. On the basis of these results, we conclude that the conversion of 1.46-g/cm³ particles to 1.42-g/cm³ particles is an integral step in the events of virus assembly.

The alteration of particles after their assembly may be necessary for the production of infectious virus. Clinton and Hayashi (1976) have presented evidence that the precursor particles interact weakly with cells when compared to the processed particles. However, when the cell-binding reaction is specifically examined, it is seen that both particles adsorb to susceptible cells at similar rates (Linser et al., this volume). For reasons which are still unclear, it would seem that the processed particle is more readily taken up by the cell after adsorption.

As is the case with most animal viruses, the role of the empty viral capsid in the MVM assembly sequence has not been defined. The empty capsids could be (1) a precursor to the 1.46-g/cm³ virus (the latter particles arising from insertion of viral DNA), (2) a degradation product of the 1.46-g/cm³ virus through extrusion of viral DNA, or (3) an independently assembled particle unrelated to full virus. The empty capsids are detected slightly before the appearance of full virus, both in electron micrographs and in extracts of synchronized infected cultures. On this basis, and because we have shown a quantitative transfer of [^{3}H]thymidine label from the 1.46-g/cm^3 to the 1.42-g/cm^3 class, it seems unlikely that the empty capsids are produced by loss of DNA from the 1.46-g/cm^3 particles.

It is possible that the empty capsids are precursors to the 1.46-g/cm^3 virus. However, assembly of the 1.46-g/cm^3 particles is immediately halted when DNA synthesis is inhibited. As there are large amounts of empty capsids present in the infected cell, either the pool of single-stranded viral DNA available for insertion must be very small or else insertion requires ongoing DNA synthesis (perhaps as a driving force for this process).

Regardless of the actual relationship between the empty capsids and the 1.46-g/cm^3 virus particle, it is clear that the presence of viral DNA has a profound effect on the surface properties of the particle. The empty capsid and the 1.46-g/cm^3 virus contain the same proteins in the same proportions, yet the 1.46-g/cm^3 particle hemagglutinates mouse red blood cells poorly and is readily processed to the 1.42-g/cm^3 form by proteolytic cleavage. On the other hand, the empty capsid particles have a high hemagglutination activity but are very resistant to proteolytic processing both in vitro and in vivo (Clinton and Hayashi 1976; Tattersall et al. 1976).

In the sequence of MVM assembly, virus is first detected in the nucleus; shortly afterwards it appears in the cytoplasm and in the culture medium. All three classes of virus particles move out of the nucleus to the cytoplasm and into the medium with equal facility. In electron micrographs we have found no evidence of any virus-specific transport process either from the nucleus to the cytoplasm or out of the cell. It appears likely that virus particles are expelled from the nucleus by some generalized process such as margination of chromatin and blistering of the nuclear membrane. Similarly, virus particles spill into the medium, probably during cellular disintegration. Membrane fragments coated with virus particles are a common feature of cultures in the last stages of infection.

ACKNOWLEDGMENTS

This investigation was supported by grants 1 K04 CA00134 and 5 R01 CA 16517 awarded by the National Cancer Institute.

REFERENCES

Clinton, G. M. and M. Hayashi. 1975. The parvovirus MVM: Particles with altered structural proteins. *Virology* **66**:261.

———. 1976. The parvovirus MVM: A comparison of heavy and light particle infectivity and their density conversion in vitro. *Virology* **74**:57.

Laemmli, U. 1970. Cleavage of structural proteins during the assembly of the head of bacteriophage T4. *Nature* **227**:680.

Richards, R., P. Linser, and R. W. Armentrout. 1977. Kinetics of assembly of a parvovirus, minute virus of mice, in synchronized rat brain cells. *J. Virol.* **22**:778.

Rose, J. A. 1974. Parvovirus reproduction. In *Comprehensive virology* (ed. H. Fraenkel-Conrat and R. R. Wagner), vol. 3, p. 1. Plenum, New York.

Singer, I. L. 1975. Ultrastructural studies of H-1 parvovirus replication. II. Induced changes in the deoxyribonucleoprotein and ribonucleoprotein components of human NB cell nuclei. *Exp. Cell Res.* **95**:205.

Singer, I. L. and H. W. Toolan. 1975. Ultrastructural studies of H-1 parvovirus replication. I. Cytopathology produced in human NB epithelial cells and hamster embryo fibroblasts. *Virology* **65**:40.

Tattersall, P. 1972. Replication of the parvovirus MVM. I. Dependence of virus multiplication and plaque formation on cell growth. *J. Virol.* **4**:872.

Tattersall, P., A. J. Shatkin, and D. C. Ward. 1977. Sequence homology between the structural polypeptides of the minute virus of mice, a parvovirus. *J. Mol. Biol.* **111**:375.

Tattersall, P., P. J. Cawte, A. J. Shatkin, and D. C. Ward. 1976. Three structural polypeptides coded for by minute virus of mice, a parvovirus. *J. Virol.* **20**:273.

Wray, W. 1975. Parallel isolation procedures for metaphase chromosomes, mitotic apparatus, and nuclei. *Methods Enzymol.* **40**:75.

Assembly of Parvovirus LuIII in Brij-58-Lysed Cells

MARKUS GAUTSCHI
GERTRUD KRONAUER
GÜNTER SIEGL

Institute of Hygiene and Medical Microbiology
University of Bern
CH 3010 Bern, Switzerland

PETER REINHARD
JOHANNES R. GAUTSCHI

Institute of Pathology
University of Bern
CH 3010 Bern, Switzerland

LuIII virus, like other parvoviruses (Rose 1974), replicates within the nucleus of the host cell (Siegl and Gautschi 1973). About 10–12 hours after infection of synchronized cell cultures, replicating viral DNA and empty capsids accumulate. These virus-specified products can be extracted in the chromatin fraction, which indicates that virus assembly may occur in close association with the host-cell chromatin (Gautschi et al. 1976). Distinct viral nucleoprotein complexes, which are probably intermediates in virus assembly, have been isolated from subnuclear fractions (Gautschi and Siegl 1975).

Here we report the use of an in vitro system, derived from Brij-58-lysed cells, to study the assembly pathway of complete LuIII virions. The advantage of this system is that nuclei are permeable to exogenously added high-molecular-weight factors (Reinhard et al. 1977; Gautschi et al. 1977).

MATERIALS AND METHODS

Cell Cultures and Virus

NB cells (SV40-transformed newborn-human kidney cells) were a generous gift of Dr. J. van der Noordaa of the University of Amsterdam. These cells were grown to a density of 4–5 × 10⁶ cells per 75-cm² plastic flask and synchronized for DNA synthesis, and were infected at release from the synchronization block as described previously (Siegl and Gautschi 1973).

Labeling of DNA

After adsorption of virus, cells were prelabeled with [¹⁴C]thymidine (TdR) (1 μCi/ml, 50 mCi/mmole) for 3 hr. ¹⁴C radioactivity incorporated into nuclear DNA was used as a reference to monitor cell number and recoveries. In vivo, viral DNA was pulse-labeled for 30 min with [³H]TdR (50 μCi/ml, 47 Ci/mmole). In vitro, viral DNA in Brij-58-lysed cells was labeled with [³H]dTTP (200 μCi/ml, 2.5 Ci/mmole).

Preparation of Brij-58-Lysed Cells

Cells from monolayer cultures were scraped off the flasks, washed, incubated for 20 min at 0°C in lysis solution (Reinhard et al. 1977) containing 0.01% or 0.1% Brij-58, centrifuged, and resuspended in the reaction mixture at a density of about 10⁷/ml. Cells lysed with 0.01% Brij-58 were still surrounded by cytoplasmic material, but cellular boundaries were no longer visible with a light microscope. In contrast, naked nuclei were produced by lysing cells with 0.1% Brij-58. Aliquots of 20 μl containing about 5 × 10⁵ nuclei were incubated at 30°C, and newly formed virus particles were isolated at various times.

Isolation of Virus Particles

After incubation, in vitro samples were diluted one- to tenfold with 20 mM Tris-HCl (pH 7.4) and stored at −70°C. To purify virus particles, samples were incubated with DNase I (final concentration 80 μg/ml) for 30 min at 37°C in the presence of 0.5 M NaCl and 50 mM MgCl₂. After addition of EDTA (final concentration 20 mM) and Sarkosyl-NL 97 (final concentration 1%), incubation was continued at 37°C for 1 hr. Samples were then loaded onto linear 10–30% sucrose gradients (0.1% Sarkosyl, 0.35 M NaCl, 10 mM Tris-HCl, 1 mM EDTA, pH 8.0). After centrifugation in a Beckman SW 50.1 rotor at 40,000 rpm for 70 min at 20°C, gradients were fractionated and the sedimentation coefficients of particles were calculated as described by Gautschi and Clarkson (1975). For buoyant-density analysis, particles were centrifuged through a 30% sucrose layer

(1 ml) and banded in the underlying CsCl gradient (3.5 ml, density 1.4 g/cm³) in an SW 50.1 rotor at 40,000 rpm for 14 hr at 15°C. Fractions of the gradients were collected from the bottom and radioactivity profiles were measured (Gautschi et al. 1976).

RESULTS

Assembly of LuIII Viruses In Vitro

Synchronized and [¹⁴C]TdR-prelabeled NB cells were pulse-labeled with [³H]TdR at 11 hr p.i. for 30 min, lysed with 0.01% Brij-58, and incubated in vitro in the presence of unlabeled 5 × 10⁻⁵ M dTTP. Virus particles were isolated after different periods of incubation in vitro, and the amount of virions newly assembled in vitro was estimated on the basis of the radioactivity profiles. The sedimentation profiles in sucrose gradients showed a time-dependent increase of DNase-resistant particles in the region of 110S (Fig. 1A). Since complete virions formed in vivo also sedimented at 110S, DNase-resistant particles formed in vitro appeared to represent complete virions. To test this notion, aliquots of samples

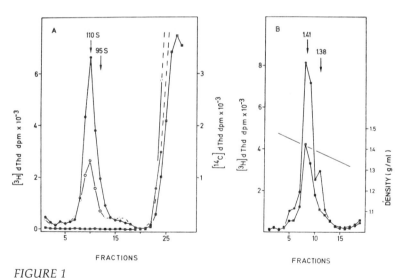

FIGURE 1

(A) Sedimentation profiles of LuIII-virus particles labeled with [³H]TdR for 30 min in vivo (o——o) and chased for 120 min in vitro (●——●). Cellular DNA of LuIII-infected and Brij-58-lysed cells was prelabeled in vivo with [¹⁴C]TdR (■——■). (B) CsCl buoyant density profiles of LuIII-virus particles labeled with [³H]TdR for 30 min in vivo (■——■) and chased for 120 min in vitro (●——●).

containing 110S structures assembled in vitro were analyzed in isopycnic CsCl gradients. Viral 110S particles formed after lysis of cells with Brij-58 were stable at high salt and banded mainly at a density of 1.41 g/cm³, i.e., at the density of mature virions (Fig. 1B). Less than 10% of the total particles had densities of either 1.44 g/cm³ or 1.38 g/cm³. Thus, the physicochemical characteristics of 110S structures assembled during incubation of Brij-58-lysed cells suggested that complete LuIII particles were formed in vitro.

To determine the kinetics of the formation of 110S particles in cells lysed with 0.01% Brij-58, the relative increase in radioactivity in the 110S peak at different times of incubation in vitro was measured. The relative number of 110S particles increased by factors of about 1.5, 2.0, and 2.4 during 15, 60, and 120 min of incubation in vitro, respectively. [14C]thymidine prelabel incorporated into cellular DNA was used to correct for variation of cell numbers.

Two control experiments were carried out to test whether the increase of radioactivity in the 110S peak really was due to an increase in the number of virus particles and not simply caused by nonspecific adsorption of ³H-label to preformed particles. First, LuIII-infected and Brij-58-lysed cells were incubated in vitro at 30°C in the presence of 10 mM EDTA to inhibit enzymatic reactions. Under these conditions no increase in radioactivity in the 110S peak could be detected. Second, different amounts of purified and labeled virus particles were mixed with infected and Brij-58-lysed cells. From these mixed samples virions were isolated and the recovery of radioactivity in the 110S region was determined. The increase in radioactivity was proportional to the number of virus particles added, and the recovery of radioactivity was about 95%. Under these conditions, therefore, radioactivity in the 110S region of sucrose gradients could be used to estimate relative numbers of complete virions containing labeled DNA. In further experiments we investigated the capacity for in vivo virus assembly of cells lysed with Brij-58 at different times in the virus replication cycle. In addition, Brij-58 concentrations were varied to study effects due to the degree of lysis. Cells were pulse-labeled in vivo for 30 min at 11, 14, and 17 hr p.i., respectively, lysed with either 0.01% or 0.1% Brij-58, and subsequently chased in vitro for 60 min. Virus particles were isolated and the numbers of virions formed in vitro were calculated as the percentage over the amount labeled during the 30-min pulse in vivo. The relative amounts of virus assembled in vitro decreased from 11 to 17 hr p.i. (Fig. 2). Cells lysed with 0.1% Brij-58 at 11 hr p.i. assembled viral particles about half as efficiently as cells lysed with 0.01% Brij-58. At later times in the virus replication cycle, however, no differences in the efficiency of assembly were detected when cells were lysed with 0.01% or 0.1% Brij-58.

Incorporation of Viral DNA Synthesized In Vitro into Complete Virions

In LuIII-virus-infected HeLa cells, empty viral capsids are present in five-
to tenfold excess over DNA-containing particles (Gautschi et al. 1976).
Similar results were obtained with NB cells. However, the ratio of 60S
empty capsids to 110S virions in NB cells decreased by approximately a
factor of 2 from 11 to 17 hr p.i. This indicated that empty capsids were
present in a large pool in the nuclei before newly synthesized progeny
viral DNA was packed into preformed capsids (for reviews see Casjens
and King 1975 and Russell and Winters 1975). It was of interest, therefore,
to test whether viral DNA synthesized in vitro was incorporated into
complete virions. At $11\frac{1}{2}$ hr p.i. LuIII-infected NB cells, ^{14}C-prelabeled in
their cellular DNA, were lysed with 0.01% Brij-58 and incubated at 30°C
in the presence of [^{3}H]dTTP. After different incubation times, incorporation of radioactivity into DNA was measured. Cells lysed with 0.01%
Brij-58 incorporated [^{3}H]dTTP for at least 2 hr. In the Hirt supernatants
prepared from lysed cells (Siegl and Gautschi, this volume), labeled DNA
of full viral genome size could be detected readily after 15 min of incubation in vitro (Fig. 3). The sedimentation spectrum of in-vitro-labeled
DNA resembled closely that found in vivo (Siegl and Gautschi, this
volume). Virus particles were isolated from these in-vitro-labeled prep-

FIGURE 2
Rate of synthesis of 110S particles in Brij-58-lysed cells
prepared at various times in
the replication cycle of LuIII
virus. Viral DNA of 110S particles was labeled for 30 min in
vivo (●——●) and chased for 60
min in vitro. Relative numbers of 110S virions formed in
cells lysed with 0.01% Brij-58
(□-----□) or 0.1% Brij-58
(■——■).

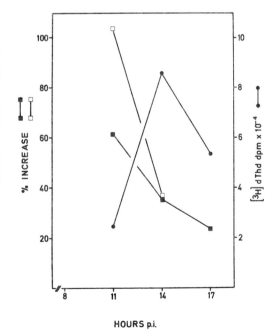

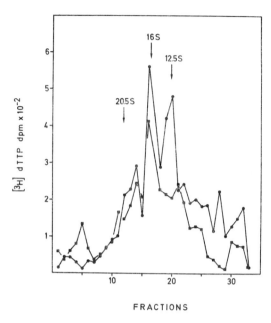

FIGURE 3
Sedimentation in alkaline gradients of virus-specific low-molecular-weight DNA labeled in vitro with [³H]dTTP for 15 min (■——■) and 120 min (●——●).

arations and analyzed in sucrose gradients (Fig. 4). At zero time and after 15 min of incubation, no ³H radioactivity was found in the 60–110S region. Thus, the labeled viral DNA which was present in the Hirt supernatants as early as 15 min after the start of in vitro incubation was not incorporated into DNase-resistant particles during that time. In addition, these results excluded unspecific trapping of labeled or unlabeled DNA by unlabeled virions already present in the incubation mixture. After 60 min of incubation in vitro, however, nuclei contained DNase-resistant and [³H]dTTP-labeled particles which sedimented as a broad peak at about 95S. Sixty minutes later, the peak had shifted to 110S with a broad shoulder in the range 70–95S. The label present in the 110S particles proved to be alkali-stable and acid-precipitable. Furthermore, particles isolated after 60 min of incubation in vitro banded in CsCl gradients mainly at 1.38 g/cm³, whereas after 120 min of incubation the majority of the labeled particles banded at 1.41 g/cm³ (Fig. 5). These results strongly suggest a precursor-product relationship between an incomplete 95S virus particle and the complete 110S virus particles.

DISCUSSION

The results presented in this paper demonstrate the in vitro assembly of LuIII in infected, Brij-58-lysed cells to complete virions. This interpretation is supported by the observations that particles formed in vitro are

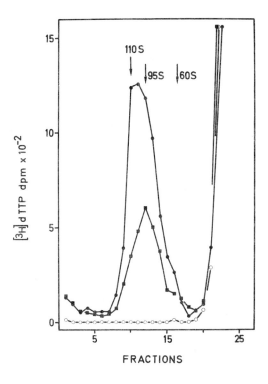

FIGURE 4
Sedimentation profiles of LuIII-virus particles labeled in vitro with [³H]dTTP during 15 min (o——o), 60 min (■——■), and 120 min (●——●) of incubation.

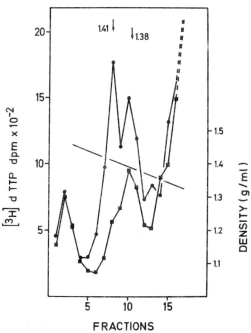

FIGURE 5
CsCl buoyant density profiles of LuIII particles labeled in vitro with [³H]dTTP during 60 min (■——■) and 120 min (●——●) of incubation.

resistant to DNase and Sarkosyl, sediment mainly at 110S, are also stable at high ionic strength, and band in CsCl at 1.41 g/cm^3. An assay of the infectivity of virions assembled in vitro is difficult, since the cells used to prepare the in vitro system already contain unlabeled infectious particles.

Different experimental conditions were used to study the assembly pathway of LuIII virus in vitro. First, viral DNA was pulse-labeled in vivo and subsequently chased into complete virions during in vitro incubation. Second, viral DNA was labeled in vitro, i.e., in Brij-58-lysed cells, and then incorporated into full particles. Viral DNA synthesized in vivo was incorporated into 110S virus particles without any delay during in vitro incubation. The rate of particle formation depended on the time in the virus replication cycle at which the cells were lysed and was also influenced by the concentration of Brij-58 used (Fig. 2). On the other hand, packaging into virions of DNA synthesized in vitro (i.e., in the absence of intact-protein synthesis) occurred only after a lag of about 30–50 min, although labeled viral DNA could be demonstrated in these nuclei after 15 min of incubation (Fig. 3). In addition, we have observed that a distinct species of incomplete particles appear to be precursors of the intact LuIII virions (Figs. 4 and 5). This interpretation is supported by results of in vivo pulse-chase experiments in HeLa cells in which labeling of 95S particles preceded that of 110S virions (Gautschi et al., unpublished results). On the basis of similar pulse-chase experiments carried out in vivo, Sundquist et al. (1973) proposed that incomplete particles represent intermediates in adenovirus assembly.

The data presented in this paper suggest strongly that the rate of LuIII-virus assembly decreased during incubation in vitro. This might explain the fact that intermediates in virus assembly were not observed in vivo or at the beginning of the in vitro incubation when the rates of assembly were still close to in vivo values. Impaired protein synthesis in the in vitro system might influence the kinetics of assembly, as has been shown in adenovirus-infected cells (Sundquist et al. 1973). Specifically, the availability of assembly factors, such as have been demonstrated in the adenovirus system (Winters and Russell 1971), might also become rate-limiting in LuIII-virus assembly.

REFERENCES

Casjens, S. and J. King. 1975. Virus assembly. *Annu. Rev. Biochem.* **44**:555.

Gautschi, J. R. and J. M. Clarkson. 1975. Discontinuous DNA replication in mouse P-815 cells. *Eur. J. Biochem.* **5**:403.

Gautschi, J. R., M. Burkhalter, and P. Reinhard. 1977. Semi-conservative DNA replication in vitro. II. Replication intermediates of mouse P-815 cells. *Biochim. Biophys. Acta* **474**:512.

Gautschi, M. and G. Siegl. 1975. Evidence for a viral maturation complex in parvovirus infected cells. *Experientia* **31**:738.

Gautschi, M., G. Siegl, and G. Kronauer. 1976. Multiplication of parvovirus LuIII in a synchronized culture system. IV. Association of viral structural poly-peptides with the host cell chromatin. *J. Virol.* **20**:29.

Reinhard, P., M. Burkhalter, and J. R. Gautschi. 1977. Semi-conservative DNA replication in vitro. I. Properties of two systems derived from mouse P-815 cells by permeabilization or lysis with Brij-58. *Biochim. Biophys. Acta* **474**:500.

Rose, J. A. 1974. Parvovirus reproduction. In *Comprehensive virology* (ed. H. Fraenkel-Conrat and R. Wagner), vol. 3, p. 1. Plenum, New York.

Russell, W. C. and W. D. Winters. 1975. Assembly of viruses. *Prog. Med. Virol.* **19**:1.

Siegl, G. and M. Gautschi. 1973. The multiplication of parvovirus LuIII in a synchronized culture system. I. Optimum conditions for virus multiplication. *Arch. gesamte Virusforsch.* **40**:105.

Sundquist, B., E. Everitt, L. Philipson, and S. Høglund. 1973. Assembly of adenoviruses. *J. Virol.* **11**:449.

Winters, W. D. and W. C. Russell. 1971. Studies on assembly of adenovirus in vitro. *J. Gen. Virol.* **10**:181.

Study of Kilham-Rat-Virus Nucleoprotein Complexes Extracted from Infected Rat Cells

MAGALI GUNTHER
BERNARD REVET

Institut de Recherches Scientifiques sur le Cancer
B.P. N° 8, Villejuif 94 800, France

The DNA of Kilham rat virus (KRV), one of the nondefective parvoviruses, has been demonstrated to be single-stranded and linear and to have a molecular weight of 1.6×10^6 (Robinson and Hetrick 1969; May and May 1970; Salzman et al. 1971).

It is now established that the replication of parvovirus single-stranded DNA involves linear double-stranded intermediates (Salzman and White 1973; Tattersall et al. 1973; Rhode 1974; Straus et al. 1976; Siegl and Gautschi 1976; Gunther and May 1976; Lavelle and Li 1977). Two replicative forms can be extracted from KRV-infected cells, one of genome length and the other of twice genome length. In a fraction of each form viral and complementary strands are covalently linked (Gunther and May 1976). It is thus likely that the replication of KRV DNA, as well as of other parvoviruses, occurs through a self-priming mechanism. In addition, little (if any) single-stranded DNA can be detected in cells infected with H-1, LuIII, or KRV (Rhode 1974; Siegl and Gautschi 1976; Gunther and May 1976); this suggests a maturation process linked to the replication of the viral strand. Such a maturation scheme would imply that virus-specific nucleoprotein complexes are formed in infected cells as maturation intermediates. The isolation and characterization of nucleo-protein complexes in KRV-infected cells was the objective of the work presented herein.

MATERIALS AND METHODS

CELL STRAINS AND VIRUS. Methods for culture of rat-embryo cells and preparation of virus stocks have been described (Gunther and May 1976). All the experiments reported here were performed in secondary cultures of rat-embryo cells.

NUCLEOPROTEIN PURIFICATION. Sixty-millimeter petri dishes, containing 3×10^6 cells, were infected with 5 plaque-forming units (PFU) of KRV per cell and then labeled with [^{3}H]thymidine (TdR) (5μCi/ml) at different times after infection. Nucleoprotein complexes were extracted by the SDS-pronase Hirt method as described previously for DNA (Gunther and May 1976) and then purified by two CsCl equilibrium density gradients and one 10–40% sucrose gradient. Fractions corresponding to nucleoproteins were dialyzed against Tris-EDTA (10 mM Tris-HCl, 1 mM EDTA, pH 8).

EQUILIBRIUM DENSITY GRADIENT. Equilibrium density gradients were carried out in 3 ml of CsCl ($\rho = 1.45$ g/cm^3) in Tris-EDTA at 35,000 rpm for 20 hr at 25°C in a Spinco SW 50.1 rotor.

SUCROSE GRADIENT SEDIMENTATION. Portions of a Hirt supernatant or of a purified nucleoprotein preparation were layered onto a 4-ml 10–40% linear sucrose gradient in 0.2 M NaCl and Tris-EDTA. Gradients were centrifuged at 40,000 rpm for 75 min at 20°C in a Spinco SW 50.1 rotor.

DNA EXTRACTION PROCEDURES AND SUCROSE GRADIENT SEDIMENTATION OF DNA. DNA was extracted from the nucleoprotein complexes by the method used for extracting viral DNA from the virions (May and May 1970). DNA to be analyzed in an alkaline sucrose gradient was extracted with 0.1 N NaOH for 20 min at room temperature. Velocity sedimentation of DNA in a 5–20% neutral (or alkaline) sucrose gradient was performed as already described except that 0.1% Sarkosyl was added in the sucrose solution (Gunther and May 1976).

ELECTRON MICROSCOPY. Purified KRV nucleoprotein complexes were analyzed by the method of Dubochet et al. (1971) or by that of Davis et al. (1971). In the first method the complexes, with DNA contents of 0.5–1.0 μg/ml, were adsorbed to charged carbon-coated copper grids. The preparations were rinsed with a 2% aqueous solution of uranyl acetate and dried by blotting on filter paper. In the second method the solution containing the complexes was made to 50% in formamide, 0.1 M Tris, 0.01 M EDTA, 0.1% cytochrome c (pH 8.5), and spread on 17% formamide,

0.01 M Tris, 0.001 M EDTA (pH 8). The film obtained was touched with a parlodion-coated grid and dried in 90% EtOH, 5×10^{-6} M uranyl acetate. In the two methods PM2 DNA and φX174 single-stranded DNA were eventually added as markers. The grids were then rotary-shadowed at an angle of 7° with platinum. They were examined at 60 kV in a Siemens Elmiskop 1A electron microscope at different magnifications. Enlargements of the plates were measured on a Hewlett-Packard digitizer.

RESULTS

Nucleoprotein Complexes Found in a Hirt Supernatant of KRV-Infected Cells

DNA of KRV-infected rat-embryo cells was labeled with [³H]TdR (5 μCi/ml) at 12 hr post-infection for 2 hr, then extracted by the Hirt procedure (Gunther and May 1976). DNA was then sedimented in a 10–40% sucrose gradient (Fig. 1a) or analyzed by band centrifugation in a CsCl solution, $\rho = 1.50$ g/cm³ (Fig. 1b). In both cases 55–60% of the radioactivity was incorporated into "free" DNA, while the remaining label was found in DNA associated with proteins. In a sucrose gradient, material sedi-

FIGURE 1
Sedimentation velocities of DNA from a Hirt supernatant. KRV-infected cells were labeled with [³H]TdR (5 μCi/ml) at 12 hr p.i. for 2 hr. DNA was then extracted as described in "Materials and Methods." (*a*) Sedimentation of 0.2 ml of a Hirt supernatant in a 10–40% sucrose gradient in 0.2 M NaCl and Tris-EDTA at 40,000 rpm for 75 min at 20°C. (*b*) Band centrifugation of 0.2 ml of a Hirt supernatant in a CsCl solution ($\rho = 1.50$ g/cm³) at 40,000 rpm for 3 hr at 25°C. D: dimer DNA duplex; M: monomer DNA duplex. Other symbols are described in the text.

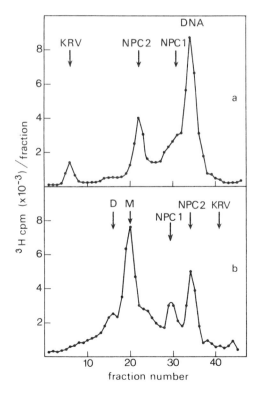

menting faster than free DNA was found (Fig. 1a). Less than 5% of the radioactivity was found at the position of the virus (110S) (the majority of the progeny virus was lost in the Hirt pellet). About 23% of the radioactivity peaked at a 60S position; this component has been termed nucleoprotein complex 2 (NPC-2). An additional component, termed nucleoprotein complex 1 (NPC-1), formed a shoulder at 35S. [Sedimentation values were calculated with purified preparations of NPC-1 and NPC-2 sedimented in 10–40% sucrose gradients with KRV added as a marker (data not shown).]

The profile of the same material after centrifugation in CsCl with a density of 1.5 g/cm³ is shown in Figure 1b. Under these conditions the migration of DNA is a function of its sedimentation coefficient (Vinograd et al. 1963), whereas the position of a nucleoprotein complex appears to depend on its buoyant density. Free DNA was resolved into the two species, monomer and dimer duplexes, already described (Gunther and May 1976), whereas the nucleoprotein complexes (NPC-1 and NPC-2) formed two bands near the top of the gradient. In an isopycnic centrifugation in CsCl ($\rho = 1.45$ g/cm³) the densities of NPC-1 and NPC-2 were found to be 1.47 and 1.42 g/cm³, respectively; the density of the virus was 1.40 g/cm³ (data not shown). The difference of 0.05 g/cm³ between the densities of the two complexes can be accounted for by different contents of DNA and/or protein. As will be shown later, NPC-1 contains twice as much DNA as NPC-2.

Velocity Sedimentation of DNA from NPC-1 and NPC-2

DNA was extracted from the purified nucleoprotein complexes as described in "Materials and Methods," then analyzed in 5–20% sucrose gradients (Fig. 2). In a neutral sucrose gradient (Fig. 2a) DNA extracted from NPC-1 behaved as double-stranded monomer-length DNA extracted from infected cells. In an alkaline sucrose gradient NPC-1 DNA cosedimented with viral DNA (Fig. 2b); less than 10% of the radioactivity was found at the position expected for a single-stranded dimer. When extracted from NPC-1, DNA therefore behaved as a double-stranded monomer without a hairpin structure. The same analysis with NPC-2 DNA showed that this DNA behaved as single-stranded viral DNA in either a neutral or an alkaline sucrose gradient (Fig. 2c,d).

Electron Microscopy

Purified nucleoprotein complexes were studied under the electron microscope. In Figure 3, structures of viruses (a), NPC-2 (b), and NPC-1 (c), spread by Dubochet's method (Dubochet et al. 1971), are shown at the same magnification (60,000). Figure 3 c and d shows that NPC-1 are

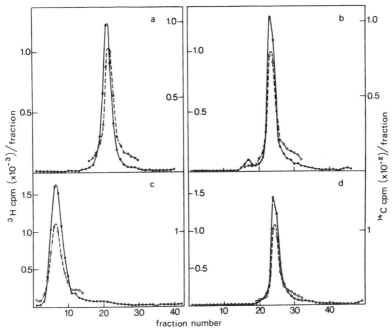

FIGURE 2

Sucrose sedimentation of DNA extracted from nucleoprotein complexes. ³H-labeled DNA was extracted from NPC-1 or NPC-2 as described in "Materials and Methods," then layered on 4 ml of a 5–20% sucrose gradient. Neutral sucrose gradients in 1 M NaCl and Tris-EDTA were centrifuged at 20°C for 150 min at 54,000 rpm in a Spinco SW 56 rotor. Alkaline sucrose gradients (pH 13) were centrifuged at 20°C for 3 hr at 54,000 rpm in a Spinco SW 56 rotor. Sedimentation of ³H-labeled DNA extracted from NPC-1 was performed in (a) neutral and (b) alkaline sucrose gradients. Sedimentation of ³H-labeled DNA extracted from NPC-2 was performed in (c) neutral and (d) alkaline sucrose gradients. ¹⁴C-labeled DNAs were added as markers in each tube. (a) Monomer double-stranded DNA; (b)–(d) KRV DNA. ●——●: ³H-labeled DNA; o-----o: ¹⁴C-labeled DNA.

double-stranded DNA molecules linked by one of their ends to a capsidlike structure. Structures of NPC-1 appeared to be very homogeneous. Very few structures like the particle shown by an arrow (Fig. 3c) could be seen—in this case the capsid appears to be associated with the interior of the DNA molecule, although a random overlap with a free capsid (Fig. 3c) cannot be excluded. The DNA molecule of NPC-1 was double-stranded (single-stranded DNA cannot be visualized with Dubochet's method) and of monomer length. The average length of NPC-1 DNA was found to be 1.59 ± 0.08 μm (86 molecules were measured); the length of PM2 DNA

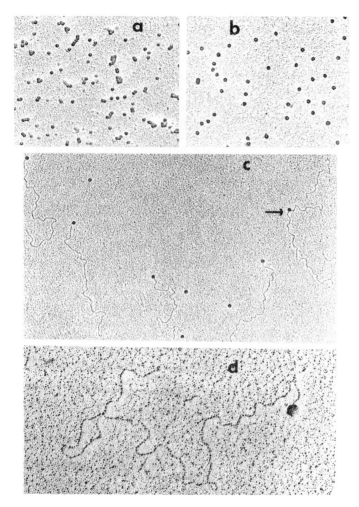

FIGURE 3
Electron micrographs of virus, NPC-2, and NPC-1 preparations
spread by Dubochet's method (see "Materials and Methods"). (a)
virus, (b) NPC-2, and (c) NPC-1 (all 30,000×). (d) NPC-1 complex
(120,000×). For an explanation of the arrow, see text.

was found to be $3.30 \pm 0.09\,\mu$m (9 molecules were measured) (Fig. 4). The
length of the double-stranded "free" monomer DNA extracted from
KRV-infected cells was found to be half the length of PM2 DNA (Hayward et al., this volume). One notices that the DNA tail of NPC-1 was 3%
smaller than half the length of PM2 DNA, but one must remember that
the measurements were done under different conditions. To confirm this
difference, NPC-1 and monomer DNA should be spread and measured

FIGURE 4
Histogram of lengths of NPC-1
DNA tails and PM2 DNA form II.

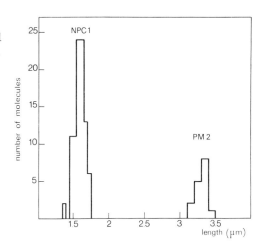

under the same conditions. When NPC-1 were spread with the Davis method, no visible single-stranded regions were found along the DNA molecule (data not shown).

DISCUSSION

In this paper we have described two species of nucleoprotein complexes extracted from KRV-infected rat cells. The sedimentation coefficients were found to be 35S for the complex NCP-1 and 60S for the complex NPC-2. Both species of nucleoprotein contained a capsidlike structure linked to a DNA molecule of genome length. [Preliminary results show that the three capsid proteins (Salzman and White 1970) are contained in both nucleoprotein complexes.] The nature of the link between proteins and DNA is not yet known, but it is resistant to our conditions of extraction (0.6% SDS, 100 μg/ml of pronase for 30 min at 37°C) and to high ionic strength (1 M NaCl or CsCl, 1.5 g/cm³).

The complex NPC-2 contains a single-stranded DNA molecule of genome length, as indicated by the sedimentation profiles of DNA in neutral and alkaline sucrose gradients. Studies of the DNA and proteins of NPC-2, as well as the buoyant density of the complex (1.42 g/cm³) in CsCl, indicate that the composition of NPC-2 is similar to that of the virus. However, the sedimentation behavior of NPC-2 (60S) is different from that of the virus (110S). Experiments are in progress to compare the structure and composition of NPC-2 and virions and to determine whether NPC-2 is a precursor of mature virus.

As shown by electron microscopy, the complex NPC-1 contains a capsidlike structure linked at one end to a double-stranded monomer. The behavior of DNA extracted from NPC-1 in neutral and alkaline sucrose gradients indicates that this DNA is double-stranded, of viral length, and without a hairpin structure. It has been proposed that progeny strands are produced by displacement of a viral strand from a replicative intermediate (Tattersall et al. 1973; Straus et al. 1976); this displacement could be symmetric for AAV and asymmetric for autonomous parvoviruses. NPC-1 complexes could represent the structures in which this displacement occurs. A possible scheme is proposed in Figure 5: An NPC-1 complex is assumed to be a structure on which a viral strand is displaced while a new viral strand is replicated by a self-priming mechanism, giving rise to an NPC-2 complex and a hairpin molecule. Preliminary pulse-chase experiments favor this hypothesis. If NPC-1 can act as a replicative intermediate, particles must exist in which the capsid is situated at positions within the duplex. However, in the electron-microscope experiments presented here such particles represent only 1% of the total number of particles. This might be due to the method used for NPC-1 purification, in which such intermediates were lost because of their fragility and/or because we selected one kind of NPC-1 complex in the CsCl density gradient. Work is in progress to determine experimental conditions allowing detection of a larger proportion of such particles. It is also of interest to determine whether the capsid is linked to one specific end of the double-stranded DNA. If so, this could explain why only one strand is packaged in KRV, a fundamental difference between the autonomous parvoviruses and AAV.

FIGURE 5
Proposed scheme for the viral strand displacement and packaging associated with a self-priming replication on an NPC-1 template.

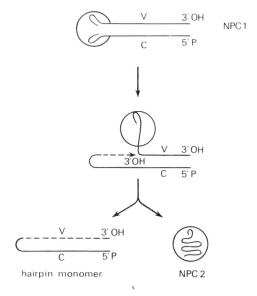

hairpin monomer

NPC 2

REFERENCES

Davis, R. W., M. Simon, and N. Davidson. 1971. Electron microscope heteroduplex method for mapping regions of base sequence homology in nucleic acids. *Methods Enzymol.* **21**:413.

Dubochet, J., M. Ducommun, M. Zollinger, and E. Kellenberger. 1971. A new preparation method for dark-field electron microscopy of biomacromolecules. *J. Ultrastruct. Res.* **35**:147.

Gunther, M. and P. May. 1976. Isolation and structural characterization of monomeric and dimeric forms of replicative intermediates of Kilham rat virus DNA. *J. Virol.* **20**:86.

Lavelle, G. and A. T. Li. 1977. Isolation of intracellular replicative forms and progeny single strands of DNA from parvovirus KRV in sucrose gradients containing guanidine hydroxychloride. *Virology* **76**:464.

May, P. and E. May. 1970. The DNA of Kilham rat virus. *J. Gen. Virol.* **6**:437.

Rhode, S. L. III. 1974. Replication process of the parvovirus H-1. II. Isolation and characterization of H-1 replicative form DNA. *J. Virol.* **13**:400.

Robinson, D. M. and F. M. Hetrick. 1969. Single-stranded DNA from the Kilham rat virus. *J. Gen. Virol.* **4**:269.

Salzman, L. A. and W. L. White. 1970. Structural proteins of Kilham rat virus. *Biochem. Biophys. Res. Commun.* **41**:1551.

―――. 1973. In vivo conversion of the single-stranded DNA of Kilham rat virus to a double-stranded form. *J. Virol.* **11**:299.

Salzman, L. A., W. L. White, and T. Kakefuda. 1971. Linear, single-stranded deoxyribonucleic acid isolated from Kilham rat virus. *J. Virol.* **7**:830.

Siegl, G. and M. Gautschi. 1976. Multiplication of parvovirus LuIII in a synchronized culture system. III. Replication of viral DNA. *J. Virol.* **17**:841.

Straus, S. E., E. D. Sebring, and J. A. Rose. 1976. Concatemers of alternating plus and minus strands are intermediates in adenovirus-associated virus DNA synthesis. *Proc. Natl. Acad. Sci.* **73**:742.

Tattersall, P., L. V. Crawford, and A. J. Shatkin. 1973. Replication of the parvovirus MVM. II. Isolation and characterization of intermediates in the replication of the viral deoxyribonucleic acid. *J. Virol.* **12**:1446.

Vinograd, J., R. Bruner, R. Kent, and J. Weigle. 1963. Band centrifugation of macromolecules and viruses in self-generating density gradients. *Proc. Natl. Acad. Sci.* **49**:902.

Electron Microscopy and Cytochemistry of H-1 Parvovirus Intracellular Morphogenesis

IRWIN I. SINGER
SOLON L. RHODE III

Institute for Medical Research
Putnam Memorial Hospital
Bennington, Vermont 05201

The parvovirus H-1 (Karasaki 1966) contains single-stranded DNA (Usategui-Gomez et al. 1969) with a molecular weight of about 1.6×10^6 (Singer and Rhode 1977). Its icosahedral capsid is composed of only two types of proteins, VP1 (m.w. 9.2×10^4) and VP2' (m.w. 7.2×10^4) (Kongsvik et al. 1974). Originally isolated from human tumors and embryos (Toolan 1972), it is capable of autonomous replication and is therefore classified as a nondefective parvovirus. The yield of H-1, however, is strikingly dependent upon the cell line in which it is propagated. While H-1 will grow in hamster-embryo fibroblasts (Rhode 1973) and in some transformed human cell lines (e.g., NB cells, an SV40-transformed newborn-human kidney epithelial line), it does not make infectious progeny in the absence of helper virus in human-embryo lung cells or human WI-26 fibroblasts (Ledinko et al. 1969). The late S or G2 phase of the host cell cycle appears to be required for H-1 replication (Rhode 1973).

In this paper we present our electron-microscope observations of the morphogenesis of H-1 virions and the accompanying damage produced in a variety of host-cell types. We also describe the use of EM immunocytochemical techniques to locate intracellular sites of accumulating capsid proteins which apparently are not incorporated into virions. Further, the possible role of these viral proteins in regulating viral DNA synthesis is investigated using simultaneous immunostaining for H-1 antigens and [³H]thymidine (TdR) autoradiography at the EM level. The above ultrastructural methods are also applied to study cells infected by various temperature-sensitive (ts) H-1 mutants with alterations in their capsid proteins (Rhode 1976). These agents are being used to probe for specific changes in normal patterns of virion morphogenesis and cytopathological damage, with the goal of defining the mechanisms of these processes more clearly.

MATERIALS AND METHODS

Cells

NB cells and first-passage hamster-embryo fibroblasts (HEF cells) were cultured as described previously (Rhode 1973; Singer and Toolan 1975). A strain of diploid human-embryo lung fibroblasts (MRC-5 cells) obtained from the American Type Culture Collection was cultured as above, except that nonessential amino acids were omitted from the growth medium.

Virus

Wild-type (wt) H-1 virus stocks prepared as before (Rhode 1973) were used to infect cover-slip cultures of the above cells at a multiplicity of 20–40 plaque-forming units (PFU) per cell at 37°C. Parasynchronous infections of NB and MRC-5 cells were obtained with a methotrexate block, and mitotically quiescent cultures of HEF cells were synchronized with 40% fetal-calf serum as already detailed (Rhode 1973; Singer and Toolan 1975). H-1 ts mutants were isolated as described previously (Rhode 1976). Adsorption of ts mutants at the permissive temperature (T_p) of 33°C was followed by incubation at the appropriate temperature(s) given below.

Electron Microscopy

FIXATION. Cells infected with wild-type or ts-mutant H-1 were grown on fluorocarbon-coated cover slips (Singer 1976) and were preserved in situ using several different fixatives with specific aims in mind. Glutaraldehyde (4%) and OsO_4 (1%) (Singer and Toolan 1975) were applied to obtain the structural preservation required for the precise characterization of H-1 virions, inclusions, and the accompanying cellular damage. Since glutaraldehyde destroys the antigenicity of intracellular H-1 antigens (Singer 1974), a solution of 2% formaldehyde, 3% dextran, and buffered saline was used as a fixative in EM immunocytochemical experiments.

This method yields reasonable morphological preservation and maintains the ability of H-1 antigens to bind anti-H-1 IgG (Singer 1976). In cases where the formaldehyde fixative did not preserve the immunological reactivity of certain ts-mutant antigens, fixation with acetone (10 min at −20°C) was appropriate. However, preservation of the ultrastructure was poor with this method.

CELL SELECTION, IN-SITU SECTIONING, AND POST-STAINING. Infected cultures were embedded in Epon in situ, and groups of cells were selected with a light microscope for subsequent in-situ ultrathin sectioning as detailed before (Singer 1976). Specimens fixed for morphological study were post-stained with uranyl acetate and lead citrate (Singer and Toolan 1975) or with the EDTA method (Bernhard 1969; Singer 1975). Immunocytochemical preparations were analyzed without heavy-metal post-staining.

Immunochemical Labeling of Intracellular H-1 Antigens

ANTIBODY PRODUCTION. Several different types of H-1 antisera made in Syrian hamsters were used. Those with the highest hemagglutination inhibition (HAI) titers were produced by deformed or "funny face" (FF) adults (Toolan, this volume) injected with 1–2 PFU/ml of H-1 at birth. It has been shown that these FF hamsters are persistently infected (Toolan 1968), so that their IgG would be expected to contain antibodies against all possible H-1 antigens, including any suspected noncapsid viral proteins. Most of our H-1-antigen-labeling experiments were performed with this FF IgG. Lower-HAI-titer H-1 antisera synthesized by normal adult hamsters injected as adults with either purified intact H-1 virus or SDS-disrupted particles (Kongsvik and Toolan 1972) were used occasionally. Since disrupted virions cannot infect the hamster being immunized, IgG to SDS-disrupted H-1 virions (anti-SDS-H-1 IgG) should only contain antibodies against capsid proteins.

ENZYME CONJUGATION AND CYTOCHEMISTRY. The IgG fractions of the above antisera were purified by precipitation with $(NH_4)_2SO_4$, and conjugated to horse-heart cytochrome c using a single-step glutaraldehyde method (Avrameas 1969) modified as before (Singer 1976). Fixation, labeling of H-1 antigens with this conjugate, and cytochemical visualization of enzyme-antibody-antigen complexes were performed as previously (Singer 1976).

HIGH-RESOLUTION EM AUTORADIOGRAPHY. Parasynchronous NB cells were infected with wt H-1 (at 37°C) or ts1 H-1 [at the restrictive temperature (T_r) of 39.5°C] and pulsed with high-specific-activity [³H]thymidine (100 Ci/mmole) for various time periods from 15 to 16 hr post-infection (p.i.). EM autoradiograms were prepared using a semi-automatic coating device and a high-resolution development process which produced compact spherical silver grains approximately 80 nm in diameter as described by Kopriwa (1973).

RESULTS

The time course of the prominent events that occur during a parasynchronous single-cycle H-1 infection in NB or HEF cells is presented in Figure 1 so that the reader can evaluate the morphological data given below in proper chronological perspective.

Morphogenesis of Wild-Type H-1 and Associated Cellular Damage

NB CELLS. H-1 particles were first observed in thin-sectioned nuclei of infected NB cells 10–12 hr p.i. A few virions were found in the euchromatic areas, while high concentrations were localized in vacuolated nucleolar fibrous centers (Fig. 2A). Most of the capsids were empty (lacked densely staining DNA centers), but a few full (DNA-containing)

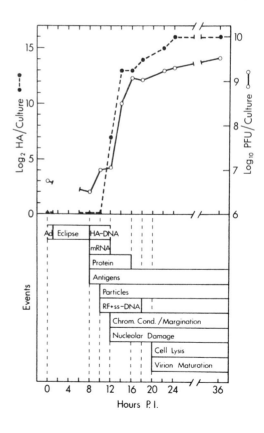

FIGURE 1
Chronology of H-1 infection. Hemagglutinin (HA) and infectious virus (PFU) synthesis occur from 10 to 16 hr p.i. concurrently with the major biochemical and morphological events of the single-cycle infection. The secondary rise in HA and PFU titers beyond 16 hr p.i. is probably due to postsynthetic modification of the H-1 capsid by cellular proteases (Kongsvik et al., this volume). Ad: adsorption of virions to plasma membranes. HA-DNA: the DNA-synthetic event required for HA synthesis (Rhode 1973). RF DNA: double-stranded replicative-form viral DNA. ssDNA: single-stranded progeny viral DNA (Rhode, this volume). (Upper portion adapted from Singer 1976.)

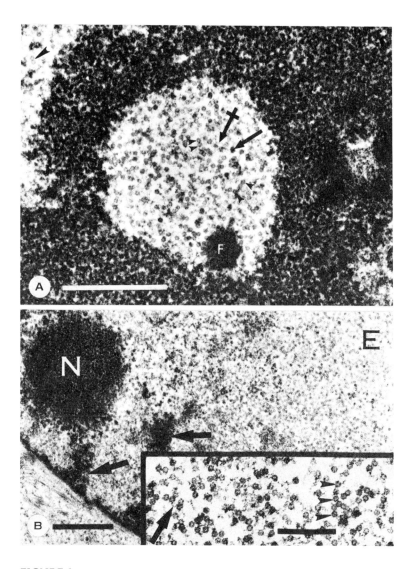

FIGURE 2

Wild-type H-1 infection in human NB cells; glut/OsO$_4$ fixation, uranyl acetate
and lead citrate staining. (A) Nucleolus with vacuolated fibrous center (12 hr
p.i.) containing many empty capsids and a few full capsids (see uncrossed
arrow) as well as condensed fibrillar component (F). Linear arrays of capsids
(arrowheads) and euchromatin fibers (crossed arrow) are also seen. Some empty
particles (arrowhead) are outside of the nucleolus. Bar = 0.5 μm; 56,900 ×.
(Adapted from Singer and Toolan 1975.) (B) Nucleus (24 hr p.i.) exhibiting many
empty capsids in the euchromatic region (E). The nucleolar remnant (N) has lost
its fibrillar components, and the heterochromatin (arrows) is very condensed.
Bar = 0.5 μm; 123,000 ×. (Inset) High magnification of euchromatic area of
above cell. Capsids may be seen attached to euchromatin fibers (arrow) and in
curvilinear aggregates (arrowheads). Bar = 0.2 μm; 79,500 ×.

particles were observed. Many of the intranucleolar capsids formed linear aggregates and appeared to be attached to euchromatinlike fibers. The fibrous nucleolar component appeared to have condensed into dense spherical masses, many of which became extruded from the nucleolus. By 18–36 hr p.i., large numbers of H-1 particles were found outside of the nucleolus in association with the euchromatin (Fig. 2B); many curvilinear aggregates of capsids seemed to be attached to euchromatin fibers (Fig. 2B, inset). About 95% of the H-1 virions synthesized by these NB cells were empty (Singer and Toolan 1975). At this stage, nucleolar remnants had totally lost their fibrillar components, and nuclear heterochromatin had condensed and marginated (Fig. 2B). No H-1 particles were ever found in the heterochromatin.

MRC-5 FIBROBLASTS. Infection of MRC-5 fibroblasts with wt H-1 leads to an abortive infection; less than 1% of the replicative-form DNA and only 5% of the progeny DNA synthesized in NB cells is produced in this system, and no full particles were detected after isopycnic centrifugation in CsCl$_2$ gradients (S. L. Rhode, unpublished). High concentrations of empty wt H-1 particles were observed in the euchromatic region of MRC-5 nuclei in cultures fixed 22 hr p.i. at 37°C (Fig. 3A). These particles tended to form numerous hollow spheres about 2 μm in diameter, similar to the ring-shaped inclusions infrequently observed in NB cells infected with wt H-1 (Singer and Toolan 1975). Empty virions were present in the lumen of this inclusion, and its wall was composed of amorphous material and particles as well (Fig. 3B). These cells are probably synthesizing altered viral proteins which are incapable of encapsidating progeny DNA despite their ability to assemble into capsids. The surface properties of these particles are probably modified also, since they exhibit an abnormal tendency to form aggregates. In addition, the ability of the temperature-sensitive mutant ts1 H-1 to form polycrystals at restrictive temperatures (see below) is not expressed in these fibroblasts. Nucleolar damage and heterochromatin condensation occur here (Fig. 3A), as they do in NB cells.

HAMSTER-EMBRYO FIBROBLASTS. Thin sections of parasynchronous HEF cells infected with wt H-1 revealed that approximately 95% of the particles synthesized are full (Fig. 4A; Singer and Toolan 1975). The cytological damage was similar to that observed in NB and MRC-5 cells, and the particles were closely associated with euchromatin (Fig. 4A). HEF cells synthesized twice the quantity of RF DNA synthesized by NB cells (S. L. Rhode, unpublished).

Immunochemistry of NB Cells Infected with Wild-Type H-1

EDTA STAINING. This technique allows one to differentiate readily between DNP-containing and RNP-containing structures (the DNP is

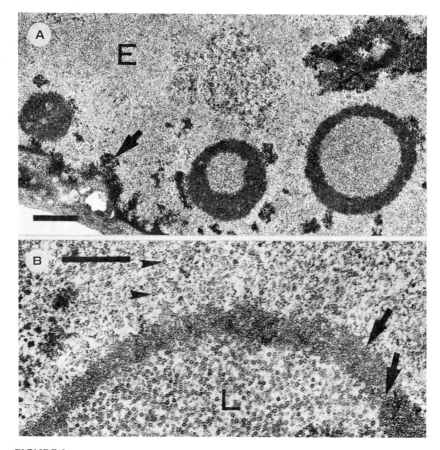

FIGURE 3
MRC-5 human fibroblasts infected by wt H-1 virus. Fixation with glut/OsO₄; staining with uranyl acetate and lead citrate. (A) Nucleus of wt-infected cell fixed 22 hr p.i. with numerous capsids in the euchromatin (E) and large spherical inclusions of H-1 particles. The nucleolus (upper right) is disrupted, and the heterochromatin (arrow) is marginated. Bar = $1.0\,\mu$m; 12,500 ×. (B) Enlargement of H-1 inclusion and adjacent euchromatin. Incomplete capsids are present in the lumen (L) and wall (arrows) of the inclusion, and in the euchromatin (arrowheads). Bar = 0.5 μm; 38,000 ×.

bleached and the RNP is heavily stained; see Bernhard 1969). We have used it to study the progress of chromatin condensation in H-1-infected NB cells. Changes occurred in the nucleolar-associated chromatin by 12 hr p.i. (Fig. 4B). Chromatin within nucleolar fibrous centers appeared condensed just after vacuolation of these centers. DNP dispersed within the nucleolonema also contracted into thickened cords which subsequently formed a cap surrounding the nucleolar remnant. Conspicuous

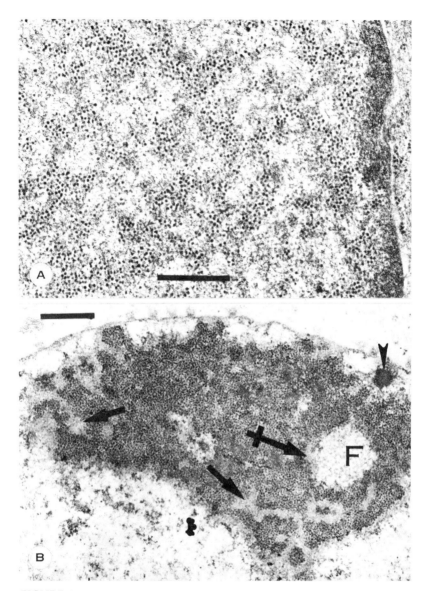

FIGURE 4

Cells infected with wt H-1. (A) Large numbers of densely stained complete H-1 virions are associated with condensing trabeculae of euchromatin in this HEF nucleus fixed 30 hr p.i. with glut/OsO₄ and stained with uranyl acetate and lead citrate. Bar = 0.5 μm; 39,400 ×. (Adapted from Singer and Toolan 1975.) (B) NB-cell nucleus fixed 12 hr p.i. using glutaraldehyde alone and stained with the EDTA method. Condensed cords of destained nucleolar-associated chromatin (uncrossed arrows) are present near the edges of the nucleolus, and within a portion (crossed arrow) of the vacuolated fibrillar center (F). An extruded mass of nucleolar fibrous material (arrowhead) is also apparent. Bar = 0.5 μm; 29,400 ×. (Reprinted, with permission, from Singer 1975.)

486

margination of extranucleolar heterochromatin was observed by 24 hr p.i. (Singer 1975).

CYTOCHROME-C ANTI-H-1-IgG STAINING. The earliest H-1-antigen-specific staining observed using cytochrome-c-conjugated FF anti-H-1 IgG was found 8 hr p.i., 2–4 hr before H-1 particles were first seen in H-1-infected cells (Singer 1976). This staining was localized in the heterochromatin bordering the nuclear envelope (Fig. 5A,B), but was not observed with the light microscope prior to thin sectioning. H-1 antigens were detected on small tufts of condensed chromatin found throughout the nucleoplasm, and at the nucleolar surface at 10 hr p.i. Starting at 12 hr p.i., coincident with an abrupt rise in HA titer (Fig. 1), the H-1-specifically stained heterochromatin condensed further and migrated to the nuclear membrane, which resulted in the appearance of extensive lightly labeled euchromatic regions (Fig. 5C). Large amounts of H-1 antigens were also located on the condensing cords of intranucleolar chromatin, and on contracted DNP at the periphery of the nucleolus from 12 to 16 hr p.i. (Figs. 4B and 5C), concurrent with the vacuolation of nucleolar fibrous centers (Fig. 2A). Extruded spheres of nucleolar fibrous material were unstained (Fig. 5C). Nucleolar-associated chromatin and nuclear heterochromatin appeared to be completely marginated by 18–36 hr p.i. and were heavily labeled with the conjugate of cytochrome c and anti-H-1 IgG (Fig. 5D,E; Singer 1975). (The same patterns of labeling were observed using cytochrome-c conjugates made with anti-H-1 IgG produced by hamsters immunized as adults with either purified *whole* or *disrupted* H-1 virus.) Many infected cells had lysed during this interval, shedding H-1 particles into the medium; extensive aggregates of these virions were observed on the plasma membranes of the remaining intact cells (Singer and Toolan 1975). The outer parts of these capsids were heavily labeled with anti-H-1 conjugate, whereas the insides were not (Fig. 5D, inset). This stain probably does not penetrate into the center of the H-1 virion.

Morphology and Immunostaining of NB Cells Infected with H-1 ts Mutants

ts2 H-1. This mutant did not form H-1 capsids at restrictive temperatures in NB cells 20–24 hr p.i.; spherical and irregular amorphous inclusions were observed instead (Fig. 6A). If the culture temperature was shifted down to the permissive temperature (24 hr T_r to 2 hr T_p) before fixation, the centers of the spherical inclusions became less dense and empty H-1 capsids appeared in them (Fig. 6B). These inclusions appeared to have transformed into masses of H-1 capsids, and isolated H-1 virions were also observed on fibers of euchromatin following an extended shiftdown period of 6 hr (Fig. 6D). However, these spherical bodies did not bind cytochrome-c-conjugated FF anti-H-1 IgG after shiftdown to T_p for 6 hr followed by either formaldehyde or acetone fixation, although marginated heterochromatin was heavily labeled and the euchromatin was slightly stained (Fig. 6C). In morphology, ts2-infected NB cells at T_p were

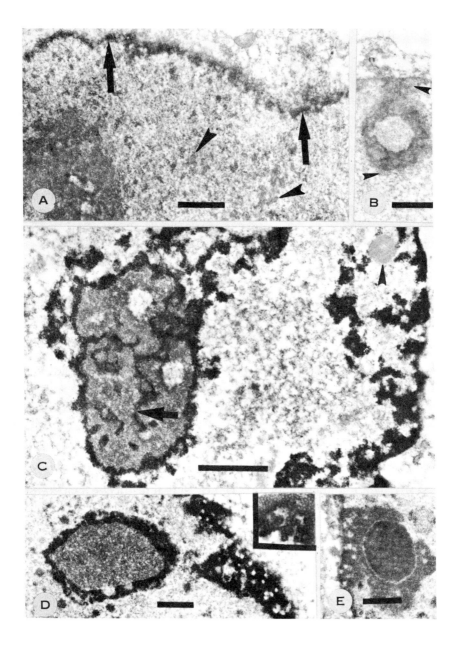

similar to wt-H-1-infected NB cells at 33°, 37°, or 39°C. Under restrictive conditions, ts2 H-1 synthesized a capsid protein incapable of HA production, showed defective progeny DNA production, and also exhibited large reductions in PFU titers (Rhode 1976).

ts1 H-1. This mutant only synthesized empty capsids which formed large globular aggregates in NB-cell nuclei 12–16 hr p.i. at T_r (Fig. 7*A,B*). These capsids did not attach to euchromatin fibers as in the case of wt H-1 infections. The capsid aggregates developed into polycrystals with linear, hexagonal, and cubic patterns of symmetry by 20–24 hr p.i. (Fig. 7*C*); one pattern could be transformed into any of the others by tilting the specimen with respect to the electron beam. Full particles were never observed at T_r, but did form at T_p. They did not engage in crystallization when infected cultures were shifted from T_p to T_r, which indicated that the surface properties of full virions differ from those of empty capsids. Most crystals dissociated into isolated capsid components after shiftdown from T_r to T_p (Fig. 7*D*). We also found that cells with the greatest pathological damage at T_r tended to have fully organized crystals rather than capsid aggregates, and that the crystals in these cells tended to be refractory to dissociation following temperature shiftdown. Further, if MRC-5 fibroblasts (see above) were infected with ts1 H-1 at 39°C, conspicuous aggregates of empty capsids formed, but they did not organize into

FIGURE 5 (see facing page)
Immunospecific labeling of wt-H-1-infected NB cells with cytochrome-c-conjugated FF anti-H-1 IgG (except *E*). Fixation with the modified formaldehyde method; sections were not stained with heavy metals. (*A*) Cell fixed 8 hr p.i.; H-1 antigens are present in heterochromatin at the nuclear membrane (arrows), but not in intranuclear tufts of condensed chromatin (arrowheads), nor within the nucleolus (lower left). Bar = 1.0 μm; 12,300 ×. (*B*) Uninfected control cell fixed 8 hr after mock infection and immunostained as above. Tufts of condensed chromatin (arrowheads) are not labeled. Bar = 1.0 μm; 11,000 ×. (*C*) Nucleus 16 hr after infection. Cords of condensed chromatin present inside (arrow) and around the nucleolus are intensely H-1-antigen positive. Marginated heterochromatin at the nuclear membrane (at right) is also heavily stained, and euchromatin is lightly labeled. Mass of extruded nucleolar fibers (arrowhead) is unlabeled. Bar = 1.0 μm; 18,500 ×. (*D*) Intensely stained nucleus prepared 36 hr p.i. Only the completely contracted nucleolar-associated chromatin and condensed nuclear heterochromatin (at right) contain large amounts of H-1 antigens. Bar = 1.0 μm; 9700 ×. (*Inset*) Aggregate of H-1 virions attached to a plasma membrane adjacent to the cell in *D*. The capsid walls are heavily stained for H-1 antigen. 32,300 ×. (*E*) Cell in stage of infection similar to that shown in *D*, but incubated with cytochrome-c-conjugated hamster IgG lacking anti-H-1 HAI activity. The concentrated chromatin surrounding the nucleolus is unlabeled. Bar = 1.0 μm; 10,400 ×. (Parts *A, B, D,* and *E* adapted from Singer 1976.)

polycrystals as in NB cells (I. I. Singer and S. L. Rhode, unpublished). These noncrystalline aggregates of empty capsids were also made by ts1 at 33°C, and by wt H-1 at 37°C or 39°C in MRC-5 cells (Fig. 3). However, ts1 H-1 capsids produced at 39°C in MRC-5 cells did not attach to euchromatin, whereas wt H-1 capsids synthesized under comparable circumstances were associated with euchromatin; this showed that ts1 and wt H-1 capsids made by this cell are different. This evidence strongly

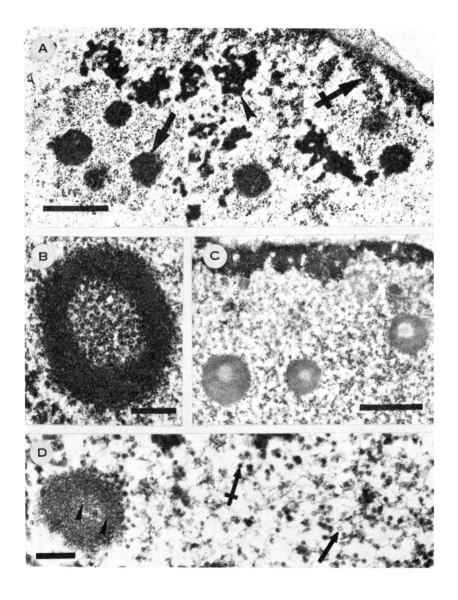

suggests that the host cell is involved in regulating the ts1 H-1 crystallization process, as well as the surface properties of the capsids it synthesizes.

Immunocytochemical preparations of ts1-H-1-infected NB cells fixed with formaldehyde 24 hr p.i. at T_r exhibited intensive staining of condensed peripheral heterochromatin and localized labeling of euchromatin, but the polycrystalline inclusions of H-1 capsids were totally unstained (Fig. 7E). These inclusions became slightly labeled as they began to dissociate following a 2-hr temperature shiftdown (24 hr T_r, 2 hr T_p), and clusters of ts1 H-1 capsids were heavily stained after 6 hr at T_p (24 hr T_r, 6 hr T_p) (Fig. 7F). However, ts1 inclusions at T_r did stain after acetone fixation, whereas post-fixation with formaldehyde inactivated the H-1 antigens of these bodies. This mutant has defects in HA, PFU, and single-stranded-DNA synthesis similar to those found for ts2 (Rhode 1976). Another mutant, ts7, possesses the same morphological and biochemical properties as ts1, and is probably the same mutant.

ts8 AND ts10H-1. Of all of the ts H-1 mutants which we studied, ts8 and ts10 appeared to be the most similar phenotypically to wt H-1. Although there was a reduction in PFU titer by a factor of about 3×10^{-4}, normal levels of HA and progeny DNA synthesis occurred at T_r. These mutants produced atypical spheroidal inclusions (Fig. 8A) composed of aggregated empty capsids (Fig. 8B). Individual full and empty particles were also found associated with the euchromatin (Fig. 8B). As in the case of ts1 and ts2, immunocytochemical preparations of ts8-infected NB cells at T_r had heavily stained heterochromatin, whereas the inclusions did not bind cytochrome-c-conjugated FF anti-H-1 IgG (Fig. 8C). Although the

FIGURE 6 (see facing page)
NB cells infected with ts2 H-1. A, B, and D were preserved with glut/OsO$_4$ and post-stained with uranyl acetate and lead citrate. C was fixed in formaldehyde and not post-stained. (A) Nucleus fixed 24 hr p.i. at T_r containing spherical (uncrossed arrow) and irregular (arrowhead) ts2 inclusion bodies. Nuclear heterochromatin is condensed (crossed arrow). Bar = 1.0 μm; 17,700 ×. (B) Spherical ts2 inclusion following a 2-hr temperature shiftdown (24 hr T_r, 2 hr T_p). Many empty capsids fill its lumen. Bar = 0.2 μm; 66,600 ×. (C) Immunochemical preparation fixed after a 6-hr shift to T_p (24 hr T_r, 6 hr T_p) and stained for H-1 antigens. The condensed heterochromatin is intensely labeled, and localized portions of the euchromatin are also stained, whereas the spherical inclusions are H-1-antigen-negative. Bar = 1.0 μm; 17,600 ×. (D) Ts2-infected nucleus prepared for optimal structural preservation after a 6-hr shiftown (24 hr T_r, 6 hr T_p). The spherical inclusion is composed of aggregated empty capsids (arrowheads). Many individual incomplete virions (uncrossed arrow) and a few full virions (crossed arrow) are attached to euchromatin fibers. Bar = 0.2 μm; 53,300 ×. (Adapted from Singer and Rhode 1977c.)

antibody conjugate may not be able to penetrate into the interior of highly ordered inclusions, no peripheral staining was observed. However, these spherical inclusions were stained intensely if they were fixed after

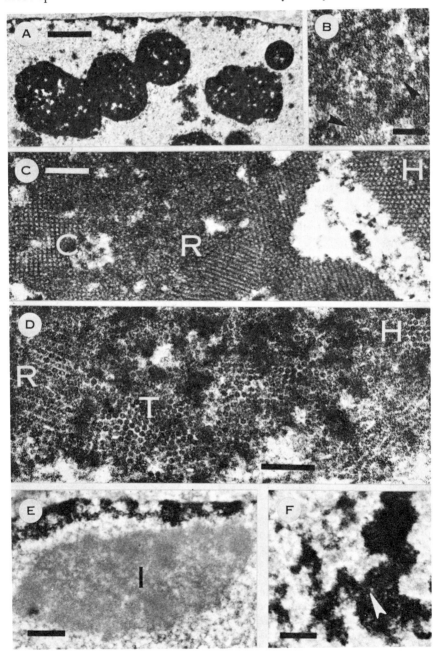

having shifted to permissive conditions for 2 hr (24 hr T_r, 2 hr T_p) (Fig. 8D).

EM Autoradiography of H-1 Viral DNA Synthesis

ANALYSIS OF SPECIFICITY. NB cells infected with wt H-1 were labeled with [^{3}H]TdR in the presence of cytosine arabinoside, an inhibitor of DNA synthesis. The number of intranuclear grains observed in these cells was very low (about 5 grains/100 μm^2), and their distribution appeared to be random. This drug also reduced the amount of radioactivity incorporated into the total extractable DNA by at least 90%. Therefore, most of the label is being incorporated in DNA. With short labeling periods, such as 10 min at 16 hr p.i., 40–60% of the total incorporation of [^{3}H]TdR into acid-insoluble material was extracted into the Hirt supernatant and was found to be almost entirely viral DNA (S. L. Rhode, unpublished data). Therefore, at least 50% of the [^{3}H]TdR is incorporated into viral DNA. To estimate the amount of viral DNA synthesis partitioning into the Hirt pellet fraction, we measured the extent to which DNA synthesis was reduced with ts14 H-1 infection in comparison to wt H-1; ts14 is an H-1 mutant with a cis-acting defect in viral DNA synthesis (see Rhode, this volume). The magnitude of the reduction in DNA synthesis caused by the mutation would be a measure of the minimum proportion of the total DNA synthesis that is viral in cultures infected with wt or ts1 H-1. It is possible, but unlikely, that wt H-1 or ts1 H-1 proteins stimulate

FIGURE 7 (see facing page)
Structure and immunostaining of ts1 H-1 inclusions in NB-cell nuclei. A–D were fixed with glut/OsO$_4$ and stained with heavy metals; E and F were preserved with formaldehyde and studied without post-staining. (A) Large dense globular ts1 inclusions in a nucleus fixed 16 hr p.i. at T_r. Bar = 2.0 μm; 6,600 ×. (B) Enlargement of inclusion shown in A reveals that it is composed of aggregated empty capsids (arrowheads). Bar = 0.2 μm; 43,400 ×. (C) Part of an individual poly-crystalline ts1 inclusion fixed 20 hr p.i. at T_r, exhibiting cubic (C), rodlike (R), and hexagonal (H) patterns. Bar = 0.2 μm; 66,600 ×. (D) A ts1 inclusion which has been shifted to T_p for 2 hr (24 hr T_r, 2 hr T_p). Now individual empty capsids may be seen in rodlike (R), toroidal (T), and hexagonal (H) arrays. Bar = 0.2 μm; 71,700 ×. (C) Preparation stained for H-1 antigens 24 hr p.i. at T_r, using cytochrome-c-conjugated FF anti-H-1 IgG. The marginated heterochromatin is heavily stained, and the euchromatin (lower right) is sparsely labeled, whereas the inclusion (I) is completely unlabeled. Bar = 0.5 μm; 20,000 ×. (F) Anti-H-1 staining performed following a prolonged shiftdown of 6 hr to T_p (24 hr T_r, 6 hr T_p). Disaggregating ts1 H-1 inclusion is now intensely H-1-antigen-positive, and labeled H-1 capsids (arrowhead) can be seen. Bar = 0.5 μm; 20,000 ×. (Parts B–D adapted from Singer and Rhode 1977b; parts E,F reprinted, with permission, from Singer and Rhode 1977c.)

unscheduled cell DNA synthesis that does not occur with ts14 infection, as the proteins of all three of these viruses support normal replicative-form DNA synthesis (Rhode, this volume). The results of this experiment indicate that approximately 90% of the total [³H]TdR uptake is due to wt H-1

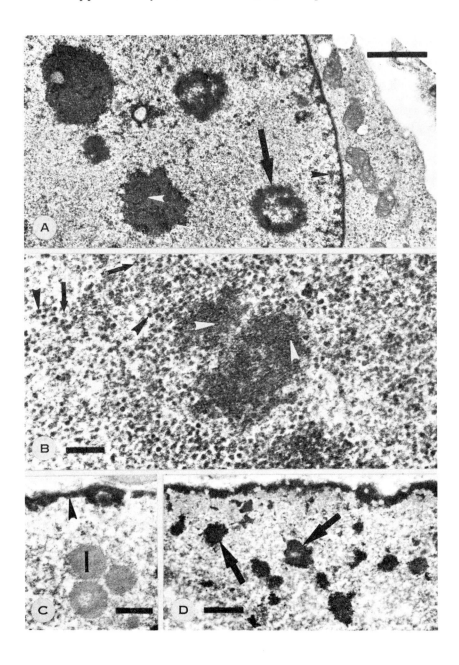

or ts1 H-1 DNA synthesis. EM autoradiograms also demonstrated that 10% of the NB-cell nuclei were synthesizing levels of DNA comparable to those of H-1-producing cells but were not H-1-infected (they lacked H-1-specific antigens and pathological damage). We therefore conclude that the grains observed in EM autoradiograms of cells infected with wt or ts1 H-1 are specific for viral DNA synthesis.

NB CELLS INFECTED WITH WILD-TYPE H-1. Although the H-1-infected cultures used in these autoradiographic experiments were fixed at the same time (16 hr after infection), early and late patterns of pathological damage were observed in addition to the middle stages of infection exhibited by the majority of infected cells. This situation is probably due to the parasynchronous nature of H-1 infection under these conditions (Rhode, this volume) and the relatively short period during which the virions are synthesized within any single virus-infected cell (approximately 6 hr). Many cells in the early stages of infection (vacuolated nucleolar fibrous centers, and small amounts of chromatin condensation) were observed. Synthesis of wt H-1 DNA was localized in the extruded masses of nucleolar fibrous material found in the euchromatin of such cells (Fig. 9A). Nuclei in combined immunocytochemical/autoradiographic preparations of similar cells exhibited tufts of condensed chromatin, which bound FF anti-H-1 IgG conjugate, and had silver grains located exclusively over extruded nucleolar fibrous components, which appeared to be H-1-antigen-negative. Autoradiograms of FF anti-H-1 immunostained cells at the mid-stage of wt H-1 infection contained dense clusters of grains over the euchromatin (Fig. 9B). The tufts of heterochromatin which were heavily stained for H-1 antigens apparently had migrated away from the regions of euchromatin synthesizing wt H-1 viral DNA. Cells in more advanced stages of infection had thickened zones of condensed chroma-

FIGURE 8 (see facing page)
NB-cell nuclei infected with ts8 H-1. *A* and *B* were fixed with glut/OsO$_4$ and stained with uranyl acetate and lead citrate. *C* and *D* were formaldehyde-fixed but not post-stained with heavy metals. (*A*) Spheroidal inclusions (arrow) are found by 16 hr p.i. at T$_r$. Nucleolar chromatin (white arrowhead) and nuclear heterochromatin (black arrowhead) are marginated. Bar = $2.0\,\mu$m; 8,000 ×. (*B*) Tangential section of ts8 inclusion (16 hr p.i. at T$_r$) shows that it is composed of aggregated empty capsids (white arrowheads). Many individual full virions (black arrowheads) and empty virions (arrows) are situated in the euchromatin. Bar = $0.2\,\mu$m; 51,700 ×. (*C*) FF anti-H-1-IgG–cytochrome-c-stained ts8-infected nucleus (24 hr p.i. at T$_r$). The inclusions (I) are unlabeled, whereas the heterochromatin (arrowhead) is heavily stained. Bar = $0.5\,\mu$m; 20,000 ×. (*D*) Nucleus stained as in *C* following a 2-hr shift to T$_p$ (24 hr T$_r$, 2 hr T$_p$); spherical inclusions (arrows) are intensely H-1-antigen-positive. Bar = $1.0\,\mu$m; 10,600 ×. (Adapted from Singer and Rhode 1977c.)

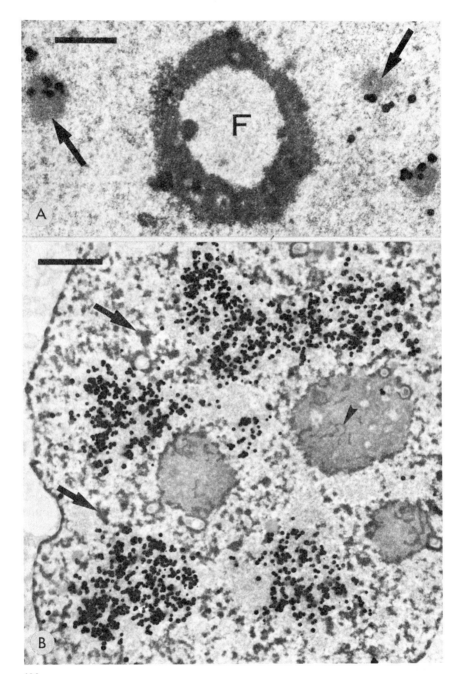

496

tin at the nuclear membrane surrounding large regions of euchromatin (Fig. 10A). These euchromatic areas were synthesizing wt H-1 DNA in a random manner; the silver grains were distributed uniformly. High concentrations of H-1 antigens were present in the marginated nuclear chromatin and nucleolar-associated chromatin (Fig. 10B). The most severely damaged cells, which had begun sloughing off their cytoplasm, were not synthesizing viral DNA. These patterns of autoradiographic labeling were not altered by varying the length of the [³H]TdR pulse from 2 to 60 minutes.

NB CELLS INFECTED WITH ts1 H-1. Autoradiograms of ts1-H-1-infected NB cells were prepared under the same conditions as those of the wt H-1 infection, except that the infection was carried out at T_r. At this temperature, most of the viral DNA synthesized would be replicative-form DNA rather than progeny DNA (Rhode 1976). The sites engaging in synthesis of ts1 H-1 DNA were essentially the same as those observed during wt H-1 infection. Patches of silver grains were found over isolated regions of euchromatin during early and intermediate stages of infection (Fig. 11A), and extruded nucleolar fibrous components were also frequently labeled as in Figure 9A. Most regions of the euchromatin were producing ts1 H-1 viral DNA in the later stages of infection (Fig. 11B). Inclusions of aggregated or crystalline ts1 capsids and H-1-antigen-positive condensed chromatin were not associated with ts1 DNA synthesis.

DISCUSSION

Role of H-1 Antigens in Nuclear Damage

Our fine-structure studies of wt H-1 morphogenesis in parasynchronous NB cells show that condensation of nuclear heterochromatin (HC) and nucleolar-associated chromatin (NAC) begin by 12 hr p.i. Disruption of

FIGURE 9 (see facing page)
Autoradiograms depicting viral DNA synthesis in wt-H-1-infected NB cells fixed 16 hr p.i. (A) Early phase of infection with vacuolated nucleolar fibrous center (F) (see Figs. 2A and 4B). Silver grains are localized over and near extruded masses of nucleolar fibers (arrows); 2-min [³H]TdR pulse, glut/OsO₄ fixation, post-staining with heavy metals. Bar = 1.0 μm; 17,000 ×. (B) Simultaneous immunological labeling of H-1 antigens and autoradiography of viral DNA synthesis in an intermediate stage of H-1 infection. Large clusters of [³H]TdR uptake are found in the euchromatin, whereas high concentrations of H-1 proteins are localized in nucleolar-associated chromatin (arrowhead) and tufts of marginating nuclear heterochromatin (arrows). 5-Min [³H]TdR pulse, formaldehyde fixation, no counterstain. Bar = 2.0 μm; 8,800 ×. (Adapted from Singer and Rhode 1978.)

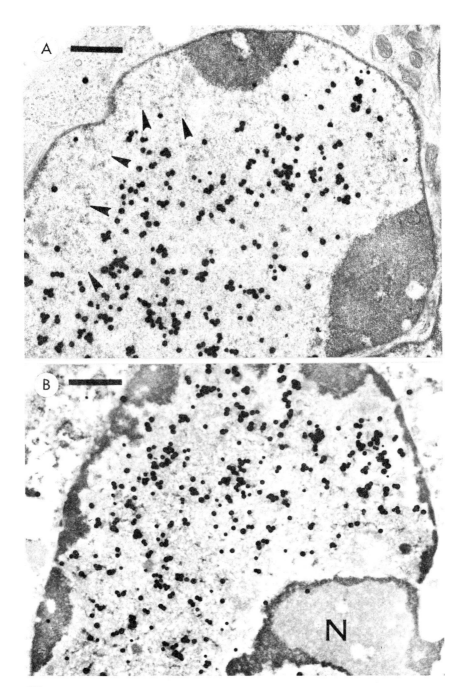

nucleolar fibrous centers and consequent formation of intranucleolar vacuoles occur at this time. None of these changes were observed in mock-infected control cultures, so these morphological alterations are induced by the viral infection (Singer and Toolan 1975). Empty H-1 capsids, first observed in thin sections at 10–12 hr p.i., accumulate preferentially in these vacuolated fibrillar areas of the nucleolus. Also, relatively low levels of H-1 antigens were localized in heterochromatin at the nuclear membrane at 8 hr p.i.; higher concentrations of them were seen in the condensing and marginating nuclear and nucleolar chromatin 12–16 hr p.i., coincident with the major hemagglutinin and infectious-virus synthesis.

We suggest that the binding of H-1 viral proteins to heterochromatin and nucleolar-associated chromatin causes them to condense, resulting in disruption of the nucleolus and heterochromatin margination. These antigens which attach to heterochromatin and nucleolar-associated chromatin are probably viral proteins which have not assembled into capsids, since H-1 virions are not found within this chromatin (Singer and Toolan 1975). It is likely that these antigens are forms of capsid proteins rather than noncapsid viral proteins, because immune sera made against SDS-disrupted H-1 virus (which are unable to promote infection of the immunized animal) produce identical patterns of immunological labeling.

Synthesis of Two Types of Antigens During H-1 Infection

The immunospecific labeling patterns of ts H-1 mutant infections indicate that H-1 capsid proteins probably form two functionally distinct intranuclear antigens. One is a thermostable chromatin-associated antigen. This antigen is found on the HC and NAC of all ts mutants that were tested, and is probably identical to the antigen responsible for immuno-

FIGURE 10 (see facing page)
Synthesis of wt H-1 DNA in advanced phases of cellular damage; NB cells fixed 16 hr p.i. (*A*) Glut/OsO$_4$ fixation; staining with uranyl acetate and lead citrate. The heterochromatin has marginated towards the nuclear membrane, forming a thick peripheral band (arrowheads). Silver grains are randomly distributed in the euchromatin. 2-Min [^{3}H]TdR pulse. Bar = 1.0 μm; 14,000 ×. (Reprinted, with permission, from Singer and Rhode 1977d.) (*B*) Formaldehyde fixation; cytochrome-c-conjugated FF anti-H-1-IgG labeling without counterstaining; 5-min [^{3}H]TdR pulse (combined immunostaining and autoradiography). Marginated nuclear heterochromatin and nucleolar-associated chromatin are strongly H-1-antigen-positive. The euchromatin is uniformly engaged in viral DNA synthesis. N = nucleolus. Bar = 1.0 μm; 14,600 ×. (Adapted from Singer and Rhode 1978.)

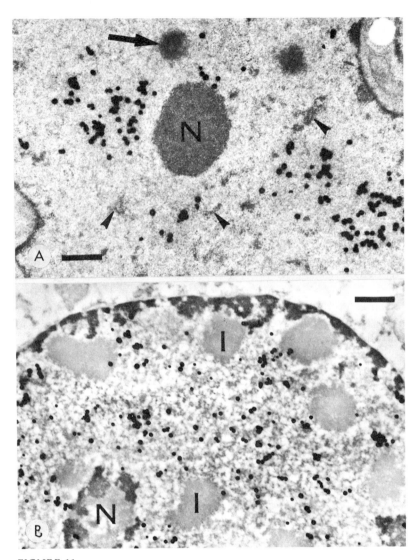

FIGURE 11

Autoradiograms of ts1-H-1-infected NB cells fixed 16 hr p.i. at T_r. (A) Intermediate phase of ts1 H-1 infection exhibiting clusters of [³H]TdR uptake localized in the euchromatin. Ts1 inclusions (arrow) and tufts of condensing heterochromatin (arrowheads) are unlabeled; N = nucleolus. 5-Min [³H]TdR pulse, glut/OsO₄ fixation, uranyl acetate-lead citrate staining. Bar = 1.0 μm; 11,300 ×. (B) Simultaneous anti-H-1-antigen staining and [³H]TdR autoradiography of an advanced stage of ts1 infection. Contracted heterochromatin at the nuclear membrane and nucleolar-associated chromatin have high concentrations of H-1 antigens (N = nucleolus), ts1 inclusions (I) are antigen-negative. Viral DNA synthesis is distributed rather randomly in the euchromatin. 5-Min [³H]TdR pulse, formaldehyde fixation, no post-staining. Bar = 1.0 μm; 10,750 ×. (Reprinted, with permission, from Singer and Rhode 1978.)

logical labeling in the wt H-1 infection. Chromatin-associated antigen remains antigenic if ts-mutant-infected cells are fixed following incubation at the restrictive temperature. The other antigen is located in the various viral inclusions induced by ts mutants in NB-cell nuclei under nonpermissive conditions, and binds IgG only if the infected cells are exposed to the permissive temperature prior to formaldehyde fixation. We call this inclusion-associated antigen (IAA). Thermoreactivation of inclusion-associated antigen occurs more readily in ts8 (by 2 hr after shift to T_p) than it does in ts1 (which requires a 6-hr incubation at T_p). However, the proteins of ts2 inclusions do not react with anti-wt-H-1 IgG under any circumstances, which suggests that inclusion-associated antigen is drastically altered in this mutant. Our morphological observations also enable us to classify the phenotypes of the ts H-1 mutants analyzed at T_r into three classes. Class-1 mutants (ts2) are unable to synthesize capsids; class-2 mutants (ts1, ts7) only produce empty capsids which form aggregates and polycrystals, but do not attach to euchromatin; class-3 (ts8, ts10) agents make full and empty capsids which are euchromatin-associated, and the empty virions also aggregate into spherical non-crystalline inclusions. The rate of IAA thermoreactivation of a given mutant is also correlated with its degree of structural alteration relative to wt H-1. All of these ts mutants exemplify how changes in the primary structure of H-1 proteins can alter the morphogenesis and properties of the capsids. In addition, our studies of HEF and MRC cells infected with wt H-1 indicate that the host-cell type may also affect the characteristics of the virions synthesized. The altered surface properties of wt capsids produced by MRC cells may be due to post-translational modification of viral proteins by the host rather than to changes in primary structure.

Regulation of H-1 DNA Synthesis by Chromatin-Associated Viral Proteins; Role of the Nucleolus

Recently (see Rhode, this volume), it has been shown that replication of H-1 replicative-form DNA requires H-1 viral proteins. H-1 ts mutants defective in progeny DNA synthesis which exhibit alterations in their capsid proteins (Rhode 1976 and above results) also indicate that H-1 proteins are required for the synthesis of this type of viral DNA. We have observed that H-1 capsids are associated with euchromatin fibers, and that unassembled capsid proteins are localized in heterochromatin and nucleolar-associated chromatin. EM autoradiography was performed on NB cells infected with wt or ts1 H-1 to determine if H-1 replicative-form and progeny DNA synthesis are localized in either of the above regions of viral protein accumulation. The sites of synthesis of replicative-form DNA (in ts1-infected cells) and progeny DNA (in wt H-1 infections) exhibited the same location. At the earliest stages of infection, dense clusters of silver grains were seen over extruded masses of fibrous nucleolar component found in the euchromatin. These regions of viral DNA synthesis appeared to increase in size with progressing cytological damage, forming larger aggregates of silver grains in the euchromatin as the H-1-antigen-positive heterochromatin marginated. Cells in relatively advanced stages of H-1 infection exhibited uniform [³H]TdR incorporation into the euchromatin. No H-1 DNA synthesis took place in the intensely H-1-antigen-positive heterochromatin and nucleolar-associated chromatin, or within the ts1 crystalline inclusions. Autoradiography was done on cells prestained with anti-H-1-IgG conjugate to ensure that cells incorporating [³H]TdR were H-1-infected. This control is especially useful when dealing with cells exhibiting early stages of pathological damage. These results suggest that H-1 DNA synthesis begins in defined regions of the euchromatin associated with extruded fibrous nucleolar components and then spreads outward. Cellular cofactors which may be required for the initiation of synthesis of H-1 parental replicative-form DNA may be located in the fibrous portion of the nucleolus. Also, input viral DNA could be sequestered within the nucleolus, and extrusion of the fibrillar component might be the mechanism required to transport it to the extranucleolar euchromatin, where viral DNA synthesis takes place. Regarding this hypothesis, it is noteworthy that the high concentrations of H-1 capsids observed within vacuolated nucleolar fibrous centers and the unassembled capsid proteins localized in nucleolar-associated chromatin in the early phases of H-1 infection are probably responsible for fibrous-center breakdown. Results from experiments in which cells are infected with various multiplicities of competing parvoviruses also suggest that the number of intranuclear sites capable of engaging in synthesis of parental replicative-form DNA is very limited—perhaps to 3–5 per nucleus (S. L. Rhode, unpublished results). Nucleolar fibrous centers would be good candidates for such sites. In view of the observations that nucleolar-associated DNA is late-replicating in human cells (Wyandt and Hecht 1973; Wyandt and Iorio 1973) and in other systems (Busch and Smetana 1970; Kuroiwa 1971), we suggest that the location of early viral DNA synthesis in nucleolar fibrous centers may also be responsible for the dependence of the replication of H-1 parental replicative-form DNA on the late S phase or the G2 phase of the cell cycle (Rhode 1973). We are conducting experiments to determine if input H-1 virions have an intranucleolar localization. In addition, once viral DNA replication has spread throughout the euchromatin, it is apparent that the type of DNA synthesis which occurs may require a morphologically different form of capsid protein. The euchromatic regions of wt-H-1-infected nuclei synthesizing progeny DNA contain numerous capsids associated with euchromatin fibers. Comparable areas in ts1-H-1-infected cells producing mostly replicative-form DNA at T$_r$

have low levels of H-1-antigen-specific staining. This labeling is due to viral protein which is not assembled into capsids since all capsids are located in ts1 inclusion bodies under restrictive conditions.

ACKNOWLEDGMENTS

We are obliged to R. L. Costantino and J. Bratton for expert technical assistance, to Dr. Helene Toolan for supplying the anti-H-1 sera used in these studies, to Dr. K. A. O. Ellem for reviewing this manuscript, and to V. Haas and J. Pratt for typing the text. We thank Dr. D. M. Parry (University of Oxford) and Dr. B. Meyrick (Brompton Hospital, London) for providing descriptions of their semiautomatic coating devices used in EM autoradiography.

This work was supported by U. S. Public Health Service Grant CA07826–12 from the National Cancer Institute, and by a generous gift from The Given Foundation.

We also thank *Experimental Cell Research, Journal of Virology, Journal of Cell Biology,* and *Virology* for permission to reproduce modified versions of our published figures.

REFERENCES

Avrameas, S. 1969. Coupling of enzymes to proteins with glutaraldehyde. *Immunochemistry* **6**:43.

Bernhard, W. 1969. A new staining procedure for electron microscopical cytology. *J. Ultrastruct. Res.* **27**:250.

Busch, H. and K. Smetana. 1970. Nucleolar DNA. In *The nucleolus*, p. 160. Academic, New York.

Karasaki, S. 1966. Size and ultrastructure of the H-viruses as determined with the use of specific antibodies. *J. Ultrastruct. Res.* **16**:109.

Kongsvik, J. R. and H. W. Toolan. 1972. Capsid components of the parvovirus H-1. *Proc. Soc. Exp. Biol. Med.* **139**:763.

Kongsvik, J. R., J. F. Gierthy, and S. L. Rhode. 1974. Replication process of the parvovirus H-1. IV. H-1-specific proteins synthesized in synchronized human NB kidney cells. *J. Virol.* **14**:1600.

Kopriwa, B. M. 1973. A reliable standardized method for ultrastructural electron microscopic radioautography. *Histochemie* **37**:1.

Kuroiwa, T. 1971. Asynchronous condensation of chromosomes from early prophase to late prophase as revealed by electron microscopic auto-radiography. *Exp. Cell Res.* **69**:97.

Ledinko, N., S. Hopkins, and H. Toolan. 1969. Relationship between potentiation of H-1 growth by human adenovirus 12 and inhibition of the "helper" adenovirus by H-1. *J. Gen. Virol.* **5**:19.

Rhode, S. L. III. 1973. Replication process of the parvovirus H-1. I. Kinetics in a parasynchronous cell system. *J. Virol.* **11**:856.

————. 1976. Replication process of the parvovirus H-1. V. Isolation and characterization of temperature-sensitive H-1 mutants defective in progeny DNA synthesis. *J. Virol.* **17**:659.

Singer, I. I. 1974. The intracellular localization of parvovirus (H-1) antigens using immunocytochrome c. In *Viral immunodiagnosis* (ed. E. Kurstak and R. Morisset), p. 101. Academic, New York.

————. 1975. Ultrastructural studies of H-1 parvovirus replication. II. Induced changes in the deoxyribonucleoprotein and ribonucleoprotein components of human NB cell nuclei. *Exp. Cell Res.* **95**:205.

————. 1976. Ultrastructural studies of H-1 parvovirus replication. III. Intracellular localization of viral antigens with immunocytochrome c. *Exp. Cell Res.* **99**:346.

Singer, I. I. and S. L. Rhode III. 1977a. Replication process of the parvovirus H-1. VII. Electron microscopy of replicative-form DNA synthesis. *J. Virol.* **21**:713.

————. 1977b. Ultrastructural studies of H-1 parvovirus replication. IV. Crystal development and structure with the temperature-sensitive mutant ts1. *J. Virol.* **24**:343.

————. 1977c. Ultrastructural studies of H-1 parvovirus replication. V. Immunocytochemical demonstration of separate chromatin-associated and inclusion-associated antigens. *J. Virol.* **24**:353.

————. 1977d. Simultaneous electron microscopic autoradiographs and immunocytochemical localization of H-1 parvovirus DNA synthesis and protein accumulation in human NB cells. In *Proceedings of the 35th Annual Meeting of the Electron Microscopy Society of America* (ed. G. W. Bailey), p. 386. Claitor's, Baton Rouge, Louisiana.

————. 1978. Ultrastructural studies of H-1 parvovirus replication. VI. Simultaneous autoradiographic and immunochemical intranuclear localization of viral DNA synthesis and protein accumulation. *J. Virol.* (in press).

Singer, I. I. and H. W. Toolan. 1975. Ultrastructural studies of H-1 parvovirus replication. I. Cytopathology produced in human NB epithelial cells and hamster embryo fibroblasts. *Virology* **65**:40.

Toolan, H. W. 1968. The picodnaviruses: H, RV and AAV. In *International review of experimental pathology* (ed. G. W. Richter and M. A. Epstein), vol. 6, p. 135. Academic, New York.

————. 1972. The parvoviruses. *Prog. Exp. Tumor Res.* **16**:410.

Usategui-Gomez, M., H. W. Toolan, N. Ledinko, F. Al-Lami, and M. S. Hopkins. 1969. Single-stranded DNA from the parvovirus H-1. *Virology* **39**:617.

Wyandt, H. E. and F. Hecht. 1973. Human Y-chromatin. II. DNA replication. *Exp. Cell Res.* **81**:462.

Wyandt, H. E. and R. J. Iorio. 1973. Human Y-chromatin. III. The nucleolus. *Exp. Cell Res.* **81**:468.

Two Populations of Infectious Virus Produced During H-1 Infection of Synchronized Transformed Cells

JOHN R. KONGSVIK
M. SUZANNE HOPKINS
KAY A. O. ELLEM

Institute for Medical Research
Putnam Memorial Hospital
Bennington, Vermont 05201

Infectious parvovirus virions sediment as a homogeneous species at 110S in sucrose velocity gradients. However, these particles are often resolved as two distinct bands when centrifuged to equilibrium in cesium chloride. In general these two bands are observed at approximately 1.46 g/cm³ and at 1.42 g/cm³. Both particle types have been reported in studies of AAV (Hoggan 1970), Haden virus (Johnson and Hoggan 1973), LuIII (Gautschi and Siegl 1973), H-1 (Rhode 1974), and MVM (Clinton and Hayashi 1975). Interestingly, "dense" forms of poliovirus (Yamaguchi-Koll et al. 1975; Wiegers et al. 1977) and of other enteroviruses (Rowlands et al. 1975) have recently been reported.

We have studied the two density types of the H-1 infectious virion produced during virus growth in synchronized human chondrosarcoma cells. In synchronized hamster cells (Rhode 1973) or human cells (Kongsvik et al. 1974a; Singer and Toolan 1975) the synthesis of viral DNA begins some 8 hr after infection. Infectious-virus production then proceeds rapidly until it levels off 20–24 hr after infection (Rhode 1973). We have examined the relationships among the various particle types produced during this period in order to gain some insight into the regulation of virus maturation in the infected cell.

MATERIALS AND METHODS

VIRUS. Radiolabeled H-1 was produced and quantitated by methods described previously (Rhode 1973).

RADIOACTIVE LABELING AND INFECTION. CS cells, originally from a human chondrosarcoma but now having only rat LDH and G6PD isozymes, were grown in Eagle's medium and synchronized by a double 16-hr methotrexate block (Hampton 1970). At the end of the block, cells were infected with H-1 at a multiplicity of 10 plaque-forming units (PFU) per cell. Double-labeled virions were obtained by continuously labeling infected CS cultures with ^{3}H-protein hydrolysate (10 μCi/ml) and [^{14}C]thymidine (1 μCi/ml, 5 μg thymidine/ml); 5-fluorodeoxyuridine (0.5 μg/ml) was added to suppress endogenous thymidine synthesis. Virions labeled only in the protein coat were obtained by continuously labeling half of the cultures with ^{3}H-protein hydrolysate (10 μCi/ml) and the other half with ^{14}C-protein hydrolysate (2–3 μCi/ml). All isotopes were obtained from New England Nuclear.

BUOYANT DENSITY CENTRIFUGATION IN CESIUM CHLORIDE. Crude virus pellets were suspended in CsCl (1.42 g/cm^3) and centrifuged for 48 hr at 35,000 rpm using the Beckman 40 fixed-angle rotor. The density gradient was monitored using a Bausch and Lomb Abbe-3L refractometer.

BUOYANT DENSITY CENTRIFUGATION IN METRIZAMIDE. In order to determine whether density differences can be detected in the absence of cesium ions, virus was centrifuged to equilibrium in gradients of metrizamide (the nonionic, tri-iodinated benzamido derivative of glucose). Densities and centrifugation times were chosen from the established values of Rickwood et al. (1973). Labeled H-1 preparations, previously established by CsCl centrifugation as heavy full (HF) particles, light full (LF) particles, or empty particles, were layered onto a solution of 58% metrizamide in 0.01 M Tris-HCl (pH 7.5), 0.001 M EDTA and centrifuged for 40–68 hr at 30,000–35,000 rpm in Beckman 40 or SW 50 rotors. Refractive-index measurements were determined and densities were calculated using the relationship: density = 3.350 × R.I. − 3.462 (Birnie et al. 1973).

POLYACRYLAMIDE GEL ELECTROPHORESIS OF VIRUS PROTEINS. Gel analyses of H-1 virus proteins were performed using (a) 10% cylindrical or 10% resolving-4% stacking vertical-slab gels (Studier 1973), (b) continuous SDS-neutral phosphate gels (Summers et al. 1965), or (c) gels employing the discontinuous buffer system of Laemmli (1970). The migration rates of the viral proteins were three times faster using the discontinuous system, and the resolution of the doublet major proteins (VP2' and VP2) was superior to that obtained in the continuous system. Molecular-weight standards used in estimating the size of H-1 proteins included phosphorylase a (94,000), bovine serum albumin (69,000), ovalbumin (43,000), and deoxyribonuclease (31,000).

AGAROSE GEL ELECTROPHORESIS OF DNA. Purified, labeled HF and LF H-1 virions were disrupted by heating for 3 min at 100°C in 2% SDS, 0.05 M Tris-HCl (pH 8.0), 0.001 M EDTA. The lysed virus was layered on 1.4% agarose gels and run for 16 hr at 35 V (3.5 mA). The running buffer was 0.08 M Tris-HCl, 0.08 M boric acid, 0.002 M EDTA (pH 8.0).

ENZYME TREATMENT OF INTACT H-1 VIRIONS. Enzymes used in this study were trypsin (2× crystallized, Worthington) dissolved in 0.075 M Tris-HCl, 0.0015 M $CaCl_2$ (pH 7.2) at 1 mg/ml, and chymotrypsin (3× crystallized, Sigma) dissolved in 0.075 M Na-phosphate buffer (pH 7.0), also at 1 mg/ml. These stock solutions were stored at −20°C in 1.0-ml aliquots.

Purified H-1 (0.1–0.2 ml) having an HA titer of 1–2 × 10^6/ml was incubated with 47–94 μg/ml of trypsin or chymotrypsin at 37°C for the intervals cited in the text. Untreated virus was incubated under the same conditions and for the same times. After incubation, aliquots of each sample were diluted 100-fold with cold phosphate-buffered saline (0.02 M Na-phosphate, 0.15 M NaCl, pH 7.2) and assayed for HA activity and infectivity as described below. The remainder of the sample was diluted with an equal volume of 2% SDS-2% mercaptoethanol in Na-phosphate (pH 7.2) or SDS-sample buffer (Laemmli 1970) and disrupted at 100°C for 3 min. The polypeptide compositions of the treated and untreated virus samples were then compared on cylindrical or vertical-slab SDS gels.

PARTICLE-INFECTIVITY RATIO. Electron-microscope particle counts of HF and LF virions were performed by Dr. I. Singer of this Institute using the method of Watson et al. (1963). Mixtures of virus and latex spheres (7.5 × 10^{12} spheres/ml) were allowed to dry on specimen grids and then examined with a JEM-7 electron microscope.

HEMAGGLUTINATION AND INFECTIVITY ASSAYS. HA assays using guinea-pig red blood cells were done as described previously (Kongsvik and Toolan 1972b). Plaque assays were done by the method of Ledinko (1967) employing a line of SV40-transformed newborn-human kidney cells (NB cells) (Shein and Enders 1962).

RESULTS

Early-Harvest vs. Late-Harvest H-1

In 12 out of 15 different H-1 preparations banded in CsCl and monitored by hemagglutination, the LF bands exceeded the amounts of HF virus by factors of 2 to 1024 (geometric mean = 24-fold). In each of the three remaining preparations the HF/LF ratio exceeded 20. This variability in the relative amounts of LF and HF virus could be due either to differences in the conditions during virus propagation or to changes in the virion population, as a function of time, during the infection process. We examined the temporal parameter by comparing the properties of virus produced early and late in a series of synchronized infections of CS cells.

Virus labeled in both the protein (^{3}H) and DNA (^{14}C) moieties was harvested and purified from cultures 26 and 96 hr after infection. The density of the virus harvested at these two times is shown in Figure 1. HF virus (56%) predominates at 26 hr (Fig. 1A), but only LF virus remains at 96 hr (Fig. 1B). There was no significant difference in the ^{3}H/^{14}C ratios of the HF or LF virus at either time (26 hr: LF = 1.09, HF = 0.960; 96 hr: LF = 0.984). In different preparations both the infectivity (PFU) and the hemagglutinating activity (HA) of the HF virus were lower than those of the LF virus (Table 1). The specific infectivities of the two particle types were compared by measuring their infectivity/radioactivity ratio (PFU/cpm in Table 1). Samples were taken from the virus peaks at each

FIGURE 1
Separation of infectious H-1 virions into two density populations by CsCl isopycnic centrifugation. Virus was harvested (*A*) 26 hr and (*B*) 96 hr after infection. Virus was spun in the 40 rotor for 48 hr at 35,000 rpm and fractions of 0.25 ml each were collected from the bottoms of the tubes. The initial CsCl concentration was 1.42 g/cm³. 20-μl aliquots were removed from each fraction and assayed for radioactivity.

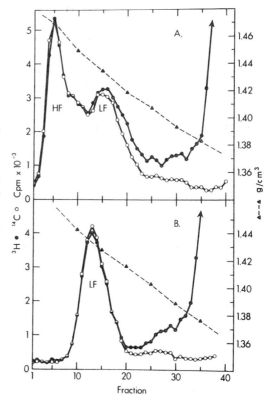

TABLE 1
Biological Activities of Virus Particles of Different Densities

Time of harvest (hr p.i.)	Type of virus particle	Specific infectivity (PFU/cpm)[a]	Specific HA (HA/cpm $\times$ 10^2)	PFU/HA ($\times$ 10^{-2})
24[b]	HF[d]	530	0.4	1,330
	LF	4,000	2.5	1,600
	E	4.4	4.2	1.1
48[b]	HF	480	1.7	282
	LF	8,400	8.6	977
	E	60	7.2	8.3
26[c]	HF	3,500	2.8	1,250
	LF	21,800	19.4	1,130
	E	286	183	1.6
36[c]	HF	1,600	3.4	471
	LF	40,000	47	851
	E	3320	46	72

[a] Radioactivity and infectivity values are the average values across the peaks.
[b] Calculations from the experiment shown in Table 2. The radiolabel was ^{3}H-amino acids in the virion proteins.
[c] Data based upon two separate experiments. The radiolabel was [^{14}C]thymidine in the virion DNA.
[d] Average densities of HF, LF, and E particles in CsCl are 1.46, 1.42, and <1.37 g/cm^3, respectively.

time. At 24 hr the specific infectivity of the LF virus was 5 times that of the HF virus, while at later times this difference was usually tenfold or more; this suggested that the HF particle may be an immature form of the LF virion. In addition, the specific HA (HA/cpm) of LF virus is 5–10 times that of HF virus. However, the infectivity/HA ratio is similar for the two classes of "full" virions. Thus, HF virions could be a population of particles with no HA or infectivity containing a small number of particles as active as LF virions. Alternatively, each HF virion may simply have an affinity for cell receptors 1/5–1/10 that of the LF particle. The infectivity of empty particles, however, is sufficiently low to be explained by contamination with LF virions, while the HA/protein cpm ratios for empty and LF particles are equivalent.

To explore the possibility of a precursor-product relationship between

HF and LF virions, the amounts of each type were quantitated at different times during synchronous infection. The results of a typical experiment are shown in Table 2. There is no significant change in the total amount of full virus between 24 and 48 hr of infection, which correlates well with the plateau of HA production after 20 hr of infection in other synchronized types (see, e.g., Rhode 1973). However, there is a change in the relative amount of HF and LF particles seen between 24 and 48 hr p.i. LF virions increase from 25% to 58% of the total particles during this period; the amount of HF virions is decreased correspondingly. The reciprocal relationship between HF and LF virus at different times is reproducible and is consistent with HF being a precursor to the LF form. The LF virus had a specific infectivity 3–18 times that of HF virus, although the infectivity/HA ratio is constant from virus band to virus band.

The lower infectivity of HF preparations is substantiated further by particle/infectivity determinations performed for us by Dr. I. Singer of this Institute on two different H-1 preparations isolated from NB cells or newborn hamsters. These ratios, although huge, still show the LF virus to have 3–8 times the infectivity of HF virus (Table 3).

Properties of HF and LF Virus Preparations

The density difference of approximately 0.04 g/cm³ between HF and LF, if due solely to the particle content of DNA, would require a considerable difference in the DNA/protein ratio of the preparations. This was not found to be the case. In several preparations of virus labeled continuously in their DNA and protein and harvested between 24 and 96 hr after

TABLE 2
Variation in Virus Particle Density During Course H–1 Infection

Time of harvest (hr p.i.)	Type of virus particle	Peak density of virus band (g/cm³)	³H-amino acid (cpm)[a]	Distribution in full virus (%)
24	HF	1.461	6060	74.8
	LF	1.431	2040	25.2
	E	≤1.390	8280	—
48	HF	1.460	3925	41.4
	LF	1.431	5560	58.6
	E	≤1.385	8360	—

[a] Radioactivity is expressed as cpm obtained under the respective density peaks. Only that virus banding between 1.417 and 1.470 g/cm³ is considered "full" virus.

TABLE 3
Particle-Infectivity Ratios of H-1

	Particle-infectivity ratio		Relative infectivity (LF/HF)
Virus/Host	"heavy fulls" (p = 1.46 g/cm^3)	"light fulls" (p = 1.42 g/cm^3)	
H-1/newborn hamsters	2.06 × 10^4 [a]	0.47 × 10^4	4.38
H-1/human NB cells	15.8 × 10^4	2.03 × 10^4	7.78
RV/newborn hamsters	121 × 10^4	38.8 × 10^4	3.12

[a] Numbers represent the number of particles counted by electron microscopy per infectious unit as determined by plaque-forming units in human NB cells.

infection, the ^{3}H/^{14}C ratios of early-harvested H-1 showed no significant differences from late-harvest H-1.

We also examined the possibility that there may be differences in the structure or size of the DNA in the LF and HF virions by comparing the migration in agarose gels of DNA extracted from both types of the particle. Comigration of DNA from the two types of virions on 1.4% agarose gels (Fig. 2) occurs in a position corresponding to genome-length, single-stranded virion DNA. Gels run in the absence of RF DNA showed that the ^{3}H-labeled material running ahead of the virion DNA is present

FIGURE 2
Coelectrophoresis of HF and LF H-1 DNA on agarose gels. ^{3}H-labeled LF DNA (30 μl) and ^{14}C-labeled HF DNA (20 μl) were layered on 1.4% agarose gels. In addition, 5 μl of double-stranded ^{3}H-labeled H-1 RF (replicative-form) DNA was added as a marker. Samples were run for 16 hr at 35 V (5 mA). The gels were sliced into 1-mm fractions and two slices per vial were counted. Recovery of counts in gel slices was 100% for ^{3}H-labeled LF and 62% for ^{14}C-labeled HF.

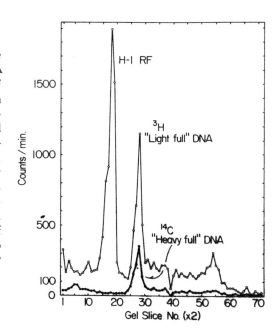

in the LF-virus preparation and is presumably due to a minority of particles containing DNA of shorter lengths. Thus, the difference between HF and LF virus does not appear to be due to a difference in DNA content or size. The same conclusion was reached by Clinton and Hayashi (1975) and Tattersall et al. (1976) in independent studies on the light and heavy forms of the parvovirus MVM (minute virus of mice).

The proteins of HF and LF virions were examined by SDS-polyacrylamide gel electrophoresis. A representative gel analysis, shown in Figure 3, compares the proteins of HF and LF virus purified from infected newborn hamsters. A difference is seen in the major-capsid-protein moiety. HF virus is generally characterized by a single major protein (VP2'), while LF virus exhibits a pronounced double band, the predominant polypeptide corresponding to VP2. This doublet formation of the major capsid protein was observed in an earlier report, but was not correlated with a particular virus type (Kongsvik et al. 1974b). SDS-disrupted H-1 virions were examined on both continuous (neutral phosphate) and discontinuous (Tris-glycine) buffer systems. The data are listed in Table 4. No major differences were found between the molecular weights determined by the three gel electrophoresis methods employed. VP2 is seen to be 3800 daltons smaller than VP2'.

FIGURE 3
Virion proteins of purified HF and LF H-1. Virions were disrupted by heating for 3 min in an equal volume of 2% SDS-2% mercaptoethanol in 0.01 M Na-phosphate (pH 7.2) and layered on cylindrical 10% polyacrylamide gels. The running buffer was 0.1 M Na-phosphate (pH 7.2) containing 0.1% SDS. The gels were then fixed and stained with Coomassie brilliant blue.

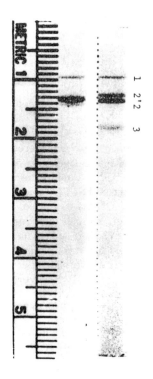

TABLE 4
Estimated Molecular Weights of Virion Proteins of H-1 Parvovirus

	Molecular weights[a] determined by	
Virion protein	individual cylinder gels[b]	slab gels[b]
VP1	94.2 ± 7.4 (16)	87.0 ± 3.4 (19)
VP2′	76.1 ± 4.0 (16)	70.0 ± 2.1 (19)
VP2	72.5 ± 3.7 (12)	66.2 ± 1.6 (8)
VP3	53.9 ± 4.5 (9)	48.8 ± 1.6 (7)

[a] m.w. × 10^{-3} ± s.d. (number).
[b] See "Materials and Methods." Standards were phosphorylase a (94K),
bovine serum albumin (69K), ovalbumin (43K), and DNase (31K).

The smallest protein observed on these gels, labeled VP3 (m.w. 48,000)
in Figure 3, was of variable occurrence and was usually associated with
virus preparations harvested after 4–5 days of infection (Kongsvik and
Toolan 1972a; Kongsvik et al. 1974a). VP3 was particularly prevalent in
H-1 isolated from baby hamsters infected at birth. At present it is not
known whether VP3 is a host-cell contaminant or a virally coded product.
A protein of similar size is associated with purified MVM when this virus
is grown in RL5E rat cells (Tattersall et al. 1976) and also in preparations of
purified Kilham rat virus (Salzman and White 1970).

While many preparations of early-harvest (26 hr) HF virions have
virtually no VP2 present in them, the LF viral forms have varying
amounts of VP2′ and VP2. The preparations illustrated in Figure 1, for
example, were processed to determine the proportion of VP2′ and VP2 by
analyzing the ratio of ^{3}H-labeled amino acid incorporated into each protein. Carrier unlabeled H-1 virus was added to the virus in the peak tubes
of each subfraction. After separation of the virion proteins in slab gels, the
stained bands were extracted and counted. The proportion of counts in
VP2′ was 75.6% of the total VP2′ + VP2 for the 26-hr HF virion, 67.1% for
the 26-hr LF virion, but only 43.9% for the 96-hr LF virion. There is thus a
considerable increase in the proportion of VP2 with virus found later in
infection.

Proteolytic Treatment of Intact Virions

Proteolytic treatment of intact H-1 converts VP2′ capsid protein to a form
that comigrates with VP2 (Kongsvik et al. 1974a). We have found that
both trypsin and chymotrypsin are effective in this cleavage reaction.

Trypsin also appears to cause a more extensive degradation of the H-1 virion with the generation of a fragment of lower molecular weight (~40,000). This fragment could originate from either VP2' or VP2 protein, but not entirely from VP1, as the loss of VP1 does not account for the larger amount of the fragment (Fig. 4). Densitometer scans of control and enzyme-treated H-1 preparations were used to calculate the total recovery of virus proteins after proteolysis. The recovery of H-1 proteins during trypsin digestion was 57% after one enzyme treatment and 40% after a further 2-hr digestion (Fig. 4, gels 2 and 3). The corresponding values for chymotrypsin-treated H-1 virus were 76% and 70%, respectively (Fig. 4, gels 5 and 6). However, only about 5% of the virion protein would be lost by removal of the 3800-dalton fragment when VP2' is cleaved to VP2.

As reported by Tattersall et al. (1976) for MVM "empty" particles, the VP2' protein of H-1 "empties" cannot be converted to the VP2 form by trypsin or chymotrypsin treatment (data not shown).

Though both trypsin and chymotrypsin routinely cleaved VP2' to VP2 in vitro, we were unable to change the density of HF to that of LF after similar incubations were followed by banding in CsCl (data not shown). This suggests that the factor responsible for the cleavage in vivo is not

FIGURE 4
Digestion of intact H-1 virus with trypsin and chymotrypsin. H-1 virus was incubated for a total of 4 hr at 37°C in the absence (gels 1 and 4) and presence of 94 μg/ml of trypsin (gel 2) and 94 μg/ml of chymotrypsin (gel 5). Two hours after the initial enzyme treatment, the same dose was added and incubation was continued for 2 hr (gels 3 and 6). Samples were then disrupted and analyzed on discontinuous SDS gels. The arrow indicates an additional cleavage product generated by trypsin action. This protein is not seen in gels of trypsin or chymotrypsin alone.

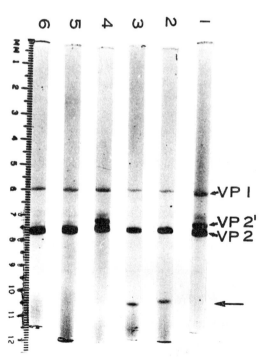

mimicked by trypsin and chymotrypsin digestions in vitro, if indeed the difference in density is a result of the in vivo cleavage. In similar studies carried out on MVM, only incubation in conditioned medium from 3-day infected cultures caused a complete change in density from HF to LF virions (Clinton and Hayashi 1976).

Nature of the In Vivo Proteolytic Activity

We sought to verify the presence of enzymatic activity which could be responsible for the HF-to-LF transition in CS cells. Infected cells were labeled with ^{3}H-amino acids, harvested after 26 hours of infection, and divided into four equal aliquots. The cell pellets were suspended in 3 ml of (*A*) TBS buffer (pH 7.4), (*B*) TBS (pH 7.4) containing 1% SDS, (*C*) 0.05 M sodium acetate (pH 5.0), and (*D*) 0.075 M Tris-HCl (pH 7.5) containing 0.0015 M CaCl$_2$. The pHs of (*C*) and (*D*) were chosen to favor the acidic and neutral proteases, respectively. After 20 sec of sonication, samples (*A*) and (*B*) were placed at 4°C, and (*C*) and (*D*) were incubated for 2 hr at 37°C. Each preparation was then diluted to 18 ml with pH 7.4 TBS and subjected to the usual purification steps of sucrose gradient centrifugation and CsCl isopycnic centrifugation. The resulting CsCl gradient profiles are shown in Figure 5. Inclusion of SDS in the sonication buffer (*B*) eliminated the very dense material sedimenting at 1.48 g/cm^3 in the absence of SDS (*A*). With the unincubated pH 7.4 virus (*A*) as the standard, sonication at pH 7.5 followed by incubation at the same pH caused a pronounced reduction (10,106 cpm) in the amount of HF virus (*D*), with a corresponding increase (6220 cpm) in LF virus.

Sonication and incubation at pH 5.0 (*C*) caused reduction in both the HF and the LF viruses, with a greater effect on the HF virus. It is possible that active acidic lysosomal enzymes are capable of degrading virus particles.

Equilibrium Centrifugation of H-1 HF and LF Virions in Metrizamide Gradients

To examine the role of the ionic environment in determining the buoyant density of HF and LF virions, the nonionic solute metrizamide was used to generate the density gradient. HF and LF viruses were prepared by banding in CsCl from infected cultures after 26 and 96 hr. They had been labeled continuously with ^{3}H-amino acids and [^{14}C]thymidine. Figure 6 shows the distribution of viral components in metrizamide gradients after 68 hr of centrifugation. LF virions harvested at 96 hr form a single peak at a density around 1.32 g/cm^3. The minor peak at 1.16 g/cm^3 appears to be free DNA. Infectivity followed the 1.32-g/cm^3 peak. Virus harvested after only 26 hr of infection shows two main peaks: one at a density of 1.32

26 hrs. P. I.

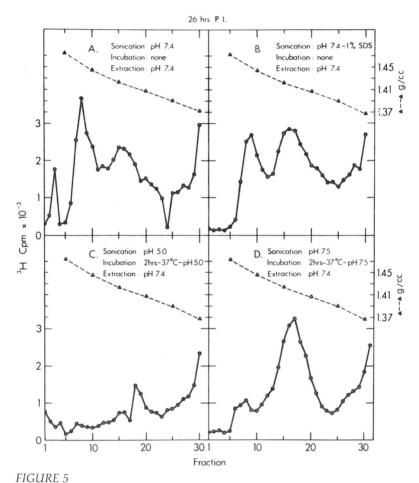

FIGURE 5

Equilibrium CsCl centrifugation of H-1 virus following sonication and incubation in cell lysates. CS cells labeled with ^{3}H-amino acids and harvested 26 hr p.i. were divided into four equal aliquots. Cell pellets were suspended in (A) 3 ml of TBS buffer (pH 7.4) (Rhode 1973), (B) TBS (pH 7.4) containing 1% SDS, (C) 0.05 M Na-acetate (pH 5.0), and (D) 0.075 M Tris-HCl (pH 7.5) containing 0.0015 M CaCl$_2$. After 20 sec of sonication, samples (A) and (B) were placed at 4°C, while (C) and (D) were incubated for 2 hr at 37°C. The samples were then subjected to purification, which culminated in CsCl centrifugation.

g/cm^3, and a second, broader peak centering on 1.20 g/cm^3. HF-virus preparations contain more of the latter, while LF virus contains more of the denser material. It can be seen that there is little infectivity associated with the less dense peak. The cpm in DNA are offset to the lighter side of the cpm in protein in the 1.20-g/cm^3 region of the gradient. This raises the

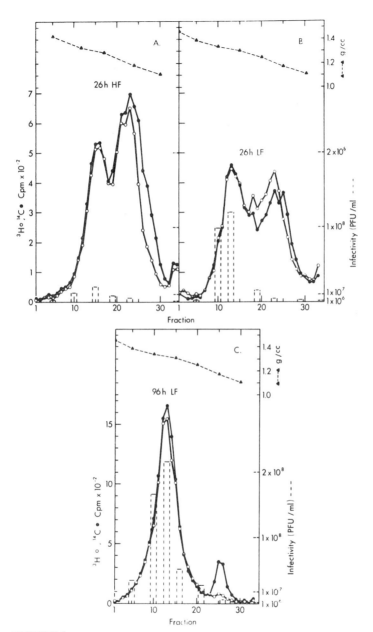

FIGURE 6
Equilibrium centrifugation of early and late H-1 virus on metrizamide gradients.
H-1 HF and LF virus, labeled in both the protein (^{3}H) and DNA (^{14}C) moieties, and
previously separated on CsCl gradients (Fig. 1), were layered on 58% metrizamide and centrifuged for 68 hr at 30,000 rpm. Selected aliquots across the
gradient were tested for infectivity. The origins of the samples analyzed on
metrizamide were 26-hr HF (fractions 3–8), 26-hr LF (fractions 14–24), and 96-hr
LF (fractions 9–20—Fig. 1). 517

possibility that particles banding at this density may break down to give empty particles, which band as a fairly symmetrical peak at 1.26 g/cm³ (data not shown)—contributing only protein cpm to the dense side of the 1.20-g/cm³ peak—and free DNA, which would band on the light side of the peak (at 1.16 g/cm³). There was no loss of infectivity during prolonged centrifugation in metrizamide, which indicated that there is no degradation of infectious virus (banding at 1.32 g/cm³) by this procedure.

DISCUSSION

The present data obtained with H-1 virus parallel closely the data from studies with MVM (Clinton and Hayashi 1975, 1976); this suggests strongly that HF virions with a density of 1.46–1.47 g/cm³ are the first form of assembled particle for both viruses. Most molecules of VP2′, the major viral protein, are then cleaved, losing some 5.4% of their mass. This event is accompanied by a large change in particle buoyant density in CsCl. In the case of MVM, adsorption of the "tailored" virus to host cells is enhanced considerably in comparison with the precursor HF virion (Clinton and Hayashi 1976). Presumably, heavy forms of both parvoviruses undergo changes in surface configuration of the coat protein which result in greater exposure of the groups involved in attachment to host cells. This configurational change in the sequence H-1 HF to H-1 LF resulted in a marked increase in the plaque-producing ability of the "tailored" virus particles. The inability of trypsin and chymotrypsin to influence the density of the HF particles, despite the loss of a similar segment of polypeptide from VP2′ as occurs in natural cleavage, indicates considerable specificity of the responsible enzyme from the infected host cell.

The buoyant density of macromolecules is markedly different when determined in CsCl or metrizamide (see, e.g., Rickwood and Birnie 1975). This has been attributed to the low hydration of metrizamide in solution compared to CsCl, so that metrizamide solutions are close to unity in water activity. Macromolecules in metrizamide should be almost fully hydrated and thus should band at a lower density than in CsCl solutions. Since HF and LF virus have only minor differences in composition, the difference in their densities presumably lies in differences in hydration. In contrast with the situation with MVM (Clinton and Hayashi 1975), rebanding of H-1 HF and LF fractions in CsCl showed little contamination of one with the other. However, both HF and LF preparations of H-1 proved to be heterogeneous in metrizamide; both contain material of 1.31 and 1.20 g/cm³, although in different proportions.

ACKNOWLEDGMENTS

We thank Dr. Helene Toolan for providing the infected newborn hamsters, Mrs. Jeanne Helft and Mr. George Edick for expert technical assistance, Mr. Rob Costantino for the graphs, and Dr. Solon Rhode for critically reviewing the manuscript. Mrs. Virginia Haas and Miss Janeen Pratt provided excellent secretarial skills.

This investigation was supported by Public Health Service Grant CA07826 from the National Cancer Institute, as well as by a generous gift from the Martin Erdmann Memorial Foundation for Scientific Research.

REFERENCES

Birnie, G. D., R. Rickwood, and A. Hill. 1973. Buoyant densities and hydration of nucleic acids, proteins, and nucleoprotein complexes in metrizamide. *Biochim. Biophys. Acta* **331**:283.

Clinton, G. M. and H. Hayashi. 1975. The parvovirus MVM: Particles with altered structural proteins. *Virology* **66**:261.

————. 1976. The parvovirus MVM: A comparison of heavy and light particle infectivity and their density conversion in vitro. *Virology* **74**:57.

Gautschi, M. and G. Siegl. 1973. Structural proteins of parvovirus LuIII. *Arch. gesamte Virusforsch.* **43**:326.

Hampton, E. G. 1970. H-1 virus growth in synchronized rat embryo cells. *Can. J. Microbiol.* **16**:266.

Hoggan, M. D. 1970. Adenovirus associated viruses. *Prog. Med. Virol.* **12**:211.

Johnson, F. B. and M. D. Hoggan. 1973. Structural proteins of Haden virus. *Virology* **51**:129.

Kongsvik, J. R. and H. W. Toolan. 1972a. Capsid components of the parvovirus H-1. *Proc. Soc. Exp. Biol. Med.* **139**:1202.

————. 1972b. Effect of proteolytic enzymes on the hemagglutinating property of the parvoviruses H-1, H-3, and RV. *Proc. Soc. Exp. Biol. Med.* **140**:140.

Kongsvik, J. R., J. F. Gierthy, and S. L. Rhode III. 1974a. Replication process of the parvovirus H-1. IV. H-1-specific proteins synthesized in synchronized human NB kidney cells. *J. Virol.* **14**:1600.

Kongsvik, J. R., I. I. Singer, and H. W. Toolan. 1974b. Studies on the red cell and antibody-reactive sites of the parvovirus H-1: Effect of fixatives. *Proc. Soc. Exp. Biol. Med.* **145**:763.

Laemmli, U. K. 1970. Cleavage of structural proteins during the assembly of the head of bacteriophage T4. *Nature* **227**:680.

Ledinko, N. 1967. Plaque assay of the effects of cytosine arabinoside and 5-iodo-2'-deoxyuridine on the synthesis of H-1 virus particles. *Nature* **214**:1346.

Rhode, S. L. III. 1973. Replication process of the parvovirus H-1. I. Kinetics in a parasynchronous cell system. *J. Virol.* **11**:856.

————. 1974. Replication process of the parvovirus H-1. II. Isolation and characterization of H-1 replicative form DNA. *J. Virol.* **13**:400.

Rickwood, D. and G. D. Birnie. 1975. Metrizamide, a new density-gradient medium. *FEBS Lett.* **50**:102.

Rickwood, D., A. Hill, and G. D. Birnie. 1973. Isopycnic centrifugation of sheared chromatin in metrizamide gradients. *FEBS Lett.* **33**:221.

Rowlands, D. J., M. W. Shirley, D. V. Sangar, and F. Brown. 1975. A high density component in several vertebrate enteroviruses. *J. Gen. Virol.* **29**:223.

Salzman, L. A. and W. L. White. 1970. Structural proteins of Kilham rat virus. *Biochem. Biophys. Res. Commun.* **41**:1551.

Shein, H. M. and J. F. Enders. 1962. Multiplication and cytopathogenicity of simian vacuolating virus 40 in cultures of human tissues. *Proc. Soc. Exp. Biol. Med.* **109**:495.

Singer, I. I. and H. W. Toolan. 1975. Ultrastructural studies of H-1 parvovirus replication. I. Cytopathology produced in human NB epithelial cells and hamster embryo fibroblasts. *Virology* **65**:40.

Studier, F. W. 1973. Analysis of bacteriophage T7 early RNA's and proteins on slab gels. *J. Mol. Biol.* **79**:237.

Summers, D. F., J. F. Maizel, Jr., and J. E. Darnell, Jr. 1965. Evidence for virus-specific non-capsid proteins in poliovirus-infected HeLa cells. *Proc. Natl. Acad. Sci.* **54**:505

Tattersall, P., P. J. Cawte, A. J. Shatkin, and D. C. Ward. 1976. Three structural polypeptides coded for by minute virus of mice, a parvovirus. *J. Virol.* **20**:273.

Watson, D. H., W. C. Russell, and P. Wildy. 1963. Electron microscopic particle counts on herpes virus using the phosphotungstate negative staining technique. *Virology* **19**:250.

Wiegers, K. J., U. Yamaguchi-Koll, and R. Drzeniek. 1977. Differences in the physical properties of dense and standard poliovirus particles. *J. Gen. Virol.* **34**:465.

Yamaguchi-Koll, U., K. J. Wiegers, and R. Drzeniek. 1975. Isolation and characterization of dense particles from poliovirus-infected HeLa cells. *J. Gen.Virol.* **26**:307.

APPENDIX

A Standardized Nomenclature for Restriction-Endonuclease Fragments

A nomenclature system for restriction-endonuclease fragments was adopted by the participants in the parvovirus meeting at Cold Spring Harbor, May 12–15, 1977. It is anticipated that all future publications dealing with the structural properties of parvovirus genomes will utilize the convention to be described below. While this scheme is simple in concept, it provides a precise description of restriction fragments in terms of their positions in the genome, rather than just their relative sizes. The system will also define unambiguously fragments resulting from double-enzyme digests and identify fragments which contain special or unusual features (for example, insertions, deletions, substitutions, point mutations within a normal cleavage site, conformational variants, etc.).

The basic principles of the system are the following:

(1) Restriction fragments shall be identified according to position, expressed in terms of percent of genome, on a physical map. The 0 position is to be the 3' terminus of the minus strand for adeno-associated viruses (and densonucleosis viruses) and the 3' terminus of the v strand for all autonomous parvoviruses. The 5' terminus of these DNA strands will be designated 100. Since palindromic sequences at the termini of parvovirus genomes result in two types of terminal structures, a normal Watson-Crick duplex and a foldback hairpin form, the 0 and 100 positions are assigned using the fully extended, nonhairpin termini only.

(2) Restriction maps will be represented in publication with the 0 position on the left and the 100 position on the right. This convention allows the direct comparison of all parvovirus genomes, in terms of both strand sense and transcription maps, as shown in Figure 1.

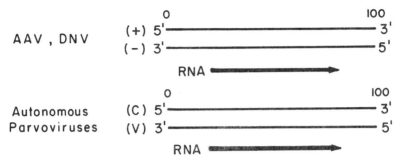

FIGURE 1
A convention that allows direct comparison of parvovirus genomes.

(3) Each restriction fragment will be designated by the limits of its map coordinates (i.e., the percent of the genome spanned relative to the 0 origin). For example, a *Hae*II fragment that is located between 0.4 and 2.5 map units would be denoted *Hae*II 0.4/2.5; an *Hha*I fragment mapping between 49 and 56 map units would be denoted *Hha*I 49/56.

(4) A restriction fragment whose left and right termini result from cleavages by two different restriction enzymes (i.e., a product of double digestion) shall be named as follows. The enzyme producing the left terminus will be designated first, and the enzyme generating the right terminus second, followed immediately by the fragment map coordinates. For example, an *Eco*RI fragment, RI 50/60, cleaved by *Hae*III at map position 57, would yield two fragments designated RI·*Hae*III 50/57 and *Hae*III·RI 57/60, respectively.

(5) To denote the appearance of a natural or inverted repetition of a restriction fragment the following convention will be followed. The map coordinates of the fragment closest to the origin (0) will be represented in the normal fashion; the map coordinates of the repetition will be denoted in brackets such that the number order indicates the presence of a natural or inverted repetition. For example, *Hae*III 2/5 (95/98) would identify a natural repetition of the *Hae*III 2/5 fragment between map coordinates 95 and 98; *Hae*III 2/5 (98/95) would identify an inverted repetition of the *Hae*III 2/5 fragment between map coordinates 95 and 98.

(6) All restriction fragments which possess special or unusual properties that result in abnormal electrophoretic mobility or an alteration in the normal cleavage pattern (i.e., a complete digest of the wild-type genome) will be identified by an asterisk (*) following the map coordinates (e.g., *Hae*III 5/11*). It will be the responsibility of the author(s) to define clearly the special properties of any such restriction fragment. Examples of special cases which will be flagged by this designation system are (1) insertions, (2) deletions, (3) substitutions, (4) point muta-

tions across cleavage sites, (4) aberrant secondary structures, and (5) conformational variants (e.g., hairpin duplexes).

(7) Restriction fragments whose precise map coordinates are unknown but which have been mapped to a localized region of the genome will be identified in the normal manner except that a question mark (?) and a fragment-size designation are to follow the known map-coordinate extremes. For example, *Hinf* 3/10?5 would identify a *Hinf* fragment that represents 5% of the genome and is located between map coordinates 3 and 10.

To further illustrate the application of these nomenclature rules, the *Hha* fragments of AAV2 DNA are listed in Table 1, using both the standard letter designation and the new map-coordinate system. The *Hha* physical map of AAV2 DNA is shown in Figure 2.

TABLE 1
Hha Fragments of AAV2 DNA: A Comparison of Nomenclature Systems

Old nomenclature		Coordinate nomenclature
fragment designation	percent of genome	fragment designation
A	28.0	68.8/97.7
B	14.5	39.5/53.6
C	12.0	20.0/32.0
D	9.0	60.7/69.8
E	6.6	53.6/60.1
F	5.3	34.0/39.5
G	4.7	11.2/15.6
H	4.3	2.3/6.5
I	3.9	6.5/10.3
K*	(2.6)[a]	0.4/2.3(99.5/97.8)*[b]
J	2.5	15.6/18.1
K	1.8	0.4/2.3(99.5/97.8)
L	1.6	32.0/34.0
M	1.5	18.1/20.0
N	0.9	10.3/11.2
O	0.6	60.1/60.7
P	0.4	0/0.4(100/99.6)

Data from L. de la Maza and B. J. Carter, *Virology* **82**:409 (1977).
[a] Parentheses indicate that the apparent molecular weight of K* varies with acrylamide concentration because of an unusual secondary structure [L. de la Maza and B. J. Carter, *Virology* **82**:409 (1977); K. Fife and K. Berns, personal communication].
[b] Parentheses indicate coordinates of inverted repeat sequence; asterisk denotes unusual structural feature described in footnote a.

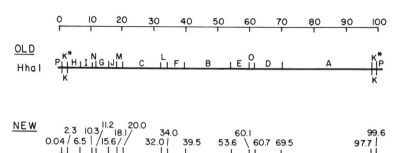

FIGURE 2
Hha physical map of AAV2 DNA.

The convention outlined above for describing restriction fragments of defined map coordinates was adopted unaminously by the meeting participants for use in publications. Although oral presentation may employ other fragment designations, the genome coordinates of the fragment should be noted where possible.

The following individuals (listed in alphabetical order) were actively involved in the evolution of the nomenclature system outlined above. Though this system was established with the particular structural properties of parvovirus DNA in mind, it can provide a much more precise and functional physical map description for any defined DNA genome. The relative merits and drawbacks of a generalized nomenclature system will be described in detail elsewhere as a joint communication from this group.

R. ARMENTROUT
R. BATES
K. BERNS
B. CARTER
M. CHOW
D. DRESSLER
K. FIFE
W. HAUSWIRTH
G. HAYWARD
G. LAVELLE
S. RHODE III
S. STRAUS
P. TATTERSALL
D. WARD

Name Index

Numbers in *italics* refer to pages on which the complete references are listed; **boldface** type designates where author's article is located in this volume.

Subject Index